CALCULUS

FOR ENGINEERING II

Second Edition

CHESTER MIRACLE

Kendall Hunt

publishing company

www.kendallhunt.com
Send all inquiries to:
4050 Westmark Drive
Dubuque, IA 52004-1840

ISBN 978-1-5249-4385-1

Published in the United States of America

Contents

5501 Review of Substitution

Suppose we are faced with a problem of finding an antiderivative. In order to find antiderivatives we classify antiderivative problems into categories. The largest category is those antiderivatives found by reversing the chain rule. The reverse of the chain rule is the Substitution Rule. This is often called u-substitution. The general Substitution Rule is

$$\int f(u(x))\frac{du}{dx}dx = \int f(u(x))du = F(u(x)) + C,$$

where $F'(u) = f(u)$. This is the formula for finding the antiderivative of a composite function

Example 1. Find the antiderivative

$$\int [3x^2 + 2\cos(6x)]^{3/2}[x - 2\sin(6x)]dx.$$

Solution. We are going to use the Substitution Rule which means that we are going to try to reverse the chain rule. The chain rule is used to differentiate composite functions. This means that the Substitution Rule is used to integrate composite functions. If we do not see the correct substitution right away, we look for the most complicated composite function in the integrand. For this example the most complicated composite function is

$$[3x^2 + 2\cos(6x)]^{3/2}.$$

We always let $u(x)$ equal the first inside function in this composite function. This means we choose $u(x)$ to be $u(x) = 3x^2 + 2\cos(6x)$. We use the substitution

$$u(x) = 3x^2 + 2\cos(6x).$$

The differential du is given by

$$du = [6x - 12\sin(6x)]dx = 6[x - 2\sin(6x)]dx.$$

Substituting these expressions for $u(x)$ and du into the integrand, we get

$$\frac{1}{6}\int u^{3/2}du = \frac{1}{6}\frac{u^{5/2}}{5/2} + C = \frac{1}{15}u^{5/2} + C.$$

1

Applying the Substitution Rule and replacing $u(x)$ with $3x^2 + 2\cos(6x)$, we have

$$\int [3x^2 + 2\cos(6x)]^{3/2}[x - 2\sin(6x)]dx = \frac{1}{15}[3x^2 + 2\cos(6x)]^{5/2} + C.$$

Example 2. Find the antiderivative

$$\int \frac{3x + 2}{6x^2 + 8x + 15}dx.$$

Solution. Suppose we do not see the correct substitution right away. At first we may not see a composite function that is part of this integrand. But let us rewrite the problem as

$$\int [6x^2 + 8x + 15]^{-1}(3x + 2)dx.$$

The composite function is $[6x^2 + 8x + 15]^{-1}$. The first inside function for this composite function is $6x^2 + 8x + 15$. Use the substitution that $u(x)$ is equal to this inside function,

$$u(x) = 6x^2 + 8x + 15.$$

The differential du is

$$du = (12x + 8)dx = 4(3x + 2)dx.$$

Substituting these expressions into the antiderivative transforms it into

$$\frac{1}{4}\int \frac{du}{u} = \frac{1}{4}\ln|u| + C.$$

Applying the Substitution Rule and replacing u with $6x^2 + 8x + 15$ gives

$$\int \frac{3x + 2}{6x^2 + 8x + 15}dx = \frac{1}{4}\ln|6x^2 + 8x + 15|dx.$$

We were able to find the antiderivative in Example 1 and in Example 2 because the integrands were actually the result of differentiating a composite function composed of elementary functions using the chain rule. It is possible to have an integrand which is almost the derivative of a function we can construct but actually is not. Consider

$$\int (x^2 + \sin x)^{1/2} (\cos x) dx.$$

If we let $u = x^2 + \sin x$, then $du = 2x + \cos x$ and so $(\cos x)dx \neq du$. We can not use substitution to find the antiderivative. In fact, we are not able to find the antiderivative for this function as an algebraic expression in terms of our usual elementary functions. For problems like this we are not able to find an algebraic expression for the antiderivative because the antiderivative does not exist as one of our known functions. This means that the antiderivative is not a function that has been defined and given a name. It is very easy to write down a function for which we can not find the antiderivative. This means that "find the antiderivative" problems can only start with "nice" functions as the integrand.

The substitution u equals the first inside function works really well when the result is an antiderivative of one of the following forms

$$\int u^n du, \quad \int \frac{du}{u}, \quad \int e^u du, \quad \int \sin u \, du, \quad \int \cos u \, du, \quad \int \sinh(u) du.$$

This method also works when the resulting antiderivative is an elementary combination of these forms.

Example 3. Find the antiderivative

$$\int \frac{x^3}{(9 + 5x^2)^{3/5}} dx.$$

Solution. The composite function is $(9 + 5x^2)^{3/5}$. The first inside function is $9 + 5x^2$. We could use the substitution $u = 9 + 5x^2$. Note that $u > 0$. However, in order to avoid fractions as powers let us use the substitution

$$u^5 = 9 + 5x^2,$$

which is the same as $u = (9+5x^2)^{1/5}$, $u > 0$. This means that $(9+5x^2)^{3/5} = u^3$. The differential is $5u^4 du = 10x dx$ or $x dx = (1/2)u^4 du$. Note that

$$x^3 dx = x^2(x dx) = (1/5)(u^5 - 9)(1/2)u^4 du.$$

3

Substituting into the integral, we get

$$\frac{1}{10} \int \frac{(u^5 - 9)u^4 du}{u^3} = \frac{1}{10} \int (u^6 - 9u) du$$

$$= \frac{1}{10} \left[\frac{u^7}{7} - \frac{9u^2}{2} \right] + C = \frac{u^7}{70} - \frac{9u^2}{20} + C.$$

Applying the Substitution Rule and substituting back $u = (9 + 5x^2)^{1/5}$, we get

$$\int \frac{x^3}{(9 + 5x^2)^{3/5}} dx = \frac{(9 + 5x^2)^{7/5}}{70} - \frac{9(9 + 5x^2)^{2/5}}{20} + C.$$

Example 4. Find the antiderivative

$$\int x^3 \sqrt{9 + x^2} dx.$$

Solution. The composite function is $\sqrt{9 + x^2}$. The first inside function is $9 + x^2$. Let us use the substitution

$$u(x) = 9 + x^2.$$

The differential is $du = 2x dx$. For this example there is a small problem of what to do with the x^3. Using $x^2 = u - 9$, we write

$$x^3 dx = x^2(x dx) = (u - 9)\frac{du}{2}.$$

Substituting into the integral, we get

$$\frac{1}{2} \int (u - 9)u^{1/2} du = \frac{1}{2} \int (u^{3/2} - 9u^{1/2}) du$$

$$= \frac{1}{2} \left[\frac{u^{5/2}}{5/2} - 9\frac{u^{3/2}}{3/2} \right] = (1/5)u^{5/2} - 3u^{3/2} + C.$$

Applying the Substitution Rule and replacing u with $9 + x^2$ we have

$$\int x^3 \sqrt{9 + x^2} dx = \frac{1}{5}(9 + x^2)^{5/2} - 3(9 + x^2)^{3/2} + C.$$

4

Given the problem of finding the antiderivative we always start by looking carefully at the integrand to see if we see a pattern. If the integrand is a product $g(x)h(x)$ or a quotient $g(x)/h(x)$, we can often see the correct u substitution right away without looking for the composite function. We look to see if $g(x)$ is the derivative of $h(x)$ except for a constant multiplier. When this is the case we know right away that the correct substitution is $u = h(x)$. Look back at Example 2. Suppose at the start of this problem we notice that $\frac{d}{dx}(6x^2 + 8x + 15) = 4(3x + 2)$. This tells us right away that $u = 6x^2 + 8x + 15$ is the correct substitution. If we fail to see this substitution right way, then we look for the most complicated composite function.

In some cases the integrand fits into one of our basic antiderivative forms, but we do not find the form by directly looking for the most complicated composite function. We need to consider that part of the integrand was obtained by differentiating $\arcsin[u(x)]$ or $\arctan[u(x)]$ using the chain rule. We need to determine if by using an appropriate function $u(x)$ we can convert the antiderivative problem to

$$\int \frac{du}{1 + u^2} = \arctan u + C \text{ and } \int \frac{du}{\sqrt{1 - u^2}} = \arcsin u + C.$$

Example 5. Find the antiderivative

$$\int \frac{dx}{\sqrt{25 - 36x^2}}.$$

Solution. Right away we see that this problem looks a lot like the antiderivative formula

$$\int \frac{du}{\sqrt{1 - u^2}} = \arcsin u + C.$$

An essential feature of this antiderivative formula which we do not have in this problem is the 1. But with a little algebra we can get a 1 in the proper position. We factor the 25 out in front as follows:

$$\sqrt{25 - 36x^2} = \sqrt{25[1 - (36x^2/25)]} = 5\sqrt{1 - (36x^2/25)}$$

5

Comparing $1 - u^2$ with $1 - (36x^2/25)$ it is now clear that we want $u^2 = (36x^2/25)$ or $u = (6x/5)$. We use the substitution

$$x = (5/6)u.$$

The differential dx is

$$dx = (5/6)du.$$

Substituting $x = (5/6)u$ and $dx = (5/6)du$ into the integral, we get

$$\int \frac{(5/6)du}{5\sqrt{1 - u^2}} = \frac{1}{6} \int \frac{du}{\sqrt{1 - u^2}} = \frac{1}{6} \arcsin u + C.$$

Applying the Substitution Rule and replacing u with $(6/5)x$, we have

$$\int \frac{dx}{\sqrt{25 - 36x^2}} = \frac{1}{6} \arcsin \left(\frac{6x}{5} \right) + C.$$

Example 6. Find the antiderivative

$$\int \frac{x}{16 + 9x^4} dx.$$

Solution. Looking at this integrand we think that this integral may have the form $\int \frac{du}{1+u^2}$. We again need a 1 in the proper position. Note that $16 + 9x^4 = 16[1 + (9x^4/(16)]$. Comparing $1 + u^2$ and $1 + (9x^4/16)$ we want $9x^4/16 = u^2$. Use the substitution

$$u = 3x^2/4.$$

The differential du is

$$du = (6x/4)dx \text{ or } xdx = (2/3)du.$$

Substituting these expressions into the integral, we get

$$\int \frac{(2/3)du}{16 + 16u^2} = \frac{1}{24} \int \frac{du}{1 + u^2} = \frac{1}{24} \arctan u + C.$$

6

Applying the Substitution Rule and replacing u with $3x^2/4$, we have

$$\int \frac{xdx}{16 + 9x^4} = \frac{1}{24} \arctan\left(\frac{3x^2}{4}\right) + C.$$

Example 7. Find the antiderivative

$$\int \frac{16x + 15}{4x^2 + 9} dx.$$

Solution. Whenever we see an integrand which can be written as a sum as the integrand can in this problem we should always consider the possibility that the correct first step is to write the integrand as a sum. The first step in finding this antiderivative is to write the integrand as a sum.

$$\int \frac{16x + 15}{4x^2 + 9} dx = \int \frac{16x}{4x^2 + 9} dx + \int \frac{15dx}{4x^2 + 9}.$$

We then evaluate each integral separately. In order to find the first antiderivative we notice right away that the derivative of $4x^2 + 9$ is a constant times x. We use the substitution $u = 4x^2 + 9$. This means $du = 8xdx$. Making this substitution the first integral is

$$\int \frac{2du}{u} = 2\ln|u| + C.$$

Replacing u with $4x^2 + 9$ gives

$$\int \frac{16x}{4x^2 + 9} dx = 2\ln(4x^2 + 9) + C.$$

The second antiderivative has no x in the numerator. We assume that it can be converted into the form

$$\int \frac{du}{1 + u^2}.$$

Note that $4x^2 + 9 = 9[1 + (4x^2/9)]$. We use the substitution $u = 2x/3$, then $dx = (3/2)du$. Substituting $x = 3u/2$ into the second integral, we get

$$\int \frac{15(3/2)du}{9(1 + u^2)} = \frac{5}{2} \int \frac{du}{1 + u^2} = \frac{5}{2} \arctan u + C.$$

Applying the Substitution Rule with $u = 2x/3$, we have

$$\int \frac{15}{4x^2 + 9} dx = \frac{5}{2} \arctan\left(\frac{2x}{3}\right) + C.$$

We are now able to write

$$\int \frac{16x + 15}{4x^2 + 9} dx = 2\ln(4x^2 + 9) + \frac{5}{2} \arctan\left(\frac{2x}{3}\right) + C.$$

Example 8. Find the antiderivative

$$\int \frac{dx}{25 + 9(x + 4)^2}$$

Solution. Looking at this integrand we think that this integral may have the form $\int \frac{du}{1+u^2}$. We need a 1 in the proper position. Note that

$$25 + 9(x + 4)^2 = 25[1 + (9/25)(x + 4)^2].$$

Comparing $1 + u^2$ and $1 + (9/25)(x + 4)^2$, we want $u^2 = (9/25)(x + 4)^2$ or

$$u = \frac{3}{5}(x + 4) \text{ or } x = (5/3)u - 4.$$

The differential is

$$du = (3/5)dx \quad \text{or} \quad dx = (5/3)du.$$

Substituting these expressions into the integral, we get

$$\int \frac{(5/3)du}{1 + u^2} = \frac{5}{3} \int \frac{du}{1 + u^2} = \frac{5}{3} \arctan u + C.$$

Applying the Substitution Rule and replacing u with $(3/5)(x + 4)$, we have

$$\int \frac{dx}{25 + 9(x + 4)^2} = \frac{5}{3} \arctan\left[\frac{3(x + 4)}{5}\right] + C.$$

Exercises

Find each of the following antiderivatives.

1. $\displaystyle \int \frac{x}{(8x^2+9)^{8/5}}\,dx$

2. $\displaystyle \int [5+3\cos(8x)]^{1/2}\sin(8x)\,dx$

3. $\displaystyle \int \frac{\cos 8x}{5+3\sin 8x}\,dx$

4. $\displaystyle \int \frac{x^3+4x}{x^4+8x^2+25}\,dx$

5. $\displaystyle \int [4\sin(6x)+12\cos(8x)+10]^{5/2}[\cos(6x)-4\sin(8x)]\,dx$

6. $\displaystyle \int [(\cos 3x)^4\sin(3x)+(\tan 3x)^2(\sec 3x)^2]\,dx$

7. $\displaystyle \int \frac{x^3}{(5x^2+8)^2}\,dx$

8. $\displaystyle \int \frac{x}{\sqrt{5x+8}}\,dx$

9. $\displaystyle \int \frac{x^3\,dx}{[9+2x^2]^{1/2}}$

10. $\displaystyle \int \frac{dx}{25+16x^2}$

11. $\displaystyle \int \frac{dx}{\sqrt{64-25x^2}}$

12. $\displaystyle \int e^{x^2+6x}(x+3)\,dx$

13. $\displaystyle \int \frac{21x+20}{9x^2+25}\,dx$

14. $\displaystyle \int \frac{12x+30}{\sqrt{25-x^2}}\,dx$

15. $\displaystyle \int \frac{dx}{25+(x+3)^2}$

16. $\displaystyle \int \frac{dx}{16+9(x+4)^2}$

5505 Integration by Parts

The formula for finding the derivative of a product is

$$\frac{d}{dx}[f(x)g(x)] = f(x)g'(x) + f'(x)g(x).$$

Taking the antiderivative of both sides of this equation, we get

$$f(x)g(x) = \int f(x)g'(x)dx + \int f'(x)g(x)dx.$$

This can be rewritten as

$$\int f(x)g'(x)dx = f(x)g(x) - \int f'(x)g(x)dx.$$

When written in this form this equation is called the "integration by parts" formula. It is the antiderivative form of the product rule for differentiation. It can be used anytime we are trying to find the antiderivative of the product of two functions. It is helpful in some cases and not helpful in others. The differentials $df = f'(x)dx$ and $dg = g'(x)dx$ can be used to write the parts formula using differential notation as

$$\int f(x)dg = f(x)g(x) - \int g(x)df.$$

This is often written using u and v instead of f and g and without mentioning the independent variable as

$$\int u\,dv = uv - \int v\,du.$$

Example 1. Find the antiderivative

$$\int x\sin(4x)dx.$$

Solution. Note that the integrand $x\sin(4x)$ is a product. This suggest that we might want to use the integration by parts formula. Let us choose the substitution

$$f(x) = x \qquad g'(x) = \sin(4x)$$
$$f'(x) = 1 \qquad g(x) = (-1/4)\cos(4x)$$

10

Substituting into the integration by parts formula, we get

$$\int x\sin(4x)dx = -\frac{x}{4}\cos(4x) - \int (1)(-1/4)\cos(4x)dx$$

$$= -\frac{x}{4}\cos 4x + \frac{1}{4}\int \cos(4x)dx.$$

This substitution is successful because we can easily find $\int \cos(4x)dx$.

$$\int x\sin(4x)dx = -\frac{x}{4}\cos(4x) + \frac{1}{16}\sin(4x) + C.$$

When we use integration by parts we must begin by choosing a function for $f(x)$ and a function for $g'(x)$. This means that there are always two possible choices. In Example 1 we choose $f(x) = x$ and $g'(x) = \sin(4x)$. Suppose we make the other possible choice which is

$$f(x) = \sin(4x) \text{ and } g'(x) = x$$

This means that

$$f'(x) = 4\cos(4x) \text{ and } g(x) = \frac{x^2}{2}.$$

Substituting into the integration by parts formula, we get

$$\int x\sin(4x)dx = \frac{x^2}{2}\sin(4x) - \int \frac{x^2}{2}(4\cos 4x)dx$$

$$= \frac{x^2}{2}\sin(4x) - 2\int x^2\cos(4x)dx.$$

This equation is correct, but it is not helpful. In order to find $\int x\sin(4x)dx$ we must now find $\int x^2\cos(4x)dx$. It is harder to find $\int x^2\cos(4x)dx$ than it is to find $\int x\sin(4x)dx$. The object in using integration by parts is to obtain a simpler integral to evaluate than the one we started with. In almost all cases one possible substitution is helpful and the other is not. In making our choices for $f(x)$ and $g'(x)$ we clearly want to choose a function for $g'(x)$ that is easy to integrate. Again every time we use integration by parts we have two choices. If we make the "not helpful" choice it is not the end of the world, we just start over by making the other choice.

Example 2. Find the antiderivative $\int x^3 (\ln x) dx$.

Solution. Note that the integrand is a product. Integration by parts works on products. It is usually best to choose $g'(x)$ first. If we try to choose $g'(x) = \ln x$, then it is hard to find $g(x)$. Therefore, the only real choice is to choose $g'(x) = x^3$. Let

$$f(x) = \ln x \qquad g'(x) = x^3$$
$$f'(x) = \frac{1}{x} \qquad g(x) = \frac{x^4}{4}.$$

The integration by parts formula is

$$\int f(x)g'(x)dx = f(x)g(x) - \int f'(x)g(x)dx.$$

Substituting into this formula, we get

$$\int x^3 (\ln x) dx = \frac{x^4}{4}(\ln x) - \int \frac{x^4}{4}\frac{1}{x}dx$$
$$= \frac{x^4}{4}(\ln x) - \frac{1}{4}\int x^3 dx$$
$$= \frac{x^4}{4}(\ln x) - \frac{x^4}{16} + C.$$

Using this substitution integration by parts was effective because the integral $\int x^3 dx$ is simpler than $\int x^3 (\ln x) dx$.

Example 3. Find $\int \arctan(3x) dx$.

Solution. At first we think that we can not use the integration by parts formula because there is no product, but there is a product. It is (1) times $\arctan(3x)$. Let

$$f(x) = \arctan(3x) \qquad g'(x) = 1$$
$$f'(x) = \frac{3}{1+9x^2} \qquad g(x) = x.$$

Substituting into the integration by parts formula, we get

$$\int \arctan(3x)\,dx = x\arctan(3x) - \int \frac{3x}{1+9x^2}\,dx.$$

In order to integrate $\int \frac{x}{1+9x^2}\,dx$ we use a u-substitution. Let $u = 1 + 9x^2$, then $du = 18x\,dx$. Using this substitution, we have

$$\int \frac{x}{1+9x^2}\,dx = \frac{1}{18}\int \frac{du}{u} = \frac{1}{18}\ln u.$$
$$= \frac{1}{18}\ln(1+9x^2).$$

Therefore,

$$\int \arctan(3x)\,dx = x\arctan(3x) - \frac{1}{6}\ln(1+9x^2) + C.$$

Example 4. Find the antiderivative $\int x^2 e^{5x}\,dx$.

Solution. Use integration by parts with

$$f(x) = x^2 \qquad g'(x) = e^{5x}$$
$$f'(x) = 2x \qquad g(x) = (1/5)e^{5x}.$$

Substituting into the parts formula, we get

$$\int x^2 e^{5x}\,dx = (1/5)x^2 e^{5x} - (2/5)\int x e^{5x}\,dx.$$

The antiderivative $\int x e^{5x}\,dx$ is simpler than $\int x^2 e^{5x}\,dx$ but still does not fit exactly into one of the antiderivative formulas. We need to apply the integration by parts formula again. In this second application of the parts formula let

$$f(x) = x \qquad g'(x) = e^{5x}$$
$$f'(x) = 1 \qquad g(x) = (1/5)e^{5x}.$$

Substituting into the parts formula, we get

$$\int x e^{5x}\,dx = (1/5)x e^{5x} - (1/5)\int e^{5x}\,dx$$
$$= (1/5)x e^{5x} - (1/25)e^{5x} + C.$$

Substituting back into the original result we get

$$\int x^2 e^{5x} dx = \frac{1}{5} x^2 e^{5x} - \frac{2}{25} x e^{5x} + \frac{2}{125} e^{5x} + C.$$

Example 5. Given that $f(0) = 18$, $f(\pi/2) = 10$ and $\int_0^{\pi/2} (\sin x) f'(x) dx = 6$, find $\int_0^{\pi/2} (\cos x) f(x) dx$.

Solution. Substitute $g'(x) = \cos x$ and $g(x) = \sin x$ into the integration by parts formula and we get

$$\int (\cos x) f(x) dx = [f(x)](\sin x) - \int [f'(x)](\sin x) dx.$$

In terms of definite integrals this says

$$\int_0^{\pi/2} (\cos x)[f(x)] dx = [f(x)](\sin x) \Big|_0^{\pi/2} - \int_0^{\pi/2} [f'(x)](\sin x) dx$$

$$= f(\pi/2) - \int_0^{\pi/2} [f'(x)](\sin x) dx.$$

Using the given values, we have

$$\int_0^{\pi/2} (\cos x)[f(x)] dx = 10 - 6 = 4$$

Exercises

Find the following antiderivatives

1. $\int x \cos(5x) dx$

2. $\int x^5 (\ln x) dx$

3. $\int x \sinh(x) dx$

4. $\int x^2 \sin(3x) dx$

5. $\int \arcsin(x) dx$

6. $\int (\ln x) dx$

7. Given that $f(0) = 3$, $f(2) = 8$, and $\int_0^2 f'(x) e^x dx = 3$ find $\int_0^2 f(x) e^x dx$.

14

5509 Integration using Trigonometric Identities

Example 1. Find the antiderivative

$$\int \frac{(\sec x)(\sin x) + (\cos x)(\tan x)^2}{\sec x + \tan x} dx.$$

Solution. This looks hopeless, but then we discover the following trigonometric identity

$$\frac{(\sec x)(\sin x) + (\cos x)(\tan x)^2}{\sec x + \tan x} = \sin x.$$

One of our antiderivative formulas is $\int (\sin x)dx = -\cos x + C$. Thus,

$$\int \frac{(\sec x)(\sin x) + (\cos x)(\tan x)^2}{\sec x + \tan x} dx = -\cos x + C.$$

Expressions involving the six basic trigonometric functions can be written in many different forms. Some forms of the same function are much easier to integrate than other forms. Given a function involving trigonometric functions we should try rewriting it using trigonometric identities. Once it is rewritten we might be able to integrate it. We will use this idea to enable us to simplify antiderivative problems.

The best known trigonometric identities are

$$\sin^2 x + \cos^2 x = 1$$
$$\cos 2x = \cos^2 x - \sin^2 x$$
$$\sin 2x = 2 \sin x \cos x$$

Combining $\sin^2 x + \cos^2 x = 1$ and $\cos^2 x - \sin^2 x = \cos 2x$ we can get

$$2 \cos^2 x = 1 + \cos(2x)$$
$$2 \sin^2 x = 1 - \cos(2x)$$

We begin our study of using trigonometric identities to find antiderivatives by finding antiderivatives of the form

$$\int \sin^n x \cos^m x \, dx.$$

Case 1. We look at how to evaluate an integral of this form in the case one or both of the powers m and n is an odd positive integer. How can we find the antiderivative? Suppose n is odd, then we write $(\sin x)^n dx$ as $(\sin x)^{n-1} \sin x \, dx$. Since $(n-1)$ is even we can convert $(\sin x)^{n-1}$ to powers of $\cos x$ using the identity $\sin^2 x = 1 - \cos^2 x$. The substitution $u = \cos x$ converts the integral into a polynomial in u which is easy to integrate.

Example 2. Find the antiderivative $\int \sin^4 x \cos^3 x dx$.

Solution. We look for odd power. Since 3 is an odd number, we write $(\cos x)^3$ as $(\cos^2 x)(\cos x) = (1 - \sin^2 x) \cos x$. This means that

$$\sin^4 x \cos^3 x = (\sin x)^4 (1 - \sin^2 x) \cos x = [(\sin x)^4 - (\sin x)^6] \cos x.$$

Using this trigonometric identity we rewrite the integral as

$$\int [(\sin x)^4 - (\sin x)^6](\cos x) dx.$$

Next, we make the substitution $u = \sin x$ and $du = \cos x dx$, and we get

$$\int (u^4 - u^6) du = \frac{u^5}{5} - \frac{u^7}{7} + C.$$

Applying the Substitution Rule and replacing u with $\sin x$, we have

$$\int \sin^4 x \cos^3 x dx = \frac{1}{5}(\sin x)^5 - \frac{1}{7}(\sin x)^7 + C.$$

This method of solution works as long as the power on the cosine is odd. We can use this same substitution to find $\int (\sin x)^{4/3} (\cos x)^3 dx$ or $\int (\sin x)^{-4} (\cos x)^3 dx$. The first step is to use the identity $(\cos x)^3 = (1 - \sin^2 x) \cos x$ to write

$$\int (\sin x)^{4/3} (\cos x)^3 dx = \int [(\sin x)^{4/3} - (\sin x)^{10/3}](\cos x) dx,$$

and to write

$$\int (\sin x)^{-4} (\cos x)^3 dx = \int [(\sin x)^{-4} - (\sin x)^{-2}](\cos x) dx.$$

Case 2. Suppose that neither $\sin x$ nor $\cos x$ have an odd number as a power. This means that both the numbers m and n are even. This includes the case where either $n = 0$ or $m = 0$. In this case we convert the integrals using the identities $\cos^2 x = (1/2)[1 + \cos(2x)]$ and $\sin^2 x = (1/2)[1 - \cos(2x)]$. We repeat the use of these identities until the only trigonometric function in the integrand is of the form $\cos(bx)$. We then have an antiderivative which is easy to find.

Example 3. Find $\int \cos^2(3x) \sin^2(3x) dx$.

Solution. When the power of both $\sin(3x)$ and $\cos(3x)$ is even we must as stated above use the following identities.

$$2 \sin^2 y = 1 - \cos(2y)$$
$$2 \cos^2 y = 1 + \cos(2y)$$

This means that

$$(\sin 3x)^2 = (1/2)(1 - \cos(6x))$$
$$(\cos 3x)^2 = (1/2)(1 + \cos(6x)).$$

These identities enable us to reduce the powers on the sine and cosine functions.

$$(\sin 3x)^2 (\cos 3x)^2 = [(1/2)(1 - \cos 6x)][(1/2)(1 + \cos 6x)]$$
$$= \frac{1}{4}\left[1 - \cos^2(6x)\right].$$

We need to get rid of the square on $\cos(6x)$. We again use the identity $\cos^2 y = (1/2)(1 + \cos 2y)$. We have

$$\cos^2 6x = (1/2)[1 + \cos(12x)].$$

$$(\sin 3x)^2 (\cos 3x)^2 = \frac{1}{4}\left[1 - \frac{1}{2}(1 + \cos 12x)\right] = \frac{1}{8} - \frac{1}{8}\cos(12x)$$

$$\int [1 - \cos(12x)]dx = x - \frac{1}{12}\sin(12x)$$

Therefore,

$$\int \sin^2(3x) \cos^2(3x) dx = \frac{x}{8} - \frac{1}{96}\sin(12x) + C.$$

17

We can, of course, write the answer in many other forms using trigonometric identities. Using either Case 1 or Case 2 we are able to find any antiderivative of the form $\int [\sin(bx)]^n [\cos(bx)]^m \, dx$, where m and n are nonnegative integers.

Example 4. Find the antiderivative

$$\int \frac{x^2}{\sqrt{9-x^2}} \, dx \text{ when } -3 < x < 3.$$

Solution. Our first try might be to use the substitution $u^2 = 9 - x^2$, but this does not convert the problem to one we can integrate. However, after a good look at the function we notice that part of it is the expression

$$\frac{1}{\sqrt{9-x^2}} = \frac{1}{3\sqrt{1-(x/3)^2}}.$$

The derivative of $\arcsin(x/3)$ contains the expression $[1-(x/3)]^{-1/2}$. Maybe the integrand in this problem was obtained by differentiating an expression involving $\arcsin(x/3)$. Let us try the substitution

$$u = \arcsin(x/3)$$
$$x = 3 \sin u.$$

Since $-3 < x < 3$, we choose $-\pi/2 < u < \pi/2$. We have

$$9 - x^2 = 9 - 9\sin^2 u = 9(1 - \sin^2 u) = 9\cos^2 u$$
$$dx = 3(\cos u)du.$$

Substituting these expressions into the given integral it becomes

$$\int \frac{(9\sin^2 u)(3\cos u)du}{3\cos u}.$$

Note that $\sqrt{\cos^2 u} = |\cos u| = \cos u$ for $-\pi/2 < u < \pi/2$. Using the identity $\sin^2 u = (1/2)(1 - \cos(2u))$, we get

$$9 \int \sin^2 u \, du = \frac{9}{2} \int [1 - \cos(2u)]du$$
$$= \frac{9}{2}u - \frac{9}{4}\sin(2u) + C$$
$$= \frac{9}{2}u - \frac{9}{2}(\sin u)(\cos u) + C.$$

18

We have $\cos u = \sqrt{1 - \sin^2 u} = \sqrt{1 - (x/3)^2} = (1/3)\sqrt{9 - x^2}$. Substituting $u = \arcsin(x/3)$, we have

$$\int \frac{x^2}{\sqrt{9 - x^2}} dx = \frac{9}{2} \arcsin(x/3) - \frac{x}{2}\sqrt{9 - x^2} + C.$$

Powers of $\sec x$ and $\tan x$

Now let us turn our attention to find integrals involving the powers of secant and tangent. We consider the problem of finding an antiderivative of the form

$$\int \tan^n x \sec^m x \, dx.$$

Case 1. Let us consider the case where m, the power on the secant, is an even positive integer. In order to integrate in this case write $(\sec bx)^m$ as $(\sec bx)^{m-2}(\sec bx)^2$ and then convert $(\sec bx)^{m-2}$ to $\tan bx$ using the identity $\tan^2 x + 1 = \sec^2 x$. Finally, make the u substitution $u = \tan bx$. Note that n may be any number as long as m is an even positive integer.

Example 5. Find the antiderivative

$$\int (\tan 3x)^4 (\sec 3x)^4 dx.$$

Solution. This is Case 1. We factor out $\sec^2(3x)$ and use the identity $\sec^2(3x) = 1 + \tan^2(3x)$ to convert the remaining powers of $\sec(3x)$ to $\tan(3x)$.

$$(\sec 3x)^4 = (\sec 3x)^2(\sec 3x)^2 = [1 + \tan^2(3x)][\sec(3x)]^2.$$

Using this trigonometric identity the integral becomes

$$\int [\tan(3x)]^4[1 + \tan^2(3x)][\sec(3x)]^2 dx.$$

When solving a Case 1 problem we always let $u = \tan(3x)$ and $du = 3\sec^2(3x)dx$. The object of using $\sec^2(3x) = 1 + \tan^2(3x)$ was to isolate $[\sec(3x)]^2$ which is derivative of $\tan(3x)$.

$$\int u^4(1 + u^2)\frac{du}{3} = \frac{1}{3}\int (u^4 + u^6)du = \frac{1}{15}u^5 + \frac{1}{21}u^7 + C.$$

Replacing u with $\tan(3x)$, we have

$$\int [\tan(3x)]^4[\sec(3x)]^4 dx = \frac{1}{15}[\tan(3x)]^5 + \frac{1}{21}[\tan(3x)]^7 + C.$$

We are also able to use a slight variation of this method when $m = 0$ to find antiderivatives of the form

$$\int \tan^n x \ dx$$

in the case n is an even positive integer. For example

$$\int \tan^4 x \ dx = \int (\sec^2 x - 1)(\tan^2 x) dx = \int (\sec^2 x \tan^2 x - \sec^2 x + 1) dx.$$

Case 2. Next we explain a method which enables us to find antiderivatives of the form $\int \tan^n x \sec^m x dx$ in the case when n is an odd positive integer and m is an odd positive integer. This method actually works for n an odd positive integer and $m \geq 1$. In order to integrate in this case, we we first factor out $(\sec x)(\tan x)$ and write

$$(\tan x)^n(\sec x)^m = (\tan x)^{n-1}(\sec x)^{m-1}(\sec x)(\tan x).$$

Since $n - 1$ is an even integer we next convert $(\tan x)^{n-1}$ to $\sec x$ using the identity $\tan^2 x = \sec^2 x - 1$. The substitution $u = \sec x$ then converts the integrand to a polynomial in u. This substitution is made in order to take advantage of $\frac{d}{dx}[\sec x] = (\sec x)(\tan x)$.

Example 6. Find the antiderivative $\int (\tan 5x)^3(\sec 5x)^3 dx$.

Solution. Note that both $\tan(5x)$ and $\sec(5x)$ have an odd power. If the power of the secant were even we could also find the antiderivative using the method of Case 1. Since the power of the tangent is odd, we write

$$(\tan 5x)^3(\sec 5x)^3 = (\tan 5x)^2(\sec 5x)^2(\tan 5x)(\sec 5x).$$

Next use the identity $\tan^2(5x) = \sec^2(5x) - 1$ to convert $\tan(5x)$ to $\sec(5x)$. The integral then is

$$\int [\sec^2(5x) - 1][\sec(5x)]^2 \sec(5x) \tan(5x) dx.$$

To convert Case 2 we always use the substitution $u = \sec(5x)$. Let $u = \sec(5x)$ and $du = 5\sec(5x)\tan(5x)dx$. Making this substitution, the integral is

$$\int (u^2 - 1)u^2 \frac{du}{5} = \frac{1}{5}\int (u^4 - u^2)du$$

$$= \frac{u^5}{25} - \frac{u^3}{15} + C.$$

Applying the Substitution Rule and replacing u with $\sec(5x)$, we have

$$\int [\tan(3x)]^3[\sec(3x)]^3 dx = \frac{1}{25}[\sec(5x)]^5 - \frac{1}{15}[\sec(5x)]^3 + C$$

We have already found $\int (\tan x)dx = \ln(\sec x) + C$. The method of Case 2 does not enable us to find $\int \tan^n x \, dx$ for n odd. For example the method of case 2 will not find $\int (\tan x)^3 dx$. However, we can find $\int (\tan x)^3 dx$ by using the identity $\tan x = \frac{\sin x}{\cos x}$ to convert $\int (\tan x)^3 dx$ into $\int (\sin x)^3(\cos x)^{-3} dx$. Indeed, this substitution will also enable us to find the antiderivative for any integral of the form $\int \tan^n x \, dx$ where n is odd.

So far we have discussed how to find an antiderivative of the form

$$\int (\tan x)^n (\sec x)^m dx$$

in the case where m is an even positive integer and in the case where both m and n are nonnegative odd integers. We have not discussed the case where n is even and m is an odd positive integer. An example of such an antiderivative is

$$\int (\tan x)^2 (\sec x)^3 dx.$$

Integrals of this type are somewhat more difficult and will not be discussed at this time except to say that all can be converted into integrals involving $\sec x$ to an odd power. However, for the record let us note that

$$\int (\sec x)dx = \ln|\sec x + \tan x| + C.$$

$$\int (\sec x)^3 dx = \frac{1}{2}[(\sec x)(\tan x) + \ln|\sec x + \tan x|] + C$$

We can easily show that this formula is correct using differentiation.

$$\frac{d}{dx}[\ln|\sec x + \tan x|] = \frac{\sec x \tan x + \sec^2 x}{\tan x + \sec x} = \sec x.$$

In our discussion so far we have considered integrals of the form $\int \sin^n(bx)\cos^m(bx)dx$ where b is the same number in both the sine function and the cosine function. We are now going to consider the problem of how to integrate where the coefficient of x is not the same in both functions.

Example 7. Find the antiderivative

$$\int \sin(4x)\sin(10x)dx.$$

Solution. In order to find the antiderivative we need the trigonometric identity

$$2(\sin A)(\sin B) = \cos(A - B) - \cos(A + B).$$

Using this identity

$$\sin(4x)\sin(10x) = \frac{1}{2}[\cos(6x) - \cos(14x)]$$

$$\int [\cos(6x) - \cos(14x)]dx = \frac{1}{6}\sin(6x) - \frac{1}{14}\sin(14x) + C$$

Therefore,

$$\int \sin(4x)\sin(10x)dx = \frac{1}{12}\sin(6x) - \frac{1}{28}\sin(14x) + C.$$

Some other trigonometric identities that are needed to work similar problems to this example are

$$2\cos A \cos B = \cos(A - B) + \cos(A + B)$$
$$2\sin A \cos B = \sin(A - B) + \sin(A + B)$$

Exercises

Find the following antiderivatives.

1. $\int \sin^3(5x)\cos^3(5x)dx$ 2. $\int \sin^3(4x)\cos^2(4x)dx$

3. $\int \sin^3(4x)dx$ 4. $\int \sin^2(5x)dx$

5. $\int \sin^2(6x)\cos^2(6x)dx$ 6. $\int [\sin(5x)]^{3/5}\cos^3(5x)dx$

7. $\int (\cos x)^{-2}(\sin x)^3 dx$ 8. $\int \dfrac{x^2}{\sqrt{16-x^2}}dx$ Hint: First let $x = 4\sin u.$

9. $\int \tan^3(4x)dx$ 10. $\int \tan^2 x \sec^4 x dx$

11. $\int [\tan(4x)]^{2/3}[\sec(4x)]^4 dx$ 12. $\int [\tan(4x)]^5[\sec(4x)]dx$

13. $\int \sin(6x)\sin(8x)dx$ 14. $\int \cos(6x)\sin(10x)dx$

15. $\int [\cos(6x)][\cos(10x)]dx$

5513 Integration using Partial Fractions

Consider the problem of finding any of the following antiderivatives.

$$\int \frac{(10x - 18)}{(x - 5)(x + 3)}dx \qquad \int \frac{12x^2 + 28x + 95}{(x^2 + 9)(x + 4)}dx \qquad \int \frac{22x^2 + 36x + 460}{(x^2 + 16)(x^2 + 25)}dx$$

The integrand in each of these problems has a product of factors as its denominator. None of our antiderivative formulas contains a product of factors in the denominator. We need to get rid of the products in the denominators of the problems in order to be able to integrate. The problem is that someone has added two fractions together. In order to add the fractions they got a common denominator which is then a product. We need to unadd these fractions.

Example 1. Express the fraction

$$\frac{10x - 18}{(x - 5)(x + 3)}$$

as the sum of two fractions in the form $\dfrac{}{x - 5} + \dfrac{}{x + 3}$.

Solution. We need to find constants A and B such that

$$\frac{A}{x - 5} + \frac{B}{x + 3} = \frac{10x - 18}{(x - 5)(x + 3)}.$$

Rewrite the two fractions on the left of the equals sign so that they have the same denominator

$$\frac{A(x + 3)}{(x - 5)(x + 3)} + \frac{B(x - 5)}{(x - 5)(x + 3)} = \frac{10x - 18}{(x - 5)(x + 3)}.$$

Since the two fractions to the left of the equals sign now have the same denominator, we can add them by adding the numerators. This gives

$$\frac{A(x + 3) + B(x - 5)}{(x - 5)(x + 3)} = \frac{10x - 18}{(x - 5)(x + 3)}.$$

Multiply both sides of this equation by $(x-5)(x+3)$ and we get

$$A(x+3) + B(x-5) = 10x - 18$$
$$(A+B)x + (3A - 5B) = 10x - 18.$$

It follows that A and B must be numbers such that

$$A + B = 10$$
$$3A - 5B = -18.$$

Multiply both sides of the first equation by 5.

$$5A + 5B = 50$$
$$3A - 5B = -18.$$

Adding these two equations gives

$$8A = 32$$
$$A = 4.$$

It follows that $B = 6$. We have shown that

$$\frac{10x - 18}{(x-5)(x+3)} = \frac{4}{x-5} + \frac{6}{x+3}.$$

We can now find the antiderivative $\int \dfrac{10x - 18}{(x-5)(x+3)}\,dx$. We have

$$\int \frac{10x - 18}{(x-5)(x+3)}\,dx = \int \left[\frac{4}{x-5} + \frac{6}{x+3}\right]\,dx = 4\ln|x-5| + 6\ln|x+3| + C.$$

Example 2. Unadd the fractions

$$\frac{12x^2 + 28x + 95}{(x^2 + 9)(x+4)}.$$

Solution. Suppose the denominator was given as $x^3 + 4x^2 + 9x + 36$ instead of $(x^2 + 9)(x+4)$. If the expression in the denominator can be factored into

25

real factors, then it should be factored before doing anything else. We need to express this fraction as the sum of two fractions, one with denominator $x^2 + 9$ and the other with denominator $x + 4$. It is important to realize that when there is a quadratic such as $x^2 + 9$ or $x^2 + 4x + 10$ as the denominator of the fraction, we must assume that the numerator is of the form $Ax + B$. In order to unadd this fraction we need to find numbers A, B, and D such that

$$\frac{Ax + B}{x^2 + 9} + \frac{D}{x + 4} = \frac{12x^2 + 28x + 95}{(x^2 + 9)(x + 4)}.$$

Rewriting the fractions on the left side so that they have the same denominator, we get

$$\frac{(Ax + B)(x + 4)}{(x^2 + 9)(x + 4)} + \frac{D(x^2 + 9)}{(x^2 + 9)(x + 4)} = \frac{12x^2 + 28x + 95}{(x^2 + 9)(x + 4)}.$$

Since the fractions on the left of the equals sign have the same denominator we can add them by adding the numerators.

$$\frac{(Ax + B)(x + 4) + D(x^2 + 9)}{(x^2 + 9)(x + 4)} = \frac{12x^2 + 28x + 95}{(x^2 + 9)(x + 4)}.$$

Since these fractions are equal and they have the same denominators, their numerators must be equal.

$$(Ax + B)(x + 4) + D(x^2 + 9) = 12x^2 + 28x + 95.$$
$$(A + D)x^2 + (4A + B)x + (4B + 9D) = 12x^2 + 28x + 95.$$

In order for this equality to be true we must have

$$A + D = 12$$
$$4A + B = 28$$
$$4B + 9D = 95.$$

Solve the first equation for A and we get $A = 12 - D$. Substitute this into the second equation and we get

$$4(12 - D) + B = 28$$
$$B - 4D = -20$$
$$B = 4D - 20.$$

Substitute $B = 4D - 20$ into the third equation and we get

$$4(4D - 20) + 9D = 95$$
$$25D = 175$$
$$D = 7.$$

Using $D = 7$ we have $B = 4(7) - 20 = 8$ and $A = 12 - 7 = 5$. We have shown that

$$\frac{5x + 8}{x^2 + 9} + \frac{7}{x + 4} = \frac{12x^2 + 28x + 95}{(x^2 + 9)(x + 4)}.$$

Example 3. Find the antiderivative

$$\int \frac{12x^2 + 28x + 95}{(x^2 + 9)(x + 4)} dx.$$

Solution. Using the result of Example 2 we can rewrite the antiderivative problem as: find the antiderivative

$$\int \left[\frac{5x + 8}{x^2 + 9} + \frac{7}{x + 4} \right] dx.$$

The first integral is a bit difficult. In order to integrate the first function we must break it into the sum of two integrals as follows:

$$5 \int \frac{x \, dx}{x^2 + 9} + 8 \int \frac{dx}{x^2 + 9} + 7 \int \frac{dx}{x + 4}.$$

We integrate the first integral $\int \frac{x dx}{x^2 + 9}$ using the substitution $u = x^2 + 9$ and $du = 2x dx$. The integral then is

$$\frac{1}{2} \int \frac{du}{u} = \frac{1}{2} \ln |u| + C = \frac{1}{2} \ln(x^2 + 9) + C.$$

We integrate the second integral $\int \frac{dx}{x^2 + 9}$ using the substitution $x = 3w$ and $dx = 3dw$. This converts the integral to

$$\frac{1}{3} \int \frac{dw}{1 + w^2} = \frac{1}{3} \arctan(w) + C = \frac{1}{3} \arctan(x/3) + C.$$

27

Integrating all three integrals, we get

$$\int \frac{12x^2 + 28x + 95}{(x^2 + 9)(x + 4)}\,dx = \frac{5}{2}\ln(x^2 + 9) + \frac{8}{3}\arctan\frac{x}{3} + 7\ln|x + 4| + C.$$

Example 4. Find the antiderivative

$$\int \frac{22x^2 + 36x + 460}{(x^2 + 16)(x^2 + 25)}\,dx.$$

Solution. We need to unadd the fraction which is the integrand. We need to find numbers $A, B, D,$ and E such that

$$\frac{Ax + B}{x^2 + 16} + \frac{Dx + E}{x^2 + 25} = \frac{22x^2 + 36x + 460}{(x^2 + 16)(x^2 + 25)}.$$

Multiplying both sides of this equation by $(x^2 + 16)(x^2 + 25)$, we get

$$(Ax + B)(x^2 + 25) + (x^2 + 16)(Dx + E) = 22x^2 + 36x + 460$$

Multiplying this becomes

$$Ax^3 + Bx^2 + 25Ax + 25B + Dx^3 + Ex^2 + 16Dx + 16E = 22x^2 + 36x + 460$$
$$(A + D)x^3 + (B + E)x^2 + (25A + 16D)x + (25B + 16E) = 22x^2 + 36x + 460.$$

In order to make this equality true we must choose $A, B, D,$ and E such that

$$A + D = 0$$
$$B + E = 22$$
$$25A + 16D = 36$$
$$25B + 16E = 460.$$

The first equation says $D = -A$. Substituting this into the third equation, we get

$$25A + 16(-A) = 36$$
$$A = 4$$
$$D = -4.$$

28

Multiply both sides of the second equation by -16 and add the result to the fourth equation.

$$-16B - 16E = -352$$
$$25B + 16E = 460$$
$$9B = 108$$
$$B = 12$$
$$E = 22 - B = 10$$

We have shown that

$$\frac{4x + 12}{x^2 + 16} + \frac{-4x + 10}{x^2 + 25} = \frac{22x^2 + 36x + 460}{(x^2 + 16)(x^2 + 25)}.$$

We can rewrite the original antiderivative problem as: find the antiderivative

$$\int \left[\frac{4x + 12}{x^2 + 16} + \frac{-4x + 10}{x^2 + 25} \right] dx.$$

In order to integrate we must split each fraction into a sum. We rewrite the integral as

$$2 \int \frac{2x\, dx}{x^2 + 16} + 12 \int \frac{dx}{x^2 + 16} - 2 \int \frac{2x\, dx}{x^2 + 25} + 10 \int \frac{dx}{x^2 + 25}$$

In order to find the first antiderivative we use the substitution $u = x^2 + 16$. In order to find the second antiderivative we use the substitution $x = 4u$. In order to find the third antiderivative we use the substitution $u = x^2 + 25$. In order to find the last antiderivative we use the substitution $x = 5u$. After finding all antiderivatives we have

$$= 2\ln(x^2 + 16) + 3\arctan\frac{x}{4} - 2\ln(x^2 + 25) + 2\arctan\frac{x}{5} + C.$$

Example 5. Find the antiderivative

$$\int \frac{2x^2 + 27x + 82}{(x+5)^2(x+4)}\, dx.$$

Solution. Since $(x + 5)^2 = x^2 + 10x + 25$ is a quadratic we should begin by assuming that the fraction is of the form $\dfrac{Ax + B}{(x + 5)^2}$. This works except that it does not yield an expression that is easily integrated. We need to go back one more step. When we have a linear factor squared as we do in this case we assume that it can be written in the form

$$\frac{A}{(x + 5)^2} + \frac{B}{x + 5} \quad \text{instead of} \quad \frac{Ax + B}{(x + 5)^2}.$$

For this integrand we want to find constants $A, B,$ and D such that

$$\frac{A}{(x + 5)^2} + \frac{B}{x + 5} + \frac{D}{x + 4} = \frac{2x^2 + 27x + 82}{(x + 5)^2(x + 4)}.$$

Adding the fractions on the left of the equal sign, we get

$$\frac{A(x + 4) + B(x + 5)(x + 4) + D(x + 5)^2}{(x + 5)^2(x + 4)} = \frac{2x^2 + 27x + 82}{(x + 5)^2(x + 4)}.$$

Since the denominators are the same, the numerators of these fractions must be equal.

$$Ax + 4A + Bx^2 + 9Bx + 20B + Dx^2 + 10Dx + 25D = 2x^2 + 27x + 82.$$

Collecting terms this becomes

$$(B + D)x^2 + (A + 9B + 10D)x + (4A + 20B + 25D) = 2x^2 + 27x + 82.$$

In order for this equation to be true the numbers $A, B,$ and D must be such that

$$B + D = 2$$
$$A + 9B + 10D = 27$$
$$4A + 20B + 25D = 82.$$

Solve the first equation for D to get $D = 2 - B$ and substitute this into the other two equations.

$$A + 9B + 10(2 - B) = 27$$
$$4A + 20B + 25(2 - B) = 82$$

These equations simplify to

$$A - B = 7$$
$$4A - 5B = 32$$

Multiply both sides of the top equation by -4.

$$-4A + 4B = -28$$
$$4A - 5B = 32$$

Adding these two equations, we get

$$-B = 4$$
$$B = -4$$
$$A = 7 + B = 3.$$

Substituting $A = 3$ and $B = -4$ into the second equation, we get

$$3 + 9(-4) + 10D = 27$$
$$D = 6.$$

We have shown that

$$\frac{3}{(x+5)^2} - \frac{4}{x+5} + \frac{6}{x+4} = \frac{2x^2 + 27x + 82}{(x+5)^2(x+4)}.$$

We can rewrite the given problem as: find the antiderivative

$$\int \left[\frac{3}{(x+5)^2} - \frac{4}{x+5} + \frac{6}{x+4} \right] dx.$$

Integrating each antiderivative separately we find

$$\int \frac{2x^2 + 27x + 82}{(x+5)^2(x+4)} dx = -\frac{3}{(x+5)} - 4\ln|x+5| + 6\ln|x+4| + C.$$

Example 6. Find the antiderivative

$$\int \frac{3x^4 + 2x^3 + 23x^2 + 10x + 50}{x^2 + 4} dx.$$

Solution. When the degree (highest power of x) of the polynomial in the numerator is greater than or equal to the degree of the polynomial in the denominator the fraction is *not* a "proper fraction". In these cases when the fraction is not a "proper fraction" we must start by dividing the numerator by the denominator using long division.

31

$$
\begin{array}{r}
3x^2 + 2x + 11 \\
x^2 + 4\sqrt{3x^4 + 2x^3 + 23x^2 + 10x + 50} \\
\underline{3x^4 \qquad + 12x^2 \qquad\qquad} \\
2x^3 + 11x^2 + 10x + 50 \\
\underline{2x^3 \qquad + 8x \qquad} \\
11x^2 + 2x + 50 \\
\underline{11x^2 \qquad + 44} \\
2x + 6
\end{array}
$$

This division tells us that

$$
\frac{3x^4 + 2x^3 + 23x^2 + 10x + 50}{x^2 + 4} = 3x^2 + 2x + 11 + \frac{2x + 6}{x^2 + 4}.
$$

This equation allows us to rewrite the given antiderivative as

$$
\int \left[3x^2 + 2x + 11 + \frac{2x + 6}{x^2 + 4} \right] dx.
$$

We then find the antiderivative as follows.

$$
\int (3x^2 + 2x + 11)dx + \int \frac{2x}{x^2 + 4} dx + 6 \int \frac{dx}{x^2 + 4}
$$

$$
= x^3 + x^2 + 11x + \ln(x^2 + 4) + 3\arctan(x/2) + C.
$$

We will not study any more methods for finding antiderivatives. There are a lot of other methods we might choose to learn that enable us to find certain antiderivatives. The real use that is made of finding antiderivatives is to evaluate definite integrals using the Fundamental Theorem of Calculus. However, with modern computer programs any really difficult to evaluate definite integrals would be evaluated using numerical methods. There is no need to be able to find difficult to evaluate antiderivatives.

Also we must be ready to use numerical methods for another reason. It is fairly easy to write an antiderivative problem for which the antiderivative can not be expressed using the collection of functions for which we have a name. Consider

$$\int \frac{x^2 dx}{[4x^2 + \cos 3x]^{3/2}}.$$

This antiderivative exists but it can not be expressed using any combination of powers, sine functions, exponential functions and so forth. In other words we have no way of writing down the antiderivative. This means that

$$\int_0^2 \frac{x^2 dx}{[4x^2 + \cos(3x)]^{3/2}}$$

can only be evaluated using numerical methods.

Exercises

Unadd the following fractions.

1. $\dfrac{3x - 11}{(2x + 9)(x + 8)}$

2. $\dfrac{3x + 57}{x^2 + 12x + 35}$

3. $\dfrac{x^2 + 38x - 27}{(x^2 + 25)(x + 8)}$

4. $\dfrac{x^3 + 33x}{x^2 + 25}$

Find the following antiderivatives.

5. $\displaystyle\int \frac{2x + 18}{(x + 4)(x + 6)}$

6. $\displaystyle\int \frac{2x + 71}{x^2 + 5x - 24} dx$

7. $\displaystyle\int \frac{x^2 + 53x + 90}{(x^2 + 9)(x + 7)} dx$

8. $\displaystyle\int \frac{7x^2 - 54x + 94}{(x^2 + 16)(x^2 + 25)} dx$

9. $\displaystyle\int \frac{43x + 144}{(x + 3)^2(x + 8)} dx$

10. $\displaystyle\int \frac{3x^3 + 8x^2 + 43x + 116}{x^2 + 16} dx$

11. $\displaystyle\int \frac{x^2 dx}{4x^2 + 25}$

12. $\displaystyle\int \frac{1}{x\sqrt{5x + 16}} dx$, Hint: Let $u^2 = 5x + 16$.

13. $\displaystyle\int \frac{1}{x\sqrt{4x + 9}} dx$

14. $\displaystyle\int \frac{6x^2 + 14x + 49}{(x - 8)(x^2 + 4x + 13)} dx$

15. $\displaystyle\int \frac{1}{6 + \sqrt{4x + 9}} dx$

5517 Substitution Rule for Definite Integrals

Let us suppose we start with the definite integral

$$\int_a^b f(x)dx,$$

and we wish to make a substitution. Let us use the substitution $x = h(u)$. Solving for u we write this as $u = m(x)$. Note that $h(m(x)) = x$ and $m(h(u)) = u$. In order to make a substitution we always need both $x = h(u)$ and $u = m(x)$. Suppose $F(x)$ is an antiderivative of $f(x)$, that is, $F'(x) = f(x)$. Using the chain rule

$$\frac{d}{du}F(h(u)) = \frac{dF}{dh}\frac{dh}{du} = F'(h(u))h'(u) = f(h(u))h'(u).$$

It follows that

$$\int f(h(u))h'(u)du = \int \frac{d}{du}[F(h(u))]du = F(h(u)) + C.$$

As a definite integral

$$\int_{m(a)}^{m(b)} f(h(u))h'(u)du = F(h(m(b))) - F(h(m(a)))$$

$$= F(b) - F(a) = \int_a^b f(x)dx.$$

We used the fact that $h(m(x)) = x$. It follows that the following theorem is true.

Theorem. Substitution Rule for Definite Integrals.

$$\int_a^b f(x)dx = \int_{m(a)}^{m(b)} f(h(u))h'(u)du$$

where $x = h(u)$ and $u = m(x)$.

34

The substitution rule says that when using a substitution in a definite integral we put everything in terms of the new variable u. Most important this includes the limits of integration. When using substitution we need to remember the definition of differential. Recall that when $x = h(u)$, then the differential is $dx = h'(u)du$.

The Substitution Rule for Definite Integrals gives one reason why we use the differential dx to indicate the variable of integration in an integral. Since the integrand in an integral may contain more than one variable we must have some way to indicate the variable of integration. We indicate that the variable of integration is x by writing dx beside the integrand. When making the substitution $x = h(u)$ we replace dx by the differential $h'(u)du$. The coefficient of the integrand behaves like a differential.

Example 1. Use the Substitution Rule for Definite Integrals and the substitution $u = 4x + 1$ to transform the integral

$$\int_0^6 \sqrt{4x + 1}\,dx.$$

Solution. Since $u = 4x + 1$, then $x = (1/4)(u - 1)$. In the notation of the Substitution Rule for Definite Integrals $x = h(u) = (1/4)(u - 1)$ and $u = m(x) = 4x + 1$. The differential is $dx = (1/4)du$.

When $x = 0$, we have $u = m(0) = 4(0) + 1 = 1$.

When $x = 6$, we have $u = m(6) = 4(6) + 1 = 25$.

Substituting these expressions into the Substitution Rule for Definite Integrals, we get

$$\int_0^6 \sqrt{4x + 1}\,dx = \int_1^{25} u^{1/2}(1/4)\,du = \frac{1}{4}\int_1^{25} u^{1/2}\,du.$$

Example 2. Use the substitution $x = (5/4)u$ and the Substitution Rule for Definite Integrals to transform the following definite integral

$$\int_{5/8}^{5\sqrt{3}/8} \frac{dx}{\sqrt{25 - 16x^2}}.$$

Solution. Since $x = h(u) = (5/4)u$, then $u = m(x) = (4/5)x$ and the differential is $dx = (5/4)du$. Also $25 - 16x^2 = 25 - 16(5u/4)^2 = 25 - 25u^2 = 25(1 - u^2)$.

When $x = 5/8$, we have $u = m(5/8) = (4/5)(5/8) = 1/2$.

When $x = 5\sqrt{3}/8$, we have $u = m(5\sqrt{3}/8) = (4/5)(5\sqrt{3}/8) = \sqrt{3}/2$.

Substituting these expressions into the Substitution Rule for Definite Integrals, we get

$$\int_{5/8}^{5\sqrt{3}/8} \frac{dx}{\sqrt{25 - 16x^2}} = \int_{1/2}^{\sqrt{3}/2} \frac{(5/4)du}{\sqrt{25(1 - u^2)}} = \frac{1}{4} \int_{1/2}^{\sqrt{3}/2} \frac{du}{\sqrt{1 - u^2}}$$

Example 3. Use the Substitution Rule for Definite Integrals and the substitution $x = (5/4)\sin u$ to transform the integral

$$\int_{5/8}^{5\sqrt{3}/8} \frac{dx}{\sqrt{25 - 16x^2}}.$$

Solution. Note that this is the same integral as in Example 2. Since $x = h(u) = (5/4)\sin u$, then $u = m(x) = \arcsin(4x/5)$. The differential is $dx = (5/4)(\cos u)du$. Also $25 - 16x^2 = 25 - 25\sin^2 u = 25(1 - \sin^2 u) = 25\cos^2 u$.

We usually do this calculation as follows. When $x = 5/8$, we have $u = \arcsin[4(5/8)/5] = \arcsin(1/2) = \pi/6$.

When $x = 5\sqrt{3}/8$, then $u = \arcsin[4(5\sqrt{3}/8)/5] = \arcsin(\sqrt{3}/2) = \pi/3$.

Substituting $x = \sin u$, $m(5/8) = \pi/6$, and $m(5\sqrt{3}/8) = \pi/3$ into the Substitution Theorem for Definite Integrals, we get

$$\int_{5/8}^{5\sqrt{3}/8} \frac{dx}{\sqrt{25 - 16x^2}} = \int_{\pi/6}^{\pi/3} \frac{(5/4)\cos u \, du}{\sqrt{25\cos^2 u}} = \frac{1}{4} \int_{\pi/6}^{\pi/3} du.$$

Example 4. Use the substitution $u = \sqrt{25 + 16x^2}$ to transform the integral

$$\int_0^3 \frac{x^3 dx}{\sqrt{25 + 16x^2}}.$$

36

Solution. An alternate and better form of $u = \sqrt{25 + 16x^2}$ is $u^2 = 25 + 16x^2$ with $u \geq 0$. Also we normally start by giving x in terms of u. We would say that the substitution is

$$x^2 = \frac{1}{16}[u^2 - 25], \text{ or } x = h(u) = (1/4)\sqrt{u^2 - 25}.$$

Taking differentials, we get

$$2x\,dx = (2/16)[u]\,du$$
$$x\,dx = (1/16)u\,du.$$

Note that $x^3\,dx = (x^2)(x\,dx) = (1/16)(u^2 - 25)(1/16)u\,du$.

When $x = 0$, we have $u^2 = 25$ or $u = 5$.

When $x = 3$, we have $u^2 = 25 + 16(9) = 169$ or $u = 13$.

Substituting these values into the Substitution Rule for Definite Integrals, we get

$$\int_0^3 \frac{x^3\,dx}{\sqrt{25 + 16x^2}} = \int_5^{13} \frac{(1/16)(u^2 - 25)(1/16)u\,du}{u} = \frac{1}{256}\int_5^{13}(u^2 - 25)\,du.$$

Since it is so easy, let us evaluate this integral

$$\frac{1}{256}\int_5^{13}(u^2 - 25)\,du = \frac{1}{256}\left[\frac{u^3}{3} - 25u\,\Big|_5^{13}\right]$$

$$= \frac{1}{256}\left[\frac{1}{3}(13^3 - 5^3) - 25(13 - 5)\right] = \frac{1472}{3(256)}.$$

Example 5. Use the substitution $x = (5/4)\tan\theta$ to transform the integral

$$\int_0^3 \frac{x^3\,dx}{\sqrt{25 + 16x^2}}.$$

Solution. Since $x = h(\theta) = (5/4)\tan\theta$, then $\theta = m(x) = \arctan[4x/5]$. The differential is $dx = (5/4)\sec^2\theta\,d\theta$. Note that $25 + 16x^2 = 25 + 16(5/4)^2\tan^2\theta = 25(1 + \tan^2\theta) = 25\sec^2\theta$.

When $x = 0$, we have $\theta = m(0) = \arctan(0) = 0$

When $x = 3$, we have $\theta = m(3) = \arctan(12/5)$.

Substituting these values into the Substitution Rule for Definite Integrals, we get

$$\int_0^3 \frac{x^3 dx}{\sqrt{25+16x^2}} = \int_0^{\arctan(12/5)} \frac{(125/64)(\tan^3 \theta)(5/4)\sec^2 \theta}{5\sec \theta} d\theta$$

$$= \frac{125}{256} \int_0^{\arctan(12/5)} \tan^3 \theta \sec \theta d\theta.$$

Note that $\sec \theta > 0$ for $0 \le \theta \le \arctan(12/5)$. Just for fun lets evaluate this integral.

$$\frac{125}{256} \int_0^{\arctan(12/5)} \tan^3 \theta \sec \theta d\theta = \frac{125}{256} \int_0^{\arctan(12/5)} (\sec^2 \theta - 1)(\tan \theta \sec \theta) d\theta$$

$$= \frac{125}{256} \left[\frac{\sec^3 \theta}{3} - \sec \theta \right] \Bigg|_0^{\arctan(12/5)}$$

Note that $\sec[\arctan(12/5)] = 13/5$ and $\sec(0) = 1$. Using these values, we have

$$= \frac{125}{256} \left[\frac{1}{3}\left(\frac{13}{5}\right)^3 - \frac{13}{5} - \frac{1}{3}(1) + 1 \right] = \frac{1472}{3(256)}.$$

Example 6. Use the substitution $x = (5/4)\sinh u$ to transform the integral

$$\int_0^3 \frac{x^3 dx}{\sqrt{25+16x^2}}.$$

Solution. Since $x = (5/4)\sinh u$, then $u = m(x) = \operatorname{arcsinh}(4x/5)$. The differential is $dx = (5/4)(\cosh u)du$. Note that $25 + 16x^2 = 25 + 16(5/4)^2 \sinh^2 u = 25[1 + \sinh^2(u)] = 25\cosh^2 u$. Note that $\sqrt{\cosh^2(u)} = \cosh(u)$ for all u.

When $x = 0$, we have $u = m(0) = \operatorname{arcsinh}(0) = 0$.

When $x = 3$, we have $u = m(3) = \operatorname{arcsinh}(12/5)$.

Substituting these values into the Substitution Rule for Definite Integrals, we get

$$\int_0^3 \frac{x^3 dx}{\sqrt{25+16x^2}} = \int_0^{\operatorname{arcsinh}(12/5)} \frac{(5/4)^3(\sinh^3 u)(5/4)(\cosh u)du}{\sqrt{25\cosh^2 u}}$$

38

$$\frac{125}{256} \int_0^{\operatorname{arc\,sinh}(12/5)} \sinh^3 u \; du.$$

Since it is so interesting, let us evaluate this integral.

$$\frac{125}{256} \int_0^{\operatorname{arc\,sinh}(12/5)} \sinh^3 u \; du = \frac{125}{256} \int_0^{\operatorname{arc\,sinh}(12/5)} (\cosh^2 u - 1)(\sinh u) du$$

$$= \frac{125}{256} \left[\frac{\cosh^3 u}{3} - \cosh u \Big|_0^{\operatorname{arc\,sinh}(12/5)} \right]$$

First $\cosh(0) = 1$. Again we need the identity $\sinh^2 u = \cosh^2 u - 1$. When $u = \operatorname{arcsinh}(12/5)$ this identity says $\sinh^2[\operatorname{arcsinh}(12/5)] + 1 = \cosh^2[\operatorname{arcsinh}(12/5)]$, on $\frac{144}{25} + 1 = \cosh^2[\operatorname{arcsinh}(12/5)]$. It follows that $\cosh[\operatorname{arcsinh}(12/5)] = 13/5$. Using these facts the above is equal to

$$= \frac{125}{256} \left[\frac{1}{3}\left(\frac{13}{5}\right)^3 - \frac{1}{3} - \frac{13}{5} + 1 \right] = \frac{1472}{3(256)}.$$

Example 7. Use the substitution $u^2 = 3x + 1$ ($u = \sqrt{3x+1}$) or $x = (1/3)(u^2 - 1)$ to transform the integral

$$\int_1^8 \frac{x}{1 - \sqrt{3x+1}} dx.$$

Solution. Note that $1 - \sqrt{3x+1} < 0$ for $1 \le x \le 8$. Since $u = m(x) = \sqrt{3x+1}$ and we have $x = h(u) = (1/3)(u^2 - 1)$. The differential is $dx = (1/3)(2u)du = (2/3)u \; du$.

When $x = 1$, we have $u^2 = 4$ or $u = 2$.

When $x = 8$, we have $u^2 = 25$ or $u = 5$.

Substituting into the Substitution Rule for Definite Integrals the value of the given integral is the same as the value of

$$\int_2^5 \frac{(1/3)(u^2 - 1)(2/3)u du}{1 - u} = -\frac{2}{9} \int_2^5 (u^2 + u) du.$$

39

Example 8. Use the transformation $u^3 = 8+13x^2$ to transform the integral

$$\int_0^3 \frac{x^3}{(8+13x^2)^{2/3}}\,dx.$$

Solution. The substitution $u^3 = 8 + 13x^2$ should actually be expressed as $x^2 = (1/13)(u^3 - 8)$ or even as $x = h(u) = \sqrt{(1/13)(u^3 - 8)}$. Note that $u \geq 2$. The differentials are $3u^2\,du = 26x\,dx$ or $x\,dx = (3/26)u^2\,du$.

When $x = 0$, we have $u^3 = 8$ or $m(0) = u = 2$.

When $x = 3$, we have $u^3 = 8 + 13(9) = 125$ or $u = 5$.

Substituting into the Substitution Rule for Definite Integrals we get that

$$\int_0^3 \frac{x^3}{(8+13x^2)^{2/3}}\,dx = \int_2^5 \frac{(1/13)(u^3 - 8)(3/26)u^2\,du}{u^2}$$

$$= \frac{3}{2(169)} \int_2^5 (u^3 - 8)\,du.$$

Exercises

1. Use the substitution $x = (5/6)u$ and the Substitution Rule for Definite Integrals to transform the following definite integral

$$\int_{5/12}^{5\sqrt{3}/12} \frac{dx}{\sqrt{25 - 36x^2}}.$$

2. Use the substitution $x = (5/6)\sin\theta$ to transform the following integral. Simplify the resulting integrand.

$$\int_{5/12}^{5\sqrt{3}/12} \frac{x^3\,dx}{\sqrt{25 - 36x^2}}.$$

3. Use the substitution $u^2 = 25 - 36x^2$ or $x = (1/6)\sqrt{25 - u^2}$ to transform the following integral. Simplify the resulting integrand.

$$\int_{5/12}^{5\sqrt{3}/12} \frac{x^3\,dx}{\sqrt{25 - 36x^2}}.$$

4. Use the substitution $u = 5x + 9$ to transform the following integral

$$\int_0^8 x^2 \sqrt{5x + 9}\, dx.$$

5. Use the substitution $u = \cos\theta$ to transform the following integral. Simplify the resulting integrand.

$$\int_0^{\pi/2} (\cos^4\theta + 4\cos\theta)(\sin\theta)\, d\theta.$$

6. Use the substitution given by $u^2 = 4x + 5$ to transform the following integral.

$$\int_1^5 \frac{x}{3 + \sqrt{4x + 5}}\, dx.$$

7. Use the substitution given by $u^2 = 7 + 2x^2$ to transform the following integral

$$\int_1^9 \frac{dx}{x\sqrt{7 + 2x^2}}.$$

8. Use the substitution $x = \sqrt{(1/2)(u^3 - 27)}$ to transform the integral

$$\int_0^7 \frac{x^3}{(27 + 2x^2)^{2/3}}\, dx.$$

9. Use the substitution $x = (3/2)\sinh u$ to transform the following definite integral:

$$\int_0^2 \frac{x^3\, dx}{\sqrt{9 + 4x^2}}$$

5521 Numerical Integration

We have discussed two operations which we called integration. The first was "find the antiderivative" or "find the indefinite integral". The second was "find the definite integral". A couple of examples of antiderivatives or indefinite integrals are

$$\int \frac{dx}{1+x^2} \qquad\qquad \int \frac{x}{1+x^3}\,dx$$

A couple of examples of definite integrals are

$$\int_0^1 \frac{dx}{1+x^2} \qquad\qquad \int_0^{10} \frac{x}{1+x^4}\,dx$$

Almost all applications that involve finding an integral require us to find the value of a definite integral rather than find an antiderivative. The most often used reason for finding antiderivatives is to use the Fundamental Theorem of Calculus to find the value of a definite integral. In the past when finding antiderivatives we have concentrated on finding the antiderivative of functions $f(x)$ whose antiderivative is some combination of our elementary functions. Recall that elementary functions are those functions such as cosine and exponential that we all know and love. There are other functions which have less well known names such as Bessel function or the error function. There are also a large number of functions which are so seldom used that they do not have names. We can easily construct functions by making combinations of our elementary functions that have as their antiderivatives these functions with no name. For such functions we have no function which is the antiderivative for the given function. If we are trying to find the definite integral of such a function we are unable to use the Fundamental Theorem of Calculus because we do not have an antiderivative function. In these cases the definite integral has a value we are just unable to find the value using the Fundamental Theorem of Calculus. A couple of examples of such definite integrals are

$$\int_0^\pi \frac{dx}{4+3\cos^3 x} \text{ and } \int_1^5 \frac{x\,dx}{\sqrt{1+x^5}}.$$

Suppose we have such a definite integral and really need to find the value of this integral. A numerical approximation to the value would be good.

Suppose we want to find an approximate value to the definite integral $\int_a^b f(x)dx$. Recall that the value of this integral is defined as the limit of a sequence of numbers. The definition of the definite integral is

$$\int_a^b f(x)dx = \lim_{n \to \infty} \frac{b-a}{n} \sum_{k=1}^{n} f\left(a + \frac{2k-1}{2n}(b-a)\right).$$

This means that as n gets larger and larger the numbers

$$\frac{b-a}{n} \sum_{k=1}^{n} f\left(a + \frac{2k-1}{2n}(b-a)\right)$$

get closer and closer to the value of $\int_a^b f(x)dx$. This clearly means that for a large enough value of n the number

$$\frac{b-a}{n} \sum_{k=1}^{n} f\left(a + \frac{2k-1}{2n}(b-a)\right)$$

is an approximation to $\int_a^b f(x)dx$. Using $\frac{b-a}{n} \sum_{k=1}^{n} f\left(a + \frac{2k-1}{2n}(b-a)\right)$ to approximate $\int_a^b f(x)dx$ is usually referred to as "approximating the value of the definite integral using the midpoint rule". Clearly in order to get a really accurate approximation to a given definite integral we would need to use a large value of n. However, when doing examples by hand we use a small value of n in order to shorten calculations. Small values of n still make the ideas clear, but keep us from doing the same calculation so many times. We save that for computers.

Example 1. Consider the integral $\int_2^8 x^3 dx$. Find the midpoint approximation

$$\frac{b-a}{n} \sum_{k=1}^{n} f\left(a + \frac{2k-1}{2n}(b-a)\right)$$

for this integral.

Solution. For this problem $a = 2$, $b = 8$, and $f(x) = x^3$. Using $a = 2$ and $b = 8$, we have

$$a + \frac{2k-1}{2n}(b-a) = 2 + \frac{2k-1}{2n}(8-2) = 2 + \frac{3(2k-1)}{n}.$$

Using $f(x) = x^3$, we get

$$f\left(a + \frac{2k-1}{2n}(b-a)\right) = \left[2 + \frac{3(2k-1)}{n}\right]^3$$

$$= 2^3 + 3(2)^2 \frac{3(2k-1)}{n} + 3(2)\left[\frac{3(2k-1)}{n}\right]^2 + \left[\frac{3(2k-1)}{n}\right]^3$$

$$= 8 + \frac{36}{n}(2k-1) + \frac{54}{n^2}(2k-1)^2 + \frac{27}{n^3}(2k-1)^3.$$

Finding the sum

$$\sum_{k=1}^{n} f\left(a + \frac{2k-1}{2n}(b-a)\right)$$

$$= \sum_{k=1}^{n} 8 + \sum_{k=1}^{n} \frac{36}{n}(2k-1) + \sum_{k=1}^{n} \frac{54}{n^2}(2k-1)^2 + \sum_{k=1}^{n} \frac{27}{n^3}(2k-1)^3$$

$$= 8n + \frac{36}{n}\sum_{k=1}^{n}(2k-1) + \frac{54}{n^2}\sum_{k=1}^{n}(2k-1)^2 + \frac{27}{n^3}\sum_{k=1}^{n}(2k-1)^3.$$

Recall that $\sum_{k=1}^{n}(2k-1) = n^2$, $\sum_{k=1}^{n}(2k-1)^2 = \frac{n}{3}(4n^2-1)$ and $\sum_{k=1}^{n}(2k-1)^3 = 2n^4 - n^2$. Using these facts the above expression is equal to

$$8n + \frac{36}{n}(n^2) + \frac{54}{n^2}\left(\frac{n}{3}\right)(4n^2 - 1) + \frac{27}{n^3}(2n^4 - n^2)$$

$$= 8n + 36n + 72n - \frac{18}{n} + 54n - \frac{27}{n} = 170n - \frac{45}{n}.$$

Finally, we have

$$\frac{b-a}{n}\sum_{k=1}^{n} f\left(a + \frac{2k-1}{2n}(b-a)\right) = \frac{6}{n}\left[170n - \frac{45}{n}\right] = 1020 - \frac{270}{n^2}.$$

The midpoint approximation for the integral

$$\int_{2}^{8} x^3\,dx \text{ for any } n \text{ is } 1020 - \frac{270}{n^2}.$$

44

The actual value of the integral is

$$\int_2^8 x^3 dx = \lim_{n \to \infty} \left[1020 - \frac{270}{n^2} \right] = 1020.$$

If $n = 25$, the midpoint approximation is

$$1020 - \frac{270}{25^2} = 1020 - 0.432 = 1019.568.$$

The error from using the midpoint approximation with $n = 25$ to approximate the value of the integral is -0.432. The actual error in using the midpoint approximation is $-\frac{270}{n^2}$ for any n. Obviously, for integrals with complicated integrands we cannot find an exact expression for the integral.

Example 2. Approximate the value of the following definite integral using the midpoint rule with $n = 6$.

$$\int_1^4 \frac{x}{1+x} dx.$$

Solution. Using $n = 6$ the approximation is

$$\frac{b-a}{6} \sum_{k=1}^6 f\left(a + \frac{2k-1}{2n}(b-a)\right).$$

Dividing the interval $1 \le x \le 4$ into 6 subintervals means that the division points are $1, 3/2, 2, 5/2, 3, 7/2$ and 4. Let $x_k^* = a + \frac{2k-1}{2n}(b-a) = 1 + \frac{2k-1}{2n}(3) = 1 + \frac{2k-1}{4}$ for $k = 1, 2, 3, 4, 5, 6$. The six midpoints of the six subintervals are

$$x_1^* = 5/4, \ x_2^* = 7/4, \ x_3^* = 9/4, \ x_4^* = 11/4, \ x_5^* = 13/4, \ x_6^* = 15/4.$$

We need the values of

$$f(x_k^*) = f\left(1 + \frac{2k-1}{4}\right) \qquad k = 1, 2, 3, 4, 5, 6.$$

45

These values are

$$f(5/4) = \frac{5/4}{1+5/4} = \frac{5}{9}, \qquad f(7/4) = \frac{7/4}{1+7/4} = \frac{7}{11}$$

$$f\left(\frac{9}{4}\right) = \frac{9}{13} \qquad f\left(\frac{11}{4}\right) = \frac{11}{15} \qquad f\left(\frac{13}{4}\right) = \frac{13}{17} \qquad f\left(\frac{15}{4}\right) = \frac{15}{19}$$

$$\frac{b-a}{6} \sum_{k=1}^{6} f\left(a + \frac{2k-1}{2n}(b-a)\right) =$$

$$\frac{4-1}{6}\left[\frac{5}{9} + \frac{7}{11} + \frac{9}{13} + \frac{11}{15} + \frac{13}{17} + \frac{15}{19}\right] = 2.0859$$

Since this is really a simple integral we know its actual value. The actual value is 2.0837. The approximate value 2.0859 differs from the actual value 2.0837 by 0.0022. The error is 0.0022.

The sequence of numbers $M_n = \frac{b-a}{n}\sum_{k=1}^{n} f\left(a + \frac{2k-1}{2n}(b-a)\right)$ is the sequence used to define the definite integral $\int_a^b f(x)dx$, that is, $\lim_{n\to\infty} M_n = \int_a^b f(x)dx$. Even so the sequence $\{M_n\}$ is not the best sequence of numbers we can use to approximate $\int_a^b f(x)dx$. That is, there are other sequences of numbers $S_1, S_2, S_3, \cdots, S_n \cdots$ such that $\lim_{n\to\infty} S_n = \int_a^b f(x)dx$ and such that for a given value of n the number S_n is closer to $\int_a^b f(x)dx$ than is M_n. Put another way, we are saying that

$$\left| S_n - \int_a^b f(x)dx \right| < \left| M_n - \int_a^b f(x)dx \right|$$

Finding such a sequence $\{S_n\}$ is, of course, a very desirable thing. It means that using S_{25} to approximate the integral gives a smaller error than using M_{25} to approximate the integral. The task of a numerical analyst is to find such a sequence $\{S_n\}$. We will not discuss how a numerical analyst might go about finding such a sequence $\{S_n\}$. Instead we will just discuss another method for finding an approximate value of the integral that seems like it should give a good approximation to a definite integral. This method is commonly called "the trapezoid rule".

The method for approximating the definite integral $\int_a^b f(x)dx$ using the trapezoid rule is as follows. We divide the closed interval $a \le x \le b$ on the x axis into n subintervals of equal length. The length of each subinterval is $\dfrac{b-a}{n}$. This means that the points where we subdivide the interval $a \le x \le b$ are

$$x_k = a + \frac{b-a}{n}k \text{ with } k = 0, 1, 2, \cdots, n.$$

Note that $x_0 = a$ and $x_n = b$. We can think of $\int_a^b f(x)dx$ as the sum of n integrals where the kth integral is over the k subinterval which is $x_{k-1} \le x \le x_k$. Using a well known property of integrals, we can write

$$\int_a^b f(x)dx = \sum_{k=1}^{n} \int_{x_{k-1}}^{x_k} f(x)dx = \int_{x_0}^{x_1} f(x)dx + \int_{x_1}^{x_2} f(x)dx + \cdots + \int_{x_{n-1}}^{x_n} f(x)dx.$$

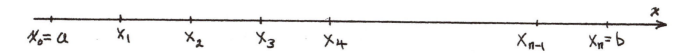

Consider the integral over a typical subinterval $x_{k-1} \le x \le x_k$. Our problem is to find an approximate value for

$$\int_{x_{k-1}}^{x_k} f(x)dx.$$

In order to make this discussion more concrete let us think of the definite integral as giving the area of a region in the xy plane. We want to find the area of

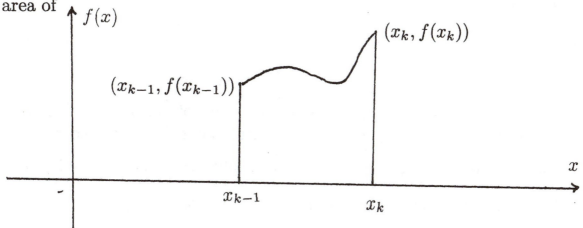

47

We need to approximate the actual curve $y = f(x)$ with a simpler curve. Nothing is simpler than a straight line. The curve $y = f(x)$ intersects the line $x = x_{k-1}$ at the point $(x_{k-1}, f(x_{k-1}))$. The curve $y = f(x)$ intersects the line $x = x_k$ at the point $(x_k, f(x_k))$. Let us find the area bounded by the straight line through the points $(x_{k-1}, f(x_{k-1}))$ and $(x_k, f(x_k))$.

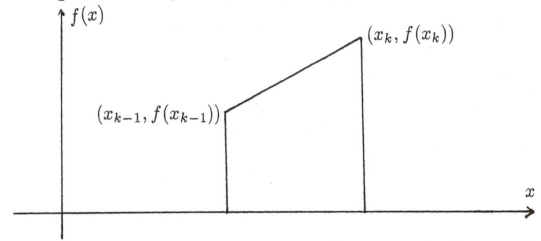

We will then use this area to approximate the area under the curve between these lines. This figure is a trapezoid. The area of this trapezoid is

$$\frac{1}{2}[f(x_{k-1}) + f(x_k)][x_{k-1} - x_k].$$

The numbers $[x_k - x_{k-1}]$, $k = 1, 2, \cdots, n$ are all the same and are equal to $(b-a)/n$. Let $\Delta x = \frac{b-a}{n}$. The area of a single typical trapezoid is

$$\frac{\Delta x}{2}[f(x_{k-1}) + f(x_k)].$$

There is a trapezoid on each subinterval $x_{k-1} \leq x \leq x_k$. We use the sum of the area of all these trapezoids as an approximation to the value of the definite integral.

$$\int_a^b f(x)dx \approx \frac{\Delta x}{2} \sum_{k=1}^{n}[f(x_{k-1}) + f(x_k)].$$

When we use this sum to approximate the value of the integral we are approximating the value of the definite integral using the trapezoid rule.

When approximating $\int_{x_{k-1}}^{x_k} f(x)dx$ we could use a more complicated curve than a straight line to approximate $y = f(x)$ for $x_{k-1} \leq x \leq x_k$. This

48

would very likely give a better approximation than the area of a trapezoid. We leave such discussions to a later time when we are discussing more advanced topics in numerical analysis.

Example 3. Find an approximate value of the integral $\int_1^5 \dfrac{x}{1+x^2}\,dx$ using the trapezoid rule with $n = 8$.

Solution. First, $\Delta x = \dfrac{b-a}{n} = \dfrac{5-1}{8} = \dfrac{1}{2}$. We divide the interval of integration $1 \le x \le 5$ into $n = 8$ subintervals using the division points

$$x_k = a + \frac{k}{n}(b-a) = 1 + \frac{k}{2}, \qquad k = 0, 1, 2, \cdots, 8.$$

$$x_0 = 1, \quad x_1 = 3/2, \quad x_2 = 2, \quad x_3 = 5/2, \cdots x_8 = 5 \qquad x$$

In order to find the approximation using the trapezoid rule we need to find the sum

$$\sum_{k=1}^{8} [f(x_{k-1}) + f(x_k)]$$

$$= [f(x_0) + f(x_1)] + [f(x_1) + f(x_2)] + [f(x_2) + f(x_3)]$$
$$+ \cdots + [f(x_6) + f(x_7)] + [f(x_7) + f(x_8)]$$
$$= f(x_0) + 2f(x_1) + 2f(x_2) + 2f(x_3) + \cdots + 2f(x_7) + f(x_8).$$

Note the pattern in this sum. We now find the value of this sum.

Using $f(x) = \dfrac{x}{1+x^2}$ we compute the function values as follows:

$$f(1) = \frac{1}{1+1} = \frac{1}{2} \quad f\left(\frac{3}{2}\right) = \frac{3/2}{1+(3/2)^2} = \frac{6}{13}$$

$$f(2) = \frac{2}{5} \quad f\left(\frac{5}{2}\right) = \frac{10}{29} \quad f(3) = \frac{3}{10} \quad f\left(\frac{7}{2}\right) = \frac{14}{53}$$

$$f(4) = \frac{4}{17} \quad f\left(\frac{9}{2}\right) = \frac{18}{85} \quad f(5) = \frac{5}{26}$$

49

$$2[f(3/2) + f(2) + f(5/2) + f(3) + f(7/2) + f(4) + f(9/2)] = 4.43515$$

$$f(1) + (4.43515) + f(5) = 5.12746$$

$$\frac{\Delta x}{2}(5.12746) = \frac{1/2}{2}(5.12746) = 1.28186$$

Using the trapezoid rule with $n = 8$ we found the approximate value of the definite integral $\int_1^5 \frac{x}{1+x^2}dx$ to be 1.28186. Since this is a simple integrand we can find the exact value of this integral using the Fundamental Theorem of Calculus. The exact value is 1.28247. The error in our approximate value is

$$|E_T| = |1.28186 - 1.28247| = 0.00061$$

We can obtain a better approximation to the integral $\int_1^5 \frac{x}{1+x^2}dx$ using the trapezoid rule by using a larger value of n.

The question of how large is the difference between the approximate value of an integral obtained by a numerical method and the actual value of the integral is a very important question. This is most important when we are dealing with an integral which we cannot evaluate using the Fundamental Theorem of Calculus. For such an integral we cannot find the exact error as we did in Example 2 and Example 3. For such integrals what we need is a bound for the error make in a numerical approximation.

Definition. A bound is a number such that we can say for sure that the error is no larger than the bound.

The actual error is usually quite a bit less than the bound and so the error is not equal to the bound. But with a bound we know for sure

$$|\text{actual error}| \leq |\text{bound}|.$$

A formula for the error bound in the trapezoid rule is given by the following theorem.

Error Bound for Trapezoid Rule. Suppose $|f''(x)| \leq K$ for $a \leq x \leq b$. If E_T denotes the actual error made in approximating the integral $\int_a^b f(x)dx$ using the trapezoid rule, then

$$|E_T| \leq \frac{K(b-a)^3}{12n^2}.$$

Using this theorem we can compute an error bound for the integral $\int_1^5 \frac{x}{1+x^2}dx$ in the case $n = 8$. This is the integral from Example 3.

Finding the value of K is the most difficult part. Note that K is the maximum value of $|f''(x)|$ on the interval $a \leq x \leq b$. We learned to find such maximum values earlier. First note that we are looking at $f''(x)$ and not $f(x)$.

$$f(x) = \frac{x}{1+x^2}$$

$$f'(x) = \frac{1-x^2}{(1+x^2)^2}$$

$$f''(x) = \frac{2x^3 - 6x}{(1+x^2)^3}.$$

We need to find the critical points of the function $f''(x)$. Our usual method is to find the derivative of $f''(x)$. This is somewhat long and we skip the details.

$$f'''(x) = \frac{-6(x^4 - 6x^2 + 1)}{(1+x^2)^4}.$$

Note that $1 + x^2 \geq 1 > 0$. The denominator is never zero. The derivative $f'''(x)$ exists for all values of x. The function $g(x) = x^4 - 6x^2 + 1$ is equal to zero for four real values of x. The values of x for which $g(x) = 0$ are $x = \pm\sqrt{3+\sqrt{8}}$ and $x = \pm\sqrt{3-\sqrt{8}}$. Note that $\sqrt{3+\sqrt{8}} = \sqrt{2}+1$ and $\sqrt{3-\sqrt{8}} = \sqrt{2}-1$. Only one of these numbers is in the interval $1 \leq x \leq 5$. That value is $x = \sqrt{3+\sqrt{8}} \approx 2.414$. The only critical value of $f''(x)$ is $x = 2.414$. The end points are $x = 1$ and $x = 5$. Note that $f'''(1) = \frac{3}{2} > 0$, and that $f'''(5) = -\frac{6(476)}{456976} < 0$. Thus, it follows that $f'''(x) > 0$ for $1 < x < 2.414$ for $f'''(x) < 0$ for $2.414 < x < 5$.

51

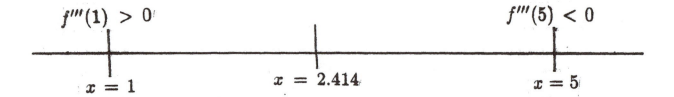

$f'''(1) > 0$ $f'''(5) < 0$

$x = 1$ $x = 2.414$ $x = 5$

Applying the first derivative test we conclude that the function $f''(x)$ has a local maximum at $x = 2.414$. Since $f'''(1) > 0$, we apply the Left End Point Theorem at a left end point and conclude that $f''(x)$ has a local minimum value at $x = 1$. Since $f'''(5) < 0$, we apply the Right End Point Theorem at a right end point and conclude that $f''(x)$ has a local minimum value at $x = 5$. The values of $f''(x)$ at these points are

$$f''(1) = \frac{-4}{8} = -0.5$$
$$f''(2.414) = 0.0429$$
$$f''(5) = \frac{220}{17576} = 0.0125$$

We need the maximum value of $|f''(x)|$ not $f''(x)$. A local maximum value of $|f''(x)|$ occurs where $f''(x)$ has a local maximum or where $f''(x)$ has a local minimum. For this function the largest maximum value of $|f''(x)|$ actually occurs where $f''(x)$ has a local minimum, that is, $|f''(x)| = |-0.5| = 0.5$ is the largest value of $|f''(x)|$ for $1 \le x \le 5$. We have

$$|f''(x)| \le 0.5 \text{ for } 1 \le x \le 5.$$

The value of K is $K = 0.5$. It is relatively easy to find the other numbers to substitute into the bound. The numbers a and b are just the limits of integration. For this integral $a = 1$ and $b = 5$. Also we are considering the case when $n = 8$. The bound is given by

$$|E_T| \le \frac{k(b-a)^3}{12n^2} = \frac{(0.5)(4)^3}{12(8)^2} = 0.04167$$

The actual error 0.00061 is much smaller than this error bound 0.04167 given by the error bound formula. This is usually the case. The actual error made in the approximation is much less than the error bound found using the formula found in Error Bound for Trapezoid Rule Theorem.

We can obtain a better approximation to the integral $\int_1^5 \frac{x}{1+x^2} dx$ using the trapezoid rule by using a larger value of n. This is reflected in the fact that the error bound is smaller when we use a larger value of n. When $n = 50$ the error bound is

$$|E_T| \le \frac{(0.5)(4)^3}{12(50)^2} < 0.0011$$

Of course, the actual error using $n = 50$ would be smaller than this error bound.

The number K in the error bound is the maximum value of $|f''(x)|$ for $a \le x \le b$. This is a maximum problem on a closed interval. As we saw for $f(x) = \frac{x}{1+x^2}$ finding the maximum value of $|f''(x)|$ for $1 \le x \le 5$ is a somewhat difficult problem. The problems are much easier when the value of K is given.

Example 4. Suppose $f(x)$ is such that $|f''(x)| \le 3$ for $0 \le x \le 5$. Find the bound on the error if we approximate the integral $\int_0^5 f(x)dx$ using the trapezoid rule with $n = 100$.

Solution. If E_T denotes the actual error made in approximating the integral using the trapezoid rule, then the error bound is given by

$$\frac{K(b-a)^3}{12n^2}.$$

and we can be sure that E_T satisfies

$$|E_T| \le \frac{K(b-a)^3}{12n^2}.$$

In this case $K = 3$, $a = 0$, $b = 5$, and $n = 100$. The error bound is

$$\frac{3(5)^3}{12(100)^2} = 0.003125.$$

In the case $n = 500$ the error bound for this same integral is

$$\frac{3(5)^3}{12(500)^2} = 0.000125.$$

Let E_T denote the actual error made in approximating this integral in case $n = 500$. Although we do not know the actual value of E_T we can be sure that $|E_T| \leq 0.000125$.

When we consider evaluating an integral using a numerical approximation we certainly want the approximation to be accurate. We know that the larger the value of n the more accurate the approximation. An important question is how large must the value of n be in order to be absolutely sure that our approximate value is correct to a certain given number of decimal places?

Example 5. Suppose $f(x)$ is such that $|f''(x)| \leq 5$ for $1 \leq x \leq 4$. What is the smallest value of n such that we can be sure that the error in approximating the integral $\int_1^4 f(x)dx$ using the trapezoid rule is less than 10^{-4}, that is, correct to 4 decimal places.

Solution. The fact that $|f''(x)| \leq 5$ for $1 \leq x \leq 4$ says that $K = 5$. The error bound is given by

$$\frac{K(b-a)^3}{12n^2} = \frac{5(4-1)^3}{12n^2} = \frac{45}{4n^2}.$$

The question is: how large must n be in order to have the error bound $\frac{K(b-a)^3}{12n^2}$ less than 10^{-4}. Clearly, we have $\frac{K(b-a)^3}{12n^2} < 10^{-4}$ when n is such that

$$\frac{45}{4n^2} < 10^{-4}.$$

This is the same as

$$\frac{4n^2}{45} > 10^4$$

$$n^2 > \frac{45}{4}(10^4)$$

$$n > \frac{10^2}{2}\sqrt{45} = 335.4.$$

Since n must be an integer it follows that $n = 336$ is the smallest value of n such that we can be sure that $\frac{K(b-a)^3}{12n^2} < 10^{-4}$. We have $n = 336$ is the smallest value of n such that the error bound is less than 10^{-4}. If the error bound is 10^{-4}, then we can be sure that the actual error E_T satisfies $|E_T| \le 10^{-4}$.

Example 6. Suppose we want to approximate the integral

$$\int_1^4 (8x + 6x^2 - x^3)\,dx$$

using the trapezoid rule. What is the smallest value of n such that we can be sure that the actual error is less than 10^{-5}, that is, such that the error bound is less than 10^{-5}?

Solution. We must find n such that $\frac{K(b-a)^3}{12n^2} < 10^{-5}$. First, we need to find the value of $K = \max |f''(x)|$ for $1 \le x \le 4$.

$$f'(x) = 8 + 12x - 3x^2$$
$$f''(x) = 12 - 6x.$$

The graph of $f''(x) = 12 - 6x$ is a straight line with a negative slope. Since $f''(1) = 6$ and $f''(4) = -12$ the line goes through the points $(1, 6)$ and $(4, -12)$. It follows that when $1 \le x \le 4$ that $-12 \le 12 - 6x \le 6$. This means that $|12 - 6x| \le 12$ for $1 \le x \le 4$. We have $|f''(x)| \le 12$ for $1 \le x \le 4$. This says that $K = 12$. It follows that

$$\frac{K(b-a)^3}{12n^2} \le \frac{12(4-1)^3}{12n^2} = \frac{27}{n^2}.$$

Clearly we have $\frac{K(b-a)^3}{12n^2} < 10^{-5}$ when n is such that

$$\frac{27}{n^2} < 10^{-5}.$$

This is the same as

$$\frac{n^2}{27} > 10^5$$
$$n^2 > 27(10^5)$$
$$n > 1643.2.$$

55

Since n must be an integer, $n = 1644$ is the smallest value of n such that the error bound $\frac{K(b-a)^3}{12n^2}$ is less than 10^{-5}. This means that although the actual error made by approximating the integral using $n = 1644$ is much less than 10^{-5}, we can be absolutely sure that the error is less than 10^{-5}.

Example 7. Consider the function $f(x) = x^2 + x^{-1}$ for $1 \leq x \leq 8$. Find $K = \max |f''(x)|$ for $1 \leq x \leq 8$.

Solution. First,

$$f'(x) = 2x - x^{-2}$$
$$f''(x) = 2 + 2x^{-3}$$
$$f'''(x) = -6x^{-4}.$$

Since $f'''(x) < 0$ for $x > 0$ and $f'''(x)$ is the derivative of $f''(x)$ it follows that $f''(x)$ is decreasing for $x > 0$. Consider $f''(x)$ as a function defined on the closed interval $1 \leq x \leq 8$. Since $f''(x)$ is a decreasing function this means that $f''(x)$ has a local maximum at $x = 1$ and a local minimum at $x = 8$. Since $f''(1) = 4$ and $f''(8) = 2 + (1/256)$ it follows that $2 + (1/256) \leq f''(x) \leq 4$ for $1 \leq x \leq 8$. It follows that $|f''(x)| \leq 4$ for $1 \leq x \leq 8$. This says $K = 4$.

Exercises

1. Find an approximate value of the following definite integral using the midpoint rule with $n = 6$:

$$\int_2^8 \frac{6-x}{4+x} dx.$$

2. Find an approximate value of the following definite integral using the trapezoid rule with $n = 8$.

$$\int_0^4 \frac{x^2}{1+x^2} dx$$

3. Find an approximate value of the following definite integral using the trapezoid rule with $n = 5$.

$$\int_1^6 \frac{x}{1+x} dx.$$

4. Let $f(x) = 4x^2 - (1/2)x^3$. Find $K = \max |f''(x)|$ for $1 \leq x \leq 6$.

5. Let $f(x) = x^4 - 6x^3 + 5x^2 + 8x + 11$. Find $K = \max |f''(x)|$ for $1 \leq x \leq 4$.

6. Suppose $|f''(x)| \leq 5$ for $1 \leq x \leq 7$. Find the error bound (the maximum possible error made) when we approximate the integral $\int_1^7 f(x)dx$ using the trapezoid rule with $n = 100$. Recall $|E_T| \leq$ error bound.

7. Find $\frac{b-a}{n} \sum_{k=1}^{n} f(a + \frac{2k-1}{2n}(b-a))$ for the following integral. What is the actual error made for a given value of n when we approximate this integral using the midpoint rule.

$$\int_0^5 x^3 dx.$$

8. Suppose $|f''(x)| \leq 15$ for all x. Find the smallest value of n such that when we approximate the following integral using the trapezoid rule the error bound 10^{-4}.

$$\int_2^6 f(x)dx.$$

9. Suppose we are estimating the definite integral $\int_2^8 f(x)dx$ using the trapezoid rule. Given that $|f''(x)| \leq 6$ for all x, what is the smallest value of n we can choose and still be sure that $|E_T| \leq 10^{-6}$, that is, the error bound is less than or equal to 10^{-6}.

5525 Improper Integrals

We have used the expression $\int_a^b f(x)dx$ to denote a proper integral.

Definition Proper Integral. The integral $\int_a^b f(x)dx$ is called a proper integral if it can be evaluated using the Fundamental Theorem of Calculus.

Fundamental Theorem of Calculus. Suppose $f(x)$ is defined and continuous for $a \le x \le b$ and that $F'(x) = f(x)$ for $a \le x \le b$, then

$$\int_a^b f(x)dx = F(b) - F(a).$$

Note that a and b are finite numbers. If $f(x)$ is not continuous for $a \le x \le b$, then $\int_a^b f(x)dx$ is not a proper integral. Except for evaluating a few integrals using the definition we have always used the Fundamental Theorem of Calculus to evaluate definite integrals. All integrals which we have considered have been proper integral, that is, they could be evaluated using the Fundamental Theorem of Calculus. In order for the definite integral $\int_a^b f(x)dx$ to be proper the numbers a and b must be finite and the function $f(x)$ must be continuous for $a \le x \le b$. A definite integral that is not a proper integral is called an improper integral. We will now discuss certain improper integrals.

The expression

$$\int_a^\infty f(x)dx$$

denotes an improper integral. The expression "∞" is not a number. We often write ∞ instead of $+\infty$ when working with improper integrals. The limit "∞" makes this not a proper integral and so it is an improper integral. It is assumed that $f(x)$ is defined and continuous for each and every x such that $x \ge a$. Because this is a new kind of integral we must define what it is in terms of ideas that we already understand.

Definition of Improper Integral of First Kind. Suppose $f(x)$ is continuous for any x such that $x \ge a$, then

$$\int_a^\infty f(x)dx = \lim_{t \to +\infty} \int_a^t f(x)dx.$$

58

The integral $\int_a^t f(x)dx$ is just a proper integral on the interval $a \leq x \leq t$. Find the limit $\lim\limits_{t \to \infty}$ is also an idea which we have discussed before. Finding the value of an improper integral involves two steps both of which we already know how to do.

Example 1. Find the value of the improper integral

$$\int_1^\infty \frac{dx}{(x+5)^2}.$$

Solution. Step 1. Note that the integrand $(x+5)^{-2}$ is defined for all x such that $x \geq 1$. We first find the value of the definite integral

$$\int_1^t \frac{dx}{(x+5)^2},$$

where t can be any real number such that $t > 1$.

$$\int_1^t \frac{dx}{(x+5)^2} = \frac{(x+5)^{-1}}{-1}\bigg|_1^t = \frac{-1}{(t+5)} + \frac{1}{6}.$$

Step 2. We need to find the limit,

$$\lim_{t \to \infty}\left[\frac{1}{6} - \frac{1}{t+5}\right] = \frac{1}{6}.$$

The limit exists. When the limit exists we say that the integral is convergent. Applying the definition of improper integral the value of this improper integral is

$$\int_1^\infty \frac{dx}{(x+5)^2} = \frac{1}{6}.$$

Example 2. Find the value of the improper integral

$$\int_2^\infty \frac{dx}{\sqrt{x+7}}.$$

Solution. Note that the integrand $(x+7)^{-1/2}$ is defined for all x such that $x \geq 2$. Since one of the limits is "∞", this is an improper integral. Step 1 is to evaluate the regular (proper) integral

$$\int_2^t \frac{dx}{\sqrt{x+7}} \text{ for } t > 2.$$

59

The value of this integral is

$$\int_2^t \frac{dx}{\sqrt{x+7}} = \frac{(x+7)^{1/2}}{1/2}\Big|_2^t = 2\sqrt{t+7} - 6.$$

Next, we find the limit

$$\lim_{t\to\infty} [2\sqrt{t+7} - 6].$$

This limit does not exist. We sometimes indicate that the limit does not exist by saying that the limit is "∞". When the limit does not exist the improper integral does not have a value. In this case, we say that the improper integral is divergent.

Example 3. Evaluate the following improper integral

$$\int_3^\infty \frac{x}{(x^2 + 16)^{3/2}} dx.$$

Solution. Note that the integrand $\frac{x}{(x^2+16)^{3/2}}$ is defined for all x such that $x \geq 3$. This integral is improper because it contains the symbol "∞" as a limit. Finding the value of an improper integral is done in two steps. First, we find the value of the usual type (proper) integral

$$\int_3^t \frac{x}{(x^2 + 16)^{3/2}} dx,$$

where t is any number such that $t > 3$.

$$\int_3^t \frac{x}{(x^2 + 16)^{3/2}} dx = \frac{-1}{\sqrt{x^2 + 16}}\Big|_3^t = \frac{-1}{\sqrt{t^2 + 16}} + \frac{1}{5}.$$

The second step is to find the limit

$$\lim_{t\to\infty} \left[\frac{1}{5} - \frac{1}{\sqrt{t^2 + 16}}\right] = \frac{1}{5}.$$

The limit exists. This improper integral is convergent. It's value is

$$\int_3^\infty \frac{x}{(x^2 + 16)^{3/2}} dx = \frac{1}{5}.$$

Example 4. Evaluate the improper integral

$$\int_0^\infty \frac{x}{x^2+9}\,dx.$$

Solution. This is an improper integral because it has "∞" as a limit. Note that $\frac{x}{x^2+9}$ is defined for all x such that $x \geq 0$. The first step is to evaluate the regular integral

$$\int_0^t \frac{x}{x^2+9}\,dx,$$

where t is any number such that $t > 0$.

$$\int_0^t \frac{x}{x^2+9}\,dx = \frac{1}{2}\ln(x^2+9)\Big|_0^t = \frac{1}{2}\ln(t^2+9) - \frac{1}{2}\ln 9.$$

Next we evaluate the limit

$$\lim_{t\to\infty}\left[(1/2)\ln(t^2+9) - \ln 3\right].$$

Since $\lim_{y\to\infty}\ln(y)$ does not exist, the limit $\lim_{t\to\infty}[\ln(t^2+9)] = +\infty$, does not exist. This limit does not exist. This improper integral is divergent. This improper integral has no value.

In order for $\int_a^b f(x)\,dx$ to be a proper integral it must be true that $f(x)$ is defined and continuous for each and every value of x such that $a \leq x \leq b$. If this is not the case, then $\int_a^b f(x)\,dx$ is not a proper integral. Let us consider the case when $f(x)$ is not defined for $x = b$. In this case the integral $\int_a^b f(x)\,dx$ is not a proper integral.

Definition of Improper Integral of the Second Kind. Suppose $f(x)$ is defined and continuous for all x such that $a \leq x < b$, but $f(b)$ is not defined. In this case we define

$$\int_a^b f(x)\,dx = \lim_{t\to b^-}\int_a^t f(x)\,dx,$$

where $\lim_{t\to b^-}$ is the limit as t approaches b using only values of t that are less than b.

Example 5. Evaluate the integral

$$\int_1^{10} \frac{dx}{\sqrt{10-x}}.$$

Solution. Note that the integrand $(10-x)^{-1/2}$ is not defined for $x = 10$. This means that this integral is an improper integral. Note that $(10-x)^{-1/2}$ is defined and continuous for $1 \le x < 10$. There are two steps to evaluating this improper integral. The first step is to evaluate the regular integral

$$\int_1^t \frac{dx}{\sqrt{10-x}},$$

where t is a number such that $1 < t < 10$.

$$\int_1^t \frac{dx}{\sqrt{10-x}} = -\frac{(10-x)^{1/2}}{1/2}\Big|_1^t = -2(10-t)^{1/2} + 6.$$

The second step is to evaluate the limit

$$\lim_{t \to 10^-} [6 - 2\sqrt{10-t}] = 6.$$

This limit exists. This improper integral is convergent. The value of this improper integral is

$$\int_1^{10} \frac{dx}{\sqrt{10-x}} = 6.$$

The symbol $\lim_{t \to 10^-}$ means that we let t get closer and closer to 10 but only using values of t such that $t < 10$. Since the expression $\sqrt{10-t}$ is continuous at $t = 10$, we can find the value of the limit by replacing t with 10 in the function $\sqrt{10-t}$.

There is a great temptation when evaluating convergent improper integrals of the second kind to pretend that the integral is not improper. An incorrect solution of this example would read as follows

$$\int_1^{10} \frac{dx}{(10-x)^{1/2}} = -2(10-x)^{1/2}\Big|_1^{10} = -2(10-10)^{1/2} + 2(10-1)^{1/2} = 6.$$

62

This solution is incorrect because it applies the Fundamental Theorem of Calculus to an improper integral. The Fundamental Theorem of Calculus does not apply to improper integrals. This so called solution yields the correct value of the integral when the integral is convergent, but it is still not a correct solution. It yields the correct result because the antiderivative is continuous at the end point and so we evaluate the limit by substituting into the antiderivative. We could prove a theorem about this, but we have not done so.

Again an improper integral may not be evaluated using the Fundamental Theorem of Calculus. The theorem does not apply. Getting the correct value of the integral does not justify incorrect steps. Some people even want to apply the Fundamental Theorem of Calculus to improper integrals of the first kind by using $+\infty$ and $-\infty$ like numbers. Please do not do this as it is not only wrong, but it causes great pain in people who read it.

Example 6. Evaluate the improper integral

$$\int_0^5 \frac{dx}{5-x}.$$

Solution. Note that the integrand $(5-x)^{-1}$ is not defined for $x = 5$. Since the integrand $(5-x)^{-1}$ is not defined for the limit of integration $x = 5$, this is an improper integral of the second kind. We find the value of this integral in two steps. First, we find the value of the proper integral

$$\int_0^t \frac{dx}{5-x}$$

where t is a number such that $0 < t < 5$.

$$\int_0^t \frac{dx}{5-x} = -\ln(5-x)|_0^t = \ln 5 - \ln(5-t).$$

The second step is to find the limit.

$$\lim_{t \to 5-} [\ln 5 - \ln(5-t)].$$

Recall that $\lim_{y \to 0+} \ln y = -\infty$. The expression $\ln(0)$ is not defined. We might like to write $\lim_{t \to 5-} [\ln 5 - \ln(5-t)] = -\infty$. This limit does not exist. This integral is divergent. This improper integral does not have a value.

Example 7. Evaluate the integral

$$\int_2^8 \frac{dx}{\sqrt{x-2}}.$$

Solution. Note that the integrand $(x-2)^{-1/2}$ is defined for x such that $2 < x \le 8$, but is not defined for $x = 2$. When the integrand is not defined at the lower limit of integration it is just the same as an integrand that is not defined at the upper limit. This is an improper integral. It is also called an improper integral of the second kind. An improper integral with the integrand not defined at the lower limit is defined in a very similar way to the way an integral with an integrand undefined at the upper limit is defined. This improper integral of the second kind is defined as

$$\int_2^8 \frac{dx}{\sqrt{x-2}} = \lim_{t \to 2+} \int_t^8 \frac{dx}{\sqrt{x-2}}.$$

The symbol $\lim_{t \to 2+}$ indicates the limit as t approaches 2 but using only values of t such that $t > 2$. The first step is to evaluate the following regular integral where $2 < t < 8$.

$$\int_t^8 \frac{dx}{\sqrt{x-2}} = 2(x-2)^{1/2}\Big|_t^8 = 2(6)^{1/2} - 2\sqrt{t-2}.$$

Next we find the limit (from the plus side)

$$\lim_{t \to 2+} [2\sqrt{6} - 2\sqrt{t-2}] = 2\sqrt{6}.$$

Since $\sqrt{t-2}$ is continuous we find the value of the limit by substituting $t = 2$ into $\sqrt{t-2}$. This limit exists. This integral is convergent. The value of this improper integral is

$$\int_2^8 \frac{dx}{\sqrt{x-2}} = 2\sqrt{6}.$$

The first step when faced with the problem to evaluate a definite integral should be to ask the question: "Is this a proper integral or is this an improper integral?" If it is a proper integral then we evaluate it using the Fundamental Theorem of Calculus. If it is an improper integral, then we evaluate it using the two step method for evaluating improper integrals.

Example 8. Evaluate the improper integral

$$\int_0^\infty \frac{dx}{\sqrt{x}}.$$

Solution. This is an improper integral because it has "∞" as a limit. It is also an improper integral because the integrand $x^{-1/2}$ is not defined when $x = 0$. This is really the sum of two improper integrals and so must be considered as a sum.

$$\int_0^\infty \frac{dx}{\sqrt{x}} = \int_0^1 \frac{dx}{\sqrt{x}} + \int_1^\infty \frac{dx}{\sqrt{x}}.$$

Consider the first integral first.

$$\int_0^1 \frac{dx}{\sqrt{x}} = \lim_{t \to 0+} \int_t^1 \frac{dx}{\sqrt{x}}.$$

We have

$$\int_t^1 \frac{dx}{\sqrt{x}} = 2x^{1/2} \Big|_t^1 = 2 - 2t^{1/2}.$$

Next evaluate the limit

$$\lim_{t \to 0+} (2 - 2t^{1/2}) = 2.$$

The limit exists and is equal to 2. Consider the second integral

$$\int_1^\infty \frac{dx}{\sqrt{x}} = \lim_{t \to +\infty} \int_1^t \frac{dx}{\sqrt{x}}.$$

We have

$$\int_1^t \frac{dx}{\sqrt{x}} = 2x^{1/2} \Big|_1^t = 2t^{1/2} - 2.$$

65

Next, evaluate the limit

$$\lim_{t \to +\infty} (2t^{1/2} - 2).$$

This limit does not exist. The first integral is convergent but the second integral is divergent. This means that the sum is divergent. The improper integral

$$\int_0^\infty \frac{dx}{\sqrt{x}}$$

does not exist. It is a divergent integral.

The problems in all the examples considered so far have been easy to solve. There are two steps in any problem to evaluate an improper integral. A problem is more difficult if either of these steps is more difficult. We can have problems in which it is more difficult to find the antiderivative of the integrand. We can have problems in which it is more difficult to find the limit.

Example 9. Evaluate the improper integral

$$\int_1^\infty \frac{\ln(x)}{x^2} dx.$$

Solution. This is an improper integral because it has "∞" as a limit. Note that the integrand $(x^{-2}) \ln x$ is defined for all x such that $x \geq 1$. We need to find the antiderivative

$$\int \frac{\ln(x)}{x^2} dx.$$

Use the substitution $u = \ln x$, then the differential is $du = (\frac{1}{x})dx$ or $dx = x\, du$. Also $x = e^u$ or $x^{-1} = e^{-u}$. Substituting we get

$$\int u e^{-u} du.$$

We integrate this last integral using the integration by parts formula.

$$f(u) = u \quad g'(u) = e^{-u}$$
$$f'(u) = 1 \quad g(u) = -e^{-u}.$$

66

Substituting these functions into the integration by parts formula, we get

$$\int ue^{-u}du = -ue^{-u} + \int e^{-u}du = -ue^{-u} - e^{-u} + C.$$

This tells us that

$$\int \frac{\ln(x)}{x^2}dx = -(\ln x)\left(\frac{1}{x}\right) - \frac{1}{x} + C.$$

The first step in finding the value of the improper integral is to evaluate

$$\int_1^t \frac{\ln(x)}{x^2}dx.$$

Since we know the antiderivative of the integrand we are able to write

$$\int_1^t \frac{\ln(x)}{x^2}dx = -\frac{\ln(x)}{x} - \frac{1}{x}\bigg|_1^t = -\frac{\ln t}{t} - \frac{1}{t} + 1.$$

Next, we need to find the limit

$$\lim_{t\to\infty}\left[-\frac{\ln t}{t} - \frac{1}{t} + 1\right].$$

Using L'Hospital's Rule

$$\lim_{t\to\infty}\frac{\ln t}{t} = \lim_{t\to\infty}\frac{1}{t} = 0.$$

Using this we get

$$\lim_{t\to\infty}\left[-\frac{\ln t}{t} - \frac{1}{t} + 1\right] = 1.$$

The limit exists. The integral is convergent. The value of the improper integral is

$$\int_1^\infty \frac{\ln(x)}{x^2}dx = 1.$$

Example 10. Evaluate the integral

$$\int_0^\infty (\cos x)dx.$$

Solution. This is an improper integral because it has "∞" as a limit. We must first evaluate the definite integral

$$\int_0^t (\cos x)dx = \sin x \big|_0^t = \sin t$$

for $t > 0$. Step 2 is to find the limit

$$\lim_{t \to +\infty} \sin t.$$

Note that as t gets larger and larger it is always true that $-1 \le \sin t \le 1$. As t gets larger and larger the value of $\sin t$ does not get close to any fixed value, but keeps going back and forth between -1 and $+1$. The limit $\lim_{u \to +\infty} \sin t$ does not exist. This is a divergent improper integral.

Remark. The value of some integrals

(a) $\int_0^\infty \frac{\sin x}{x}dx = \frac{\pi}{2}$ (b) $\int_0^\infty \frac{dx}{(x^2+1)^2} = \frac{\pi}{4}$

(c) $\int_0^\infty \frac{\cos x}{(x^2+1)^2}dx = \frac{\pi}{2e}$ (d) $\int_0^\infty \frac{dx}{x^3+1} = \frac{2\pi}{3\sqrt{3}}$

(e) $\int_0^\infty \frac{dx}{x^4+1} = \frac{\pi}{2\sqrt{2}}$ (f) $\int_0^\pi \sin^{2n}(x)dx = \frac{(2n)!}{2^{2n}(n!)^2}\pi$

Exercises

1. $\int_1^\infty \frac{dx}{x+5}$

2. $\int_1^\infty \frac{dx}{(x+1)^2}$

3. $\int_3^\infty \frac{x}{(x^2+9)^{3/2}}dx$

4. $\int_0^\infty \frac{x\,dx}{\sqrt{x^2+1}}$

5. $\int_1^\infty \frac{dx}{x^2+1}$

6. $\int_0^\infty (\sin x)dx$

7. $\int_0^5 \frac{dx}{\sqrt{5-x}}$

8. $\int_0^8 \frac{dx}{8-x}$

9. $\int_{-2}^4 \frac{dx}{\sqrt{16-x^2}}$

10. $\int_1^{10} \frac{dx}{\sqrt{x-1}}$

11. $\int_1^\infty (x \ln x)dx$

12. $\int_1^\infty \frac{4x+6}{x^2+2x}dx$

5527 Arc Length and Surface Area

Consider a segment of a straight line. We know how to find the length of a line segment. If this is a line segment in a coordinate plane connecting the point with coordinates (x_1, y_1) to the point with coordinates (x_2, y_2), then we even know a formula for the length of the segment. The length of the segment is

$$\sqrt{(x_2 - x_1)^2 + (y_2 - y_1)^2}$$

Suppose we wish to find the length of a curve in the coordinate plane which is not a line segment. How do we find the length? First, we need some way of defining this curve. Suppose we have a curve in the plane which is the graph of $y = f(x)$ for $a \leq x \leq b$, where $f(x)$ is continuous. What is the length of this curve? We approach this problem as follows. Divide the interval $a \leq x \leq b$ on the x axis into smaller subintervals each of length $\Delta x = \frac{b-a}{n}$. Denote the division points by $a = x_0, x_1, x_2, \cdots, x_n = b$. Locate the points on the curve with coordinates $(x_0, f(x_0))$, $(x_1, f(x_1))$, $(x_2, f(x_2)), \cdots, (x_n, f(x_n))$.

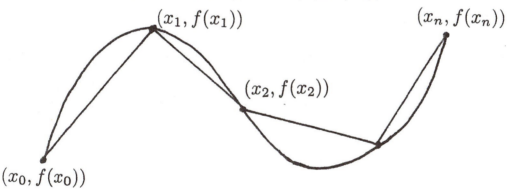

$(x_1, f(x_1))$ $(x_n, f(x_n))$ $(x_2, f(x_2))$ $(x_0, f(x_0))$

The length of the line segment connecting the points $(x_0, f(x_0))$ and $(x_1, f(x_1))$ is less than the length of the section of the curve connecting the points $(x_0, f(x_0))$ and $(x_1, f(x_1))$. The length of the line segment is

$$\sqrt{[x_1 - x_0]^2 + [f(x_1) - f(x_0)]^2}.$$

The length of the line segment connecting the points $(x_1, f(x_1))$ and $(x_2, f(x_2))$ is

$$\sqrt{[x_2 - x_1]^2 + [f(x_2) - f(x_1)]^2}.$$

The length of a line segment connecting two points is less than the length of a curve connecting the two points. The sum of the lengths of all the line segments is

$$\sum_{k=1}^{n} \sqrt{[x_k - x_{k-1}]^2 + [f(x_k) - f(x_{k-1})]^2}.$$

Since $x_k - x_{k-1} = \Delta x$, the sum of the lengths of the line segments is

$$\sum_{k=1}^{n} \sqrt{1 + \left[\frac{f(x_k) - f(x_{k-1})}{x_k - x_{k-1}}\right]^2} \Delta x.$$

The sum of the lengths of all the line segments is less than the length of the curve.

It is clear that the more division points that we use the larger the value of this sum. The limit of this sum as the number of divisions of the interval $a \le x \le b$ becomes infinite is the length of the curve. Also the limit of this sum as the number of divisions becomes infinite is a definite integral. This means that the length of this curve which is called its "arc length" is given by a definite integral.

Theorem 1. The Arc Length Formula. Consider the curve which is the graph of $y = f(x)$ for $a \le x \le b$. Assume $f'(x)$ exists for $a < x < b$. The arc length of this curve is given by the definite integral

$$\int_a^b \sqrt{1 + [f'(x)]^2} \, dx = \int_a^b \sqrt{1 + \left(\frac{dy}{dx}\right)^2} \, dx.$$

Example 1. Find the arc length of the curve given by $y = x^{3/2}$ between the points $(1,1)$ and $(9,27)$.

Solution. Since $f(x) = x^{3/2}$, we have $f'(x) = (3/2)x^{1/2}$. It follows that

$$\sqrt{1 + [f'(x)]^2} = \sqrt{1 + (9/4)x} = (1/2)\sqrt{4 + 9x}.$$

Substituting into the integral formula, the arc length is given by

$$L = \int_1^9 (1/2)\sqrt{4 + 9x} \, dx.$$

70

We transform this integral using the substitution $u = 4 + 9x$. The differential is $du = 9dx$ or $dx = (1/9)du$.

When $x = 1$, we have $u = 4 + 9(1) = 13$.
When $x = 9$, we have $u = 4 + 9(9) = 85$.

The substitution rule for definite integrals says that

$$L = \int_{13}^{85} (1/2)u^{1/2}(1/9)du = (1/18)\int_{13}^{85} u^{1/2}du$$

$$= \frac{1}{18}\frac{2}{3}u^{3/2}\Big|_{13}^{85} = \frac{1}{27}[(85)^{3/2} - (13)^{3/2}] = 27.29.$$

In this problem we see one of the characteristics of arc length problems. Arc length problems never have nice answers.

Surface Area

Next we are going to consider the problem of how to find the area of the surface of an object. Suppose we have a cylinder with no top and no bottom. The area of the surface is $A = 2\pi rh$. Suppose we have a sphere. The area of the surface of a sphere is $A = 4\pi r^2$. A cylinder and a sphere are examples of a special kind of surface called a surface of revolution. A surface of revolution is formed when a curve is rotated about a line. We are going to find a formula for finding the area of a surface of revolution. We form a surface of revolution as follows. We start with a curve in the xy plane which is the graph of $y = f(x)$ for $a \le x \le b$.

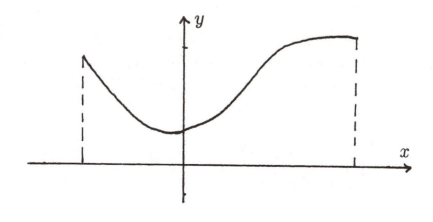

We then revolve this curve about the x axis generating a surface. The resulting surface encloses a figure which is symmetric about the x axis.

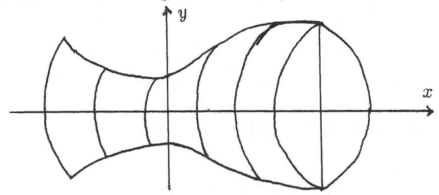

We then find the area of the surface of this figure.

Theorem 2. Surface Area Formula. Suppose $f(x)$ is a continuous function with $f(x) > 0$ for $a \leq x \leq b$. The surface area of the surface obtained by rotating the curve $y = f(x)$ for $a \leq x \leq b$ about the x axis is

$$S = 2\pi \int_a^b f(x)\sqrt{1 + [f'(x)]^2}\,dx.$$

Example 2. The arc (curve segment) of the parabola $y = \sqrt{6x + 7}$ from $(3, 5)$ to $(7, 7)$ is rotated about the x axis. Find the area of the surface generated.

Solution. A sketch of the graph of this curve segment is

However, in finding surface area a sketch of the graph is not necessary. All we need to do is substitute into the formula for surface area which is an integral. We have $f(x) = \sqrt{6x + 7}$ and so $f'(x) = (1/2)(6x + 7)^{-1/2}(6) = 3(6x + 7)^{-1/2}$.

$$1 + [f'(x)]^2 = 1 + \left[\frac{3}{\sqrt{6x + 7}}\right]^2 = 1 + \frac{9}{6x + 7} = \frac{6x + 16}{6x + 7}.$$

72

$$S = 2\pi \int_3^7 \sqrt{6x+7}\sqrt{\frac{6x+16}{6x+7}}\,dx = 2\pi \int_3^7 \sqrt{6x+16}\,dx.$$

We transform this integral using the substitution $u = 6x+16$, and $du = 6dx$.

When $x = 3$, we have $u = 6(3) + 16 = 34$.

When $x = 7$, we have $u = 6(7) + 16 = 58$.

$$S = 2\pi \int_3^7 \sqrt{6x+16}\,dx = \frac{2\pi}{6} \int_{34}^{58} u^{1/2}\,du = \frac{\pi}{3} \left[\frac{2}{3} u^{3/2} \Big|_{34}^{58} \right]$$

$$= \frac{2\pi}{9} \left[58^{3/2} - 34^{3/2} \right] = 169.97.$$

Exercises

1. Find the arc length of the curve which is the graph of $f(x) = (1/24)(x^2 + 16)^{3/2}$ for $0 \le x \le 4$.

2. Find the arc length of the curve which is the graph of $f(x) = (1/4)x^4 + (1/8)x^{-2}$ for $1 \le x \le 5$. Arc length $= 156.12$.

3. Find the arc length of the curve which is the graph of $f(x) = (1/3)x^{3/2} - x^{1/2}$ for $0 \le x \le 4$.

4. Find the arc length of the curve which is the graph of $f(x) = \cosh x$ for $0 \le x \le 1$.

5. Find the surface area of the surface obtained by rotating the curve $y = \sqrt{2x+8}$ for $0 \le x \le 8$ about the x axis.

6. Find the surface area of the surface obtained by rotating the curve $y = \frac{x^3}{6} + \frac{1}{2x}$ for $1 \le x \le 3$ about the x axis.

7. Find the surface area of the surface obtained by rotating the curve $y = x^3$ for $0 \le x \le 2$ about the x axis.

5529 First Moments and Centroids

We are going to discuss first moments. The idea behind moments is based on the following type of situation. Two children are on a seesaw. A seesaw is a plank resting on a fulcrum (wooden horse). Suppose one child weighs 80 lbs and the other child weighs 50 lbs. In order for the seesaw to be balanced the child whose weight is smaller, 50 lbs, must be seated farther from the fulcrum than the child weighing 80 lbs.

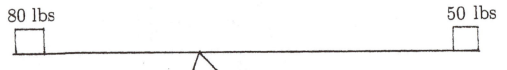

The question is: What must be the relative distances? We find the relative distances by computing the first moment for each child. The first moment is weight times distance. In order for the seesaw to balance the first moments must be equal. Let d_1 equal the distance from the pivot point to the 80 lb child and d_2 equal the distance from the pivot point to the 50 lb child. To say that the moments for each child must be equal is to say that

$$80d_1 = 50d_2.$$

This means that in order for the seesaw to balance we must have $d_1 = (5/8)d_2$. In order for the seesaw to balance the moment for the heavy child must equal the moment for the lighter child.

When we talk about the first moment of an object we must talk about the first moment of the object about a specific line. Suppose the object is a point mass located at the point P, then we find the first moment of this object about the line L. Let d denote the perpendicular distance from the point P to the line L, then the moment is given by

$$\text{First Moment} = (\text{distance})(\text{mass}) = (d)(m).$$

If we have several objects of masses $m_1, m_2, \cdots, m_k$ then the first moment of all objects taken together about the line L is

$$d_1 m_1 + d_2 m_2 + \cdots + d_k m_k,$$

where d_1 is the distance from the line to the object of mass m_1, and so forth.

74

When computing moments we will assume that all the objects and the line are all in the same plane. Allowing objects to be in 3-space is more complicated. Also if objects are on both sides of the line then we need to discuss negative first moments. We will not discuss the case when the objects are on both sides of the line. We will always be working in a coordinate plane. We will only consider cases where the objects in question are on the positive side of the line. Also we can discuss these ideas by either saying first moment = (mass)(distance) or by saying first moment = (weight)(distance). In the case of (mass)(distance) we would need to know the mass of the object and in the case of (weight)(distance) we would need to know the weight of the object. We will not make a big issue of this point.

As everyone really knows we do not want to discuss objects with all their mass concentrated at a single point, that is, point masses. We will discuss something called a lamina. We think of this as a thin flat plate. In our discussions the plate will be part of the coordinate plane. We will assume that the plate (lamina) has a constant area density, usually denoted by ρ. When we say that a lamina has constant density ρ, what does this mean? Suppose we take some part of the total lamina and the area of this part is A, then for the lamina to have constant area density ρ means that the mass of this part of the lamina is ρA.

Theorem 1. The moment of a long thin rectangle of constant width, w, of length, L, and constant density ρ about a line perpendicular to the rectangle and through one end of the rectangle is

$$(1/2)(\text{length})(\text{density})(\text{area}).$$

Proof. Suppose the rectangle is along the segment of the x axis given by $0 \leq x \leq L$, and the line is the y axis. Divide the interval $0 \leq x \leq L$ into n short subintervals each of length Δx and cut the rectangle into pieces of length Δx. Let w equal the width of the rectangle. The small piece of width w and length Δx is essentially a point mass with mass = density times area $= \rho w(\Delta x)$. The first moment equals the distance from the y axis to the small section. The distance from the y axis to this small section is x. The moment of the small section (distance) times (mass) = is $(x)[\rho w \Delta x]$. Add up the moments of all the small pieces

$$\sum_{k=1}^{n} \rho w x (\Delta x).$$

Take the limit as the number of subdivisions become infinite and we get that the moment of the long thin rectangle is given by the integral

$$\int_0^L \rho w x \, dx = \rho w \frac{x^2}{2}\bigg|_0^L = \frac{1}{2}\rho w L^2 = \rho(L/2)(\text{Area}).$$

Example 1. Consider the region bounded by the x axis and the parabola $y = 8x - x^2$. Suppose this region is covered by a lamina of density ρ, find the moment of this lamina about the y axis.

Solution. Note that there is no real need to specify units. The density ρ may be given in either mass density units or weight density units. In order to find the total moment we divide the region into small pieces such that we can easily find the moment for each piece and then add the contributions of the pieces.

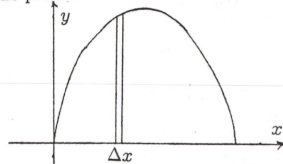

Divide the interval $0 \le x \le 8$ on the x axis into subintervals of length Δx. Slice the lamina into long thin rectangles of width Δx and length from the x axis to the parabola. For any x the length of the rectangle is $8x - x^2$. Any point mass inside the long thin rectangle is the same distance from the y axis. This distance is equal to the x coordinate of the point. The moment of the long thin rectangle about the y axis is given by

$$\Delta M_y = \rho x(\text{area}) = \rho x(y)\Delta x = \rho x(8x - x^2)\Delta x.$$

Adding the moments for all these rectangles together and taking the limit as the number becomes infinite we get that the moment for the entire lamina is given by the integral

$$M_y = \int_0^8 \rho x(8x - x^2)\,dx = \rho \int_0^8 (8x^2 - x^3)\,dx$$

$$= \rho \left[\frac{8x^3}{3} - \frac{x^4}{4}\bigg|_0^8\right] = \rho\frac{8^4}{12} = \frac{1024}{3}\rho.$$

76

If x is given in feet and density in slugs/ft^2, then the units of moment are slugs-ft. If x is given in meters and density in kilograms/m^2, then the units of moment are kilogram-meters. If x is given in feet and the density in lbs/ft^2, then the units of moment are lbs-ft.

Example 2. Consider the region bounded by the x axis and the parabola $y = 8x - x^2$. Suppose this region is covered by a lamina of density ρ, find the moment of this lamina about the x axis.

Solution. Divide the interval $0 \le x \le 8$ on the x axis into subintervals of length Δx. Slice the lamina into long thin rectangles of width Δx and length from the x axis to the parabola $y = 8x - x^2$.

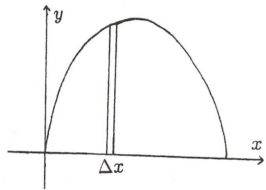

We use Theorem 1 to find the moment of this long thin rectangle about the x axis. Note that the x axis is a line perpendicular to the rectangle which passes through one end of the rectangle. Theorem 1 says

$$\Delta M_x = (1/2)(\text{length})(\text{density})(\text{Area})$$
$$= (y/2)(\rho)(y\Delta x) = (\rho/2)y^2(\Delta x) = (\rho/2)(8x - x^2)^2\Delta x.$$

Adding the moments of all these rectangles together and taking the limit as the number becomes infinite we get that the moment for the entire lamina is given by the integral

$$M_x = \frac{\rho}{2}\int_0^8 (8x - x^2)^2 dx = \frac{\rho}{2}\int_0^8 (64x^2 - 16x^3 + x^4)dx$$

$$= \frac{\rho}{2}\left[\frac{64x^3}{3} - 4x^4 + \frac{x^5}{5}\Big|_0^8\right] = \frac{(8^5)(\rho)}{60} = \frac{8192}{15}\rho.$$

Example 3. Consider the region bounded by the x axis and the parabola $y = 8x - x^2$. Suppose that the region is covered by a lamina of constant density ρ, find the mass of the lamina.

Solution. Since the area density ρ is a constant the mass is given by density times area.

$$\text{Area} = \int_0^8 (8x - x^2)dx = 4x^2 - \frac{x^3}{3}\Big|_0^8 = \frac{256}{3}.$$
$$\text{Mass} = \frac{256}{3}\rho.$$

Definition. The centroid of a lamina is a point. Let M_x denote the moment of the lamina about the x axis and let M_y denote the moment of the lamina about the y axis, then the centroid of the lamina is the point with coordinates $(\bar{x}, \bar{y})$ where

$$\bar{x} = \frac{M_y}{\text{Mass}} \text{ and } \bar{y} = \frac{M_x}{\text{Mass}}.$$

The number $\bar{x}$ is the distance from the y axis to the centroid and $\bar{y}$ is the distance from the x axis to the centroid. A point mass of the same magnitude as the total mass of the lamina concentrated at the centroid $(\bar{x}, \bar{y})$ has the same moment about the x axis and the same moment about the y axis as does the given lamina. If the entire mass of a lamina is concentrated at the center of mass then its moment about any line remains unchanged. The line may not pass through the lamina. Recall that all our discussion assumes the lamina is on the positive side of the line. The centroid is also called the center of mass. If the lamina is placed on the end of a long thin rod perpendicular to the lamina whose end is at the center of mass, then the lamina is balanced and will not tip.

Example 4. Consider the region bounded by the x axis and the parabola $y = 8x - x^2$. Suppose the region is covered by a lamina of density ρ, find the centroid of this lamina.

Solution. We have already found the moments and weight or mass of this lamina. They are

$$M_x = \frac{8192}{15}\rho, \quad M_y = \frac{1024}{3}\rho, \quad \text{and Mass} = \frac{256}{3}\rho.$$

If follows that

$$\bar{x} = \frac{M_y}{\text{Mass}} = \left(\frac{1024\rho}{3}\right)\left(\frac{3}{256\rho}\right) = 4$$

$$\bar{y} = \frac{M_x}{\text{Mass}} = \left(\frac{8192\rho}{15}\right)\left(\frac{3}{256\rho}\right) = \frac{32}{5}.$$

Example 5. Consider the region bounded by the line $y = x + 4$ and the parabola $y = x^2 - 4x + 4$. Suppose this region is covered by a lamina of constant density ρ, find the moment about the x axis, the moment about the y axis, and the weight of this lamina. Assume that weight units are given this time.

Solution. The line $y = x + 4$ and the parabola $y = x^2 - 4x + 4$ intersect at the points $(0, 4)$ and $(5, 9)$. We make a sketch of this region. We divide the region into long thin rectangles of width Δx whose lower end is on the parabola and whose upper end is on the line.

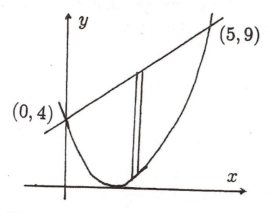

We need to know the first moment of this long thin rectangle about the x axis. The bottom end of this rectangle is not on the x axis. We know how to find the moment of a rectangle about a line through one end of the rectangle. With this in mind, we draw two long thin rectangles, one from the x axis to the line and one from the x axis to the parabola.

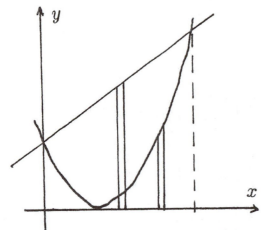

The moment about the x axis of the lamina covering the rectangle which extends from the x axis to the line is $(\rho/2)(y_{\text{line}})^2\Delta x$. The moment about the x axis of the lamina covering the rectangle which extends from the x axis to the parabola is $(\rho/2)(y_{\text{parabola}})^2\Delta x$. The rectangle we want is the rectangle from the x axis to the line take away the rectangle from the x axis to the parabola. Therefore the moment about the x axis for the desired rectangle is

$$\Delta M_x = (\rho/2)\left[(y_{\text{line}})^2 - (y_{\text{parabola}})^2\right]\Delta x$$
$$= (\rho/2)\left[(x+4)^2 - (x^2 - 4x + 4)^2\right]\Delta x.$$

Adding all the moments ΔM_x for their rectangles together and taking the limit as the number of rectangles becomes infinite, we get that the moment of the whole lamina about the x axis is given by the integral

$$M_x = \frac{\rho}{2}\int_0^5 [(x+4)^2 - (x^2 - 4x + 4)^2]dx.$$

$$= \frac{\rho}{2}\int_0^5 [-x^4 + 8x^3 - 23x^2 + 40x]dx$$

$$= \frac{\rho}{2}\left[-\frac{5^5}{5} + 8\frac{5^4}{4} - 23\frac{5^3}{3} + 40\frac{5^2}{2}\right] = \frac{250}{3}\rho.$$

Next let us find the moment of the lamina about the y axis. Return to our large thin rectangle with bottom end on the parabola and top end on the line. All the particles in this long thin rectangle are the same distance from the y axis. That distance is x. This means that the first moment of

80

this thin rectangle about the y axis is

$$\Delta M_y = \text{(distance)(density)(length)(width)}$$
$$= \rho[x][y_{\text{line}} - y_{\text{parabola}}]\Delta x$$
$$= \rho x[(x+4) - (x^2 - 4x + 4)]\Delta x$$

Adding all the moments ΔM_y and taking the limit as the number of rectangles becomes infinite. We get that the moment of the whole lamina about the y axis is given by the integral

$$M_y = \rho \int_0^5 x(-x^2 + 5x)dx = \rho\left[-\frac{5^4}{4} + \frac{5^4}{3}\right] = \frac{625}{12}\rho.$$

The area of this region is given by

$$\text{Area} = \int_0^5 [(x+4) - (x^2 - 4x + 4)]dx = \frac{125}{6}.$$

The weight of the lamina is $\frac{125}{6}\rho$. The center of weight $(\bar{x}, \bar{y})$ of this lamina is given by

$$\bar{x} = \frac{M_y}{\text{weight}} = \left(\frac{625}{12}\rho\right)\left(\frac{6}{125\rho}\right) = \frac{5}{2},$$

$$\bar{y} = \frac{M_x}{\text{weight}} = \left(\frac{250}{3}\rho\right)\left(\frac{6}{125\rho}\right) = 4.$$

The center of weight is $(5/2, 4)$.

So far we have found the first moment of a region about the x axis, M_x, and the first moment about the y axis, M_y. What happens if we find the first moment about some other line?

Example 6. Consider the region bounded by the x axis and the parabola $y = 8x - x^2$. Suppose this region is covered by a lamina of density ρ, find the first moment of this lamina about the line $x = -5$. Note that the lamina is on the positive side of the line.

Solution. Divide the interval on the x axis into subintervals of length Δx. Slice the lamina into long thin rectangles of width Δx and length from the x axis to the parabola.

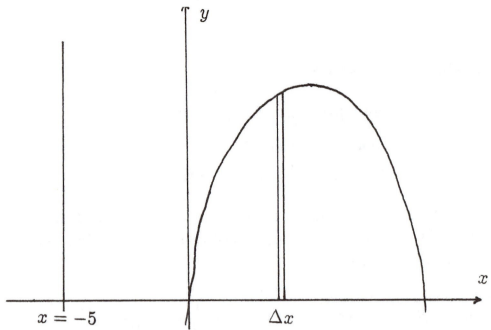

For any x the length of the rectangle is $8x - x^2$. Any point mass in the rectangle is the same distance from the line $x = -5$. The distance is $x + 5$. The moment of the long thin rectangle about $x = -5$ is

$$\rho(x+5)(\text{Area}) = \rho(x+5)(y)(\Delta x) = \rho(x+5)(8x - x^2)\Delta x.$$

Adding the moments of all these rectangles together and taking the limit as the number becomes infinite, we get that the moment for the entire lamina is given by

$$\int_0^8 \rho(x+5)(8x - x^2)dx = \int_0^8 \rho(x)(8x - x^2)dx + 5\rho \int_0^8 (8x - x^2)dx$$
$$= M_y + 5(\text{density})(\text{Area})$$
$$= M_y + 5(\text{Mass}).$$

Thus we see that if we want to find the moment of this lamina about a line parallel to the y axis but 5 more units distant from the lamina, then the moment is $M_y + 5(\text{Mass of lamina})$. This is actually a general rule. If we want to know the moment of this lamina about the line $x = -10$, then the moment is

$$M_y + 10(\text{Mass of lamina}).$$

If we want to know the moment of this lamina about the line $y = -6$, then the moment is

$$M_x + 6(\text{Mass of lamina}).$$

82

Recall the formula for center of mass

$$\bar{x} = \frac{M_y}{\text{Mass}} \text{ and } \vec{y} = \frac{M_x}{\text{Mass}}.$$

These can be written as

$$M_y = (\text{Mass})(\bar{x}) \text{ and } M_x = (\text{mass})(\bar{y}).$$

Now $\bar{x}$ is the distance from the y axis to the center of mass and $\bar{y}$ is the distance from the x axis to the center of mass. This means that

$$M_x = (\text{mass})(\text{distance from } x \text{ axis to center mass})$$
$$M_y = (\text{mass})(\text{distance from } y \text{ axis to center of mass}).$$

In fact this is a general rule. This general rule is why we like to find the center of mass.

General rule. Suppose we have a region R and a line that does not pass through the region, but the region is on the positive side of the line. Suppose the region is covered by a lamina. The first moment of this lamina about the line is

(mass of lamina)(distance from line to center of mass of the lamina).

Example 7. Consider the region bounded by the x axis and the parabola which is the graph of $y = 8x - x^2$. Suppose this region is covered by a lamina of density ρ, find the first moment of this lamina about the line $y = -x$.

Solution. We found the center of mass of this region in Example 4. It is $(4, 32/5)$. We need the equation of the line through the point $(4, 32/5)$ which is also perpendicular to the line $y = -x$. This is a line of slope 1 through the point $(4, 32/5)$. The equation of this line is $y - (32/5) = 1(x - 4)$. The line $y - (32/5) = x - 4$ is perpendicular to the line $y = -x$ and also contains the point $(4, 32/5)$. The lines $y = -x$ and $y = x + (12/5)$ intersect at the point $(-6/5, 6/5)$. The length of the line segment from $(-6/5, 6/5)$ to $(4, 32/5)$ is

$$\sqrt{[4 + 6/5]^2 + [(6/5) - (32/5)]^2} = (26/5)\sqrt{2}.$$

83

The perpendicular distance from the line $y = -x$ to the center of mass $(4, 32/5)$ is $(26/5)\sqrt{2}$. The mass of this lamina was found in Example 4 to be $(256/3)\rho$. The first moment of this lamina about the line $y = -x$ is

$$\text{moment} = (\text{distance})(\text{mass}) = [(26/5)\sqrt{2}][(256/3)\rho]$$
$$= \frac{6656}{15}\sqrt{2}\rho = (627.5)\rho.$$

When working with first moments in a practical situation the complete discussion of first moments is a little more complicated than is the discussion given here. In a physics or engineering class there would be a more general introduction to the ideas of moment using the term torque. A torque that causes a rotation in one direction would be labeled as positive while a torque that causes a rotation in the other direction would be labeled negative. When working with the vectors $\vec{i}, \vec{j}$ and $\vec{k}$ in a right handed coordinate system the positive directions are already determined. We will discuss this a bit as part of our discussion of the vectors $\vec{i}, \vec{j}$, and $\vec{k}$.

Let us return to the example of the seesaw for a moment. We said that the seesaw was balanced when $80d_1 = 50d_2$. This is the same as $80d_1 - 50d_2 = 0$. In practice we usually say that one moment is $-50d_2$, that is, negative. When the seesaw is balanced the sum of all moments (torques) is zero. Having both negative and positive moments (torques) results from considering first moments (torques) as vectors. The bottom line is that in a more general discussion the laminas one side of the line result in positive torque while the laminas on the other side of the line result in negative torque. In our discussion the problems are given in such a way that all moments are positive. While this simplifies the discussion it still retains all the calculus aspects of the problems. In order to illustrate this let us outline an example where the axis of rotation passes through the lamina.

Extra Example. Suppose a lamina of density ρ covers the region bounded by the curve $y = 8x - x^2$ and the x axis. Find the first moment of this lamina about the line $x = 3$. Compare this with Example 1.

$$\text{Moment} = -\rho \int_0^3 (3-x)(8x-x^2)dx + \rho \int_3^8 (x-3)(8x-x^2)dx$$

$$\text{Moment} = \rho \int_0^8 (x-3)(8x-x^2)dx$$

$$= \rho \int_0^8 x(8x-x^2)dx - 3\rho \int_0^8 (8x-x^2)dx$$

$$= M_y - 3(\text{mass}).$$

Exercises

1. Let R denote the region bounded by the curve $y = \sin x$, the x axis, the line $x = 0$, and the line $x = \pi/2$. The region R is covered by a lamina of constant density ρ. Find M_x, the first moment about the x axis, and M_y, the first moment about the y axis, for this lamina.

2. The region bounded by the parabola $y = 6x - x^2$ and the line $y = -x$ is covered by a lamina of constant density ρ. Find M_y, the moment of this lamina about the y axis.

3. The region bounded by the parabola $y = x^2 - 4x + 4$ and the line $y = x + 1$ is covered by a lamina of density ρ. Find M_x, the first moment of this lamina about the x axis.

4. The region bounded by the parabola $y = 6x - x^2$ and the line $y = x$ is covered by a lamina of constant mass density ρ. Find M_x, the moment about the x axis, and M_y, the moment about the y axis. Find the mass of the lamina and its center of mass. Also find the first moment of this lamina about the line $x = -10$ and about the line $y = -4$.

5. The region bounded by the parabola $y = (x - 3)^2$ and the line $y = 4$ is covered by a lamina of mass density ρ. Find M_x and M_y. Find the mass of this lamina and its center of mass. Also find the first moment of this lamina about the line $y = -4$.

6. The region bounded by the parabola $y = 10x - x^2$ and the x axis is covered by a lamina of density ρ. Find M_x, M_y, and the center of mass for this lamina. Find the first moment of this lamina about the line $y = -x$.

5531 Hydrostatic Force

We are going to discuss the force that is exerted by a liquid on the walls of a container. One of the best ways to think of this is to consider the side of a large aquarium full of water. Suppose that the side is vertical, made of glass, and the aquarium is full of water. The water exerts a force on the glass in the direction toward the outside of the aquarium. This force is called hydrostatic force. Also the water exerts pressure against the glass. Pressure is defined as force per unit area. At any given point in the water the pressure is the same in all directions. Water pressure increases the deeper you go. In order to compute water pressure we need to know the weight density of water. In the engineering system the weight density of water is 62.5 lb/ft^3. In the mks system the weight density of water is 9800 newtons/m^3. Suppose we have a very small section of glass at the end of the aquarium. Suppose the section is below the surface of the water, then the pressure on this section is equal to weight density times the depth. More importantly, since force equals pressure times area, the force exerted by the water against this very small section of glass plate is given by

$$\text{Force} = (\text{weight density})(\text{depth})(\text{Area of section})$$

Example 1. There is a larger glass plate in the shape of a rectangle in the end of a large tank. The top edge of the rectangle is parallel to the surface of the water and is 4 feet below the surface. The tank is full of water to the top which is 4 feet above the top edge of the rectangle. The length of the side of the rectangle parallel to the surface is 20 feet. The length of the side perpendicular to the surface is 8 feet. Find the hydrostatic force on the glass plate.

Solution. This is not a completely easy problem because the pressure is not constant, but increases as the depth increases.

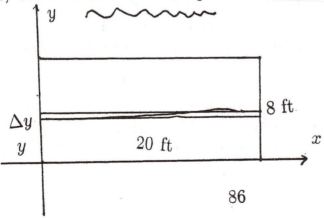

Draw the x axis along the bottom of the plate and a y axis along the left edge of the plate. Our strategy is to break up the problem into a large number of small parts, approximate the force on each part, add the results, take the limit, and then evaluate the resulting integral.

Divide the section $0 \le y \le 8$ of the y axis into subintervals of length Δy. Draw a long thin rectangle of width Δy parallel to the x axis and perpendicular to the y axis. The most important thing about this rectangle is that it is parallel to the ground. This means that all points in this rectangle are basically the same distance below the surface of the water. The water pressure at all points in this rectangle is the same. The distance from the x axis to the surface is 12 feet. The distance from the x axis to the long thin rectangle is y feet. This rectangle is $12 - y$ feet under the surface of the water. The hydrostatic force on this rectangle is

$$\Delta F = (\text{density})(\text{depth})(\text{length})(\Delta y)$$
$$= \rho(12 - y)(x)\Delta y = \rho(12 - y)(20)\Delta y.$$

Adding the force on all the little rectangles together and taking the limit as the number of rectangles becomes infinite, the total force on the plate is given by

$$F = \int_0^8 20\rho(12 - y)dy = 20\rho(8^2) = 1280\ \rho = 80,000\text{lbs}.$$

Example 2. The end of a large tank is in the shape of the isosceles trapezoid shown below. The tank is full of water to the top of the trapezoid. Find the hydrostatic force on the end of the tank. All lengths are given in feet.

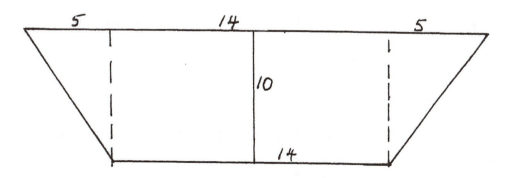

Solution. We begin by drawing a coordinate system on this end of the tank.

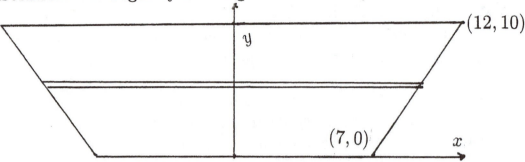

We could draw the coordinate system other places and with positive in different directions, but this seems to be the best choice. Divide the interval $0 \leq y \leq 10$ on the y axis into short subintervals of length Δy. Draw a long thin rectangle of width Δy parallel to the ground. Note that basically all points in this long thin rectangle are the same distance below the surface of the water. This distance is $10 - y$. The hydrostatic force on this long thin rectangle is

$$\Delta F = (\text{density})(\text{depth})(\text{length})(\text{width})$$
$$= \rho(10 - y)(2x)(\Delta y).$$

The hard part of this problem is that the length of the rectangle $(2x)$ is not a constant for all values of y. We need a relation (equation) that gives the value of x for a given value of y. The right edge of the trapezoid is a line through the points $(7, 0)$ and $(12, 10)$. The slope of this line is

$$m = \frac{y_2 - y_1}{x_2 - x_1} = \frac{10 - 0}{12 - 7} = 2.$$

The general equation of a line through the point (x_1, y_1) with slope m is

$$y - y_1 = m(x - x_1).$$

Substituting $(x_1, y_1) = (7, 0)$ and $m = 2$ into this general equation, we get

$$y - 0 = 2(x - 7)$$
$$y = 2x - 14.$$

This is the equation of the line which forms the right side of the trapezoid. Solving for $2x$, we have

$$2x = y + 14.$$

Substituting for $2x$ in ΔF we get an expression for ΔF involving only y and not x.

$$\Delta F = \rho(10 - y)(y + 14)\Delta y.$$

Adding up the forces on all the thin rectangles and taking the limit as the number of rectangles becomes infinite, we get that the total hydrostatic force is given by the integral

$$F = \rho \int_0^{10} (10 - y)(y + 14)dy = \rho \int_0^{10} (-y^2 - 4y + 140)dy$$

$$= \rho\left[-\frac{10^3}{3} - 2(10^2) + 140(10)\right] = \frac{2600}{3}\rho = 54,167 \text{ lbs.}$$

The solution of a hydrostatic force problem involves more than finding the correct answer for the force. The solution of a hydrostatic force problem results in the evaluation of an integral. The solution is not complete unless there is a clear and complete description of what the variable of integration represents.

Example 3. The edge of a large tank has the equation $y = (x - 2)^2$ where the edge of the tank is shown. The top of the water is at $y = 4$.

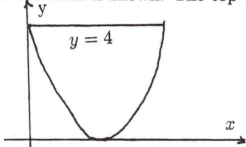

Find the hydrostatic force on the end of the tank. All lengths are given in feet.

Solution. There is water from $y = 0$ to $y = 4$. We begin by dividing the interval $0 \leq y \leq 4$ into subintervals each of length Δy. Draw a long thin rectangle perpendicular to the y axis and parallel to the ground. The long thin rectangle extends from the left side of the parabola to the right side of the parabola. Each point inside the rectangle is basically the same depth under the surface of the water.

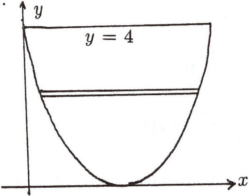

The long thin rectangle is $4 - y$ feet below the surface of the water. The hydrostatic force exerted by the water on the long thin rectangle is given by

$$\Delta F = (\text{density})(\text{depth})(\text{length})(\Delta y).$$

We have a little difficulty finding the length of the rectangle. The rectangle goes from the left side of the parabola to the right side of the parabola. The equation of the parabola is

$$y = (x - 2)^2.$$

Solving this equation for x, we get

$$x = 2 + \sqrt{y} \text{ and } x = 2 - \sqrt{y}.$$

The equation whose graph is the left half of the parabola is $x = 2 - \sqrt{y}$. The equation whose graph is the right half of the parabola is $x = 2 + \sqrt{y}$. The length of the long thin rectangle for a given value of y is

$$(x_{\text{right}} - x_{\text{left}}) = (2 + \sqrt{y}) - (2 - \sqrt{y}) = 2\sqrt{y}.$$

It follows that

$$\Delta F = \rho(4 - y)(2\sqrt{y})\Delta y.$$

90

Adding up the forces on all the long thin rectangles and taking the limit as the number of rectangles becomes infinite, we get that the hydrostatic force on the whole parabolic face is given by the following integral.

$$F = \rho \int_0^4 (4-y)(2\sqrt{y})dy = \rho \int_0^4 (8y^{1/2} - 2y^{3/2})dy$$

$$= \rho \left[\frac{2(8)(4^{3/2})}{3} - \frac{4(4^{5/2})}{5} \right] = \frac{256}{15}\rho = 1067 \text{ lbs.}$$

Exercises

1. The end of a large tank is a rectangle of horizontal length 15 meters and height 12 meters. Suppose the tank is full of water to a depth of 12 meters. Find the hydrostatic force on the end of the tank.

2. The end of a large tank is in the shape of the isosceles trapezoid shown below. The tank is full of water to the top of the trapezoid. Find the hydrostatic force on the end of the tank. The lengths are given in feet.

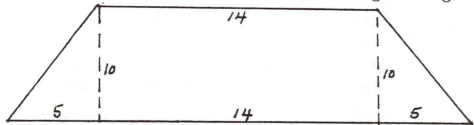

3. There is a glass plate in the end of a large tank in the shape of the isosceles triangle shown below. The top edge of the plate is parallel to the surface of the water and is 5 feet below the surface of the water. The tank is full of water to the top which is 5 feet above the top edge of the triangle. Find the hydrostatic force on the glass plate. The lengths are given in feet.

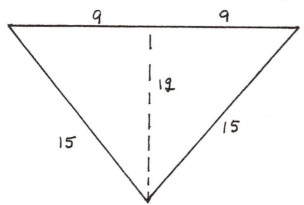

91

4. The edge of one end of a large tank has the equation $y = (x-3)^2$ where the end of the tank is shown. The top of the water is at $y = 9$. Find the hydrostatic force on the end of the tank. All lengths are given in feet.

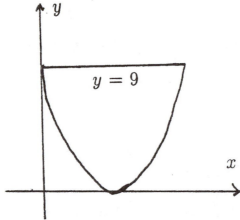

5. There is a glass plate in the end of a large tank in the shape of the isosceles triangle shown below. The top vertex of the triangle is 5 meters below the surface of the water. The base of the triangle is parallel to the ground. The tank is full of water to the top which is 5 meters above the top vertex of the triangle. Find the hydrostatic force on the glass plate. The lengths are given in meters.

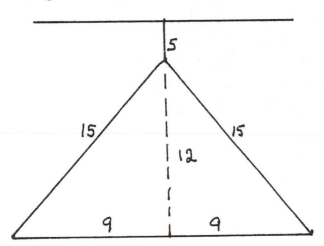

6. There is an odd shaped glass plate at the end of a large tank. Using a well chosen set of coordinates the left edge of the plate has the equation $y = 3x$ and the right edge of the glass plate has the equation $y = x^2$. The water level is $y = 10$. Find the hydrostatic force on the glass plate. The lengths are given in feet.

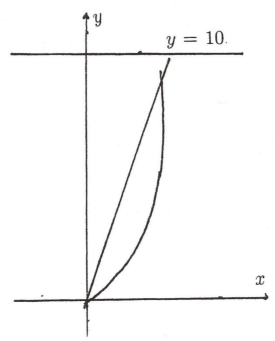

5537 Variables Separable

We are going to discuss differential equations. A differential equation is an equation involving the independent variable x, the dependent variable y, and one or more derivatives of the dependent variable. The complete solution of a differential equation is to find an expression for the dependent variable as a function of the independent variable. The problem of finding an antiderivative can be viewed as a very simple differential equation. The problem given

$$\frac{dy}{dx} = 4x^3 + 8x + 6\cos(2x),$$

find the antiderivative can be viewed as a differential equation. We integrate and find that $y = x^4 + 4x^2 + 3\sin(2x) + C$ is the solution. The thing that makes this differential equation so easy to solve is that it does not contain the dependent variable y. An example of an even simpler differential equation is the following.

Example 1. Find y as a function of x such that

$$\frac{dy}{dx} = 2x - 6.$$

Solution. We just find the antiderivative in order to get the solution

$$y = x^2 - 6x + C,$$

where C is the constant of integration and can be any real number. Since C can be any real number, the differential equation has infinitely many solutions. All the solutions of this equation taken together form "a family of functions". The graphs of the family form a family of curves (parabolas). We have a different function for each value of the constant of integration C. Some of the solution functions in this family are

$$y = x^2 - 6x$$
$$y = x^2 - 6x + 9$$
$$y = x^2 - 6x + 5$$
$$y = x^2 - 6x - 7.$$

94

If we graph all the curves in this family of solutions, the graphs would fill the entire xy plane. This is a typical situation. The solution of every first order differential equation is a family of curves. Even though we get different functions by changing the value of C we still call the solution of a differential equation containing the arbitrary constant C, the general solution (singular) of the differential equation. Below is a sketch of the members of this family of solutions obtained by letting $C = 9, 5, 0$, and -7.

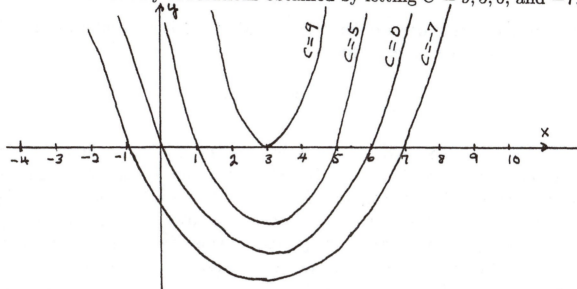

Besides finding the general solution of a differential equation, we also solve initial value problems involving the differential equation. If in addition to the differential equation we require an initial condition that picks out a single solution from the family of solutions, then the problem is called an initial value problem. The solution of an initial value problem is unique.

Example 2. Solve the initial value problem

$$\frac{dy}{dx} = 2x - 6 \text{ and } y(0) = 5.$$

Solution. We already found the general solution of this differential equation. It is

$$y = x^2 - 6x + C.$$

The initial condition says that we must find the member of the family of solutions such that $y = 5$ when $x = 0$. Substituting we get

$$5 = 0 + C.$$

The solution of the initial value problem is the unique function

$$y = x^2 - 6x + 5.$$

We will now advance to consider nontrivial differential equations. We consider equations that contain the dependent variable y in addition to x and $\frac{dy}{dx}$.

Definition. A first order differential equation is called variables separable if it can be written in the form

$$\frac{dy}{dx} = f(x)g(y).$$

This says that we can express the right side as a product of an expression in x times an expression in y. It follows that for a variables separable differential equation we can get all the y's on one side of the equal sign and all the x's on the other side. In order to solve a variables separable differential equation we get all the y's on one side of the equals sign and all the x's on the other side and write the equation in differential form. We then integrate both sides.

Example 3. Solve the following differential equation. Find y as a function of x such that
$$\frac{dy}{dx} = \frac{3(y+6)}{3x+8}.$$

Solution. This equation is variables separable and so we rewrite it with the y's on one side of the equals sign and the x's on the other side of the equals sign. We could also say that we are writing this equation in differential form. The result is
$$\frac{dy}{y+6} = \frac{3dx}{3x+8}.$$

We find the antiderivative of both sides of the equation. In order to find $\int \frac{3dx}{3x+8}$, we let $u = 3x + 8$ and $du = 3dx$. The integral then has the form $\int \frac{du}{u} = \ln|u|$. Using this idea we integrate the expression on each side of the equals and get

$$\ln|y+6| = \ln|3x+8| + \ln C.$$

Recall that $\ln[C|3x + 8|] = \ln|3x + 8| + \ln C$. Using this we rewrite the result of the integration as

$$\ln|y + 6| = \ln[C|3x + 8|].$$

If the logs are equal, then the functions are equal. Taking antilogs we get

$$|y + 6| = C|3x + 8|.$$

Somewhere during the solution we should notice that a lot of our arithmetic is not correct if $x = -8/3$. Also the original differential equation is not defined for $x = -8/3$. We need to require that $x \neq -8/3$. The constant of integration is written as $\ln C$. The constant of integration can be any real number. The expression $\ln C$ takes on all real values for $C > 0$. We should really say $C > 0$. The expression $|y + 6| = C|3x + 8|$, with $C > 0$, and the expression $y + 6 = C(3x + 8)$, C any real number define the same family of functions. This means that $|y + 6| = C|3x + 8|$, $x \neq -8/3$, and $y + 6 = C(3x + 8)$ $x \neq -8/3$ have the same meaning. Since this is true, we usually write the general solution of the differential equation as

$$y = -6 + C(3x + 8).$$

This is a family of rays that start at the point $(-8/3, -6)$, but not including the point $(-8/3, -6)$ and go to infinity. The family includes the ray $y = -6$ for $x > -8/3$ and the ray $y = -6$ for $x < -8/3$ both of which are solutions of the differential equation as given. This family includes the ray $y = 3x + 2$ defined for $x < -8/3$ and the ray $y = 3x + 2$ defined for $x > -8/3$. Note that the differential equation is satisfied since $\frac{d}{dx}(3x + 2) = 3$ and $\frac{3(3x+2+6)}{3x+8} = 3$ for $x \neq -8/3$ but $\frac{3(y+6)}{3x+8}$ is undefined for $x = -8/3$.

Example 4. Find the solution of the initial value problem

$$\frac{dy}{dx} = \frac{3(y + 6)}{3x + 8} \quad \text{and} \quad y(0) = 10.$$

Solution. The solution is a function which satisfies both equations. We already have the general solution of the differential equation. It is

$$y = -6 + C(3x + 8).$$

97

We need the unique member of this family of solutions such that $y = 10$, when $x = 0$. Substituting we get

$$10 = -6 + C(8) \text{ or } C = 2.$$

The solution of the initial value problem is

$$y = -6 + 2(3x + 8), \text{ for } x > -8/3.$$

Next let us solve a related initial value problem:

$$\frac{dy}{dx} = \frac{3(y + 6)}{3x + 8} \text{ and } y = -8 \text{ when } x = -3.$$

Substituting the values $y = -8$ and $x = -3$ into $y = -6 + C(3x + 8)$, we get $-8 = -6 + C(-1)$ or $C = 2$. The solution is

$$y = -6 + 2(3x + 8) \text{ for } x < -8/3.$$

Example 5. Find the general solution of the differential equation

$$\frac{dy}{dx} = \frac{3(y^2 + 9)}{y(2x + 5)}.$$

Solution. This equation is variables separable. We need to get the y's on one side of the equals sign and the x's on the other side. We rewrite the equation in differential form as

$$\frac{y\,dy}{y^2 + 9} = \frac{3\,dx}{2x + 5}.$$

Note that $\frac{d}{dy}[y^2 + 9] = 2y$ and $\frac{d}{dx}[2x + 5] = 2$. With this in mind, we multiply both sides of the equation by 2 to get

$$\frac{2y\,dy}{y^2 + 9} = \frac{3(2\,dx)}{2x + 5}.$$

Integrating both sides we get

$$\ln(y^2 + 9) = 3\ln|2x + 5| + \ln C.$$

98

Recall that $3 \ln |2x+5| = \ln |2x+5|^3$ and $\ln |2x+5|^3 + \ln C = \ln[C|2x+5|^3]$. Therefore, we have

$$\ln(y^2 + 9) = \ln[C|2x + 5|^3].$$

If the logs are equal, then the functions are equal. Taking antilogs we get

$$y^2 + 9 = C|2x + 5|^3.$$

This is the same family of functions as

$$y^2 + 9 = C(2x + 5)^3.$$

Note that we assume $x \neq -5/2$ which should have been stated as part of the original problem.

Suppose we have the solution of a differential equation which can be solved for y as an explicit function of x then we should solve for y. If, on the other hand, suppose we have an equation which defines y implicitly as a function of x in such a way that we can not solve for y to the first power, then we do not usually solve for y in the general solution. This is sometimes not the case for an initial value problem. The expression $y^2 + 9 = C(2x+5)^3$ defines y implicitly as a function of x. Therefore, we do not solve the equation for y. We just assume that a detailed analysis of the functions defined implicitly by the equation $y^2 + 9 = C(2x+5)^3$ is another question and that we are not interested in doing a detailed analysis at this time. We have an expression for the function which is the solution of the differential equation and so we are done. We could, for example, consider the case $C = 1$, that is, $y^2 = -9 + (2x+5)^3$. This equation defines two functions implicitly. These functions are the plus and minus square root of $-9 + (2x+5)^3$. These functions are defined for $x \geq (1/2)(\sqrt[3]{9} - 5) \approx -1.46$.

We are able to find the differential equation for a mixing problem because we can add rates. Simple mixing problems lead to differential equations that are variables separable.

Example 6. (A mixing problem) Initially a large tank contains 50 lbs of salt dissolved in 800 gallons of water. Brine that contains 1/2 lbs of salt per gallon enters the tank at the rate of 10 gallons per minute. The well mixed brine is drained from the tank at the same rate of 10 gal/min. Find an expression for the amount of salt in the tank at time t.

Solution. First, the unknown is the number of pounds of salt in the tank at time t. Let $y(t)$ denote the number of pounds of salt in the tank at time t. Note that the number of gallons of brine in the tank is constant and is always 800 gallons. The derivative $\frac{dy}{dx}$ is the rate of change of the amount of salt in the tank with respect to time. The rate of change of the amount of salt in the tank at time t can also be expressed as

$$\frac{dy}{dt} = (\text{rate in}) - (\text{rate out}).$$

The rates for salt are as follows:

$$\text{Rate in} = (1/2 \text{ lb/gal})(10 \text{ gal/min}) = (5 \text{ lb/min})$$

Note that $\frac{y(t)}{800}$ is the total amount of salt in the tank divided by the total number of gallons of water in the tank.

$$\text{Rate out} = \left(\frac{y(t)}{800} \text{lb/gal} \right) (10 \text{gal/min}) = \left(\frac{y(t)}{80} \text{ lb/gal} \right).$$

The differential equation is

$$\frac{dy}{dt} = 5 - \frac{y(t)}{80}.$$
$$\frac{dy}{dt} = \frac{400 - y}{80}$$

This equation is variables separable. We rewrite it in differential form.

$$\frac{dy}{400 - y} = \frac{dt}{80}$$

Integrating we get

$$-\ln|400 - y| = (1/80)t + \ln C'$$
$$\ln|400 - y| = -(t/80) + \ln C.$$

Note that $C = 1/C'$. Recall that $y \equiv \ln(e^y)$.

$$\ln|400 - y| = \ln[e^{-t/80}] + \ln C$$
$$\ln|400 - y| = \ln[Ce^{-t/80}].$$

Taking antilogs, we have

$$400 - y = Ce^{-t/80}$$

$$y = 400 - Ce^{-t/80}.$$

This is an initial value problem. The initial condition is that the number of pounds of salt in the tank at time $t = 0$ is $y = 50$. Substituting these values into the general solution, we get

$$50 = 400 - C \text{ or } C = 350.$$

The solution of the initial value problem is

$$y = 400 - 350e^{-t/80}.$$

This is an expression for the amount of salt in the tank at any time t. The number of pounds of salt in the tank when $t = 40$ minutes is

$$y = 400 - 350e^{-1/2} = 187.7 \text{ lbs.}$$

Exercises

1. Solve the initial value problem

$$\frac{dy}{dx} = \frac{4(y+5)}{2x+3} \text{ and } y(1) = 45. \text{ Note that } x \neq -3/2.$$

Also find the solution when the initial condition is $y(-4) = 45$.

2. Solve the initial value problem

$$\frac{dy}{dx} = \frac{9x(y+4)}{3x^2+4} \text{ and } y(0) = 36.$$

3. Find the general solution of the differential equation

$$\frac{dy}{dx} = \frac{2x(y^3+8)}{3y^2(x^2+5)}.$$

4. Initially a large tank contains 40 lbs of salt dissolved in 600 gallons of water. Brine that contains 1/3 lbs of salt per gallon of water enters the tank at the rate of 8 gallons per minute. The well mixed brine is drained from the tank at the same rate of 8 gal/min. Find an expression for the amount of salt in the tank at time t.

5. For what values of x does the solution of the following differential equation have either a maximum value or a minimum value?

$$\frac{dy}{dx} = \frac{x^2 - 8x + 12}{3y^2 + 4}$$

6. Find the equation of the curve $y = f(x)$ whose graph passes through the point $(2, 1)$ and whose graph at (x, y) always has the slope x^3/y^2, $y \neq 0$.

7. Initially a very large tank contains 100 lbs of salt dissolved in 500 ft^3 of water. Brine that contains 2 lbs of salt per ft^3 of water enters the tank at the rate of 20 ft^3 per minute. The well mixed brine is drained from the tank at the same rate of 20 ft^3 per minute. Find an expression for the amount of salt in the tank at time t.

8. Initially a large tank contains 2,000 newtons of salt dissolved in 200 m^3 (cubic meters) of water. Brine that contains 16 newtons of salt per m^3 of water enters the tank at the rate of 5 m^3 per minute. The well mixed brine is drained from the tank at the same rate of 5 m^3 per minute. Find an expression for the amount of salt in the tank at time t.

5541 Exponential Growth

The simplest model of population growth makes the assumption that the population grows at a rate proportional to the size of the population. If $N(t)$ denotes the number of individuals present in a population at time t, then

$$\text{Rate of change} = \frac{dN}{dt}.$$

The phrase "is proportional to" means "equals some unknown constant times". When we say "the rate of change is proportional to the number present" we are saying

$$\frac{dN}{dt} = kN,$$

where k is a constant to be determined. If the amount $N(t)$ of a quantity present obeys $N'(t) = kN$, then we also say that $N(t)$ obeys "the law of exponential growth". Although it is somewhat true to say that a population of animals obeys the law of exponential growth, we assume that it is more nearly true about a population of bacteria. We assume that the rate at which bacteria in certain populations grow obeys the law of exponential growth.

Example 1. The rate at which a certain culture of bacteria grows is proportional to the number of bacteria present at time t. That is, the population obeys the law of exponential growth. For a certain population there are 9000 bacteria present at a certain moment in time. After 4 hours the number of bacteria is found to be 12000. How many bacteria are there after 12 hours?

Solution. Let $N(t)$ denote the number of bacteria at time t. The rate of change of N is $\frac{dN}{dt}$. To say that the rate of change is proportional to the amount N is to say that $\frac{dN}{dt}$ equals some unknown constant times N.

$$\frac{dN}{dt} = kN$$

This is a variables separable differential equation.

$$\frac{dN}{N} = kdt.$$

Integrating both sides, we get

$$\ln N = kt + \ln C = \ln(e^{kt}) + \ln C$$
$$\ln N = \ln[Ce^{kt}].$$

Taking antilogs, we have

$$N = Ce^{kt}.$$

Note that $N > 0$ and $C > 0$. There are two constants in the expression for $N(t)$. There is the constant of proportionality, k, and the constant of integration, C. We first apply the initial condition that $N = 9000$ when $t = 0$ in order to find the constant of integration.

$$9000 = Ce^o = C.$$

We have another piece of data which we use to determine the constant of proportionality. The other data is that $N = 12000$ when $t = 4$. Substituting these values into $N = 9000e^{kt}$, we get

$$12000 = 9000e^{4k}.$$
$$e^{4k} = 4/3.$$

Taking the fourth root of both sides of this equation, we have

$$e^k = (4/3)^{1/4}$$

Raise both sides of this equation to the t power, we get

$$e^{kt} = (4/3)^{t/4}.$$

Therefore, the solution is

$$N(t) = 9000(4/3)^{t/4}.$$

When $t = 12$, the number of bacteria is

$$N(12) = 9000(4/3)^{12/4} = 9000(64/27) = 21,333.$$

Example 2. A certain colony of bacteria grows at a rate proportional to the number of bacteria present. At a certain time there are 12 mg of bacteria present. Exactly 20 hours later there are 18 mg. When will there be exactly 40.5 mg present?

Solution. Let $Q(t)$ denote the mass of the bacteria at time t. We assume $Q > 0$. The rate at which the bacteria increase in mass is $Q'(t)$. The phrase "is proportional to" translates as "is equal to some unknown constant times". The mathematical equation is

$$\frac{dQ}{dt} = kQ,$$

where k is the proportionality constant. This is a differential equation and it is variables separable. We write the differential equation in differential form as

$$\frac{dQ}{Q} = kdt.$$

Integrating both sides, we get

$$\ln Q = kt + \ln C.$$

This is an implicit equation for Q. We usually write this in exponential form because that solves for Q. Recall that $\ln[\exp(kt)] \equiv kt$. Using this we rewrite the equation as

$$\ln Q = \ln[\exp(kt)] + \ln C.$$

The log of a product is the sum of the logs. Using this we rewrite this equation as

$$\ln Q = \ln[Ce^{kt}].$$

Taking antilogs, we get

$$Q = Ce^{kt}.$$

The initial condition $Q = 12$ when $t = 0$ enables us to evaluate the constant of integration

$$12 = Ce^o = C.$$

Also we are given that $Q = 18$ when $t = 20$. Substituting these values into $Q = 12e^{kt}$ gives

$$18 = 12e^{20k},$$

$$(3/2) = e^{20k}.$$

We can solve this equation to get $k = (1/20)\ln(3/2)$. However, we can also find e^{kt} by first raising both sides to $1/20$ the power.

$$(3/2)^{1/20} = (e^{20k})^{1/20} = e^k.$$

Then raise both sides to t power.

$$(3/2)^{t/20} = e^{tk}.$$

Using this the mass of the bacteria at any time t is given by

$$Q(t) = 12(3/2)^{t/20}.$$

The question is: for what value of t is $Q(t) = 40.5$?

$$40.5 = 12(3/2)^{t/20}$$

$$\left(\frac{27}{8}\right) = \left(\frac{3}{2}\right)^{t/20}$$

$$\left(\frac{3}{2}\right)^3 = \left(\frac{3}{2}\right)^{t/20}$$

$$t/20 = 3$$

$$t = 60.$$

There will be 40.5 mg of bacteria in 60 hours.

Example 3. At a certain instant in time there is 25 grams of a certain radioactive substance present. After 8 hours 20 grams are present. How much is present after 24 hours? What is the half life of this substance?

Solution. The rate at which the amount of a radioactive substance changes (decays) is proportional to the amount present. The decay of a radioactive substance follows the law of exponential decay. We say exponential decay instead of exponential growth when the constant of proportionality is a negative number. Let $A(t)$ denote the amount of the substance present at time t. The rate $A'(t)$ at which the amount changes is proportional to the amount present, $A(t)$. The rate $A'(t)$ equals some unknown constant times $A(t)$. The differential equation is

$$\frac{dA}{dt} = kA.$$

We have solved this differential before. The solution is

$$A = Ce^{kt}.$$

Applying the initial condition that $A = 25$ when $t = 0$, we get

$$A = 25e^{kt}.$$

The other piece of data which is given is that $A = 20$ when $t = 8$. Substituting these numbers into $A = 25e^{kt}$, we get

$$20 = 25e^{8k}, \text{ or } (4/5) = e^{8k}.$$

Taking the 8th root of both sides, gives

$$(4/5)^{1/8} = e^k.$$

Note that $k = (1/8)\ln(4/5) < 0$. Raising both sides to the t-th power, we have

$$(4/5)^{t/8} = e^{kt}.$$

The amount of radioactive substance at any time t is given by

$$A(t) = 25\left(\frac{4}{5}\right)^{t/8}.$$

When $t = 24$ this gives

$$A(24) = 25\left(\frac{4}{5}\right)^{24/8} = 25\left(\frac{64}{125}\right) = \frac{64}{5} = 12.8.$$

The term "half life" is defined as follows. If at $t = 0$ the amount of a radioactive substance present is A_0, then the half life is that amount of time later when the amount of radioactive substance present is $A_0/2$. The initial amount in this example is 25 so the half life is that time when the amount is 12.5. We can solve for this value of t is the equation

$$12.5 = 25 \left(\frac{4}{5}\right)^{t/8}$$

$$\frac{1}{2} = \left(\frac{4}{5}\right)^{t/8}$$

$$-\ln 2 = (t/8)\ln(4/5)$$

$$t = \frac{-8\ln 2}{\ln(4/5)} = 24.85.$$

The half life is 24.85 hours.

Exercises

1. The rate at which a population of bacteria grows is proportional to the size of the population. In a certain population of bacteria the number present at time $t = 0$ is $12,000$. After exactly 5 hours had passed the population was found to be $15,000$. How many bacteria were there after 10 hours?

2. The rate at which a population of bacteria grows is proportional to the size of the population. For a certain population of bacteria the number present initially was 5400. After exactly 20 hours had passed the population was found to be 9000. How many hours later will there be 25,000 bacteria?

3. At a certain instant in time there is 32 grams of a certain radioactive substance present. After 10 hours 24 grams are present. How much is present after 60 hours?

4. After the administration of a certain drug is stopped the quantity remaining in the patients body decreases at a rate proportional to the amount present. The half life of the drug in the body is 20 hours. How many hours does it take for the drug level in the body to be reduced to 25% of its original level.

108

5545 Logistic Equation

Suppose $y(t)$ denotes the number of individuals in a population such as a population of wolves or bacteria. Our first attempt to model population growth was to assume that the rate of growth of the population is proportional to the number of individuals present. This gives the differential equation

$$\frac{dy}{dt} = ky \text{ with the solution } y(t) = Ce^{kt}.$$

This is the equation of exponential growth. This model says that the population grows without bound. Exponential growth may occur in the initial stages of a population growth, but it can not continue indefinitely. Population growth levels off as a result of factors such as limited food supply and over crowding. We need a model that takes into account the fact that there is a limit to how large a population can become. We need a model that says that when the population reaches a certain number of individuals, then the environment can no longer support any more individuals. This number is called the carrying capacity and we will denote it by N. We want to say something like the rate of growth is proportional to $N - y$. We will not discuss models that do not really work. The model that works pretty well is to assume that the population $y(t)$ satisfies the equation

$$\frac{dy}{dt} = ky(N - y),$$

where k is a constant to be determined by the growth rate and N is the carrying capacity. Note that when $y(t) \equiv N$, then $\frac{dy}{dt} = 0$. This says that when the population reaches the carrying capacity, then the rate at which the population changes is zero. This then is a more reasonable model. The equation $y'(t) = ky(N - y)$ is usually called the logistic equation. Note that $ky(N - y)$ is a quadratic expression in y. In order to solve a logistic equation we must be able to solve equations of the form

$$\frac{dy}{dt} = ay^2 + by + c$$

where a, b, c are constants. A differential equation of this form is called a logistic equation. We will now consider the solutions of differential equations of this form. These differential equations are variables separable.

Example 1. Find the general solution of the differential equation

$$\frac{dy}{dt} = y^2 - 4y - 12.$$

Solution. Note that the independent variable t does not appear in the equation. All logistic type equations are variables separable. We should always begin by factoring the quadratic in y if possible.

$$\frac{dy}{dt} = (y+2)(y-6).$$

In differential form this equation is

$$\frac{dy}{(y+2)(y-6)} = dt.$$

We want to find the antiderivative of both sides of this equation. In order to integrate the left side we must use partial fractions to unadd. We need to find constants A and B such that

$$\frac{A}{y+2} + \frac{B}{y-6} = \frac{1}{(y+2)(y-6)}.$$

Multiplying both sides of this equation by $(y+2)(y-6)$, we get

$$A(y-6) + B(y+2) = 1$$
$$(A+B)y + (-6A+2B) = 1.$$

We must choose constants A and B such that

$$A + B = 0$$
$$-6A + 2B = 1.$$

The solution is $A = -1/8$ and $B = 1/8$. Thus

$$\frac{-1/8}{y+2} + \frac{1/8}{y-6} = \frac{1}{(y+2)(y-6)}.$$

Using this equality we can rewrite the differential equation as

$$\left(\frac{-1/8}{y+2} + \frac{1/8}{y-6}\right) dy = dt.$$

Multiply both sides of this equation by 8 and we get

$$\left(\frac{1}{y-6} - \frac{1}{y+2}\right) dy = 8dt.$$

Integrating both sides gives

$$\ln|y-6| - \ln|y+2| = 8t + \ln C \text{ with } C > 0.$$

Recall that $8t \equiv \ln(e^{8t})$. Using this we rewrite the last equation as

$$\ln|y-6| - \ln|y+2| = \ln(e^{8t}) + \ln C.$$

Recall that $\ln(A) + \ln(B) = \ln(AB)$ and $\ln(A) - \ln(B) = \ln(A/B)$. Using these two properties of logs, we rewrite this equation as

$$\ln\left|\frac{y-6}{y+2}\right| = \ln[Ce^{8t}].$$

If the logs are equal, then the functions are equal, that is,

$$\left|\frac{y-6}{y+2}\right| = Ce^{8t} \text{ with } C > 0.$$

This is the general solution which is a family of functions. This same family of functions is described by the equation

$$\frac{y-6}{y+2} = Ce^{8t} \text{ when } C \neq 0.$$

This equation defines y implicitly as a function of t. Note that $y \equiv 6$ is also a solution of the differential equation. Hence, we can allow $C = 0$. Since we can solve this equation explicitly for y we should do that. Multiplying both sides of this equation by $(y+2)$ we get

$$y - 6 = (y+2)Ce^{8t} = Cye^{8t} + 2Ce^{8t}$$

$$y - Cye^{8t} = 6 + 2Ce^{8t}$$

$$y(1 - Ce^{8t}) = 6 + 2Ce^{8t}$$

$$y = \frac{6 + 2Ce^{8t}}{1 - Ce^{8t}}.$$

This is the general solution of the differential equation. This means that every solution of the given differential equation can be expressed by

$$y(t) = \frac{6 + 2Ce^{8t}}{1 - Ce^{8t}}$$

by choosing an appropriate value of C. This is a true statement except for one solution. The function $y(t) \equiv -2$ is a solution of $y'(t) = y^2 - 4y - 12$, but there is no value of C such that

$$y = \frac{6 + 2Ce^{8t}}{1 - Ce^{8t}}$$

becomes $y = 2$. Note that $C = 0$ gives $y(t) \equiv 6$.

A slightly different path for our solution would cause us to write the general solution as

$$y = \frac{-2 - 6Ce^{-8t}}{1 - Ce^{-8t}}.$$

Substituting $C = 0$ into this form of the general solution does give $y(t) \equiv -2$ as a solution. However, using this form of the solution there is no value of C which gives the solution $y(t) \equiv 6$. The solutions $y(t) = -2$ and $y(t) \equiv 6$ are special in some sense. Note that the first step in our solution of $y'(t) = (y + 2)(y - 6)$ was to divide by $(y + 2)(y - 6)$. In order for this to be correct we must assume $y \neq -2$ and $y \neq 6$. When solving logistic differential equations we will just find the the usual general solution at this time and not concern ourselves with these special solutions. We will discuss these special solutions in our future study of differential equations.

Example 2. Solve the initial value problem

$$\frac{dy}{dt} = y^2 - 4y - 12 \text{ and } y(0) = 3.$$

Solution. We already found the general solution of this differential equation in Example 1. It is

$$y = \frac{6 + 2Ce^{8t}}{1 - Ce^{8t}}.$$

Substituting $t = 0$ and $y = 3$ into the general solution, we get

$$3 = \frac{6 + 2C}{1 - C}.$$

Solving for C gives $C = -3/5$. The solution of the initial value problem is

$$y = \frac{6 + 2(-3/5)e^{8t}}{1 - (-3/5)e^{8t}} = \frac{30 - 6e^{8t}}{5 + 3e^{8t}}.$$

Example 3. Solve the initial value problem

$$\frac{dy}{dt} = y^2 + 16, \text{ and } y(0) = 4.$$

Solution. Note that $y^2 + 16$ does not factor and $y^2 + 16 \neq 0$. We write this equation in differential form

$$\frac{dy}{y^2 + 16} = dt.$$

Rewrite this as

$$\frac{4dy}{16 + y^2} = 4dt.$$

Integrating both sides, we get

$$\arctan y/4 = 4t + C.$$

Recall that $\tan[\arctan(A)] = A$. Take the tangent of both sides and we have

$$\frac{y}{4} = \tan(4t + C)$$

$$y = 4\tan(4t + C).$$

Substituting the initial condition $y = 4$ and $t = 0$ into the general solution, we get

$$4 = 4 \tan C, \text{ or } \tan(C) = 1$$
$$C = \pi/4.$$

The solution of the initial value problem is

$$y = 4 \tan(4t + \pi/4).$$

Exercises

1. Solve the initial value problem:

$$\frac{dy}{dt} = y^2 - 8y \text{ and } y(0) = 2.$$

2. Solve the initial value problem

(a)
$$\frac{dy}{dt} = (2 + y)(8 - y) \text{ and } y(0) = 2.$$

(b)
$$\frac{dy}{dt} = (2 + y)(8 - y) \text{ and } y(0) = 10.$$

3. Solve the initial value problem:

$$\frac{dy}{dt} = y^2 + 9 \text{ and } y(0) = 3\sqrt{3}.$$

4. Find the general solution of

$$\frac{dy}{dt} = (2 - y)(6 - y).$$

5553 Euler Method

We have been considering differential equations of the form

$$\frac{dy}{dx} = f(x, y).$$

Most such equations can not be solved by the elementary methods like the ones we have discussed. With this in mind, and with modern computers available, we would like to find some methods for finding a numerical approximation to the solution of a given differential equation. Of course, we would need to pick one solution at a time from the family of solutions to the differential equation. This means that we would need to try to solve an initial value problem. Suppose we have an initial value problem in the following form:

$$\frac{dy}{dx} = f(x, y) \text{ and } y(x_0) = y_0.$$

We will start with the goal of computing an approximate value for the solution at the finite set of points $x_1, x_2, x_3, \cdots, x_n$, with $x_k = x_0 + kh$ where h is fixed and $k = 1, 2, 3, \cdots, n$. The constant difference h between successive values of x is called the step size. If $y(x)$ denotes the exact solution of the initial value problem, then we want to find approximate values of $y(x)$ at the points where $x = x_k$, $k = 0, 1, 2, \cdots, n$. The values of h and n are usually decided before the computation starts.

As part of the problem we are given that $y = y_0$ when $x = x_0$. As a start we want to find the next value of $y(x)$, the value of $y(x)$ when $x = x_1$. Let y_1 denote the approximate value of the actual solution $y(x)$ at $x = x_1$. The question is: how shall we find the approximate value y_1 with the information given? Note that the information given is enough to uniquely determine the solution. Also the given information is enough to determine the linear approximation of the solution at $x = x_0$. Let us use the linear approximation to approximate the actual solution. Recall that the linear approximation $L_a(x)$ of a function $y(x)$ at the point where $x = a$ is

$$L_a(x) = y(a) + y'(a)(x - a).$$

The linear approximation $L_a(x)$ is very nearly equal to the function $y(x)$ for values of x near a. The linear approximation of the solution $y(x)$ near

$x = x_0$ is

$$L_{x_0}(x) = y(x_0) + y'(x_0)(x - x_0).$$

It follows that $L_{x_0}(x) \simeq y(x)$ for x near x_0. Since x_1 is a value of x close to x_0 it follows that $L_{x_0}(x_1) \simeq y(x_1)$. Therefore, let us choose the value y_1 to be $y_1 = L_{x_0}(x_1)$, then $y_1 \simeq y(x_1)$.

$$y_1 = y(x_0) + y'(x_0)(x_1 - x_0)$$
$$y_1 = y_0 + hy'(x_0) = y_0 + h[f(x_0, y(x_0))].$$

This means that (x_1, y_1) is an approximate point on the solution curve. Suppose we have found several approximate points on the solution curve, how do we find the next approximate point? Suppose we have found the point (x_{k-1}, y_{k-1}). We proceed from the approximate point (x_{k-1}, y_{k-1}) on the solution curve to the next approximate point (x_k, y_k) using this same idea for approximating the solution curve, that is, using the linear approximation to approximate the actual solution. First, we find the linear approximation to the solution curve which passes through the point (x_{k-1}, y_{k-1}) at the point (x_{k-1}, y_{k-1}). This linear approximation is

$$L(x) = y(x_{k-1}) + y'(x_{k-1})(x - x_{k-1}).$$

The differential equation says that $y'(x_{k-1}) = f(x_{k-1}, y_{k-1})$. Using this the linear approximation is

$$L(x) = y_{k-1} + f(x_{k-1}, y_{k-1})(x - x_{k-1}).$$

The value of the solution at $x = x_k$ is $y(x_k)$. Let y_k denote the approximate value of $y(x_k)$. That is, y_k is the y coordinate of the next approximate solution point. We choose the value y_k to be $y_k = L(x_k)$. This is using the linear approximation to approximate the actual curve.

$$y_k = L(x_k) = y_{k-1} + f(x_{k-1}, y_{k-1})(x_k - x_{k-1})$$
$$y_k = y_{k-1} + hf(x_{k-1}, y_{k-1}).$$

We used the fact that $(x_k - x_{k-1}) = h$. When we use the formula

$$y_k = y_{k-1} + hf(x_{k-1}, y_{k-1}) \quad k = 1, 2, \cdots, n$$

116

to find the values of y_k, we are using Euler's method for finding an approximate solution to an initial value problem. Euler's method generates much larger errors than more advanced methods, but it illustrates the basic ideas behind numerical methods. We can get a smaller error by using a smaller value of h. We study Euler's method because it is simple and it illustrates the basic ideas.

Example 1. Use Euler's method to find an approximate solution to the initial value problem

$$\frac{dy}{dx} = 2x + y \text{ and } y(1) = 2.$$

Using $h = 0.2$ and $n = 5$.

Solution. Note that we will find an approximate solution on the interval $1 \leq x \leq 2$. Instead of saying $h = 0.2$ and $n = 5$ the problem could have said find the solution on the interval $1 \leq x \leq 2$ using $n = 5$. The general formula for the Euler method is

$$y_k = y_{k-1} + hf(x_{k-1}, y_{k-1}).$$

For this problem we have $h = 0.2$ and $f(x, y) = 2x + y$. For this problem the general formula is

$$y_k = y_{k-1} + (0.2)[2x_{k-1} + y_{k-1}].$$

In order to find the first value y_1 we replace k with 1 in the general formula.

$$y_1 = y_0 + (0.2)[2x_0 + y_0].$$

We need values for x_0 and y_0. These values are given by the initial condition, $x_0 = 1$ and $y_0 = 2$. Thus,

$$y_1 = 2 + (0.2)[2(1) + 2] = 2.8.$$

The first computed approximate coordinates of a point on the solution curve are $(1.2, 2.8)$.

In order to find y_2 we replace k with 2 in the general formula. This gives

$$y_2 = y_1 + (0.2)[2x_1 + y_1].$$

The numbers x_1 and y_1 are the previously computed values $x_1 = 1.2$ and $y_1 = 2.8$.

$$y_2 = 2.8 + (0.2)[2(1.2) + 2.8] = 3.84$$

In order to find y_3 we replace k with 3 in the general formula. This gives

$$y_3 = y_2 + (0.2)[2x_2 + y_2].$$

The numbers x_2 and y_2 are $x_2 = 1.4$ and $y_2 = 3.84$.

$$y_3 = 3.84 + (0.2)[2(1.4) + 3.84] = 5.168.$$

In order to find y_4 we replace k with 4 in the general formula. This gives

$$y_4 = y_3 + (0.2)[2x_3 + y_3].$$

The numbers x_3 and y_3 are $x_3 = 1.6$ and $y_3 = 5.168$.

$$y_4 = (5.168) + (0.2)[2(1.6) + 5.168] = 6.8416.$$

In order to find y_5 we replace k with 5 in the general formula. This gives

$$y_5 = y_4 + (0.2)[2x_4 + y_4].$$

The numbers x_4 and y_4 are $x_4 = 1.8$ and $y_4 = 6.8416$.

$$y_5 = (6.8416) + (0.2)[2(1.8) + 6.8416] = 8.9299.$$

We have found the five approximate points on the solution curve to be

$$(1.2, 2.8), \ (1.4, 3.84), \ (1.6, 5.168), \ (1.8, 6.8416) \text{ and } (2, 8.9299).$$

Since this initial value problem is easy to solve we can solve it and find the value of y corresponding to $x = 2$. The value is 10.3097. The approximating value is 8.9299. Not very good. We should have used a smaller value of h and a larger value of n, say $n = 100$.

Exercises

1. Find an approximate solution of the following initial value problem on the interval $1 \leq x \leq 2$ using $n = 5$.

$$\frac{dy}{dx} = 3x + 2y \text{ and } y(1) = 2.$$

2. Find an approximate solution of the following initial value problem on the interval $1 \leq x \leq 1.6$ using $n = 6$.

$$\frac{dy}{dx} = \frac{2y + 3}{x + 5} \text{ and } y(1) = 2.$$

3. Find an approximate solution of the following initial value problem on the interval $(1/2) \leq x \leq 1$ using $n = 5$.

$$\frac{dy}{dx} = 3x - \sqrt{y} \text{ and } y(1/2) = 2.$$

5549 First Order Linear Differential Equations

A differential equation is called first order linear if it can be written in the form

$$a(x)\frac{dy}{dx} + b(x)y = h(x),$$

where $a(x), b(x)$, and $h(x)$ are functions of the independent variable x only. All first order linear differential equations are solved using exactly the same sequence of steps. We will illustrate these steps by solving an example.

Example 1. Solve the differential equation

$$x\frac{dy}{dx} + 3y = 24x^5.$$

Solution. This equation fits exactly into the general form using $a(x) = x$, $b(x) = 3$, and $h(x) = 24x^5$. This shows beyond doubt that this differential equation is first order linear. The "standard form" for a first order linear differential equation is

$$\frac{dy}{dx} + f(x)y = g(x).$$

Notice that when written in the standard form the coefficient of $\frac{dy}{dx}$ is one, and y and $\frac{dy}{dx}$ are on the same side of the equals sign.

Step 1. Write the differential equation in standard form. In order to put this equation in standard form we must divide both sides by x. We get

$$\frac{dy}{dx} + \frac{3}{x}y = 24x^4.$$

Step 2. Integrate the coefficient of y. This could be a difficult step.

$$p(x) = \int \frac{3}{x}dx = 3\ln x.$$

Step 3. Take the exponential, that is, find $e^{p(x)}$. The expression $e^{p(x)}$ is called the integrating factor. The integrating factor for this problem is

$$e^{p(x)} = e^{3\ln x} = e^{\ln x^3} = x^3.$$

Step 4. Consider the differential equation as it was written in standard form (see Step 1). Multiply both sides of the differential equation in standard form by the integrating factor.

$$x^3 \left[\frac{dy}{dy} + \frac{3}{x}y \right] = x^3 [24x^4]$$

$$x^3 \frac{dy}{dx} + (3x^2)y = 24x^7.$$

After this step the left side of the equation is always the derivative of the integrating factor times y. Use the product rule for differentiating to check that this is actually true for this problem.

$$\frac{d}{dx}[x^3 y] = x^3 \frac{dy}{dx} + 3x^2 y.$$

If the left side of the equation is not the derivative of the integrating factor times y, then a mistake has been made. Since they are equal we replace $x^3 y' + 3x^2 y$ with $\frac{d}{dx}[x^3 y]$ in the differential equation.

$$\frac{d}{dx}[x^3 y] = 24x^7.$$

Step 5. Integrate both sides of the equation. Note that it is always true that

$$\int \frac{d}{dx}[f(x)]dx = f(x).$$

Using this fact to integrate the left side we integrate both sides of the last equation to get

$$x^3 y = 3x^8 + C.$$

$$y = 3x^5 + Cx^{-3} \quad \text{for } x \neq 0.$$

This is a family of functions. Note that we are able to find an explicit expression for the solution. The solution of an initial value problem involving this differential equation would be a single member of this family. Suppose that in addition to the differential equation $xy' + 3y = 24x^5$ we require that the initial condition $y(1) = 8$ be satisfied. This means that C must be selected such that

$$8 = 3 + C \text{ or } C = 5.$$

The solution of the initial value problem is

$$y = 3x^5 + 5x^{-3}.$$

Example 2. A very large tank contains 50 lbs of salt dissolved in 600 gallons of water. Brine that contains 2/3 lbs of salt per gallon of water enters the tank at the rate of 6 gallons per minute. The mixture leaves the tank at the slower rate of 4 gallons per minute. How many pounds of salt are in the tank after 30 minutes?

Solution. Let $A(t)$ denote the number of pounds of salt in the tank at time t. At the start of time there are 600 gallons of brine in the tank. Every minute 6 gallons of brine enters the tank and only 4 gallons leave the tank. This means that the amount of brine in the tank increases at the rate of 2 gal/min. The number of gallons of brine in the tank after t minutes is

$$600 + 2t.$$

The rate at which the amount of salt in the tank is changing with respect to time is given by the rate at which salt enters the tank minus the rate at which salt leaves the tank. The rate at which salt enters the tank is

$$\text{Rate in } = (2/3 \text{ lb/gal})(6 \text{ gals/min}) = 4 \text{ lbs/min}.$$

The rate at which the salt is leaving the tank is

$$\text{Rate out } = \left(\frac{A(t)}{600 + 2t} \text{ lbs/gal}\right)(4 \text{ gals/min}) = \frac{2}{300 + t} \text{ lbs/min}.$$

The differential equation for the amount of salt in the tank is

$$\frac{dA}{dt} = 4 - \frac{2}{300 + t}A.$$

Rewrite the equation in the standard form for a first order linear differential equation. It is

$$\frac{dA}{dt} + \frac{2}{300 + t}A = 4.$$

Since this equation is in standard form we are ready for Step 2. Step 2 is to integrate the coefficient of A.

$$p(t) = \int \frac{2}{300 + t} dt = 2\ln(300 + t).$$

Step 3. We find $e^{p(t)}$, the integrating factor, using the properties of logarithms as follows

$$e^{p(t)} = e^{2\ln(300+t)} = e^{\ln(300+t)^2} = (300 + t)^2.$$

Step 4. Multiply both sides of the differential equation in standard form by the integrating factor $(300 + t)^2$.

$$(300 + t)^2 \left[\frac{dA}{dt} + \frac{2}{300 + t} A \right] = 4(300 + t)^2$$

$$(300 + t)^2 \frac{dA}{dt} + 2(300 + t)A = 4(300 + t)^2$$

Check to see that the left hand side is the derivative of the integrating factor times the dependent variable.

Using the rule for differentiating a product we see that

$$\frac{d}{dt} \left[(300 + t)^2 A(t) \right] = (300 + t)^2 \frac{dA}{dt} + 2(300 + t)A$$

Using this equality we can rewrite the differential equation as

$$\frac{d}{dt} \left[(300 + t)^2 A \right] = 4(300 + t)^2.$$

Integrating both sides, we get

$$(300 + t)^2 A = \frac{4}{3}(300 + t)^3 + C.$$

$$A(t) = (4/3)(300 + t) + C(300 + t)^{-2}.$$

We evaluate the constant of integration C using the initial condition that $A = 50$ when $t = 0$.

$$50 = (4/3)(300) + C(300)^{-2}$$
$$C = -350(300)^2$$

$$A(t) = (4/3)(300 + t) - 350(300)^2(300 + t)^{-2}.$$

This is the amount of salt in the tank at any time t. The question was: what is A when $t = 30$?

$$A(30) = (4/3)(300 + 30) - 350(300)^2(330)^{-2}$$
$$= 151 \text{ lbs.}$$

Exercises

1. Find the general solution of the following differential equation:

$$x\frac{dy}{dx} - 2y = 12x^3 \cos 3x.$$

2. Solve the initial value problem:

$$x\frac{dy}{dx} + 3y = 8x \text{ and } y(1) = 6.$$

3. Find the general solution of $x\frac{dy}{dx} = 4y + 20x^5 \sin(5x)$.

4. Solve the initial value problem:

$$\frac{dy}{dt} = 12 - \frac{6y}{(100 + 2t)} \text{ and } y(0) = 100.$$

5. A tank contains 25 kg of salt dissolved in 4,000 liters of water. Brine that contains (1/10) kg of salt per liter of water enters the tank at the rate of 20 liters/min. Brine is drained from the tank at the slower rate of 15 liters/min. Find an expression for the amount of salt in the tank at time t.

6. A tank contains 80 lbs of salt dissolved in 600 gallons of water. Brine that contains (1/3) lbs of salt per gallon of water enters the tank at the rate of 15 gals/min. Brine is drained from the tank at the slower rate of 10 gals/min. Find an expression for the amount of salt in the tank at time t.

5575 Polar Coordinates

In order to analyze a geometric figure in a plane the figure needs to be in a coordinate plane. In the past, this has always been a cartesian coordinate plane. In a cartesian plane we have two perpendicular reference lines. These lines are the x axis and the y axis. The location of a given point in the plane is specified by giving its coordinates relative to these axes. The x coordinate of a point is 2 if the point is on the line perpendicular to the x axis (parallel to the y axis) that passes through the point on the x axis where $x = 2$. The y coordinate of a point is 3 if the point is on the line perpendicular to the y axis (parallel to the x axis) that passes through the point on the y axis where $y = 3$. This determines the point since the intersection of two straight lines is a unique point.

We are now going to locate points in the plane using coordinates called polar coordinates. Fundamental to polar coordinates is the pole(origin) and the polar axis. The pole is a point in the plane. The polar axis is a half line starting at the pole and going to infinity. In all our discussion of polar coordinates there is always an associated set of rectangular coordinates. The pole (origin) of the polar coordinates is always the same as the origin of the rectangular system. The polar axis is always the same as the positive x axis, and one unit on the polar axis is always the same length as one unit on the x axis. We actually locate points in the plane using polar coordinates by saying that the point is at the intersection of a certain circle with a certain line. However, our description of this location is usually as follows. Given the polar coordinates (r, θ) the point located by these coordinates is found as follows. The number θ is an angle. Suppose the angle whose measure is θ has its initial side the same as the polar axis, then the point in question is on the terminal side of that angle. If $\theta > 0$, then the angle is obtained by turning in the counterclockwise direction. If $\theta < 0$, then the angle is found by turning in the clockwise direction. The first number, r, is the polar distance. If $r > 0$, then r is the distance from the origin (pole) to the point in question along the terminal side of the angle.

Example 1. Plot the point in the plane whose polar coordinates are $(2, \pi/3)$.

Solution. We turn the angle $\pi/3$ in the counterclockwise direction. We find the point on the terminal side of this angle that is 2 units from the origin (pole).

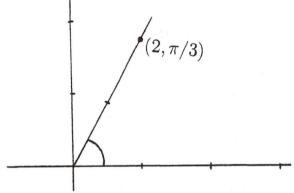

$(2, \pi/3)$

Example 2. Plot the point in the plane whose polar coordinates are $(3, 5\pi/4)$.

Solution. We turn the angle $5\pi/4$ in the counterclockwise direction. We find the point on the terminal side of this angle that is 3 units from the origin.

$(3, 5\pi/4)$

Example 3. Plot the point in the plane whose polar coordinates are $(3, -5\pi/4)$.

Solution. We turn the angle of magnitude $5\pi/4$ but in the clockwise direction. We find the point on the terminal side that is 3 units from the origin.

$(3, -5\pi/4)$

We also allow the value of r to be negative. First, we extend through the origin the ray which is the terminal side of the angle with measure θ. This gives a ray which is an extension of the terminal side of the angle. Negative values of r are marked off on this negative extension ray. If $r < 0$, then the point located is a distance $|r|$ down the negative extension from the origin.

Example 4. Plot the point in the plane with polar coordinates $(-3, \pi/4)$.

Solution. We turn the angle of magnitude $\pi/4$ in the counterclockwise direction. Extend the terminal side of this angle through the origin. The point with coordinates $(-3, \pi/4)$ is on the negative extension of the terminal side 3 units from the origin.

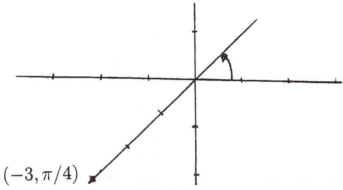

Example 5. Plot the point in the plane with polar coordinates $(-3, -3\pi/4)$.

Solution. We turn the angle of magnitude $3\pi/4$ in the clockwise direction. Extend the terminal side of this angle through the origin to obtain the negative extension ray. The point with coordinates $(-3, -3\pi/4)$ is on the negative extension of the terminal side 3 units from the origin.

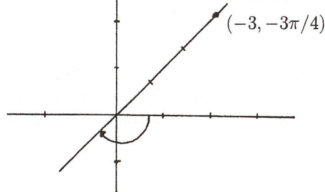

Given a set of polar coordinates (r, θ) there is only one point in the coordinate plane with these coordinates. However, the reverse is not true. Given a point in the coordinate plane, the point has many sets of polar coordinates. The polar coordinates $(2, 3\pi/4)$ and $(2, 11\pi/4)$ locate the same point because the terminal side of the angle $11\pi/4$ is the same as the terminal side of $3\pi/4$. The polar coordinates $(3, 3\pi/2)$ and $(-3, \pi/2)$ locate the same point because the terminal side of the angle $3\pi/2$ and the negative extension of the angle $\pi/2$ are the same.

Example 6. Given the point with rectangular coordinates $(-2,2)$ find four different sets of polar coordinates for this point such that $-2\pi < \theta < 2\pi$.

Solution. The ray from the origin through the point $(-2,2)$ is the terminal side of the angle $3\pi/4$. The distance from the origin to the point $(-2,2)$ is $\sqrt{(-2)^2 + 2^2} = 2\sqrt{2}$. Thus one set of polar coordinates is $(2\sqrt{2}, 3\pi/4)$.

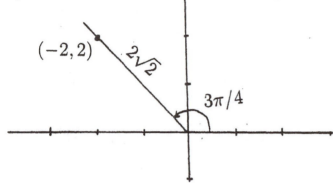

The ray from the origin through the point $(-2,2)$ is also the terminal side of the angle $-5\pi/4$. Another set of polar coordinates for this point is $(2\sqrt{2}, -5\pi/4)$.

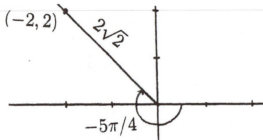

The ray from the origin through the point $(-2,2)$ is the negative extension of the terminal side of the angle $-\pi/4$. The coordinate on the negative extension is $-2\sqrt{2}$. A set of polar coordinates for this point is $(-2\sqrt{2}, -\pi/4)$.

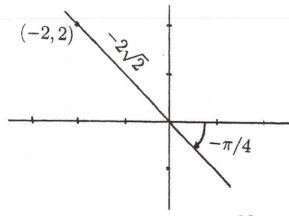

128

The ray from the origin through the point $(-2, 2)$ is the negative extension of the terminal side of the angle $7\pi/4$.

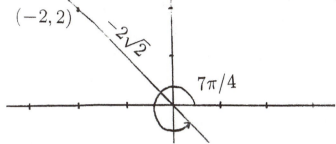

The coordinate on the negative extension is $-2\sqrt{2}$. A set of polar coordinates for this point is $(-2\sqrt{2}, 7\pi/4)$.

Suppose (r, θ) is a set of polar coordinates for a point. Let (x, y) denote the unique set of rectangular coordinates for this same point. It is easy to make a sketch showing all these quanities if the point is located in the first quadrant.

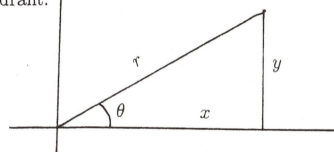

For a given values of r and θ the definition of trigonometric functions tells us that x and y are always given by

$$x = r\cos\theta \text{ and } y = r\sin\theta.$$

On the other hand, for given values of x and y the corresponding values of r and θ are not uniquely determined. We have

$$x^2 + y^2 = r^2.$$

Given x and y we have two possible values of r. These values are $r = +\sqrt{x^2 + y^2}$ and $r = -\sqrt{x^2 + y^2}$. Given x and y the angle θ can have any values such that

$$\tan\theta = \frac{y}{x}.$$

Given x and y there are an infinite number of possible values for θ. This causes problems.

We make two kinds of graphs using rectangular coordinates. Most rectangular graphs that we construct are the graphs of functions. For example we graph the function $f(x) = x^2(1+x^2)^{-1}$ or the function $f(x) = 1 + \cos x$. Calculators will only make these kinds of graphs. The other kind of graph we construct is the graph of an equation. For example, we graph the equation $4x^2 + 9y^2 + 8x + 54y = 27$. We are now going to discuss the construction of graphs using a polar coordinate plane. We only graph equations using polar coordinates. We construct the graph of a polar equation as follows. Suppose we have an equation involving the polar coordinates r and θ which can be written in the form $r = f(\theta)$. This says that for each value of θ there corresponds a unique value of r. The graph of this equation consists of all points whose polar coordinates (r, θ) satisfy the equation $r = f(\theta)$.

Example 7. Sketch the graph of $r = 2 \sin \theta$.

Solution. First, we select some values of θ and compute the corresponding values of r. When $\theta = \pi/6$, we have $r = 2\sin(\pi/6) = 1$. When $\theta = \pi/4$, we have $r = 2\sin(\pi/4) = \sqrt{2}$. Continuing in this manner we construct the following chart of values. Note that we have a somewhat odd situation in making this chart. When writing polar coordinates we write (r, θ) with the value of r first, but it is traditional when making the chart for a graph to put the value of θ first in the left most column.

θ	r
0	0
$\pi/6$	1
$\pi/4$	$\sqrt{2}$
$\pi/3$	$\sqrt{3}$
$\pi/2$	2
$2\pi/3$	$\sqrt{3}$
$3\pi/4$	$\sqrt{2}$
$5\pi/6$	1
π	0
$7\pi/6$	-1
$5\pi/4$	$-\sqrt{2}$

In order to construct the graph we first plot the points with these coordinates. We then connect successive points with a smooth curve. The assumption is that should we plot more points then these points would also be on this smooth curve.

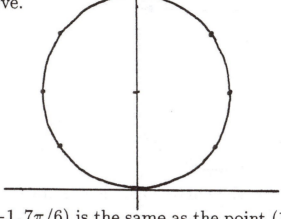

Note that the point $(-1, 7\pi/6)$ is the same as the point $(1, \pi/6)$. The point $(-\sqrt{2}, 5\pi/4)$ is the same point as $(\sqrt{2}, \pi/4)$. As we use values of θ larger than π we are starting to trace the curve again. The graph is tracing itself over again. We obtain the entire circle by plotting points such that $0 \le \theta \le \pi$. If we plot the points with $\pi \le \theta \le 2\pi$, we go around the circle a second time. This happens very often with polar graphs. Very often the graph for $2\pi \le \theta \le 4\pi$ is the same as the graph for $0 \le \theta \le 2\pi$. The graph of a polar equation $r = f(\theta)$ consists of all points that have at least one polar representation (r, θ) that satisfies the equation $r = f(\theta)$. This means that some of the many other coordinates of this point may not satisfy the equation $r = f(\theta)$.

Example 8. Suppose we have a polar equation $r = f(\theta)$ with the following chart of values. Sketch the graph of this equation.

θ	r
0	2
$\pi/6$	0.73
$\pi/4$	0
$\pi/3$	-0.73
$\pi/2$	-2
$2\pi/3$	-2.73
$3\pi/4$	-2.83
$5\pi/6$	-2.73
π	-2

Solution. First, note that $(2, 0)$ and $(-2, \pi)$ are the same point.

131

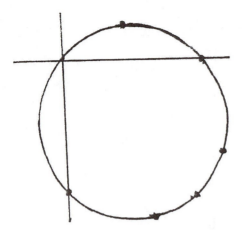

Example 9. Consider the curve which is the graph of the polar equation $r = 3(\cos\theta + \sin\theta)$. Find the rectangular equation whose graph is the same curve.

Solution. Multiply both sides of $r = 3(\cos\theta + \sin\theta)$ by r and we get

$$r^2 = 3(r\cos\theta + r\sin\theta).$$

Next we substitute using the relations $x = r\cos\theta$, $y = r\sin\theta$, and $r^2 = x^2 + y^2$.

$$x^2 + y^2 = 3(x + y).$$

Example 10. Consider the curve which is the graph of the rectangular equation $x^2 + y^2 = 5x$. Find the polar equation whose graph is this same curve.

Solution. If (x, y) denotes a point on this curve, then $x^2 + y^2 = 5x$. This point must have polar coordinates (r, θ) for some value of r and θ. There is some value of r and θ such that $x = r\cos\theta$ and $y = r\sin\theta$. For this value of r and θ, we have

$$r^2\cos^2\theta + r^2\sin^2\theta = 5r\cos\theta.$$
$$r^2(\cos^2\theta + \sin^2\theta) = 5r\cos\theta$$

$$r = 5\cos\theta$$

This value of r and θ must satisfy the equation $r = 5\cos\theta$. Note that while there is only one solution of the equation $x^2 + y^2 = 5x$ that locates the point with coordinates $(5/2, 5/2)$ on this graph there are many solutions

132

of the equation $r = 5\cos\theta$ that plot as this same point. For example, $(\pi/4, (5/2)\sqrt{2})$ and $(5\pi/4, -(5/2)\sqrt{2})$ both plot as this same point.

Example 11. Suppose a curve is the graph of the polar equation

$$3r\cos\theta + 4r\sin\theta = 5, \text{ or } r = 5(3\cos\theta + 4\sin\theta)^{-1}.$$

What equation in Cartesian coordinates has the same graph?

Solution. Each point on this curve has a set of polar coordinates (r, θ) such that $3r\cos\theta + 4r\sin\theta = 5$. Using these values of (r, θ) the x coordinates of this point is given by $x = r\cos\theta$ and the y coordinate is given by $y = r\sin\theta$. Therefore, the Cartesian coordinates of the point satisfy the equation

$$3x + 4y = 5.$$

Example 12. Suppose a certain curve is the graph of the polar equation

$$r^2\cos 2\theta = 1, \text{ for } -\pi/4 < \theta < \pi/4, \ 3\pi/4 < \theta < 5\pi/4, \text{ etc.}$$

What equation in Cartesian coordinates has this same graph? The expression $r^2\cos 2\theta = 1$ means $r = \sqrt{\sec(2\theta)}$ or $r = -\sqrt{\sec(2\theta)}$ and is not of the form $r = f(\theta)$.

Solution. We need to rewrite this equation using the identity $\cos 2\theta = \cos^2\theta - \sin^2\theta$.
$$r^2[\cos^2\theta - \sin^2\theta] = 1. \text{ Note } |x| \geq 1.$$

Using $x = r\cos\theta$ and $y = r\sin\theta$ we convert this to the Cartesian equation

$$x^2 - y^2 = 1.$$

Exercises

1. Plot the points with the following polar coordinates

a) $(4, 5\pi/4)$ b) $(2, -\pi/3)$ c) $(-3, \pi/4)$

d) $(3, 4)$ e) $(-3, -5\pi/4)$ f) $(-4, -5\pi/6)$.

2. Given the point with rectangular coordinates $(-2, -2)$ find four different sets of polar coordinates for this point such that $-2\pi < \theta < 2\pi$.

3. Given the point with Cartesian coordinates $(-2\sqrt{3}, 2)$ find four different sets of polar coordinates for this point such that $-2\pi < \theta < 2\pi$.

4. Suppose we have a polar equation $r = f(\theta)$ with the following chart of values. Sketch the graph of this equation.

θ	r
0	2
$\pi/6$	$\sqrt{3}$
$\pi/4$	$\sqrt{2}$
$\pi/3$	1
$\pi/2$	0
$2\pi/3$	-1
$3\pi/4$	$-\sqrt{2}$
$5\pi/6$	$-\sqrt{3}$
π	-2

5. Suppose we have a polar equation $r = f(\theta)$ with the following chart of values. Sketch the graph of this equation.

θ	r
0	-2
$\pi/6$	-0.73
$\pi/4$	0
$\pi/3$	0.73
$\pi/2$	2
$2\pi/3$	2.73
$3\pi/4$	2.83
$5\pi/6$	2.73
π	2

6. Consider the curve which is the graph of the following Cartesian equations, find the polar equation whose graph is the same curve.

a) $x^2 + y^2 = 6y$

b) $3x + 4y = 6$.

c) $x^2 + y^2 = 36$

7. Consider the curve which is the graph of the following polar equation. Find the Cartesian equation whose graph is the same curve.

a) $r = 4(\cos\theta + \sin\theta)$

b) $r^2 \sin 2\theta = 1$.

c) $r = \tan\theta \sec\theta$

5579 Graphing Polar Equations

We will only construct the graph of polar equations that can be written in the form $r = f(\theta)$. It is too difficult to graph equations which can not be written in this form.

The graph of a polar equation of the form $r = (c)\cos\theta$ or $r = (c)\sin\theta$ is a circle. We have already constructed the graph of an example of such a circle. The next most common polar curve is called a cardioid. The graphs of all equations of the form $r = c(1+\sin\theta)$ and $r = c(1+\cos\theta)$ are cardioids.

Example 1. Sketch the graph of the equation $r = 2(1 + \cos\theta)$ which is a cardioid.

Solution. First, we make a chart of values using $0 \le \theta \le 2\pi$.

θ	r
0	4
$\pi/6$	$2 + \sqrt{3} = 3.7$
$\pi/4$	$2 + \sqrt{2} = 3.4$
$\pi/3$	3
$\pi/2$	2
$3\pi/4$	$2 - \sqrt{2} = 0.6$
π	0
$5\pi/4$	$2 - \sqrt{2} = 0.6$
$3\pi/2$	2
$7\pi/4$	$2 + \sqrt{2} = 3.4$
2π	4

In order to construct the graph we first plot these points and then connect them with a smooth curve.

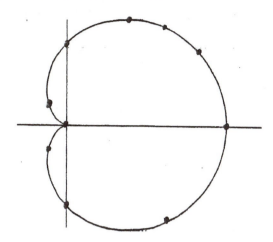

As we all know a graphing calculator will draw graphs of equations in polar coordinates. This is usually very helpful, but it can also lead to wrong conclusions. If we draw the graph of $r = 2(1 + \cos\theta)$, $0 \le \theta \le 2\pi$, using a calculator we see the cardioid sketched above. If we draw the graph of $r = 2(1 + \cos\theta)$, $2\pi \le \theta \le 4\pi$, using a calculator we see the same cardioid. This means that the graph of $r = 2(1 + \cos\theta)$, $0 \le \theta \le 4\pi$, is the cardioid drawn twice. This can be a critical issue if we only want to draw the graph once. Also we can be pretty sure that if we use $0 \le \theta \le 6\pi$ then we are going to draw the graph three times. On the other hand consider the case of the circle whose equation is $r = 2\cos\theta$. The graph of $r = 2\cos\theta$, $0 \le \theta \le \pi$, draws the circle once while the graph of $r = 2\cos\theta$, $0 \le \theta \le 2\pi$, draws the circle twice.

When we look at the graph of the equation $y = f(x)$ in cartesian coordinates it is very easy to decide the range of values of the independent variable x. We just project the graph onto the x axis and we clearly see the values of x.

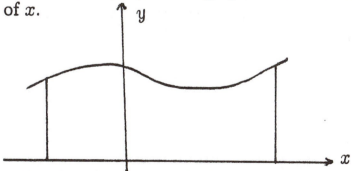

On the other hand, when we look at the graph of a polar equation $r = f(\theta)$ there is no easy method to determine the range of values of the independent variable θ. In fact there are usually several ranges of values for θ which graph as the same section of curve. In the cartesian coordinate system each point has only one set of coordinates, but in the polar coordinate system each point has many sets of coordinates. In the applications of graphing in polar coordinates it is essential that we know the exact values of θ which give the section of curve under consideration.

The graph of the cardioid $r = 1 + \cos\theta$, $0 \leq \theta \leq 2\pi$, is shown below.

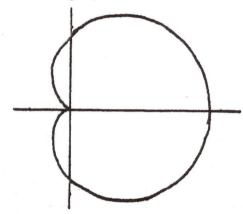

The graph of the polar equation $r = \cos\theta - 1$, $0 \leq \theta \leq 2\pi$, is the exact same cardioid. Suppose we are interested in the section of this cardioid which is in the second quadrant. The graph of $r = 1 + \cos\theta$ for $\pi/2 \leq \theta \leq \pi$ is the section in the second quadrant. On the other hand graph of $r = \cos\theta - 1$ for $3\pi/2 \leq \theta \leq 2\pi$ is the same section of the same cardioid in the second quadrant. It takes more than a casual look at the curve to determine which values of θ would graph as a given section of the curve.

Example 2. The graph of $r = \cos\theta - 1$ is a cardioid. What values of θ graph as the section of the cardioid in the first quadrant.

Solution. The most accurate method for determining the values of θ which give a certain section of a graph is to make a chart of a few values. A short chart of values for the equation $r = \cos\theta - 1$ follows.

θ	r
0	0
$\pi/2$	-1
π	-2
$3\pi/2$	-1
2π	0

Locate these points on the graph.

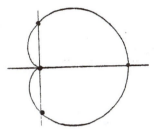

Checking values of θ we can now see that the section of graph in the first quadrant is given by $\pi \leq \theta \leq 3\pi/2$. The section of curve in the second quadrant is given by $3\pi/2 \leq \theta \leq 2\pi$.

Example 3. Sketch the graph of the polar equation $r = 1 + 2\sin\theta$.

Solution. Recall that the graph of $r = (c)(1 + \sin\theta)$ is a cardioid. The graph of $r = 1 + 2\sin\theta$ will look something like a cardioid, but will not be exactly a cardioid. We can be reasonably sure that we get the complete curve using $0 \leq \theta \leq 2\pi$. Before we construct the chart of values let us note that we can have $r = 0$ for some values of θ.

$$1 + 2\sin\theta = 0$$
$$\sin\theta = -1/2$$
$$\theta = 7\pi/6 \text{ and } \theta = 11\pi/6.$$

We want to be sure to include these values of θ on our chart.

θ	r
0	1
$\pi/4$	$1 + \sqrt{2} = 2.4$
$\pi/3$	$1 + \sqrt{3} = 2.7$
$\pi/2$	3
$3\pi/4$	$1 + \sqrt{2} = 2.4$
π	1
$7\pi/6$	0
$5\pi/4$	$1 - \sqrt{2} = -0.4$
$3\pi/2$	-1
$7\pi/4$	$1 - \sqrt{2} = -0.4$
$11\pi/6$	0
2π	1

Plotting these points and connecting them with a smooth curve we get the graph.

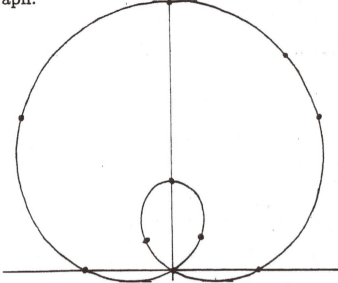

We should, of course, check that our graph is correct by graphing $r = 1 + 2\sin\theta$, $0 \leq \theta \leq 2\pi$, using the calculator. Note that the small inside loop is given by $r = 1 + 2\sin\theta$, $7\pi/6 \leq \theta \leq 11\pi/6$. Check this also with the calculator. We should learn to sketch curves of this type as they can be useful later.

Example 4. Sketch the graph of the polar equation $r = \cos 2\theta$.

Solution. At first we might think that the entire curve is given by $0 \leq \theta \leq \pi$ since we have 2θ instead of θ. But as we construct the graph we will see that this is not the case. It is often helpful to first determine the values of θ such that $r = 0$. For this graph we have

$$\cos 2\theta = 0$$
$$2\theta = \frac{\pi}{2}, \frac{3\pi}{2}, \frac{5\pi}{2}, \frac{7\pi}{2}$$
$$\theta = \frac{\pi}{4}, \frac{3\pi}{4}, \frac{5\pi}{4}, \frac{7\pi}{4}.$$

This indicates that there are several parts to this curve. Since there are several parts to this curve let us begin by constructing the section of curve given by $r = \cos 2\theta$, $0 \leq \theta \leq \pi/2$. We begin by making the following chart:

ϑ	r
0	1
$\pi/8$	0.7
$\pi/4$	0
$3\pi/8$	−0.7
$\pi/2$	−1

Plotting these points we get the following curve

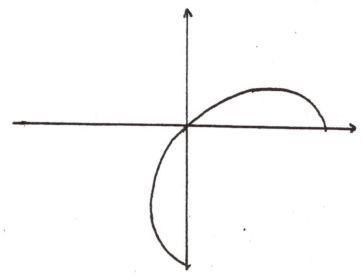

The next section of the curve is given by $r = \cos 2\theta$, $\pi/2 \leq \theta \leq \pi$. Make a chart using these value of θ.

θ	r
$\pi/2$	−1
$5\pi/8$	−0.7
$3\pi/4$	0
$7\pi/8$	0.7
π	1

Using these points we sketch the following section of the curve.

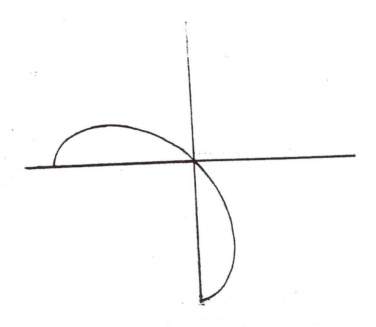

Continuing in this manner we obtain the four leaf rose.

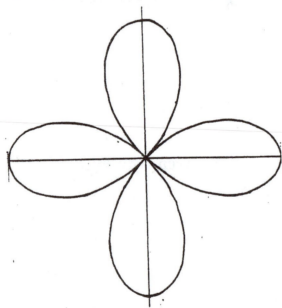

What values of θ give the top leaf of the rose? This is the leaf entirely above the x axis. In order to answer this question it is helpful to look at the values of θ in $r = \cos 2\theta$ for which $r = 0$. Recall that these value are

$$\theta = \frac{\pi}{4}, \frac{3\pi}{4}, \frac{5\pi}{4}, \frac{7\pi}{4}.$$

Each leaf starts and ends at one of these values of θ. Go back and look at the steps in the construction of this rose. Note that when $\theta = \frac{3\pi}{2}$, we have $r = -1$. The point $(-1, 3\pi/2)$ is at the top of the leaf in question. The top leaf is the graph of $r = \cos(2\theta)$ for $5\pi/4 \leq \theta \leq 7\pi/4$.

142

It is possible to draw very complicated graphs using polar coordinates. For example, the graph of $r = \sin(8\theta/5)$ for $0 \le \theta \le 10\pi$ is shown below. This is a rose with 16 loops.

We will not study such complicated graphs. Analyzing such graphs takes us too far from the ideas that we really want to study.

When studying curves that are the graphs of cartesian equations in x and y we often talked about the slope of the curve and the equation of the tangent line to the curve. These ideas are not so natural when studying the graphs of curves given by polar equations. We may discuss how to find the slope of a polar curve later. For now let us be clear, the slope of a polar curve is *not* given by $\frac{dr}{d\theta}$.

Exercises

1. The following is the value chart for a certain polar equation $r = f(\theta)$. Sketch the graph of this equation.

θ	r
0	2
$\pi/4$	3.4
$\pi/2$	4
$3\pi/4$	3.4
π	2
$5\pi/4$	0.6
$3\pi/2$	0
$7\pi/4$	0.6
2π	2

2. The following is the value chart for a certain polar equation $r = f(\theta)$. Sketch the graph of this equation.

θ	r
0	-3
$\pi/6$	-2.6
$\pi/4$	-2.1
$\pi/3$	-1.5
$\pi/2$	0
$2\pi/3$	1.5
$3\pi/4$	2.1
$5\pi/6$	2.6
π	3

3. Sketch the graph of the cardioid $r = \sin\theta - 1$ for $0 \le \theta \le 2\pi$. What values of θ give the points on this curve which are in the third quadrant?

4. Sketch the graph of the equation $r = \theta \sin\theta$, for $0 \le \theta \le 2\pi$.

5. Sketch the graph of the polar equation $r = 1 - 2\cos\theta$ for $0 \le \theta \le 2\pi$. What values of θ give the points on this curve that are in the inside part of the curve.

6. Sketch the graph of the polar equation $r = 3\sin 2\theta$ for $0 \le \theta \le 2\pi$. What values of θ give the points on the leaf of this rose which is in the second quadrant?

5583 Area of a Simple Polar Region

We are going to find the area of a region which has the graph of a polar equation as part of its boundary. We have already discussed how to find the area of a region where part or all of the boundary is the graph of a cartesian equation. In the cartesian case we begin with a region bounded by the x axis, the line $x = a$, the line $x = b$, and the graph of $y = f(x)$. We then moved on to more complicated regions which were constructed using these fundamental regions. We use the same plan when considering regions bounded by polar curves. We begin with a fundamental region and first learn to find the area of such regions. The following theorem describes the fundamental region and tells us how to find its area.

Theorem. Area Theorem for Polar Curves. Suppose R is the region bounded by the rays $\theta = \alpha$ and $\theta = \beta$ which start at the origin. Suppose the third side of the region is bounded by the graph of the polar equation $r = f(\theta)$ for $\alpha \le \theta \le \beta$. The area of the region is given by the definite integral

$$\text{Area of } R = \frac{1}{2} \int_\alpha^\beta [f(\theta)]^2 d\theta.$$

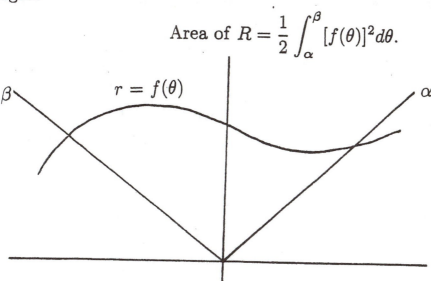

We will not prove this theorem. We must always select the angles α and β such that $\alpha < \beta$. The region described is basically a three sided region. The region is shaped like a piece of pie. The graph of $r = f(\theta)$ is called the outer boundary. The sharp point of the piece of pie must be the origin. The rays must start at the origin, but the rays may be the negative extension as well as the positive extension. For example, we may be using the negative of the ray $\theta = 5\pi/4$. The negative extension of $\theta = 5\pi/4$ is a half line in the first quadrant and not in the third quadrant. In view of all the different coordinates a single point may have we must be careful to find the exact values of α and β such that the curve in question is the graph of $r = f(\theta)$ for $\alpha \leq \theta \leq \beta$.

Example 1. Find the area of that part of the circle $r = 4\cos\theta$ that is in the first quadrant.

Solution. A sketch of the graph of $r = 4\cos\theta$.

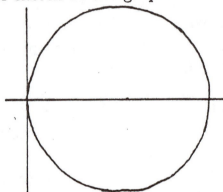

Let us check the coordinates of a few points. When $\theta = 0$, we get $r = 4$. When $\theta = \pi/4$, we get $r = 2\sqrt{2} = 2.8$. When $\theta = \pi/2$, we get $r = 0$. This should make it clear that the upper half of the circle is given by $r = 4\cos\theta$ for $0 \leq \theta \leq \pi/2$. The upper half of the circle is between the rays $\theta = 0$ and $\theta = \pi/2$. We do not want to make the mistake of thinking it is between $\theta = 0$ and $\theta = \pi$. Once we are sure that the curve which is the outer boundary is given by $r = 4\cos\theta$ for $0 \leq \theta \leq \pi/2$, the area is an easy application of the Area Theorem for Polar Curves.

$$\text{Area} = \frac{1}{2}\int_0^{\pi/2}(4\cos\theta)^2\,d\theta$$

$$= 4\int_0^{\pi/2}(1 + \cos 2\theta)\,d\theta$$

$$= 4\left[\theta + \frac{1}{2}\sin 2\theta\,\Big|_0^{\pi/2}\right] = 2\pi.$$

We, of course, already knew the answer since the area of the whole circle equals $\pi r^2 = \pi(2)^2 = 4\pi$.

Example 2. Find the area enclosed by the inside loop of the curve $r = 1 + 2\sin\theta$.

146

Solution. We begin with a sketch of the curve. Note that in order to get the entire curve we must use $0 \leq \theta \leq 2\pi$.

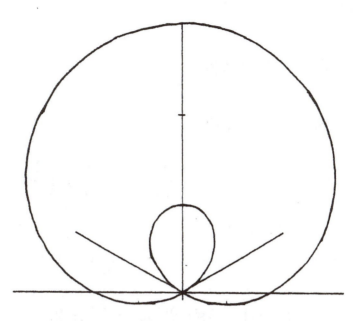

We want the area of the region enclosed by the small inner loop. We must find two rays such that the region in question is between these rays. We must find α and β such that the inner loop is the graph of $r = 1 + 2\sin\theta$ for $\alpha \leq \theta \leq \beta$. A quick guess might lead us to think that the rays are $\theta = 0$ and $\theta = \pi$, but this is not the case. We must study the graph of $r = 1 + 2\sin\theta$ in some detail. First, the loop starts at the origin, that is, $r = 0$. We are looking for values of θ such that $r = 0$.

$$1 + 2\sin\theta = 0$$
$$\sin\theta = -1/2$$
$$\theta = \frac{7\pi}{6} \text{ and } \theta = \frac{11\pi}{6}.$$

Half way between $7\pi/6$ and $11\pi/6$ is $3\pi/2$. When $\theta = 3\pi/2$, we have $r = 1 + 2\sin(3\pi/2) = -1$. The point $(-1, 3\pi/2)$ is the point on the curve at the top of the small loop. This makes it clear that the small loop is the graph of $r = 1 + 2\sin\theta$ for $\frac{7\pi}{6} \leq \theta \leq \frac{11\pi}{6}$. If in doubt we can check more values of θ. The small loop is bounded by the rays $\theta = 7\pi/6$ and $\theta = 11\pi/6$. The outer boundary of the region is the small loop. The Area Theorem for Polar Curves says that the area is given by

$$\frac{1}{2}\int_{7\pi/6}^{11\pi/6}[1+2\sin\theta]^2 d\theta$$

$$=\frac{1}{2}\int_{7\pi/6}^{11\pi/6}[1+4\sin\theta+4\sin^2\theta]d\theta$$

$$=\int_{7\pi/6}^{11\pi/6}\left[\frac{3}{2}+2\sin\theta-\cos 2\theta\right]d\theta$$

$$=\frac{3}{2}\left[\frac{4\pi}{6}\right]-2\left[\cos\left(\frac{11\pi}{6}\right)-\cos\left(\frac{7\pi}{6}\right)\right]-\frac{1}{2}\left[\sin\left(\frac{11\pi}{3}\right)-\sin\left(\frac{7\pi}{3}\right)\right]$$

$$=\pi-(3/2)\sqrt{3}=0.5435.$$

The area of the region inside the outer loop and also in the first quadrant is given by

$$\frac{1}{2}\int_0^{\pi/2}[1+2\sin\theta]^2 d\theta=\frac{3\pi}{4}+2.$$

This would also include the region which is also inside the inner loop.

Example 3. Find the area of the region which is inside the graph of the polar curve $r=2(\cos\theta-1)$ and also in the first quadrant.

Solution. A sketch of the curve $r=2(\cos\theta-1)$ for $0\le\theta\le 2\pi$ is a cardioid.

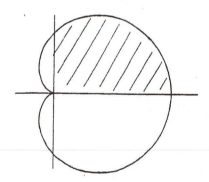

Clearly the region in the first quadrant is between the rays $\theta=0$ and $\theta=\pi/2$, but are these the correct values of θ that give that section of graph. Is the section of cardioid in the first quadrant the graph of $r=2(\cos\theta-1)$ for $0\le\theta\le\pi/2$. In order to determine if this really is the case let us check a few points. When $\theta=0$, we get $r=0$. When $\theta=\pi/4$, we get $r=\sqrt{2}-2=-0.6$. When $\theta=\pi/2$, we get $r=-1$. Plot these points

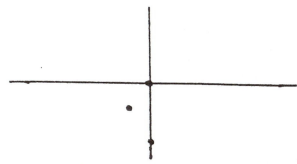

These are not points on the section of curve in the first quadrant. These points are actually points on the section of curve in the third quadrant. We need to consider some different values of θ. When $\theta = \pi$, we get $r = 2(-1-1) = -4$. When $\theta = 5\pi/4$, we get $r = -\sqrt{2} - 2 = -3.14$. When $\theta = 3\pi/2$, we get $r = -2$. Let us plot the points with these coordinates.

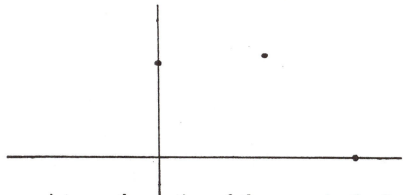

These are points on the section of the curve in the first quadrant. The section of curve in the first quadrant is the graph of $r = 2(\cos \theta - 1)$ for $\pi \leq \theta \leq 3\pi/2$. The region in the first quadrant is between the negative extension of the ray $\theta = \pi$ and the negative extension of the ray $\theta = 3\pi/2$. The Area Theorem for Polar Curves tells us that the area in the first quadrant is given by

$$\text{Area} = \frac{1}{2} \int_{\pi}^{3\pi/2} [2(\cos \theta - 1)]^2 d\theta$$

$$= \int_{\pi}^{3\pi/2} [3 + \cos 2\theta - 4 \cos \theta] d\theta$$

$$= 3\theta + \frac{1}{2} \sin 2\theta - 4 \sin \theta \Big|_{\pi}^{3\pi/2}$$

$$= \frac{3\pi}{2} + 4.$$

Example 4. Find the area of the region bounded by the graph of the polar equation $r = \cos \theta - 2$ and the rays $\theta = \pi/4$ and $\theta = 3\pi/4$.

Solution. First, we make a sketch of the region.

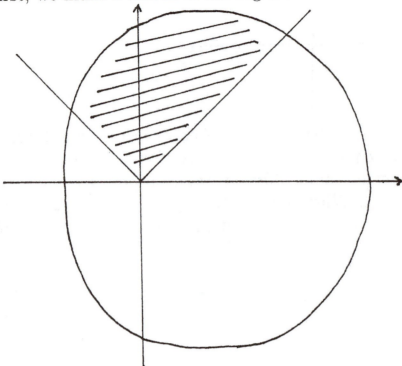

The region in question is between the rays $\theta = \pi/4$ and $\theta = 3\pi/4$, but the important question is: is the section of the curve which is the outer boundary of this region the graph of $r = \cos\theta - 2$ for $\pi/4 \leq \theta \leq 3\pi/4$? Looking back at Example 3 we are pretty sure that the answer to this question is "no". Do we need to use the negative extension of some rays? Maybe we need to use the negative extension of the ray $\theta = 5\pi/4$ instead of the ray $\theta = \pi/4$. We answer these questions by a careful look at the graph of $r = \cos\theta - 2$. We need to find angles α and β such that the outer boundary of the region is the graph of $r = \cos\theta - 2$ for $\alpha \leq \theta \leq \beta$. We keep checking values until we find that when $\theta = 5\pi/4$, we get $r = (-\sqrt{2}/2) - 2 = -2.7$. When $\theta = 3\pi/2$, we get $r = -2$. When $\theta = 7\pi/4$, we get $r = (\sqrt{2}/2) - 2 = -1.3$. Let us plot these points.

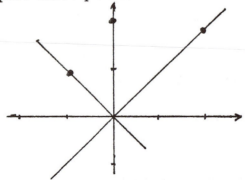

It is clear that the outer boundary of the region is the graph of $r = \cos\theta - 2$ for $5\pi/4 \le \theta \le 7\pi/4$. The region is between the negative extension of $\theta = 5\pi/4$ and the negative extension of $\theta = 7\pi/4$. We can now apply the Area Theorem for Polar Curves to get the area of the region.

$$\text{Area} = \frac{1}{2}\int_{5\pi/4}^{7\pi/4} [\cos\theta - 2]^2 d\theta$$

$$= \int_{5\pi/4}^{7\pi/4} [\frac{9}{4} + \frac{1}{4}\cos 2\theta - 2\cos\theta] d\theta$$

$$= \frac{9}{4}\theta + \frac{1}{8}\sin 2\theta - 2\sin\theta \Big|_{5\pi/4}^{7\pi/4}$$

$$\frac{9}{4}\left[\frac{\pi}{2}\right] + \frac{1}{8}\left[\sin(7\pi/2) - \sin(5\pi/2)\right] - 2\left[\sin(7\pi/4) - \sin(5\pi/4)\right]$$

$$= \frac{9\pi}{8} + \frac{1}{8}[-1 - 1] - 2\left[-\frac{\sqrt{2}}{2} + \frac{\sqrt{2}}{2}\right] = \frac{9\pi}{8} - \frac{1}{4}.$$

We would use integration by parts to find the following useful antiderivative formulas.

$$\int x\sin x\, dx = -x\cos x + \sin x + C$$

$$\int x\cos x\, dx = x\sin x + \cos x + C$$

$$\int x^2\sin^2 x\, dx = \frac{x^3}{6} - \frac{x^2}{4}\sin 2x - \frac{x}{4}\cos 2x + \frac{1}{8}\sin 2x + C$$

$$\int x^2\cos^2 x\, dx = \frac{x^3}{6} + \frac{x^2}{4}\sin 2x + \frac{x}{4}\cos 2x - \frac{1}{8}\sin 2x + C$$

Example 5. The polar curve $r = \theta\sin\theta - \pi$, $0 \le \theta \le 2\pi$, encloses a region in the plane. Find the area of that part of this region which is in the second quadrant.

151

Solution. Although it is not necessary in this case we begin by sketching a graph of this curve.

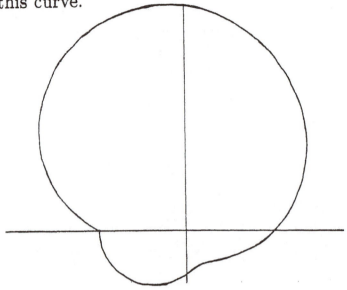

We need to find the section of this curve which is the second quadrant. Our first guess is that the section of curve is given by $\pi/2 \leq \theta \leq \pi$. Let us make a short chart

θ	r
$\pi/2$	$-\pi/2$
$3\pi/4$	$(3\sqrt{2}\pi/8) - \pi = -1.47$
π	$-\pi$

When we plot these points we see that they are all points in the fourth quadrant. These points are not part of the boundary of the region in the second quadrant. The points in the second quadrant must be on the negative extension of the terminal sides of angles. Look at some values of θ such that $3\pi/2 \leq \theta \leq 2\pi$.

θ	r
$3\pi/2$	$(3\pi/2)(-1) - \pi = -5\pi/2 = -7.85$
$7\pi/4$	$(7\pi/4)(-1/\sqrt{2}) - \pi = -7.03$
2π	$2\pi(0) - \pi = -\pi.$

The points with coordinates $(-7.85, 3\pi/2)$, $(-7.03, 7\pi/4)$ and $(-\pi, 2\pi)$ are in the second quadrant. The area of the region enclosed by the curve and also in the second quadrant is given by

$$\frac{1}{2} \int_{\frac{3\pi}{2}}^{2\pi} [\theta \sin\theta - \pi]^2 \, d\theta.$$

152

We hope that problems of this kind just say "set up the integral" as this is a long integral to evaluate. But just for fun let us evaluate this integral in this case. Squaring we have

$$\frac{1}{2} \int_{\frac{3\pi}{2}}^{2\pi} \left[\theta^2 \sin^2 \theta - 2\pi\theta \sin\theta + \pi^2 \right] d\theta.$$

Using the formulas given above this definite integral is equal to

$$\frac{\theta^3}{12} - \frac{\theta^2}{8} \sin 2\theta - \frac{\theta}{8} \cos 2\theta + \frac{1}{16} \sin 2\theta + \pi\theta \cos\theta - \pi \sin\theta + \frac{\pi^2}{2} \theta \Big|_{3\pi/2}^{2\pi}.$$

$$= \frac{1}{12} \left[(2\pi)^3 - \left(\frac{3\pi}{2} \right)^3 \right] - \frac{1}{8} \left[4\pi^2 \sin 4\pi - \frac{9\pi^2}{4} \sin 3\pi \right]$$

$$- \frac{1}{8} \left[2\pi \cos 4\pi - \frac{3\pi}{2} \cos 3\pi \right] + \frac{1}{16} \left[\sin 4\pi - \sin 3\pi \right]$$

$$+ \pi \left[2\pi \cos 2\pi - \frac{3\pi}{2} \cos \frac{3\pi}{2} \right] - \pi \left[\sin 2\pi - \sin \frac{3\pi}{2} \right] + \frac{\pi^2}{2} \left(\frac{\pi}{2} \right)$$

$$= \frac{\pi^3}{12} \left[\frac{64 - 27}{8} \right] - \frac{1}{8} \left[2\pi + \frac{3\pi}{2} \right] + \pi \left[2\pi \right] - \pi + \frac{\pi^3}{4}$$

$$= \frac{61\pi^3}{96} + 2\pi^2 - \frac{23\pi}{16} = 34.9251.$$

There are several applications of finding the area of a region when working with cartesian xy coordinates. For example, if the function is velocity, then the area is position. If the function is force, then the area is work. If the function is linear density, then the area is mass. However, there are not really as many applications to finding area in polar coordinates. Most of these applications have to do with rotational motion which is not usually discussed in detail in a course of calculus of one independent variable.

Exercises

1. Find the area of the region bounded by the graph of the polar equation $r = 2(1 - \cos\theta)$ which is in the second quadrant.

2. Find area of the region bounded by the graph of the polar equation $r = 3 + 2\sin\theta$ which is also in the third quadrant.

3. Set up the integral which when evaluated will give the area of the region enclosed by the outer curve which is the graph of the polar equation $r = 2\sin\theta - 1$.

4. The polar curve which is the graph of $r = (3\pi/2) + \theta\sin\theta$, $0 \le \theta \le 2\pi$, encloses a region in the plane. Set up the integral which when evaluated will equal to the area of this region which is also in the upper half plane.

5. Find the area of the region which is inside the curve which is the graph of the polar equation $r = 2(\sin\theta - 1)$ and also in the first quadrant.

6. Find the area of the region bounded by the graph of the polar equation $r = 4(\sin\theta - 1)$, the ray $\theta = \pi/4$ and the ray $\theta = \pi/2$ which is also in the first quadrant.

7. Find the area of the region bounded by the graph of the polar equation $r = -2\cos\theta$ which is below the ray $\theta = 7\pi/6$ and also in the third quadrant.

8. The polar curve $r = \theta(\cos\theta - 1)$, $0 \le \theta \le 2\pi$, encloses a region in the plane. Set up the integral which when evaluated will equal to the area of the part of this region which is in the first quadrant.

9. Consider the graph of $r = 4\theta$, $\theta > 0$. This graph is a spiral. Find the area of the region between the first and second turn of the spiral and also in the second quadrant.

5585 Area Between Polar Curves

We are now going to find the area of regions in the polar coordinate plane where part or all of the boundary of the region is the graph of two separate polar equations.

Example 1. Find the area of the region which is both inside the graph of $r = 2 + 2\sin\theta$, and inside the graph of $r = 6\sin\theta$, and also in the first quadrant.

Solution. The graph of the polar equation $r = 2 + 2\sin\theta$ is a cardioid and the graph of the polar equation $r = 6\sin\theta$ is a circle. What are the coordinates of the points where these two curves intersect?

$$6\sin\theta = 2 + 2\sin\theta$$
$$4\sin\theta = 2$$
$$\sin\theta = 1/2$$
$$\theta = \pi/6 \text{ and } \theta = 5\pi/6.$$

All corresponding r values are $r = 3$. A sketch of the graph of these two curves

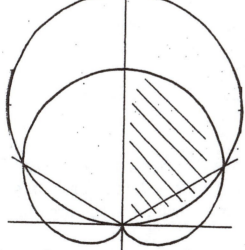

The region whose area we must find is the sum of two parts. One part is a rather long thin region between the circle $r = 6\sin\theta$ and the ray $\theta = \pi/6$. The second part is bounded by the ray $\theta = \pi/6$, the ray $\theta = \pi/2$, and the cardioid $2 + 2\sin\theta$. We find the total area by finding each area separately and adding the results together.

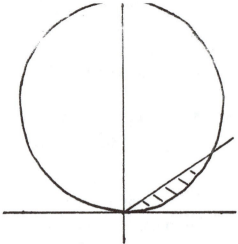

The long thin part between the circle $r = 6 \sin \theta$ and the line $\theta = \pi/6$ is between the rays $\theta = 0$ and $\theta = \pi/6$. The outer boundary of this part is given by $r = 6 \sin \theta$ for $0 \le \theta \le \pi/6$. Applying the Area Theorem for Polar Coordinates, the area is

$$\frac{1}{2} \int_0^{\pi/6} [6 \sin \theta]^2 d\theta.$$

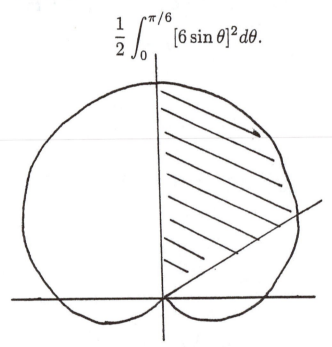

The larger section is between the rays $\theta = \pi/6$ and $\theta = \pi/2$. The outer boundary of this region is given by $r = 2 + 2 \sin \theta$ for $\pi/6 \le \theta \le \pi/2$. Applying the Area Theorem for Polar Coordinates the area of this region is

$$\frac{1}{2} \int_{\pi/6}^{\pi/2} [2 + 2 \sin \theta]^2 d\theta$$

The total area is the sum of these two areas

$$\frac{1}{2} \int_0^{\pi/6} [6 \sin \theta]^2 d\theta + \frac{1}{2} \int_{\pi/6}^{\pi/2} [2 + 2 \sin \theta]^2 d\theta$$

156

$$= 18 \int_0^{\pi/6} \sin^2 \theta d\theta + 2 \int_{\pi/6}^{\pi/2} (1 + 2 \sin \theta + \sin^2 \theta) d\theta$$

$$= 9 \int_0^{\pi/6} (1 - \cos 2\theta) d\theta + \int_{\pi/6}^{\pi/2} (3 + 4 \sin \theta - \cos 2\theta) d\theta$$

$$= 9 \left[\frac{\pi}{6} - \frac{1}{2} \frac{\sqrt{3}}{2} \right] + \left[3 \left(\frac{\pi}{2} - \frac{\pi}{6} \right) \right] - 4 \left[0 - \frac{\sqrt{3}}{2} \right] - \frac{1}{2} \left[0 - \frac{\sqrt{3}}{2} \right]$$

$$= \frac{5\pi}{2}.$$

Example 2. Find the area of the region which is inside the graph of the circle $r = 5 \cos \theta$ and outside the graph of $r = 3 + \cos \theta$.

Solution. First, what are the coordinates of the points where these two curves intersect?

$$3 + \cos \theta = 5 \cos \theta$$
$$\cos \theta = 3/4$$
$$\theta = 0.7227 \text{ and } \theta = -0.7227.$$

We make a sketch of the graph of these two curves.

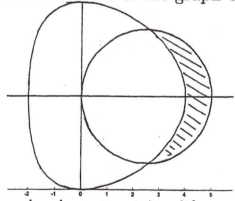

The region is symmetric with respect to the x axis. The area of that part of the region in the fourth quadrant is the same as the area of that part of the region in the first quadrant. We will find the area of the part of the region in the first quadrant and multiply by 2. The region in the first quadrant is the difference between two regions. Let R_1 denote the region between the rays $\theta = 0$ and $\theta = 0.7227$ with outer boundary given by $r = 5 \cos \theta$ for $0 \leq \theta \leq 0.7227$.

157

The area of this region is

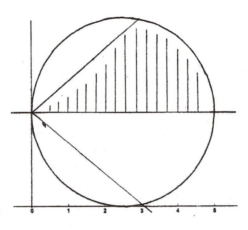

$$\frac{1}{2}\int_0^{0.7227}(5\cos\theta)^2\,d\theta.$$

Let R_2 denote the region between the rays $\theta = 0$ and $\theta = 0.7227$ with outer boundary given by $r = 3 + \cos\theta$ for $0 \le \theta \le 0.7227$.

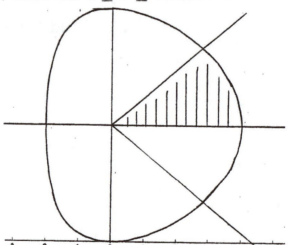

The area of this region is

$$\frac{1}{2}\int_0^{0.7227}(3+\cos\theta)^2\,d\theta.$$

The area of the region we want is two times the area of region R_1 minus

158

two times the area of region R_2. This area is equal to

$$\int_0^{0.7227} (25\cos^2\theta)d\theta - \int_0^{0.7227} (3+\cos\theta)^2 d\theta$$

$$= \frac{25}{2}\int_0^{0.7227}(1+\cos 2\theta)d\theta - \int_0^{0.7227}(\frac{19}{2}+6\cos\theta+\frac{1}{2}\cos 2\theta)d\theta$$

$$= \frac{25}{2}\left[\theta + \sin\theta\cos\theta\Big|_0^{0.7227}\right] - \left[\frac{19}{2}\theta + 6\sin\theta + \frac{1}{2}\sin\theta\cos\theta\Big|_0^{0.7227}\right]$$

$$= \frac{25}{2}\left[0.7227 + \frac{\sqrt{7}}{4}\frac{3}{4}\right] - \left[\frac{19}{2}(0.7227)\right] - 6\left[\frac{\sqrt{7}}{4}\right] - \frac{1}{2}\frac{\sqrt{7}}{4}\frac{3}{4}$$

$$= 3(0.7227) + \frac{75\sqrt{7}}{32} - \frac{6\sqrt{7}}{4} - \frac{3\sqrt{7}}{32}$$

$$= 4.1525.$$

Example 3. Find the area of the region which is inside the circle $r = 6\sin\theta$, but is outside the cardioid $r = 2\sin\theta - 2$.

Solution. Let us try to find some coordinates of points where these two curves intersect.

$$6\sin\theta = 2\sin\theta - 2$$

$$\sin\theta = -1/2$$

$$\theta = 7\pi/6 \text{ and } \theta = 11\pi/6.$$

When $\theta = 7\pi/6$, we get $6\sin(7\pi/6) = -3$ and $2\sin(7\pi/6) - 2 = -3$. When $\theta = 11\pi/6$, we get $6\sin(11\pi/6) = -3$ and $2\sin(11\pi/6) - 2 = -3$. The points $(-3, 7\pi/6)$ and $(-3, 11\pi/6)$ are on both curves. Let us draw a sketch of the graphs of both curves.

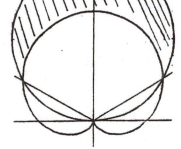

Our first idea is that the ray which passes through the intersection of the two curves is the ray $\theta = \pi/6$. But since $(-3, 11\pi/6)$ satisfies both equations it is more likely that this ray is the negative extension of the ray $11\pi/6$. Whatever the case we very much need to express the region in question in terms of simple regions. Let R_1 denote the region bounded by the two rays and the circle $r = 6\sin\theta$.

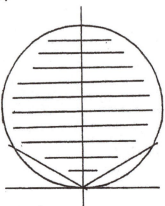

Let R_2 denote the region bounded by the two rays and the cardioid $r = 2\sin\theta - 2$.

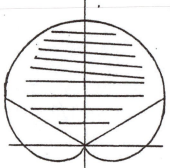

Clearly the region whose area we wish to find is the region R_1 take away the region R_2. We find the area of the region R_1. We find the area of the region R_2. We subtract one result from the other. The outer boundary of the region R_1 is the graph of $r = 6\sin\theta$ for $\pi/6 \leq \theta \leq 5\pi/6$. The region R_1 is bounded by the ray $\theta = \pi/6$ and the ray $\theta = 5\pi/6$. Applying the Area Theorem for Polar Coordinates the area of R_1 is

$$\text{Area of } R_1 = \frac{1}{2}\int_{\pi/6}^{5\pi/6} [6\sin\theta]^2\,d\theta = \frac{1}{2}\int_{7\pi/6}^{11\pi/6} [6\sin\theta]^2\,d\theta$$

$$= \int_{\pi/6}^{\pi/2} [6\sin\theta]^2\,d\theta.$$

160

We must be careful with the cardioid. What are the angles α and β such that the graph of the section of cardioid in question is given by $r = 2\sin\theta - 2$ for $\alpha \le \theta \le \beta$? We need to check the coordinates of some points. When $\theta = 7\pi/6$, we get $r = -3$. When $\theta = 3\pi/2$, we get $r = -4$. When $\theta = 11\pi/6$, we get $r = -3$. Plot these points.

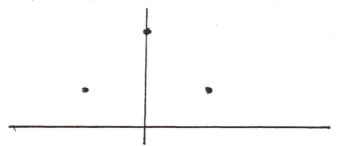

The outer boundary of R_2 is the section of cardioid given by $r = 2\sin\theta - 2$ for $7\pi/6 \le \theta \le 11\pi/6$. This is the important fact. We can also say that the region R_2 is between the negative extension of the ray $r = 7\pi/6$ and the negative extension of the ray $\theta = 11\pi/6$. Applying the Area Theorem for Polar Coordinates the area of R_2 is

$$\text{Area of } R_2 = \frac{1}{2}\int_{7\pi/6}^{11\pi/6} (2\sin\theta - 2)^2 d\theta.$$

The area of the region in question is the area of R_1 subtract the area of R_2. The area of the region which is inside the circle $r = 6\sin\theta$ and outside the cardioid $r = 2\sin\theta - 2$ is

$$\frac{1}{2}\int_{\pi/6}^{5\pi/6} [6\sin\theta]^2 d\theta - \frac{1}{2}\int_{7\pi/6}^{11\pi/6}(2\sin\theta - 2)^2 d\theta.$$

Example 4. Find the area of the region which is inside the cardioid $r = 2 - 2\sin\theta$ and outside the circle $r = -6\sin\theta$ and also inside the first or fourth quadrant $(x \ge 0)$.

Solution. First, let us try to find some values of θ for which these curves intersect.

$$-6\sin\theta = 2 - 2\sin\theta$$

$$\sin\theta = -1/2$$

$$\theta = 7\pi/6 \text{ and } \theta = 11\pi/6.$$

Next, let us sketch the graph of the polar equation $r = -6\sin\theta$ and the polar equation $r = 2 - 2\sin\theta$ on the same graph.

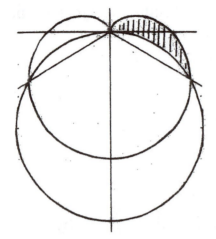

Let R_1 denote the region bounded below by the ray $\theta = 11\pi/6$ and above by the cardioid $r = 2 - 2\sin\theta$. A sketch of R_1 is

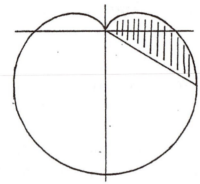

Let R_2 denote the region bounded below by the ray $r = 11\pi/6$ and above by the circle $r = -6\sin\theta$. A sketch of R_2 is

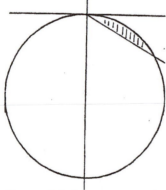

The region for which we want to find the area is the region R_1 take away the region R_2. The area of the region for which we must find the area is the area of R_1 subtract the area of R_2. We consider finding the area of R_1 and finding the area of R_2 as two separate problems. We have a slight problem with the outer boundary of the region R_1. As we normally draw the graph the part of the cardioid $r = 2 - 2\sin\theta$ in the first quadrant using values the part of the graph in the first quadrant we use values of θ given by $0 \le \theta \le \pi/2$. We draw the part of the graph in the fourth quadrant using values of θ given by $11\pi/6 \le \theta \le 2\pi$. We could just use two separate intervals of values of θ. But also notice that the entire section of the cardioid is the graph of $r = 2 - 2\sin\theta$ for $-\pi/6 \le \theta \le \pi/2$. The region R_1 is between the ray $\theta = -\pi/6$ and the ray $\theta = \pi/2$. Applying the Area Theorem for Polar Coordinates the area of R_1 is

$$\text{Area of } R_1 = \frac{1}{2}\int_{-\pi/6}^{\pi/2}[2 - 2\sin\theta]^2 d\theta$$

The section of the circle which is the outer boundary of the region R_2 is the graph of $r = -6\sin\theta$ for $5\pi/6 \le \theta \le \pi$. The region R_2 is between the negative extension of the ray $r = 5\pi/6$ and the negative extension of the ray $\theta = \pi$. Applying the Area Theorem for Polar Coordinates the area of R_2 is

$$\text{Area of } R_2 = \frac{1}{2}\int_{5\pi/6}^{\pi}[-6\sin\theta]^2 d\theta.$$

The area of the region in question is the area of R_1 subtract the area of R_2. The area of the region which is both inside the cardioid $r = 2 - 2\sin\theta$ and outside the circle $r = -6\sin\theta$ is

$$\frac{1}{2}\int_{-\pi/6}^{\pi/2}[2 - 2\sin\theta]^2 d\theta - \frac{1}{2}\int_{5\pi/6}^{\pi}[-6\sin\theta]^2 d\theta$$

$$= 2\int_{-\pi/6}^{\pi/2}(1 - 2\sin\theta + \sin^2\theta)d\theta - 18\int_{5\pi/6}^{\pi}\sin^2\theta d\theta$$

$$= \int_{-\pi/6}^{\pi/2}(3 - 4\sin\theta - \cos 2\theta)d\theta - 9\int_{5\pi/6}^{\pi}(1 - \cos 2\theta)d\theta$$

$$= (3\theta + 4\cos\theta - (1/2)\sin 2\theta)\big|_{-\pi/6}^{\pi/2} + [-9\theta + (9/2)\sin 2\theta]\big|_{5\pi/6}^{\pi}$$

163

$$= 3\left[\frac{\pi}{2} + \frac{\pi}{6}\right] + 4\left[\cos\frac{\pi}{2} - \cos\frac{\pi}{6}\right] - \frac{1}{2}\left[\sin\pi + \sin\frac{\pi}{3}\right]$$

$$- 9\left[\pi - \frac{5\pi}{6}\right] + \frac{9}{2}\left[\sin 2\pi - \sin\frac{5\pi}{3}\right]$$

$$= 3\left[\frac{2\pi}{3}\right] + 4\left[0 - \frac{\sqrt{3}}{2}\right] - \frac{1}{2}\left[0 + \frac{\sqrt{3}}{2}\right] - 9\left(\frac{\pi}{6}\right) + \frac{9}{2}\left[0 + \frac{\sqrt{3}}{2}\right]$$

$$= \frac{\pi}{2}.$$

Exercises

1. Find the definite integrals which when evaluated will equal to the area of the region which is both inside the curve which is the graph of the polar equation $r = 2 + 2\cos\theta$ and inside the curve which is the graph of the polar equation $r = 2(1 + \sqrt{2})\cos\theta$.

2. Find the area of the region which is both inside the circle which is the graph of the polar equation $r = 3\sin\theta$ and also inside the cardioid which is the graph of the polar equation $r = 2 - 2\sin\theta$.

3. Find the area of the region which is both inside the curve $r = 2 + \sin\theta$ and inside the curve $r = 2 - \cos\theta$.

4. Find the area of the region which is both inside the curve which is the graph of the polar equation $r = 2\cos\theta - 2$ and outside the curve which is the graph of the polar equation $r = 6\cos\theta$.

5. Find the definite integrals which when evaluated will equal to the area of the region which is both inside the cardioid $r = 3 - 3\cos\theta$ and outside the circle $r = -8\cos\theta$ and also in either the first or second quadrant.

6. Find the area of the region which is both inside the curve $r = \sin\theta - \sqrt{2}$ and the circle $r = -\sin\theta$, and also in the third quadrant.

6751 Introduction to Vectors

Quantities such as force, displacement, and velocity can not be represented by a single number. In order to completely describe these quantities we must give not only their magnitude but also their direction. This causes us to introduce the idea of vectors.

The geometric way to define vectors is to define a vector as a directed line segment. Directed means that the line segment has an initial point at one end and a terminal point at the other end. The direction is the direction from the initial point toward the terminal point. The magnitude of the vector is the length of the line segment. For example, we can represent the velocity of an object as a vector. The length of the vector is the same as the speed of the object and the direction of the vector indicates the direction of the movement.

It is very difficult to discuss vectors using just old fashion geometry. It is even difficult to determine the length of a given vector in such a discussion. We need a method for locating points. We are going to discuss vectors in a plane. We will always assume that there is a given coordinate system for this plane. This enables us to specify any point in the plane by giving its coordinates. This enables us to discuss vectors algebraically.

Definition. The vector $\vec{v} = a\vec{i} + b\vec{j}$ in standard position is the directed line segment in the plane with initial point $(0,0)$ and terminal point (a,b).

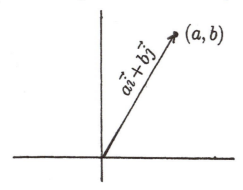

Other notations sometimes used for the vector $\vec{v}$ are $[a,b]$, $< a,b >$, and (a,b). The vector $\vec{v}$ need not always be in standard position. If two line segments have the same direction and the same length, then they represent the same vector. The vector $\vec{i}$ is of length one from $(0,0)$ to $(1,0)$. The vector $\vec{j}$ is of length one from $(0,0)$ to $(0,1)$.

165

Note that we have two ways of expressing a vector. There is the algebraic way which is $a\vec{i} + b\vec{j}$. There is the geometric way which is to draw a line segment in the required direction and of the required length.

Example 1. The vector $-3\vec{i} + 5\vec{j}$ is shown below in standard position.

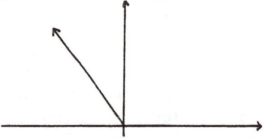

Example 2. Suppose the vector $4\vec{i} - 3\vec{j}$ is placed with its initial point at $(5,2)$. What is the terminal point?

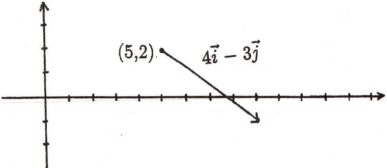

Solution. $5 + 4 = 9$ and $-3 + 2 = -1$. The terminal point is $(9, -1)$.

Example 3. Find the vector $\vec{v}$ if the initial point is $(-4, 3)$ and the terminal point is $(5, -1)$.

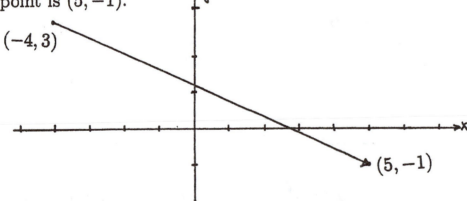

Solution. $[5 - (-4)]\vec{i} + [-1 - 3]\vec{j} = 9\vec{i} - 4\vec{j}$.

Theorem. We denote the length of the vector $\vec{v} = a\vec{i} + b\vec{j}$ by $|\vec{v}|$ or $||\vec{v}||$. When $\vec{v}$ is placed in standard position it is an easy application of the

Pythagorean Theorem to see that

$$\|\vec{v}\| = \sqrt{a^2 + b^2}$$

Example 4. Find the length of the vector $12\vec{i} - 5\vec{j}$.

Solution. $\sqrt{12^2 + (-5)^2} = 13$. The length is 13.

Given the vector $\vec{v} = a\vec{i} + b\vec{j}$ the number a is called the component of $\vec{v}$ in the direction of $\vec{i}$ or in the x direction and b is called the component of $\vec{v}$ in the $\vec{j}$ direction or in the y direction. Sometimes we refer to a as the horizontal component and b as the vertical component. We also refer to the vector $a\vec{i}$ as the vector component of $\vec{v}$ in the direction of $\vec{i}$ and the vector $b\vec{j}$ as the vector component of $\vec{v}$ in the direction of $\vec{j}$.

Definition. A vector $\vec{u}$ is called a unit vector if it is of length 1, that is, $\|\vec{u}\| = 1$.

Example 5. The vector $\dfrac{12}{13}\vec{i} - \dfrac{5}{13}\vec{j}$ is a unit vector because

$$\left(\frac{12}{13}\right)^2 + \left(-\frac{5}{13}\right)^2 = 1.$$

Exercises

1. Find the vector $\vec{v}$ if the initial point is $(-5, -3)$ and the terminal point is $(2, 7)$.

2. Suppose the vector $-5\vec{i} + 3\vec{j}$ is placed with its initial point $(2,4)$ what is its terminal point? Sketch a graph of the vector $-5\vec{i} + 3\vec{j}$ with its initial point at $(2, 4)$.

3. Find the length of the vector $-16\vec{i} + 12\vec{j}$. Find a vector the same length as $-16\vec{i} + 12\vec{j}$ but in the opposite direction. Find a unit vector in the same direction as $-16\vec{i} + 12\vec{j}$. What is the component of $-16\vec{i} + 12\vec{j}$ in the x direction? What is the vector component of $-16\vec{i} + 12\vec{j}$ in the direction of $\vec{i}$?

Definition of Addition of Vectors. If $\vec{v} = a\vec{i} + b\vec{j}$ and $\vec{w} = c\vec{i} + d\vec{j}$ are vectors positioned so that the initial point of $\vec{w}$ is at the terminal point of $\vec{v}$, then the sum $\vec{v} + \vec{w}$ is the vector from the initial point of $\vec{v}$ to the terminal point of $\vec{w}$.

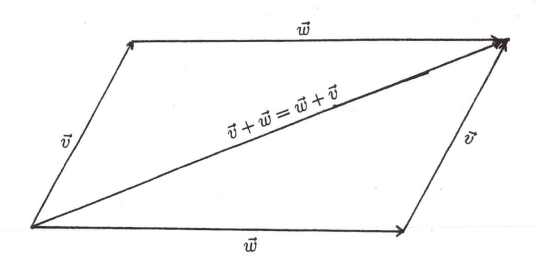

Parallelogram Law. Suppose the initial point of $\vec{w}$ is at the terminal point of $\vec{v}$. Let $\vec{p}$ denote the vector from the initial point of $\vec{v}$ to the terminal point of $\vec{w}$. Second, suppose the initial point of $\vec{v}$ is at the terminal point of $\vec{w}$. Let $\vec{q}$ denote the vector from the initial point of $\vec{w}$ to the terminal point of $\vec{v}$, then $\vec{p} = \vec{q}$. We might also express this as $\vec{v} + \vec{w} = \vec{w} + \vec{v}$.

Theorem 1. If $\vec{v} = a\vec{i} + b\vec{j}$ and $\vec{w} = c\vec{i} + d\vec{j}$ then $\vec{v} + \vec{w} = (a+c)\vec{i} + (b+d)\vec{j}$.

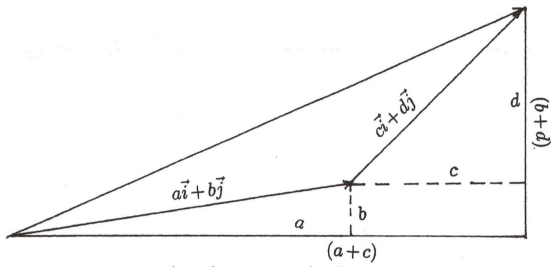

Example 1. If $\vec{v} = -3\vec{i} + 4\vec{j}$ and $\vec{w} = 5\vec{i} + \vec{j}$, then $\vec{v} + \vec{w} = 2\vec{i} + 5\vec{j}$. Below we see the vectors added graphically.

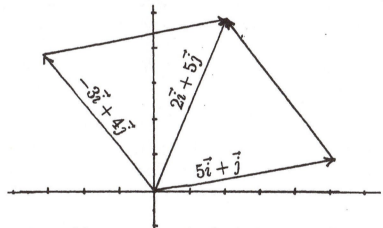

Theorem 1 enables us to easily find the sum of two vectors by just adding their components. When working with vectors it is customary to refer to real numbers as scalars.

Definition of scalar multiplication. If c is a scalar and $\vec{v} = a\vec{i} + b\vec{j}$ is a vector, then $c\vec{v}$ is a vector of length given by $|c|\|\vec{v}\|$. If $c > 0$, then the direction of $c\vec{v}$ is the same as the direction of $\vec{v}$. If $c < 0$, then the direction of $c\vec{v}$ is a vector in the opposite direction of $\vec{v}$.

169

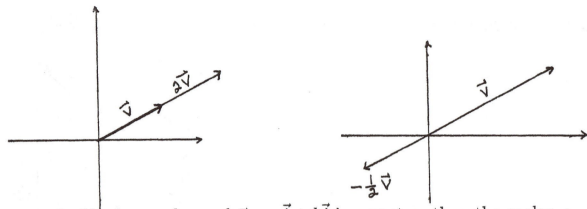

Theorem 2. If c is a scalar and $\vec{v} = a\vec{i} + b\vec{j}$ is a vector, then the scalar c times the vector $\vec{v}$ is the vector $c\vec{v} = (ca)\vec{i} + (cb)\vec{j}$.

Example 2. If $\vec{v} = 3\vec{i} - 5\vec{j}$ and $\vec{w} = -7\vec{i} + 4\vec{j}$ find $6\vec{v} - 8\vec{w}$.

Solution. $6\vec{v} - 8\vec{w} = 6(3\vec{i} - 5\vec{j}) - 8(-7\vec{i} + 4\vec{j}) = (18\vec{i} - 30\vec{j}) + (56\vec{i} - 32\vec{j}) = 74\vec{i} - 62\vec{j}$.

Theorem 3. If two vectors are parallel then one is a scalar multiple of the other. If $\vec{v}$ and $\vec{w}$ are parallel than there is a scalar c such that $\vec{v} = c\vec{w}$.

Example 3. Find a vector parallel to $3\vec{i} - 4\vec{j}$ of length 30. Find a unit vector parallel to $3\vec{i} - 4\vec{j}$.

Solution. The length of $3\vec{i} - 4\vec{j}$ is $\sqrt{9 + 16} = 5$. Any vector parallel to $3\vec{i} - 4\vec{j}$ must be of the form $c(3\vec{i} - 4\vec{j})$. Clearly we need $c = 6$. The vector $18\vec{i} - 24\vec{j}$ is parallel to $3\vec{i} - 4\vec{j}$ and of length 30. The vector $(1/5)(3\vec{i} - 4\vec{j})$ is one unit long and parallel to $3\vec{i} - 4\vec{j}$.

Example 4. Forces with magnitudes of 500 pounds and 400 pounds act at a point which we label as $(0,0)$. The first force acts with an angle of $30°$ and the second force with an angle of $-45°$ with the x axis. Find the resultant force.

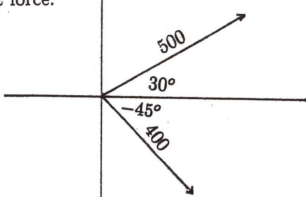

Solution. Let the vector $\vec{F}_1$ denote the first force and the vector $\vec{F}_2$ denote the second force. The direction of the first force is the same as the direction of the unit vector $(\cos 30°)\vec{i} + (\sin 30°)\vec{j}$. The direction of the second force is the same as the unit vector $(\cos 45°)\vec{i} - (\sin 45°)\vec{j}$. The first force is

$$500[(\cos 30°)\vec{i} + (\sin 30°)\vec{j}] = (250\sqrt{3})\vec{i} + (250)\vec{j}.$$

The second force is

$$400[(\cos 45°)\vec{i} - (\sin 45°)\vec{j}] = (200\sqrt{2})\vec{i} - (200\sqrt{2})\vec{j}.$$

The resultant force is the sum of these two vectors.

$$[(250\sqrt{3})\vec{i} + (250)\vec{j}] + (200\sqrt{2})\vec{i} - (200\sqrt{2})\vec{j}]$$

$$= [250\sqrt{3} + 200\sqrt{2}]\vec{i} + [250 - 200\sqrt{2}]\vec{j}$$

$$= (715.86)\vec{i} - (32.84)\vec{j}.$$

The force acting on an object has both magnitude and direction and so must be represented as a vector. If an object is not in motion, then the sum of the forces acting on the object must be zero. A force vector points in the direction in which the force acts and its length is a measure of the force's strength.

Example 5. A 200 pound weight hangs from two wires as shown below. Find the forces $\vec{F}_1$ and $\vec{F}_2$ that the wires exert on the weight.

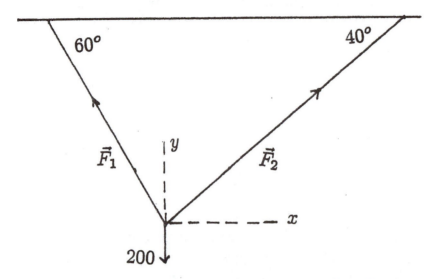

Solution. The first thing we must do is to decide on a coordinate system. As indicated in the above diagram we place the origin of the coordinate system on the object. We place the x axis parallel to the ground and the y axis vertical. In this coordinate system we can easily specify the direction of all forces. We do know the direction of the forces $\vec{F}_1$ and $\vec{F}_2$. The two things that we do not know are the magnitudes of the forces $\vec{F}_1$ and $\vec{F}_2$. Let us use the notation $T_1 = \|\vec{F}_1\|$ and $T_2 = \|\vec{F}_2\|$. From the given geometry we easily see that

$$\vec{F}_1 = T_1[(-\cos 60°)\vec{i} + (\sin 60°)\vec{j}]$$
$$= T_1[(-1/2)\vec{i} + (\sqrt{3}/2)\vec{j}],$$
$$\vec{F}_2 = T_2[(\cos 40°)\vec{i} + (\sin 40°)\vec{j}]$$

Gravity exerts a force of $-200j$ on the object. The sum of the forces acting on the object is zero.

$$\vec{F}_1 + \vec{F}_2 - 200\vec{j} = \vec{0}$$
$$\vec{F}_1 + \vec{F}_2 = 200\vec{j}.$$

Therefore,

$$T_1[(-\cos 60°)\vec{i} + (\sin 60°)\vec{j}] + T_2[(\cos 40°)\vec{i} + (\sin 40°)\vec{j}]$$
$$= 0\vec{i} + 200\vec{j}$$

172

If two vectors are equal, then each of their components must be equal.

$$T_1(-\cos 60^o) + T_2(\cos 40^o) = 0$$
$$T_1(\sin 60^o) + T_2(\sin 40^o) = 200.$$

This is a system of equations with two unknowns T_1 and T_2. We need to solve this system. We solve the first equation for T_2,

$$T_2 = T_1 \frac{\cos 60^o}{\cos 40^o} \qquad (*)$$

Substitute this into the second equation.

$$T_1(\sin 60^o) + T_1 \frac{\cos 60^o}{\cos 40^o}(\sin 40^o) = 200.$$

Multiply both sides by $(\cos 40^o)$.

$$T_1(\sin 60^o)(\cos 40^o) + T_1(\cos 60^o)(\sin 40^o) = 200(\cos 40^o)$$
$$T_1(\sin 100^o) = 200(\cos 40^o)$$

$$T_1 = 200 \frac{\cos 40^o}{\sin 100^o} = 155.57.$$

Substituting this value in $(*)$, we get

$$T_2 = 200 \frac{\cos 40^o}{\sin 100^o} \frac{\cos 60^o}{\cos 40^o} = 200 \frac{\cos 60^o}{\sin 100^o} = 101.54.$$

It follows that

$$\vec{F_1} = (155.57)[(-\cos 60^o)\vec{i} + (\sin 60^o)\vec{j}]$$
$$= -77.79\vec{i} + 134.73\vec{j}$$
$$\vec{F_2} = (101.54)[(\cos 40^o)\vec{i} + (\sin 40^o)\vec{j}]$$
$$= 77.79\vec{i} + 65.27\vec{j}.$$

Note that $134.73 + 65.27 = 200$ and $-77.79 + 77.79 = 0$.

Exercises

1. Let $\vec{v} = -4\vec{i} + 3\vec{j}$ and $\vec{w} = 5\vec{i} + 2\vec{j}$. On the same set of axis draw $\vec{v}, \vec{w}$, and $\vec{v} + \vec{w}$.

2. Let $\vec{v} = -2\vec{i} + 6\vec{j}$. On the same set of axis draw $\vec{v}$, $2\vec{v}$, and $(-1/2)\vec{v}$.

3. Find a vector parallel to $-4\vec{i} + 3\vec{j}$ that is 20 units long.

4. If $\vec{v} = 5\vec{i} - 2\vec{j}$ and $\vec{w} = 3\vec{i} + 7\vec{j}$ find $6\vec{v} - 4\vec{w}$ and $-3\vec{v} + 10\vec{w}$. No need to give a sketch.

5. Three forces all in the same plane of magnitude 300 newtons, 180 newtons, and 250 newtons act on an object located at the origin at angles of -30°, 45°, and 135° respectively with the positive x axis. Forces act in the direction away from the origin. Find the resultant force.

6. In order to carry a 100 pound weight two workers lift at the end of short ropes tied to an eyelet at the top of the weight. One rope makes an angle of 25° with the vertical and the other rope makes an angle of 30° with the vertical. Find the tension in each rope. Find the force exerted on the eyelet by each rope. Recall that tension is the magnitude of the force vector.

7. A 500 pound weight hangs from two wires as shown below. Find the forces $\vec{F}_1$ and $\vec{F}_2$ that the wires exert on the weight. The weight is in equilibrium.

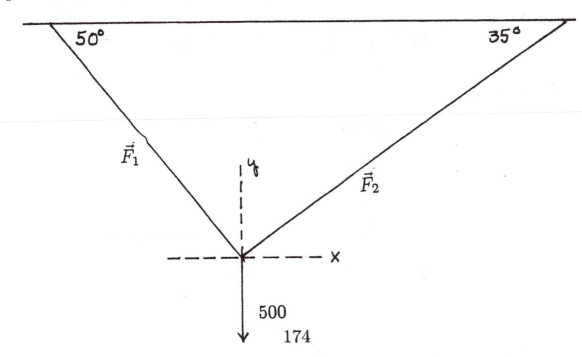

174

6755 The Dot Product

Geometric definition of dot product. The dot product of two vectors is a scalar which may be either positive or negative. When discussing the dot product we define the angle θ between two vectors with the same initial point to be the angle satisfying $0 \leq \theta \leq \pi$. If $\vec{v}$ and $\vec{w}$ are two vectors, then the dot product $\vec{v} \cdot \vec{w}$ is given by

$$\vec{v} \cdot \vec{w} = ||\vec{v}|| \, ||\vec{w}|| \cos \theta,$$

where θ is the angle between $\vec{v}$ and $\vec{w}$.

Algebraic definition of dot product. Given the vectors $\vec{v} = a\vec{i} + b\vec{j}$ and $\vec{w} = c\vec{i} + d\vec{j}$, the dot product is

$$\vec{v} \cdot \vec{w} = ac + bd.$$

Now we should, of course, prove that these two definitions are equivalent, that is,

$$||\vec{v}|| \, ||\vec{w}|| \cos \theta = ac + bd.$$

We will not give the proof here. It is an exercise in trigonometry involving the identity $\cos(A - B) = \cos A \cos B + \sin A \sin B$.

When asked to find the dot product of two vectors $a\vec{i} + b\vec{j}$ and $c\vec{i} + d\vec{j}$ we would always use the algebraic form $(a\vec{i} + b\vec{j}) \cdot (c\vec{i} + d\vec{j}) = ac + bd$. On the other hand the geometric definition can be very useful. It says

$$\cos \theta = \frac{\vec{v} \cdot \vec{w}}{||\vec{v}|| \, ||\vec{w}||}.$$

This formula can be used to find the angle between two vectors.

Example 1. Find the angle in radians between $\vec{v} = 5\vec{i} + 7\vec{j}$ and $\vec{w} = -3\vec{i} + 4\vec{j}$.

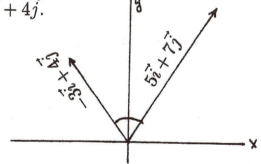

Solution. We have $\vec{v} \cdot \vec{w} = -15 + 28 = 13$. $||\vec{v}|| = \sqrt{5^2 + 7^2} = \sqrt{74}$, and

175

$||\vec{w}|| = \sqrt{(-3)^2 + 4^2} = 5$. Therefore,

$$\cos\theta = \frac{13}{5\sqrt{74}} = 0.3022$$

$$\theta = 1.2638 \text{ or } 72.41°.$$

In math we usually measure this angle in radians, but in some practical problems we might want to measure it in degrees.

Theorem. If $\vec{v} \neq \vec{0}$ and $\vec{w} \neq \vec{0}$, but $\vec{v} \cdot \vec{w} = 0$, then the vectors $\vec{v}$ and $\vec{w}$ are perpendicular.

Example 2. Given the vector $\vec{v} = 5\vec{i} - 4\vec{j}$, find a vector $\vec{w}$ perpendicular to $\vec{v}$.

Solution. We need to find a vector $\vec{w} = a\vec{i} + b\vec{j}$ such that $\vec{v} \cdot \vec{w} = 0$. We need to find a and b such that $(5\vec{i} - 4\vec{j}) \cdot (a\vec{i} + b\vec{j}) = 5a - 4b = 0$. This means $5a = 4b$ or $b = (5/4)a$. We have lots of choices for a since the length of $\vec{w}$ is not given. Choose $a = 4$, then $b = 5$. This means $(5\vec{i} - 4\vec{j}) \cdot (4\vec{i} + 5\vec{j}) = 20 - 20 = 0$. The vector $4\vec{i} + 5\vec{j}$ is perpendicular to $5\vec{i} - 4\vec{j}$. In fact the vector $b\vec{i} - a\vec{j}$ is always perpendicular to $a\vec{i} + b\vec{j}$.

Given the vector $\vec{v} = 3\vec{i} - 4\vec{j}$, as stated earlier the coefficient 3 of $\vec{i}$ is called the component of $\vec{v}$ in the direction of $\vec{i}$ and the coefficient -4 of $\vec{j}$ is called the component of $\vec{v}$ in the direction of $\vec{j}$. We also refer to 3 and -4 as the components of the vector $\vec{v}$. We also use the word component in a more general way.

Definition 3. Given vectors $\vec{v}$ and $\vec{w}$ we say that the component of $\vec{v}$ in the direction of $\vec{w}$ is given by

$$\text{Comp}_{\vec{w}}\vec{v} = \frac{\vec{v} \cdot \vec{w}}{||\vec{w}||}.$$

Note that $\text{Comp}_{\vec{w}}\vec{v}$ is a scalar which is also given by $||\vec{v}|| \cos\theta$.

Example 3. Let $\vec{v} = 7\vec{i} - 8\vec{j}$ and $\vec{w} = -3\vec{i} + 4\vec{j}$ find $\text{Comp}_{\vec{w}}\vec{v}$.

Solution. We have $\vec{v} \cdot \vec{w} = -21 - 32 = -53$ and $||\vec{w}|| = \sqrt{(-3)^2 + 4^2} = 5$. Therefore,

$$\text{Comp}_{\vec{w}}\vec{v} = \frac{-53}{5} = -10.6.$$

176

Definition 4. Given vectors $\vec{v}$ and $\vec{w}$, we define the projection of $\vec{v}$ on $\vec{w}$ as the vector

$$\text{Proj}_{\vec{w}}\vec{v} = \left(\frac{\vec{v}\cdot\vec{w}}{||\vec{w}||}\right)\frac{\vec{w}}{||\vec{w}||} = \left[\frac{\vec{v}\cdot\vec{w}}{||\vec{w}||^2}\right]\vec{w}.$$

Note that $\frac{\vec{w}}{||\vec{w}||}$ is a vector one unit long in the same direction as $\vec{w}$. The length of the vector $\text{Proj}_{\vec{w}}\vec{v}$ is $\frac{|\vec{v}\cdot\vec{w}|}{||\vec{w}||}$. Recall that $\text{Comp}_{\vec{w}}\vec{v} = \frac{\vec{v}\cdot\vec{w}}{||\vec{w}||}$. This means $\text{Proj}_{\vec{w}}\vec{v} = [\text{Comp}_{\vec{w}}\vec{v}]\frac{\vec{w}}{||\vec{w}||}$. Also note that $(\text{Proj}_{\vec{w}}\vec{v})\cdot\vec{w} = \vec{v}\cdot\vec{w}$. Finally, the vector $\text{Proj}_{\vec{w}}\vec{v}$ is also called the vector component of $\vec{v}$ in the direction of $\vec{w}$.

It is easy to picture geometrically the relationship between $\vec{v}$, $\vec{w}$, and $\text{Proj}_{\vec{w}}\vec{v}$ when all vectors are in standard position. Draw a line perpendicular to $\vec{w}$ through the tip of $\vec{v}$. The tip of vector $\text{Proj}_{\vec{w}}\vec{v}$ is where this line intersects the line containing the vector $\vec{w}$. When the angle between $\vec{v}$ and $\vec{w}$ is greater than $90°$ the vector $\text{Proj}_{\vec{w}}\vec{v}$ is in the opposite direction to $\vec{w}$. Note that if $\vec{u}$ and $\vec{w}$ are parallel then $\text{Proj}_{\vec{w}}\vec{v} = \text{Proj}_{\vec{u}}\vec{v}$.

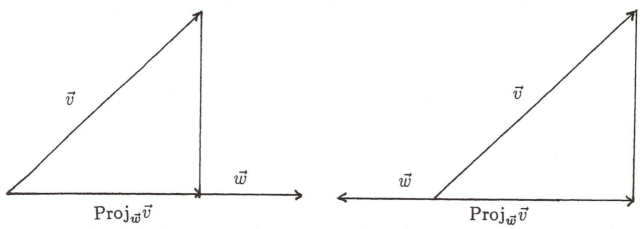

Example 4. Given the vectors $\vec{v} = -4\vec{i} + 3\vec{j}$ and $\vec{w} = 7\vec{i} + 2\vec{j}$, find $\text{Proj}_{\vec{w}}\vec{v}$. Draw a sketch of these three vectors in standard position.

Solution. We have $\vec{v}\cdot\vec{w} = -28 + 6 = -22$, $||\vec{v}|| = 5$ and $||\vec{w}|| = \sqrt{53}$.

$$\text{Proj}_{\vec{w}}\vec{v} = \frac{-22}{53}(7\vec{i} + 2\vec{j}) = -2.91\vec{i} - 0.83\vec{j}.$$

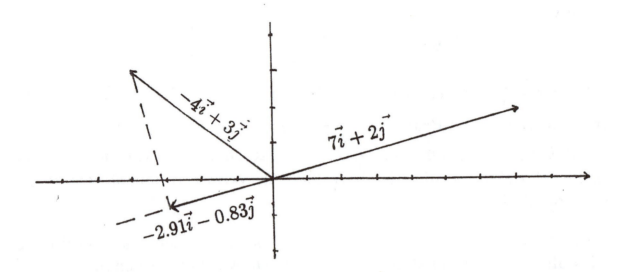

The vector $\text{Proj}_{\vec{w}}\vec{v}$ is also called the component vector of vector $\vec{v}$ in the direction of $\vec{w}$. It is often useful when working with a vector $\vec{v}$ to resolve the vector $\vec{v}$ into two component vectors. One component vector of $\vec{v}$ would be the component vector of $\vec{v}$ in the direction of some given vector $\vec{w}$ and the other component vector would be a vector perpendicular to the given vector $\vec{w}$. This means that we want to express the given vector $\vec{v}$ as

$$\vec{v} = \text{Proj}_{\vec{w}}\vec{v} + (\text{Another vector perpendicular to } \text{Proj}_{\vec{w}}\vec{v})$$

How do we find a vector $\vec{z}$ such that

$$\vec{v} = \text{Proj}_{\vec{w}}\vec{v} + \vec{z}?$$

Certainly we can use $\vec{z} = \vec{v} - \text{Proj}_{\vec{w}}\vec{v}$. But we also want $\vec{z}$ to be perpendicular to the vector $\text{Proj}_{\vec{w}}\vec{v}$ (perpendicular to $\vec{w}$). Let us take the dot product

$$\vec{z} \cdot (\text{Proj}_{\vec{w}}\vec{v}) = (\vec{v} - \text{Proj}_{\vec{w}}\vec{v}) \cdot (\text{Proj}_{\vec{w}}\vec{v})$$

$$= \vec{v} \cdot (\text{Proj}_{\vec{w}}\vec{v}) - (\text{Proj}_{\vec{w}}\vec{v}) \cdot (\text{Proj}_{\vec{w}}\vec{v})$$

$$= \vec{v} \cdot \left(\frac{\vec{v} \cdot \vec{w}}{\|\vec{w}\|^2}\right) \vec{w} - \|\text{Proj}_{\vec{w}}\vec{v}\|^2$$

$$= \frac{(\vec{v} \cdot \vec{w})^2}{\|\vec{w}\|^2} - \frac{(\vec{v} \cdot \vec{w})^2}{\|w\|^2} = 0.$$

Since $\vec{z} \cdot (\text{Proj}_{\vec{w}} \vec{v}) = 0$, it follows that $\vec{z}$ and $\text{Proj}_{\vec{w}} \vec{v}$ are perpendicular. Also $\vec{z}$ and $\vec{w}$ are perpendicular.

Example 5. Express the vector $\vec{v} = -5\vec{i} + 4\vec{j}$ as the sum of a component vector parallel to $\vec{w} = 8\vec{i} + 3\vec{j}$ and a component vector perpendicular to $8\vec{i} + 3\vec{j}$.

Solution. The component vector of $\vec{v} = -5\vec{i} + 4\vec{j}$ in the direction of $\vec{w} = 8\vec{i} + 3\vec{j}$ is given by

$$\text{Proj}_{\vec{w}} \vec{v} = \frac{\vec{v} \cdot \vec{w}}{\|\vec{w}\|^2} \vec{w} = \frac{-28}{73}(8\vec{i} + 3\vec{j}) = -\frac{224}{73}\vec{i} - \frac{84}{73}\vec{j}.$$

The component vector perpendicular to this vector is

$$\vec{z} = -5\vec{i} + 4\vec{j} - \left[-\frac{224}{73}\vec{i} - \frac{84}{73}\vec{j}\right] = -\frac{141}{73}\vec{i} + \frac{376}{73}\vec{j}.$$

Note that

$$\left(-\frac{224}{73}\vec{i} - \frac{84}{73}\vec{j}\right) + \left(-\frac{141}{73}\vec{i} + \frac{376}{73}\vec{j}\right) = -5\vec{i} + 4\vec{j}.$$

Also note that

$$\left(-\frac{224}{73}\vec{i} - \frac{84}{73}\vec{j}\right) \cdot \left(-\frac{141}{73}\vec{i} + \frac{376}{73}\vec{j}\right) = 0.$$

The most famous application of the dot product and projection vector is work. Suppose we are moving a box along a line and applying a force of 80 Newtons in the direction parallel to the line of motion, then the work done is equal to the force times the displacement. If we move this box 30 meters applying this force, then the work done is

$$\text{work} = (80 \text{ Newtons})(30 \text{ meters}) = 2400 \text{ Newton Meters}.$$

However, if the force is not parallel to the line of displacement, then all the force does not count in computing the work. Suppose we move a box in a straight line on a level floor. Suppose the box is not as tall as you. When you push the box across the floor part of the force you exert pushes down on the box because you are touching the box at a level below your shoulders. The downward part of this force actually increases the weight of the box and makes it harder to push. Only the vector component of the force vector in the direction of the displacement vector counts when computing the work done. We can simplify calculations for work if we use vectors. Suppose a box is moved along the line segment given by the vector $\vec{D}$. Suppose while it is moving a constant force $\vec{F}$ is exerted on the box, then the work done by the force in moving the box through the displacement $\vec{D}$ is

$$\vec{F} \cdot \vec{D} = (\mathrm{Proj}_{\vec{D}}\vec{F}) \cdot \vec{D}.$$

Note that $\mathrm{Proj}_{\vec{D}}\vec{F}$ is the vector component of $\vec{F}$ parallel to $\vec{D}$. All the motion which we will consider takes place in a plane with a given coordinate system. If no coordinate system is given, then we must make one up.

Example 6. A box is moved 120 feet along a line using a force of 70 lbs. The direction of the force makes a 30° angle with the horizontal. Find the amount of work done.

Solution. We introduce a coordinate system with the origin at the initial point of motion of the box. The x axis is along the line of displacement and the y axis is vertical. The displacement vector goes from (0,0) to (120,0) and so is $120\vec{i}$.

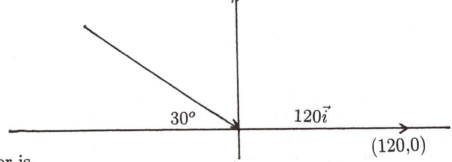

The force vector is

$$70(\cos 30^\circ)\vec{i} + 70(-\sin 30^\circ)\vec{j} = (35\sqrt{3})\vec{i} - 35\vec{j}.$$

$$\text{work} = [(35\sqrt{3})\vec{i} - 35\vec{j}] \cdot [120\vec{i}] = 4200\sqrt{3} \text{ ft lbs.}$$

Exercises

1. Find the angle in radians between the vectors $\vec{v} = -5\vec{i} + 6\vec{j}$ and $\vec{w} = 2\vec{i} + 7\vec{j}$.

2. Given $\vec{v} = 4\vec{i} - 3\vec{j}$, $\vec{w} = 2\vec{i} - 5\vec{j}$, find $\text{Comp}_{\vec{w}}\vec{v}$ and $\text{Proj}_{\vec{w}}\vec{v}$.

3. Given $\vec{v} = -3\vec{i} + 5\vec{j}$, $\vec{w} = 6\vec{i} + 2\vec{j}$, find $\text{Proj}_{\vec{w}}\vec{v}$.

4. Given $\vec{v} = 7\vec{i} - 2\vec{j}$ and $\vec{w} = -3\vec{i} + 8\vec{j}$, find $\text{Proj}_{\vec{w}}\vec{v}$. Sketch a graph of the three vectors. Find $\text{Comp}_{\vec{w}}\vec{v}$ and $(\cos\theta)\|\vec{v}\|$.

5. Find two vectors perpendicular to each of the following vectors:

(a) $12\vec{i} - 5\vec{j}$

(b) $-5\vec{i} + 7\vec{j}$

(c) $-3\vec{i} - 4\vec{j}$.

6. Express the vector $\vec{v} = 8\vec{i} - 5\vec{j}$ as the sum of a component vector parallel to $2\vec{i} + 9\vec{j}$ and a component vector perpendicular to $2\vec{i} + 9\vec{j}$.

7. A box is moved 150 feet along a line using a force of 90 lbs. The direction of the force makes an angle of 40° with the horizontal. Find the amount of work done.

6757 The Position Vector and Lines

There are two common ideas we may be thinking about when we talk about graphing vector functions. One idea is to graph a vector field. In a vector field there is a vector associated with every point. For example, at every point near the surface of the earth there is associated a vector which is the force of gravity at that point. This is a vector field. We are not going to discuss vector fields as such at this time.

The other idea we are thinking about when we graph vector functions is as follows. We also use vectors to locate points. This is done in the plane and in space. We are going to discuss using vectors to locate points in the coordinate plane. When a vector is being used to locate points we always assume that the vector is in standard position with its initial point at the origin $(0,0)$. We will only use rectangular coordinates. The vector $a\vec{i} + b\vec{j}$ locates the point with coordinates (a, b).

The vector $3\vec{i} + 4\vec{j}$ locates the point with coordinates $(3, 4)$. The vector $3\vec{i} + 4\vec{j}$ and the point $(3, 4)$ are indicated below.

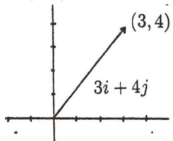

We are going to consider vector valued functions $\vec{r}(t)$ of one independent variable t. We will often shorten the term "vector valued function" to "vector function". For example, consider vector function

$$\vec{r}(t) = (2t + 3)\vec{i} + t^2\vec{j}.$$

When $t = 1$ we get the vector $5\vec{i} + \vec{j}$, that is, $\vec{r}(1) = 5\vec{i} + \vec{j}$. When $t = 2$ we have $\vec{r}(2) = 7\vec{i} + 4\vec{j}$. Letting $t = -1$ and $t = 0$, we get $\vec{r}(-1) = \vec{i} + \vec{j}$, and $\vec{r}(0) = 3\vec{i}$. The vector $5\vec{i} + \vec{j}$ locates the point $(5, 1)$. The vector $7\vec{i} + 4\vec{j}$ locates the point $(7, 4)$. The phrase "plot the graph of $\vec{r}(t) = (2t+3)\vec{i} + t^2\vec{j}$" means to plot the points located by the vectors. In order to sketch a graph of $\vec{r}(t)$ let us make a table of values for $\vec{r}(t) = (2t + 3)\vec{i} + t^2\vec{j}$.

t	$\vec{r}(t)$	x	y
-3	$-3\vec{i} + 9\vec{j}$	-3	9
-2	$-\vec{i} + 4\vec{j}$	-1	4
-1	$\vec{i} + \vec{j}$	1	1
0	$3\vec{i}$	3	0
1	$5\vec{i} + \vec{j}$	5	1
2	$7\vec{i} + 4\vec{j}$	7	4
3	$9\vec{i} + 9\vec{j}$	9	9

Graphing $\vec{r}(t) = (2t + 3)\vec{i} + t^2\vec{j}$ means to plot all the points located by these vectors. This is the same as plotting the points with the given (x, y) coordinates. We then connect the points with a smooth curve to obtain the graph of $\vec{r}(t)$. The assumption is that plotting more points would only locate more points on this smooth curve.

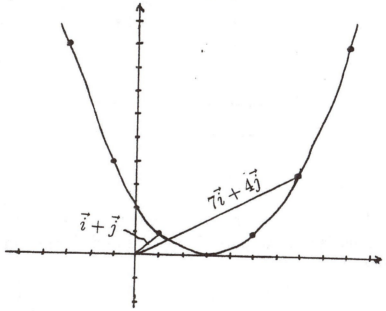

Example 1. Plot the graph of the vector valued function $\vec{r}(t) = [t(t^2 - 4)]\vec{i} + [t^2 - 1]\vec{j}$

Solution. First, we make a table of values for $\vec{r}(t)$ as follows. When $t = -3, \vec{r}(-3) = [(-3)(9 - 4)]\vec{i} + [9 - 1]\vec{j} = -15\vec{i} + 8\vec{j}$. This means that corresponding to $t = -3$ the point is $(-15, 8)$. We continue in this manner to construct a table of values for x and y.

183

t	x	y
-3	-15	8
-2	0	3
-1	3	0
0	0	-1
1	-3	0
2	0	3
3	15	8

We plot the points with these (x, y) coordinates and connect these points using a smooth curve.

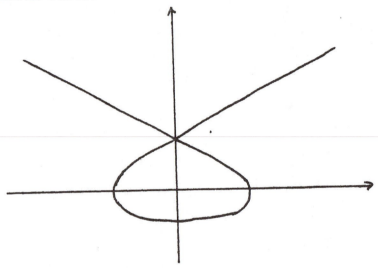

Example 2. The curve $\vec{r}(t)$ has the following value chart. Make a sketch of the curve. This means we plot the points and then connect them using a smooth curve.

t	x	y
-3	-15	-12
-2	0	-2
-1	3	4
0	0	6
1	-3	4
2	0	-2
3	15	-12

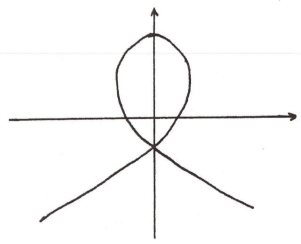

184

Example 3. Find the vector function $\vec{r}(t)$ whose graph is the line through the point $(-2, 5)$ and parallel to the vector $3\vec{i} - 2\vec{j}$.

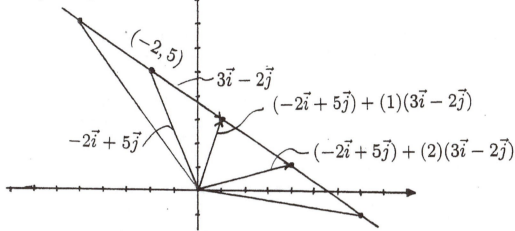

Solution. The point $(-2, 5)$ is on the line and is located by the vector $-2\vec{i} + 5\vec{j}$. Place the given vector $3\vec{i} - 2\vec{j}$ with its initial point at $(-2, 5)$. Since the vector $3\vec{i} - 2\vec{j}$ is parallel to the given line the terminal point of $3\vec{i} - 2\vec{j}$ will be on the line. This says

$$(-2\vec{i} + 5\vec{j}) + (1)(3\vec{i} - 2\vec{j})$$

locates a point on the line. Next place the vector $(2)(3\vec{i} - 2\vec{j})$ with its initial point at $(-2, 5)$. Since vector $(2)(3\vec{i} - 2\vec{j})$ is parallel to the given line the terminal point of $(2)(3\vec{i} - 2\vec{j})$ will then be on this same line. This says that

$$(-2\vec{i} + 5\vec{j}) + (2)(3\vec{i} - 2\vec{j})$$

locates a point on the line. In fact place the vector $(t)(3\vec{i} - 2\vec{j})$, t any scalar, with its initial point at $(-2, 5)$. Since the vector $(t)(3\vec{i} - 2\vec{j})$ is parallel to the line the terminal point of $(t)(3\vec{i} - 2\vec{j})$ will be on the line. This says that

$$(-2\vec{i} + 5\vec{j}) + (t)(3\vec{i} - 2\vec{j})$$

locates a point on the line for any value of t. All the points on this line are located by

$$\vec{r}(t) = (-2\vec{i} + 5\vec{j}) + t(3\vec{i} - 2\vec{j}),$$

when we use different values of t. The vector $3\vec{i} - 2\vec{j}$ is called the direction vector of the line. The line through $(-2, 5)$ and parallel to the vector $3\vec{i} - 2\vec{j}$ is the graph of the vector function $\vec{r}(t) = (-2\vec{i} + 5\vec{j}) + t(3\vec{i} - 2\vec{j})$.

185

Theorem. The graph of $\vec{r}(t) = f(t)\vec{i} + g(t)\vec{j}$ is a straight line if and only if both $f(t)$ and $g(t)$ are linear.

Example 4. Find the direction vector for the line which is the graph of the vector function $\vec{r}(t) = (3t - 8)\vec{i} + (-5t + 7)\vec{j}$.

Solution. We have $\vec{r}(t) = -8\vec{i} + 7\vec{j} + t(3\vec{i} - 5\vec{j})$. The direction vector for this line is $3\vec{i} - 5\vec{j}$. This line is parallel to the vector $3\vec{i} - 5\vec{j}$.

It is clear that in order to find the vector function $\vec{r}(t)$ whose graph is a line we need the coordinates of one point on the line and a direction vector for the line. If the direction vector is not given, then we must find it before we can find the vector function whose graph is the line.

Example 5. Find the vector function $\vec{r}(t)$ whose graph is the line through the two points $(-2, 7)$ and $(4, 3)$.

Solution. We all recall that two points determine a line. Let us sketch this line.

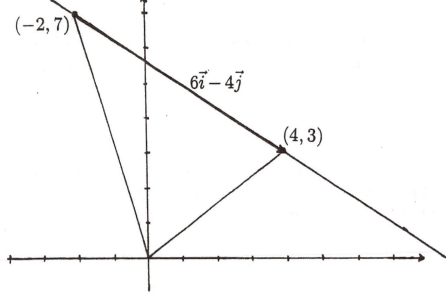

In order to find the vector function whose graph is a given line we need a point on the line and a direction vector. We actually have two points given for this line. We are not given the direction vector. Clearly a direction vector for this line is the vector with initial point at $(-2, 7)$ and terminal point at $(4, 3)$. This vector is $[4 - (-2)]\vec{i} + [3 - 7]\vec{j} = 6\vec{i} - 4\vec{j}$. Using the first point $(-2, 7)$ and the direction vector $6\vec{i} - 4\vec{j}$ the vector function of the line is

$$\vec{r}(t) = (-2\vec{i} + 7\vec{j}) + t(6\vec{i} - 4\vec{j}).$$

Note that $\vec{r}(1) = 4\vec{i} + 3\vec{j}$. Using the second point the vector function which graphs as the line is

$$\vec{R}(t) = (4\vec{i} + 3\vec{j}) + t(6\vec{i} - 4\vec{j}).$$

Both $\vec{r}(t)$ and $\vec{R}(t)$ are correct functions whose graph is this same line. This brings up a point we do not like to think about. Every curve in the plane has more than one way to express it using a vector function.

Example 6. Let $\vec{r}(t) = (-\cos t)\vec{i} + (\sin t)\vec{j}$ for $0 \le t \le \pi$. Let $\vec{R}(t) = t\vec{i} + [(1 - t^2)^{1/2}]\vec{j}$ for $-1 \le t \le 1$. Both $\vec{r}(t)$ and $\vec{R}(t)$ graph as the upper half of a circle of radius 1 with center at $(0,0)$.

Example 7. Show that $\vec{r}(t) = (-7t + 3)\vec{i} + (-3t + 5)\vec{j}$ and $\vec{R}(t) = (7t - 4)\vec{i} + (3t + 2)\vec{j}$ graph as the same line.

Solution. Since the coefficients are linear the graph of both $\vec{r}(t)$ and $\vec{R}(t)$ are lines. Start with $\vec{r}(t) = (-7t + 3)\vec{i} + (-3t + 5)\vec{j}$. The points $\vec{r}(0) = 3\vec{i} + 5\vec{j}$ and $\vec{r}(1) = -4\vec{i} + 2\vec{j}$ are two points on this line. Are these two points on $\vec{R}(t)$? Is there a value of t such that

$$(7t - 4)\vec{i} + (3t + 2)\vec{j} = 3\vec{i} + 5\vec{j}.$$

This is the same question as: is there a solution to the simultaneous equations:

$$7t - 4 = 3$$
$$3t + 2 = 5?$$

Clearly $t = 1$ is a solution, that is, $\vec{R}(1) = 3\vec{i} + 5\vec{j}$. The point $(3,5)$ is on the line $\vec{R}(t)$.

Is there a value of t such that

$$(7t - 4)\vec{i} + (3t + 2)\vec{j} = -4\vec{i} + 2\vec{j}?$$

Without doing any work we see that the value of t is $t = 0$, that is, $\vec{R}(0) = -4\vec{i} + 2\vec{j}$. The two points $(3,5)$ and $(-4, 2)$ are on both the lines $\vec{r}(t)$ and

$\vec{R}(t)$. Since two points determine a line this two lines must be the same line.

Example 8. At what point do the two lines $\vec{r}(t) = (3t+4)\vec{i} + (-2t+11)\vec{j}$ and $\vec{R}(s) = (s-5)\vec{i} + (4s+3)\vec{j}$ intersect?

Solution. Since we are discussing two different lines, in order to keep from being confused we use the letter s in $\vec{R}(s)$ instead of the letter t which we would usually use. The problem may be restated as: can we find a value of s and a value of t such that

$$(3t+4)\vec{i} + (-2t+11)\vec{j} = (s-5)\vec{i} + (4s+3)\vec{j}?$$

This vector equation is equivalent to the two simultaneous equations

$$3t + 4 = s - 5$$
$$-2t + 11 = 4s + 3.$$

Simplifying
$$3t - s = -9$$
$$-2t - 4s = -8.$$

Multiply both sides of the first equation by -4.

$$-12t + 4s = 36$$
$$-2t - 4s = -8.$$

Adding these two equations together, we get

$$-14t = 28$$

This gives $t = -2$. Substituting $t = -2$ into $3t - s = -9$, we get $3(-2) - s = -9$ or $s = 3$.

$$\vec{r}(-2) = (-6+4)\vec{i} + (4+11)\vec{j} = -2\vec{i} + 15\vec{j}$$
$$\vec{R}(3) = (3-5)\vec{i} + (12+3)\vec{j} = -2\vec{i} + 15\vec{j}.$$

The point $(-2, 15)$ is on both lines.

Example 9. Given the line which is the graph of the vector function $\vec{r}(t) = (3t+5)\vec{i} + (-4t+11)\vec{j}$, find the function $\vec{R}(s)$ whose graph is the line perpendicular to the given line and which contains the point $(-8, 13)$.

Solution. We are given one point $(-8, 13)$ on the line $\vec{R}(s)$. We must find the direction vector. The direction vector for $\vec{R}(s)$ is perpendicular to the direction vector for $\vec{r}(t)$. The direction vector of $\vec{r}(t) = (5\vec{i}+11\vec{j}) + t(3\vec{i}-4\vec{j})$ is $3\vec{i}-4\vec{j}$. The vector $4\vec{i}+3\vec{j}$ is perpendicular to $3\vec{i}-4\vec{j}$ since the dot product $(3\vec{i} - 4\vec{j}) \cdot (4\vec{i} + 3\vec{j}) = 0$. A direction vector for $\vec{R}(s)$ is $4\vec{i} + 3\vec{j}$. A vector function whose graph is the line perpendicular to $\vec{r}(t)$ is

$$\vec{R}(s) = (-8\vec{i} + 13\vec{j}) + t(4\vec{i} + 3\vec{j})$$
$$= (4t - 8)\vec{i} + (3t + 13)\vec{j}.$$

Let us look again at the fact that two different vector functions can graph as the same line. Consider the vector function $\vec{r}(t) = (3\vec{i}+5\vec{j}) + t(-7\vec{i}-3\vec{j})$. We know that $\vec{r}(t)$ graphs as a line through the point $(3, 5)$ with direction vector $-7\vec{i} - 3\vec{j}$. But we can select other points and other direction vectors to represent this line. Since $\vec{r}(-2) = 17\vec{i} + 11\vec{j}$ the point $(17, 11)$ is on this line. The vector $14\vec{i} + 6\vec{j}$ is parallel to the vector $-7\vec{i} - 3\vec{j}$. Therefore

$$\vec{R}(s) = (17\vec{i} + 11\vec{j}) + s(14\vec{i} + 6\vec{j})$$

graphs as the same line as

$$\vec{r}(t) = (3\vec{i} + 5\vec{j}) + t(-7\vec{i} - 3\vec{j}).$$

Example 10. Consider the line which is the graph of the vector equation $\vec{r}(t) = (-4\vec{i} + 7\vec{j}) + t(8\vec{i} - 5\vec{j})$. Find three other vector representations of this same line.

Solution. The vectors $\vec{r}(-1) = -12\vec{i}+12\vec{j}$, $\vec{r}(1) = 4\vec{i}+2\vec{j}$ and $\vec{r}(3) = 20\vec{i}-8\vec{j}$ all locate points on this line. The vectors $-16\vec{i} + 10\vec{j}$, $4\vec{i} - (5/2)\vec{j}$, and $40\vec{i} - 25\vec{j}$ are all parallel to the direction vector $8\vec{i} - 5\vec{j}$.

Any of these may be used to form a vector function whose graph is this same line. The vector functions

$$\vec{R}(s) = (-12\vec{i} + 12\vec{j}) + s(-16\vec{i} + 10\vec{j})$$
$$\vec{R}(s) = (4\vec{i} + 2\vec{j}) + s(4\vec{i} - (5/2)\vec{j})$$
$$\vec{R}(s) = (20\vec{i} - 8\vec{j}) + s(40\vec{i} - 25\vec{j})$$

all graph as the same line as does the vector function $\vec{r}(t) = (-4\vec{i} + 7\vec{j}) + t(8\vec{i} - 5\vec{j})$.

In fact, if we really want to be creative we can replace t by t^2. Such a replacement does, however, make some changes in the graph. Note that when $-\infty < t < \infty$, we have $t^2 \geq 0$, but t^2 does take on all positive real numbers. Clearly the graphs of the vector functions

$$\vec{r}(t) = (-4\vec{i} + 7\vec{j}) + t(8\vec{i} - 5\vec{j}) \text{ for } t \geq 0$$
$$\vec{R}(t) = (-4\vec{i} + 7\vec{j}) + t^2(8\vec{i} - 5\vec{j}) \text{ for } t \geq 0$$

are the same one half of a line. This ray starts at $(-4, 7)$ and goes to infinity in one direction only. The graph of

$$\vec{R}(t) = (-4\vec{i} + 7\vec{j}) + t^2(8\vec{i} - 5\vec{j}) \text{ for } t \leq 0$$

is the same as the graph of

$$\vec{r}(t) = (-4\vec{i} + 7\vec{j}) + t(8\vec{i} - 5\vec{j}) \text{ for } t \geq 0.$$

The graph of

$$\vec{R}(t) = (-4\vec{i} + 7\vec{j}) + t^2(8\vec{i} - 5\vec{j}) \text{ for } -\infty < t < \infty$$

does not include the whole line but rather does one half the line twice. On the other hand, the graph of

$$\vec{R}(t) = (-4\vec{i} + 7\vec{j}) + t^3(8\vec{i} - 5\vec{j}) \text{ for } -\infty < t < \infty$$

is exactly the same as the graph of

$$\vec{r}(t) = (-4\vec{i} + 7\vec{j}) + t(8\vec{i} - 5\vec{j}) \text{ for } -\infty < t < \infty.$$

We often wish to talk about a finite segment of a curve rather than the whole curve. The graph of

$$\vec{r}(t) = (-4\vec{i} + 7\vec{j}) + t(8\vec{i} - 5\vec{j}) \text{ for } 1 \leq t \leq 5$$

can be expressed as the vector $\vec{r}(5) - \vec{r}(1)$. Since $\vec{r}(1) = 4\vec{i} + 2\vec{j}$ and $\vec{r}(5) = 36\vec{i} - 18\vec{j}$ this is the line segment connecting the point $(4, 2)$ and the point $(36, -18)$.

Example 11. Consider the parabola which is the graph of $\vec{r}(t) = (t+2)\vec{i} + (t^2 - 1)\vec{j}$ for $-\infty < t < \infty$. The vertex is at $(2, -1)$. A sketch of the graph is

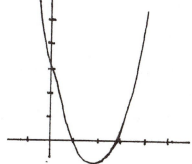

The graph of $\vec{r}(t) = (t+2)\vec{i} + (t^2 - 1)\vec{j}$ for $t \geq 0$ is the right most one half of the parabola. The graph of $\vec{R}(t) = (t^2 + 2)\vec{i} + (t^4 - 1)\vec{j}$ for $t \geq 0$ is also the graph of the right most one half of the parabola. The graph of $\vec{R}(t) = (t^2 + 2)\vec{i} + (t^4 - 1)\vec{j}$ is the right half of the parabola done twice. The graph of $\vec{r}(t) = (t+2)\vec{i} + (t^2 - 1)\vec{j}$ for $-1 \leq t \leq 1$ is the part of the parabola below the y axis. The graph of $\vec{R}(t) = (-\cos t + 2)\vec{i} + (\cos^2 t - 1)\vec{j}$ for $0 \leq t \leq \pi$ is the part of the parabola below the x axis.

There is another notation which is often used to express the ideas about graphing vector functions. Graphing using this idea is called graphing parametric equations. Given the vector function $\vec{r}(t) = f(t)\vec{i} + g(t)\vec{j}$, we say that $x = f(t)$ and $y = g(t)$ are the parametric equations for a curve.

Example 12. Graphing the vector function $\vec{r}(t) = (3t^2 + 5t)\vec{i} + (t \sin t)\vec{j}$, for $0 \leq t \leq 1$, is the same as graphing the parametric equations

$$x = 3t^2 + 5t \qquad y = t \sin t \qquad 0 \leq t \leq 1.$$

Example 13. Suppose the parametric equations for a curve are

$$x = \frac{4t}{t^2 + 1} \qquad y = t^2(3t + 2) \qquad 0 \leq t \leq 5$$

Then the vector equation for this same curve is

$$\vec{r}(t) = \frac{4t}{t^2 + 1}\vec{i} + [t^2(3t + 2)]\vec{j}, \quad 0 \leq t \leq 5.$$

191

In order to graph a curve given by the parametric equations $x = f(t)$ and $y = g(t)$ we do exactly the same thing we would do in order to graph the vector function $\vec{r}(t) = f(t)\vec{i} + g(t)\vec{j}$.

Graphing calculators graph vector functions as graphing with parametric equations. In order to graph the vector function $\vec{r}(t) = (t^3 - 4t)\vec{i} + (t^2 - 1)\vec{j}$ we set the calculator in parametric graphing mode. Let $x = t^3 - 4t$ and $y = t^2 - 1$. Set the t values at about $t_{min} = -3$ and $t_{max} = 3$ and press graph.

Exercises

1. Plot the graph of the vector valued function $\vec{r}(t) = [t^2 - 9]\vec{i} + [t(t^2 - 4)]\vec{j}$ using $-3 \le t \le 3$.

2. The curve $\vec{r}(t)$ has the following value chart. Make a sketch of the curve.

t	x	y
-3	-5	-5
-2	0	0
-1	1	3
0	0	4
1	-1	3
2	0	0
3	5	-5

3. Find a vector valued function whose graph is of the line through the point $(-8, 7)$ and parallel to the vector $5\vec{i} - 3\vec{j}$.

4. Find the direction vector for the line $\vec{r}(t) = (-7t + 8)\vec{i} + (9t + 11)\vec{j}$. Find a vector function $\vec{R}(s)$ whose graph is the line through the point $(5, -8)$ and is parallel to this line.

5. Find a vector equation for the line through the two points $(-7, 11)$ and $(3, 5)$.

6. At what point do the lines that are the graphs of the vector functions $\vec{r}(t) = (5t + 6)\vec{i} + (-3t + 11)\vec{j}$ and $\vec{R}(s) = (-4s + 7)\vec{i} + (7s - 8)\vec{j}$ intersect?

7. Find the point where the graph of the two following vector functions intersect.

$$\vec{r}(t) = (3t + 13)\vec{i} + (4t - 3)\vec{j}$$
$$\vec{R}(s) = (5s - 8)\vec{i} + (-2s - 5)\vec{j}$$

8. Show that the vector function $\vec{r}(t) = (14t + 3)\vec{i} + (8t + 9)\vec{j}$ and the vector function $\vec{R}(s) = (-7s - 4)\vec{i} + (-4s + 5)\vec{j}$ both graph the same line.

9. Show that the following vector functions each graph as the same line

$$\vec{r}(t) = (-2t + 3)\vec{i} + (7t - 5)\vec{j}$$
$$\vec{R}(s) = (4s + 1)\vec{i} + (-14s + 2)\vec{j}$$

10. Given the line which is the graph of the vector function $\vec{r}(t) = (7t - 13)\vec{i} + (-5t + 8)\vec{j}$, find the function $\vec{R}(s)$ whose graph is the line perpendicular to the given line through the point $(-7, 11)$.

11. Are the three points $(-1, 19)$, $(5, 11)$ and $(14, -1)$ on the same line?

12. Consider the line which is the graph of the vector function

$$\vec{r}(t) = (8\vec{i} - 5\vec{j}) + t(-4\vec{i} + 7\vec{j})$$

Find three other vector functions whose graph is this same line.

13. Sometimes when we have a problem to graph a vector function $\vec{r}(t)$ we like to convert it into rectangular coordinates. Consider $\vec{f}(t) = (3\cos t)\vec{i} + (4\sin t)\vec{j}$. Stated using parametric equations this would be $x = 3\cos t$ and $y = 4\sin t$. Since $\cos t = (x/3)$ and $\sin t = (y/4)$ this says

$$\left(\frac{x}{3}\right)^2 + \left(\frac{y}{4}\right)^2 = \cos^2 t + \sin^2 t = 1$$

The xy equation is the ellipse $16x^2 + 9y^2 = 144$. Find the xy equation which has the same graph as the following vector function

(a) $\vec{r}(t) = (t - 4)\vec{i} + (t^2 - 3t - 4)\vec{j}$

(b) $\vec{r}(t) = (3\cosh t)\vec{i} + (4\sinh t)\vec{j}$.

6759 Calculus of Vector Functions

Theorem 1. If $\vec{r}(t) = f(t)\vec{i} + g(t)\vec{j}$, then the derivative $\vec{r}\,'(t)$ is given by $\vec{r}\,'(t) = f'(t)\vec{i} + g'(t)\vec{j}$.

Example 1. Let $\vec{r}(t) = (t^3 + 8t)\vec{i} + (\sin 3t)\vec{j}$, then $\vec{r}\,'(t) = (3t^2 + 8)\vec{i} + (3\cos 3t)\vec{j}$.

Theorem 2. Consider the curve which is the graph of $\vec{r}(t)$. Assume that $\vec{r}\,'(b)$ exists, then $\vec{r}\,'(b)$ is a vector tangent to this curve at the point located by the vector $\vec{r}(b)$.

Proof. The vector $\vec{r}(b+h) - \vec{r}(b)$ is the vector whose initial point is the point located by the vector $\vec{r}(b)$ and whose terminal point is the point located by the vector $\vec{r}(b+h)$. For ease in graphing assume that $h > 0$.

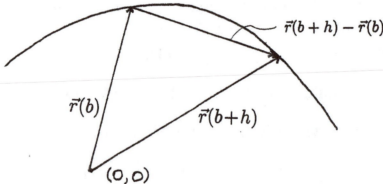

The vector $\vec{r}(b+h) - \vec{r}(b)$ is a secant line or chord for the curve $\vec{r}(t)$. The number h is a scalar and so $1/h$ is a scalar. The vector $\frac{1}{h}[\vec{r}(b+h) - \vec{r}(b)]$ is a vector with initial point the point located by $\vec{r}(b)$. It is parallel to the vector (chord) $\vec{r}(b+h) - \vec{r}(b)$. Suppose $\vec{r}\,'(b)$ exists. It follows that

$$\lim_{h \to 0+} \frac{\vec{r}(b+h) - \vec{r}(b)}{h}$$

exists and is a vector with initial point the point located by the vector $\vec{r}(b)$. We see from the graph that this vector is tangent to the curve $\vec{r}(t)$ at the point $\vec{r}(b)$. If $\vec{r}\,'(b)$ exists, then it follows that

$$\lim_{h \to 0-} \frac{\vec{r}(b+h) - \vec{r}(b)}{h} = \lim_{h \to 0+} \frac{\vec{r}(b+h) - \vec{r}(b)}{h}$$

Hence, $\vec{r}\,'(b) = \lim\limits_{h \to 0} \dfrac{\vec{r}(b+h) - \vec{r}(b)}{h}$ is a tangent vector to curve at point located by $\vec{r}(b)$.

194

Example 2. Sketch a graph of the curve which is the graph of the vector function

$$\vec{r}(t) = (2t + 1)\vec{i} + (t^2 - 4)\vec{j} \text{ for } -3 \le t \le 3,$$

and then sketch the tangent vectors $\vec{r}\,'(-2)$, $\vec{r}\,'(0)$, $\vec{r}\,'(1)$, and $\vec{r}\,'(2)$ on the same graph.

Solution. First, we make a chart of values and use these values to plot some points on the curve. We then connect these points using a smooth curve. The result is the graph of the function $\vec{r}(t)$.

t	x	y
-3	-5	5
-2	-3	0
-1	-1	-3
0	1	-4
1	3	-3
2	5	0
3	7	5

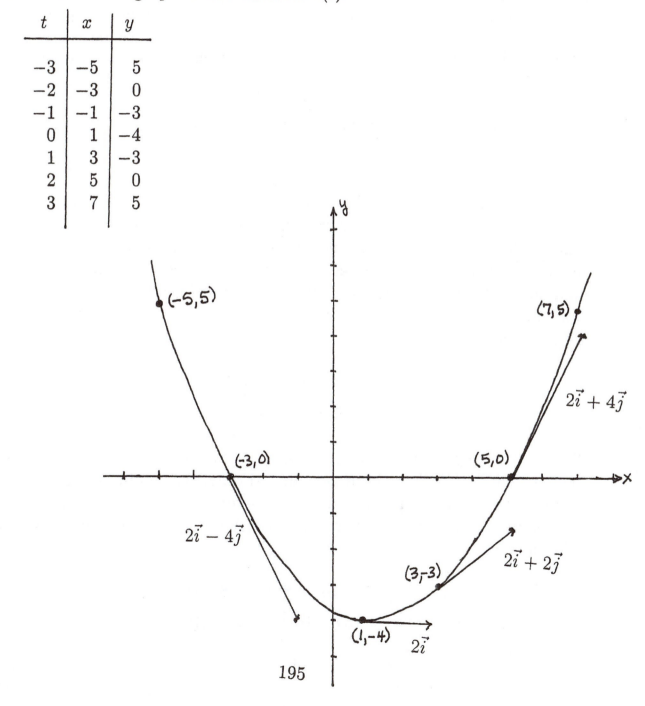

195

Next we find the tangent vectors. Differentiating $\vec{r}(t)$, we get

$$\vec{r}\,'(t) = 2\vec{i} + (2t)\vec{j}.$$

Let us find the tangent vectors to the curve $\vec{r}(t)$ at the points corresponding to $t = -2, 0, 1$, and 2. Note that $\vec{r}(-2) = -3\vec{i}$, $\vec{r}(0) = \vec{i} - 4\vec{j}$, $\vec{r}(1) = 3\vec{i} - 3\vec{j}$, and $\vec{r}(2) = 5\vec{i}$. These vectors locate the points $(-3, 0)$, $(1, -4)$, $(3, -3)$, and $(5, 0)$. Substituting the values for t we find the tangent vectors $\vec{r}\,'(-2) = 2\vec{i} - 4\vec{j}$, $\vec{r}\,'(0) = 2\vec{i}$, $\vec{r}\,'(1) = 2\vec{i} + 2\vec{j}$ and $\vec{r}\,'(2) = 2\vec{i} + 4\vec{j}$. We then plot these vectors on the above graph. From the graph we see that while all these vectors are of different lengths they are all tangent to the curve.

Example 3. Find the vector function whose graph is the tangent line to the graph of the vector function

$$\vec{r}(t) = t(t^2 - 9)\vec{i} + t(t + 2)\vec{j}$$

at the point where $t = 2$.

Solution. Since $\vec{r}(2) = 2(4 - 9)\vec{i} + 2(2 + 2)\vec{j} = -10\vec{i} + 8\vec{j}$, the point on the curve corresponding to $t = 2$ has coordinates $(-10, 8)$. This point is also a point on the tangent line. A tangent line is a line. In order to find a vector function $\vec{R}(s)$ whose graph is the tangent line we need a point on the line and a direction vector. We need a direction vector for the tangent line. Any vector tangent to the curve $\vec{r}(t)$ at this point would be a direction vector for the tangent line.

$$\vec{r}\,'(t) = (3t^2 - 9)\vec{i} + (2t + 2)\vec{j}$$
$$\vec{r}\,'(2) = 3\vec{i} + 6\vec{j}.$$

The vector $3\vec{i} + 6\vec{j}$ is a tangent vector to the curve $\vec{r}(t)$ at the point $(-10, 8)$. The tangent line to the curve is a line through the point $(-10, 8)$ with direction vector the same as a tangent vector. A direction vector of the tangent line is the tangent vector $3\vec{i} + 6\vec{j}$. A vector function whose graph is the tangent line is

$$\vec{R}(s) = (-10\vec{i} + 8\vec{j}) + s(3\vec{i} + 6\vec{j})$$
$$= (3s - 10)\vec{i} + (6s + 8)\vec{j}.$$

Note that $\vec{r}\,'(2) = 3\vec{i} + 6\vec{j}$ is not the only vector which is tangent to the curve $\vec{r}(t)$ at $(10, 8)$. Any vector which is parallel to $3\vec{i} + 6\vec{j}$ would also be a tangent vector. For example, the vectors $\vec{i} + 2\vec{j}$, $-3\vec{i} - 6\vec{j}$, and $12\vec{i} + 24\vec{j}$ are all tangent vectors. These tangent vectors either have the same direction as $\vec{r}\,'(2) = 3\vec{i} + 6\vec{j}$ or they are in the opposite direction. The vector $\vec{r}\,'(2) = 3\vec{i} + 6\vec{j}$ is called the principal tangent vector. The principal tangent vector given by the derivative $\vec{r}\,'(t)$ always points along the curve in the direction of increasing values of t. When asked to find a tangent vector we are expected to find the principal tangent vector $f'(t)\vec{i} + g'(t)\vec{j}$.

There are two vectors which are perpendicular to the tangent vector and which are easy to find. These vectors are $g'(t)\vec{i} - f'(t)\vec{j}$ and $-g'(t)\vec{i} + f'(t)\vec{j}$. These are called normal vectors.

Definition. If $\vec{r}(t) = f(t)\vec{i} + g(t)\vec{j}$, then

$$\int \vec{r}(t)dt = \left[\int f(t)dt\right]\vec{i} + \left[\int g(t)dt\right]\vec{j}$$

and

$$\int_a^b \vec{r}(t)dt = \left[\int_a^b f(t)dt\right]\vec{i} + \left[\int_a^b g(t)dt\right]\vec{j}$$

Example 4. Let $\vec{r}(t) = [4t^3 - 10t]\vec{i} + [3t^2 + 4t]\vec{j}$ then

$$\int \vec{r}(t)dt = [t^4 - 5t^2]\vec{i} + [t^3 + 2t^2]\vec{j} + a\vec{i} + b\vec{j}$$

where $a\vec{i} + b\vec{j}$ is a constant vector.

$$\int_0^5 \vec{r}(t)dt = [5^4 - 5(5)^2]\vec{i} + [5^3 + 2(5)^2]\vec{j}$$

$$= 500\vec{i} + 175\vec{j}.$$

The arc length problem. Let us consider the problem of finding the length of the curve which is the graph of the vector function $\vec{r}(t)$, $a \le t \le b$.

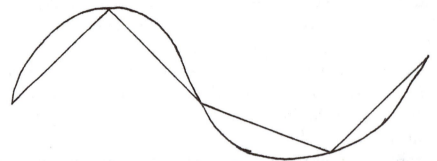

Draw a separate t number line. Divide the interval $a \le t \le b$ into n subintervals of length $\Delta t = h = \frac{b-a}{n}$. Let $t_{k+1} = t_k + h$, $a = t_0$ and $t_n = b$, then the division points on the t number line are $a = t_0, t_1, t_2, \cdots t_n = b$. That number line is divided into n intervals of equal length by these points. Plot the points $\vec{r}(t_0), \vec{r}(t_1), \vec{r}(t_2), \cdots, \vec{r}(t_n)$ on the curve. Let $\vec{r}(c)$ and $\vec{r}(c + h)$ denote two successive points on the curve. How long is the chord (vector) from $\vec{r}(c)$ to $\vec{r}(c+h)$. This is the vector $\vec{r}(c+h) - \vec{r}(c)$. The length of this vector is

$$||\vec{r}(c+h) - \vec{r}(c)|| = ||[f(c+h) - f(c)]\vec{i} + [g(c+h) - g(c)]\vec{j}||$$

$$= \sqrt{[f(c+h) - f(c)]^2 + [g(c+h) - g(c)]^2}$$

$$= h\sqrt{\left[\frac{f(c+h) - f(c)}{h}\right]^2 + \left[\frac{g(c+h) - g(c)}{h}\right]^2}$$

The sum of the lengths of all the chords is

$$\sum_{k=1}^{n} h\sqrt{\left[\frac{f(c+h) - f(c)}{h}\right]^2 + \left[\frac{g(c+h) - g(c)}{h}\right]^2}$$

The sum of the lengths of all the chords is approximately equal to the length of the curve and is actually equal to the length of the curve in the limit as $h \to 0$. Note that $\lim_{h \to 0}$ is the same as $\lim_{n \to \infty}$. Note that

$$\lim_{h \to 0} \frac{f(c+h) - f(c)}{h} = f'(c).$$

Also recall that the limit as $n \to \infty$ of a sum like the above sum is a definite integral. Taking the limit of the above sum as $n \to \infty$ ($h \to 0$) it follows that the length of the curve $\vec{r}(t)$ for $a \le t \le b$ is given by

$$\int_a^b \sqrt{[f'(t)]^2 + [g'(t)]^2}\, dt = \int_a^b \|\vec{r}\,'(t)\|\, dt.$$

Example 5. Find the arc length of the curve $\vec{r}(t) = (12t - t^3)\vec{i} + (6t^2)\vec{j}$ for $0 \le t \le 3$.

Solution. We have $\vec{r}\,'(t) = (12 - 3t^2)\vec{i} + (12t)\vec{j}$.

$$\|\vec{r}\,'(t)\| = \sqrt{(12 - 3t^2)^2 + (12t)^2}$$
$$= \sqrt{144 + 72t^2 + 9t^4} = 12 + 3t^2$$

$$\int_0^3 [12 + 3t^2]\, dt = 12t + t^3 \Big|_0^3 = 63.$$

If the curve $\vec{r}(t)$ for $a \le t \le b$ is on the positive side of the x axis, that is, $\vec{r}(t) \ge 0$, then a surface is generated when this curve is revolved about the x axis. An interesting problem is to find the surface area of such a surface. The formula for the area of such a surface is

$$2\pi \int_a^b y(t)\|\vec{r}(t)\|\, dt.$$

However, we are not going to discuss this idea here.

Consider the curve which is the graph of the vector function $\vec{r}(t)$. We may ask the question: how curvy is this curve? Suppose this curve is a highway and you are driving along the highway at a constant speed, would you slide off the highway because the curves are too sharp? It turns out that we have a measure of the sharpness of a curve.

Definition. Suppose a curve C is the graph of the vector function $\vec{r}(t) = f(t)\vec{i} + g(t)\vec{j}$ for $a \le t \le b$, and that $f''(t)$ and $g''(t)$ exist and are continuous for $a \le x \le b$. The curvature $k(t)$ for this curve for any value of t is given by

$$k(t) = \frac{|f'(t)g''(t) - g'(t)f''(t)|}{[(f'(t))^2 + (g'(t))^2]^{3/2}}$$

199

Note that $k(t) \geq 0$. The curvature of a curve at a point is a measure of how much the curve changes direction at that point. The idea of curvature is based on the fact that a circle has constant curvature.

Example 6. Let $\vec{r}(t) = (b\cos t)\vec{i} + (b\sin t)\vec{j}$ for $0 \leq t \leq 2\pi$. The graph of the vector function $\vec{r}(t)$ is a circle of radius b. Find the curvature of this circle for any value of t.

Solution. We have

$$\vec{r}\,'(t) = f'(t)\vec{i} + g'(t)\vec{j} = (-b\sin t)\vec{i} + (b\cos t)\vec{j}$$
$$\vec{r}\,''(t) = f''(t)\vec{i} + g''(t)\vec{j} = (-b\cos t)\vec{i} + (-b\sin t)\vec{j}.$$

The curvature is

$$k(t) = \frac{|(-b\sin t)(-b\sin t) - (-b\cos t)(b\cos t)|}{[(-b\sin t)^2 + (-b\cos t)^2]^{3/2}}$$
$$= \frac{b^2\sin^2 t + b^2\cos^2 t}{[b^2\sin^2 t + b^2\cos^2 t]^{3/2}} = \frac{b^2}{b^3} = \frac{1}{b}.$$

The curvature of a circle of radius b at any point is equal to $1/b$. The curvature of a circle is constant. A circle with a small radius has a large curvature while a circle with a large radius has a small curvature.

Example 7. Let $\vec{r}(t) = (5t + 8)\vec{i} + (7t - 13)\vec{j}$. The graph of the function $\vec{r}(t)$ is a line. Find the curvature of this line for any value of t.

Solution. We have
$$\vec{r}\,'(t) = 5\vec{i} + 7\vec{j}$$
$$\vec{r}\,''(t) = 0\vec{i} + 0\vec{j}.$$

The curvature is given by

$$k(t) = \frac{|(5)(0) - (0)(7)|}{[5^2 + 7^2]} = 0.$$

The curvature of a line is zero. A straight line has zero curvature.

When discussing the definite integral $\int_a^b f(x)dx$ we spent a lot of time finding the area of the region bounded by the curve $y = f(x)$, the line $x = a$,

200

the line $x = b$, and the x axis. The situation is more complicated when discussing area enclosed by the graph of $\vec{r}(t)$. The curve $y = f(x)$ intersects the line $x = a$ exactly one time. The curve $\vec{r}(t)$ may intersect the line $x = a$ any number of times. This makes it more difficult to exactly describe the region we are discussing. We will not discuss how to find the area of a region partially bounded by the curve which is the graph of $\vec{r}(t)$. This is usually discussed in a course in multivariable calculus.

Exercises

1. If $\vec{r}(t) = [4t^3 + 8t]\vec{i} + [3t(t+4)]\vec{j}$, find $\vec{r}\,'(t)$, $\int \vec{r}(t)dt$, and $\int_0^2 \vec{r}(t)dt$.

2. Sketch the graph of $\vec{r}(t) = [t(t^2 - 9)]\vec{i} + [9 - t^2]\vec{j}$ for $-3 \leq t \leq 3$. On the same graph sketch the tangent vectors $\vec{r}\,'(-2)$, $\vec{r}\,'(0)$, $\vec{r}\,'(1)$, and $\vec{r}\,'(2)$.

3. Consider the curve $\vec{r}(t) = (2t^3 + t)\vec{i} + (4t^2 + t)\vec{j}$. Find the point on this curve corresponding to $t = 2$. Find the vector function whose graph is the tangent line to this curve at the point where $t = 2$.

4. Consider the curve which is the graph of the vector function $\vec{r}(t) = (t^2 - 4t)\vec{i} + (2t^3 - 8)\vec{j}$. Find the vector function whose graph is the tangent line to this curve at the point $(-4, 8)$.

5. Consider the curve which is the graph of the vector function

$$\vec{r}(t) = (t^3 - t^2 + 4)\vec{i} + (-t^2 + 3t + 3)\vec{j}.$$

Find the vector function whose graph is the tangent line to this curve at the point corresponding to $t = 2$.

6. Find the arc length of the curve $\vec{r}(t) = (t^2)\vec{i} + (t^3)\vec{j}$ for $1 \leq t \leq 4$.

7. A particle is moving in space such that its position at time t is given by

$$\vec{r}(t) = [(1/2)t^4 - t^2]\vec{i} + [(4/3)t^3]\vec{j} \text{ for } 1 \leq t \leq 3.$$

Find the total distance traveled by the particle along this curved path during the time interval $1 \leq t \leq 3$.

8. Find the curvature of the curve which is the graph of the vector function $\vec{r}(t) = (t^3 - 3t^2)\vec{i} + (t^2 + 1)\vec{j}$ at the point where $t = 2$.

201

An Optional Remark

Many problems about polar coordinates can be resolved by reference to the same problem using vector functions. Suppose we wish to graph the polar equation $r = g(\theta)$. The formulas for polar coordinates are $x = r\cos\theta$ and $y = r\sin\theta$. When $r = g(\theta)$, this is $x = g(\theta)\cos\theta$ and $y = g(\theta)\sin\theta$. Suppose a certain curve is obtained by graphing $r = g(\theta)$, $\alpha \le \theta \le \beta$, using polar coordinates. The same curve is obtained by graphing

$$\vec{r}(\theta) = [g(\theta)\cos\theta]\vec{i} + [g(\theta)\sin\theta]\vec{j} \text{ for } \alpha \le \theta \le \beta.$$

Graphing the polar equation $r = 2\cos\theta$, $0 \le \theta \le \pi$, is the same as graphing the vector function $[(2\cos\theta)\cos\theta]\vec{i} + [(2\cos\theta)\sin\theta]\vec{j}$, $0 \le \theta \le \pi$. In either case we get the circle with center (1,0) and radius 1.

Suppose we want to find the arc length of $\vec{r}(\theta) = [g(\theta)\cos\theta]\vec{i} + [g(\theta)\sin\theta]\vec{j}$, $\alpha \le \theta \le \beta$.

$$\vec{r}\,'(\theta) = [-g(\theta)\sin\theta + g'(\theta)\cos\theta]\vec{i} + [g(\theta)\cos\theta + g'(\theta)\sin\theta]\vec{j}.$$

After a little work, we get

$$||\vec{r}\,'(\theta)||^2 = [g(\theta)]^2 + [g'(\theta)]^2.$$

Therefore, the arc length for $r = g(\theta)$, $\alpha \le \theta \le \beta$, is

$$\int_{\alpha}^{\beta} \sqrt{[g(\theta)]^2 + [g'(\theta)]^2}\, d\theta.$$

6761 Velocity and Acceleration

Suppose that the position of an object moving in the coordinate plane is given at time t by $\vec{r}(t)$. For example, if $\vec{r}(t) = (t^3 + 2)\vec{i} + (2t^2 + 5t)\vec{j}$ then when $t = 1$ the object is at $\vec{r}(1) = 3\vec{i} + 7\vec{j}$ or the object is located at (3,7). When $t = 2$ the object is at $\vec{r}(2) = 10\vec{i} + 18\vec{j}$ or the object is located at (10,18). In general if $h > 0$, then $\vec{r}(b+h) - \vec{r}(b)$ is the vector from the position of the object at time $t = b$ to the position of the object at time $t = b + h$. The vector $\vec{r}(b+h) - \vec{r}(b)$ gives the change in position. Change in position divided by change in time is average velocity.

$$\text{Average velocity} = \frac{\vec{r}(b+h) - \vec{r}(b)}{h}.$$

The instantaneous velocity is obtained by taking the limit as $h \to 0$.

Theorem 1. If the position of an object at time t is given by $\vec{r}(t)$, then the instantaneous velocity of the object is given by the derivative $\vec{r}\,'(t)$. Speed $= \|\vec{r}\,'(t)\|$.

Example 1. A particle is moving in the coordinate plane such that its position at time t is given by $\vec{r}(t) = (2t^2 + 5)\vec{i} + (t^2 - 3t)\vec{j}$. Find the velocity of this particle at $t = 1$, $t = 2$, and $t = 4$.

Solution. We have $\vec{r}\,'(t) = (4t)\vec{i} + (2t - 3)\vec{j}$.

$$\vec{r}\,'(1) = 4\vec{i} - \vec{j} \quad \vec{r}\,'(2) = 8\vec{i} + \vec{j} \quad \vec{r}\,'(4) = 16\vec{i} + 5\vec{j}$$

If we graph the vector function $\vec{r}(t)$ we obtain the path along which the particle is moving in the plane. Recall that the vector $\vec{r}\,'(b)$ is a tangent vector to this graph at the point located by $\vec{r}(b)$. Thus we see that the vector representing the velocity of a particle is tangent to the path of the particle.

Theorem 2. If a particle is moving in the plane such that its position at time t is given by $\vec{r}(t)$, then the acceleration of the particle at time t is $\vec{r}\,''(t)$.

Example 2. An object is moving in the plane such that its position at time t is given by

$$\vec{r}(t) = (t^4)\vec{i} + (t^3 + 4t)\vec{j}.$$

Find the acceleration of this object at time $t = 2$. At time $t = 2$ find the vector component of acceleration in the direction of velocity and the vector component of acceleration which is perpendicular to velocity.

Solution. The velocity at any time t is given by

$$\vec{r}\,'(t) = (4t^3)\vec{i} + (3t^2 + 4)\vec{j}.$$

The acceleration at any time t is given by

$$\vec{r}\,''(t) = (12t^2)\vec{i} + (6t)\vec{j}.$$

At time $t = 2$ the velocity and acceleration are

$$\vec{v} = \vec{r}\,'(2) = 32\vec{i} + 16\vec{j}$$
$$\vec{a} = \vec{r}\,''(2) = 48\vec{i} + 12\vec{j}.$$

The vector component of acceleration in the direction of velocity is given by

$$\text{Proj}_{\vec{v}}\vec{a} = \frac{(48\vec{i} + 12\vec{j}) \cdot (32\vec{i} + 16\vec{j})}{32^2 + 16^2}(32\vec{i} + 16\vec{j})$$
$$= \frac{108}{5}(2\vec{i} + \vec{j}) = \frac{216}{5}\vec{i} + \frac{108}{5}\vec{j}.$$

Note that this vector is parallel to velocity $\vec{v} = 16(2\vec{i} + \vec{j})$. The vector component of acceleration in the direction perpendicular to velocity is given by

$$\vec{a} - \text{Proj}_{\vec{v}}\vec{a} = 48\vec{i} + 12\vec{j} - \left[\frac{216}{5}\vec{i} + \frac{108}{5}\vec{j}\right]$$
$$= \frac{24}{5}\vec{i} - \frac{48}{5}\vec{j}.$$

Note that $\frac{108}{5}(2\vec{i}+\vec{j}) + \frac{24}{5}(\vec{i}-2\vec{j}) = 48\vec{i}+12\vec{j}$ and $\left[\frac{108}{5}(2\vec{i} + \vec{j})\right] \cdot \left[\frac{24}{5}(\vec{i} - 2\vec{j})\right] = 0$. The vector component of acceleration which is parallel to velocity, $\frac{108}{5}(2i + j)$ is causing the speed of the object to increase. The vector component of acceleration which is perpendicular to velocity, $\frac{24}{5}(\vec{i} - 2\vec{j})$, is causing the direction of motion of the object to change. From this we can easily compute the force required to keep the object on the path.

Example 3. Suppose a particle is moving around a circle of radius 4 such that its position at time t is given by $\vec{r}(t) = 4(\cos t)\vec{i} + 4(\sin t)\vec{j}$. Find the velocity and acceleration at $t = \pi/4$ and $t = \pi/2$.

Solution. The derivative is $\vec{r}\,'(t) = (-4\sin t)\vec{i} + (4\cos t)\vec{j}$ so that the velocities are $\vec{r}\,'(\pi/4) = (-2\sqrt{2})\vec{i} + (2\sqrt{2})\vec{j}$ and $\vec{r}\,'(\pi/2) = -4\vec{i}$. Note that speed $= \|\vec{r}\,'(t)\|$ is always 4. We have $\vec{r}\,''(t) = (-4\cos t)\vec{i} + (-4\sin t)\vec{j} = -\vec{r}(t)$. The accelerations are $\vec{r}\,''(\pi/4) = (-2\sqrt{2})\vec{i} + (-2\sqrt{2})\vec{j}$ and $\vec{r}\,''(\pi/2) = -4\vec{j}$. Let us find the vector component of acceleration in the direction of velocity. Note that $\vec{a} \cdot \vec{v} = 16[(-\cos t)\vec{i} + (-\sin t)\vec{j}] \cdot [(-\sin t)\vec{i} + (\cos t)\vec{j}] = 0$. These vectors are perpendicular. This means $\text{Proj}_{\vec{v}}\vec{a} = 0\vec{i} + 0\vec{j}$. There is no vector component of acceleration in the direction of velocity. The acceleration vector always points toward the origin.

Example 4. A particle is moving in the plane such that its velocity vector is given by $\vec{r}\,'(t) = (6t^2+5)\vec{i} + (4t+3)\vec{j}$. It's initial position is $\vec{r}(0) = 7\vec{i} + 11\vec{j}$. Find $\vec{r}(t)$ the position of the particle at time t.

Solution. Since $\vec{r}\,'(t)$ is the velocity the position is given by

$$\vec{r}(t) = \int \vec{r}\,'(t)dt = \int [(6t^2 + 5)\vec{i} + (4t + 3)\vec{j}]dt$$
$$= (2t^3 + 5t)\vec{i} + (2t^2 + 3t)\vec{j} + a\vec{i} + b\vec{j}.$$

We must choose the constant vector $a\vec{i} + b\vec{j}$ such that $\vec{r}(0) = 7\vec{i} + 11\vec{j}$. This means $\vec{r}(0) = a\vec{i} + b\vec{j} = 7\vec{i} + 11\vec{j}$. Therefore,

$$\vec{r}(t) = (2t^3 + 5t)\vec{i} + (2t^2 + 3t)\vec{j} + 7\vec{i} + 11\vec{j}$$
$$= (2t^3 + 5t + 7)\vec{i} + (2t^2 + 3t + 11)\vec{j}$$

Example 5. A cannon located at $(0, 0)$ fires a projectile with muzzle velocity of 480 ft/sec with an angle of elevation of $\pi/6$. Find the path $\vec{r}(t)$ of this projectile.

Solution. We assume that the path of the projectile lies in a plane, that the ground is level, the x axis is horizontal, and the y axis is vertical.

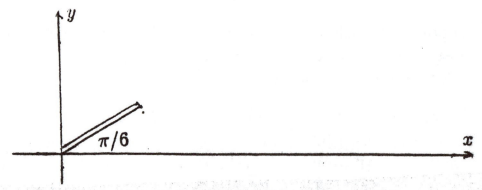

We assume that the only force acting on the projectile is gravity, that is,

$$\text{Force} = -32m\vec{j}.$$

Newton's law says that Force $= m\vec{a}$, where $\vec{a}$ is acceleration. Therefore, $m\vec{a} = -32m\vec{j}$, or

$$\vec{a} = -32\vec{j}.$$

Taking the indefinite integral of both sides we get

$$\vec{v}(t) = -32t\vec{j} + a\vec{i} + b\vec{j}$$

where $\vec{v}(t) = $ velocity and $a\vec{i} + b\vec{j}$ is a constant vector. The muzzle speed is 480 ft/sec. The elevation of the muzzle is 30°. This means that the initial velocity is

$$\vec{v}(0) = (480)\left[\frac{\sqrt{3}}{2}\vec{i} + \frac{1}{2}\vec{j}\right] = (240\sqrt{3})\vec{i} + (240)\vec{j}.$$

Substituting $t = 0$ into $\vec{v}(t) = (-32t)\vec{j} + a\vec{i} + b\vec{j}$, we get $a\vec{i} + b\vec{j} = (240\sqrt{3})\vec{i} + (240)\vec{j}$. Therefore,

$$\vec{v}(t) = (-32t)\vec{j} + (240\sqrt{3})\vec{i} + (240)\vec{j}$$
$$= (240\sqrt{3})\vec{i} + (-32t + 240)\vec{j}.$$

Taking the indefinite integral of both sides we get the position vector

$$\vec{r}(t) = 240\sqrt{3})(t)\vec{i} + (-16t^2 + 240t)\vec{j} + a\vec{i} + b\vec{j}.$$

The projectile starts at $(0,0)$ so that $\vec{r}(0) = \vec{0}$. This means $a\vec{i} + b\vec{j} = 0$. Therefore,

$$\vec{r}(t) = (240\sqrt{3})(t)\vec{i} + (-16t^2 + 240t)\vec{j}.$$

Question. When does the projectile hit the ground? The projectile hits the ground when the component of position in the $\vec{j}$ direction is zero, that is, when

$$-16t^2 + 240t = 0$$
$$-16t(t - 15) = 0$$

The projectile starts at ground level when $t = 0$ and hits the ground when $t = 15$.

How far does the projectile travel during this 15 seconds?

$$\vec{r}(15) = (240\sqrt{3}(15)\vec{i} = 6235\vec{i}.$$

The projectile travels 6235 feet. We usually answer this question by giving the length of the line segment from the starting position to the finishing position. That is how we answered the question here. We could also answer the question by finding the arc length of the path of the projectile. That distance is given by

$$\int_0^{15} \|\vec{r}\,'(t)\| dt = \int_0^{15} \sqrt{(240\sqrt{3})^2 + (-32t + 240)^2}\, dt.$$

Example 6. An object of mass 2 kg is moving in the plane such that the total force on the object is given by $\vec{F}(t) = (6t - 2)\vec{i} + 12t^2\vec{j}$. The initial position of the object is $(0,0)$ and the initial velocity is $5\vec{i} + 2\vec{j}$. Find the position of the object at any time t.

Solution. Let $\vec{r}(t)$ denote the position of the object at time t. Recall that force equals mass times acceleration. Since acceleration equals $\vec{r}\,''(t)$ this says that $\vec{F}(t) = 2\vec{r}\,''(t)$.

$$\vec{r}\,''(t) = (1/2)[(6t - 2)\vec{i} + 12t^2\vec{j}] = (3t - 1)\vec{i} + 6t^2\vec{j}.$$

The indefinite integral is

$$\vec{r}\,'(t) = [(3/2)t^2 - t]\vec{i} + 2t^3\vec{j} + a\vec{i} + b\vec{j},$$

where $a\vec{i} + b\vec{j}$ is the constant of integration. We are given that $\vec{r}\,'(0) = a\vec{i} + b\vec{j} = 5\vec{i} + 2\vec{j}$. Thus,

$$\vec{r}\,'(t) = [(3/2)t^2 - t]\vec{i} + 2t^3\vec{j} + 5\vec{i} + 2\vec{j}.$$

Integrating

$$\vec{r}(t) = [(1/2)t^3 - (1/2)t^2]\vec{i} + (1/2)t^4\vec{j} + 5t\vec{i} + 2t\vec{j} + c\vec{i} + d\vec{j}.$$

Using $\vec{r}(0) = 0\vec{i} + 0\vec{j}$, we get $c\vec{i} + d\vec{j} = 0\vec{i} + 0\vec{j}$. The position of the object at any time t is given by

$$\vec{r}(t) = (1/2)[t^3 - t^2]\vec{i} + (1/2)t^4\vec{j} + 5t\vec{i} + 2t\vec{j}$$
$$= (1/2)[t^3 - t^2 + 10t]\vec{i} + (1/2)[t^4 + 4t]\vec{j}.$$

Exercises

1. If the position of a particle in the coordinate plane is given by

$$\vec{r}(t) = (t^4 + 8t^2)\vec{i} + (3t^3 + 7t)\vec{j}$$

a) Find the velocity and acceleration of this particle at time $t = 2$.

b) At time $t = 2$ find the vector component of acceleration in the direction of velocity and the vector component of acceleration which is perpendicular to velocity.

2. A particle is moving in the plane such that its velocity vector is given by

$$\vec{r}\,'(t) = (8t^3 + 7)\vec{i} + (9t^2 + 4t)\vec{j}.$$

It's initial position is $\vec{r}(0) = 5i - 3j$. Find the expression $\vec{r}(t)$ for its position at any time t.

3. A cannon located at $(0,0)$ fires a projectile with muzzle velocity of 384 ft/sec with an angle of elevation of $\pi/6$. Find the path $\vec{r}(t)$ of this projectile.

4. An object of mass 3kg is moving in the plane such that the total force on the object is given by $\vec{F}(t) = (72t^2)\vec{i} + (36t - 9)\vec{j}$. The initial position of the object is $(0,0)$ and the initial velocity is $8\vec{i} + 3\vec{j}$. Find the position of the object at any time t.

6701 Limits of Sequences

A sequence is an ordered list of numbers that goes on forever. There are two common ways to denote a sequence. The first way is illustrated by the following example:

$$\left\{ \frac{1}{2}, \frac{2}{3}, \frac{3}{4}, \frac{4}{5}, \frac{5}{6}, \ldots \right\}.$$

This is an ordered list and is enclosed by the brackets { } to make it clear that the list is a sequence. The third term in the sequence is 3/4, the fifth term is 5/6. The three dots after the 5/6 indicate "and continuing on in this pattern". It is assumed that the reader can see the pattern. This means that the reader could add more terms to the list if he wanted to. Some more terms in this sequence are:

$$\left\{ \frac{1}{2}, \frac{2}{3}, \frac{3}{4}, \frac{4}{5}, \frac{5}{6}, \frac{6}{7}, \frac{7}{8}, \frac{8}{9}, \ldots \right\}.$$

The second method of indicating a sequence is to give a formula for the terms of the sequence. For example,

$$a_n = \frac{n+2}{2n+1}$$

is such a formula. Using this formula, we can find any term (number) in this sequence by substituting into the formula. The sixth term in this sequence is found by replacing n by 6 as follows:

$$a_6 = \frac{6+2}{2(6)+1} = \frac{8}{13}.$$

The tenth term is obtained by replacing n by 10 in the formula,

$$a_{10} = \frac{10+2}{2(10)+1} = \frac{12}{21}.$$

Given a sequence represented in one of these two forms we should be able to represent the sequence in the other form. The first sequence is given as a list. How would we represent the first sequence using the other method, the formula? We find the formula for the nth term after a careful inspection of the given terms. The formula for the nth term of the first sequence is

$$a_n = \frac{n}{n+1}.$$

Note that this formula gives $a_5 = 5/6$ and $a_8 = 8/9$. Since we have the formula for the second sequence we can easily make a list to represent it. We just substitute $n = 1, 2, 3, \ldots$ into the formula. A few terms for the second sequence are:

$$\left\{ \frac{3}{3}, \frac{4}{5}, \frac{5}{7}, \frac{6}{9}, \frac{7}{11}, \frac{8}{13}, \ldots \right\}.$$

Once we have learned the notation for sequences we would like to determine which sequences have a limit. For the sequence $\{a_n\}$ we want to find

$$\lim_{n \to \infty} a_n.$$

What does this mean? As n get larger and larger do the terms in the sequence a_n get closer and closer to some fixed number L? If the answer is yes, then L is called the limit of the sequence and we write

$$\lim_{n \to \infty} a_n = L.$$

Example 1. Find $\lim_{n \to \infty} \dfrac{2n + 7}{5n + 3}$.

Solution. First, note that

$$\frac{2n + 7}{5n + 3} = \frac{(2n + 7)(1/n)}{(5n + 3)(1/n)} = \frac{2 + 7/n}{5 + 3/n}.$$

As we substitute larger and larger integers for n into the fraction, the expressions $7/n$ and $3/n$ get closer and closer to zero. When $n = 1000$ then $7/n = .007$, but when $n = 1,000,000$ then $7/n = .000007$. The bigger the value of n the closer $7/n$ gets to 0. As n gets larger and larger, the expression

$$\frac{2 + 7/n}{5 + 3/n} \text{ gets closer and closer to } \frac{2 + 0}{5 + 0} = \frac{2}{5}.$$

Therefore, we say $\lim\limits_{n\to\infty} \dfrac{2n+7}{5n+3} = \dfrac{2}{5}$.

Example 2. Let $a_n = 2n+3$, find $\lim\limits_{n\to\infty} a_n$.

Solution. It is clear that as we substitute in larger and larger values for n into $2n+3$ the value of $2n+3$ gets larger and larger. The values of $2n+3$ do not get close to any particular number. The limit $\lim\limits_{n\to\infty} (2n+3)$ does not exist. This sequence does not have a limit. We sometimes write $\lim_{n\to\infty}(2n+3) = +\infty$, but this does not mean that $+\infty$ or ∞ is a super large number. This is just a short hand way of indicating to another person the way in which the limit failed to exist. We still should write "the limit does not exist".

Example 3. Find $\lim\limits_{n\to\infty} (-1)^n \dfrac{3n+8}{5n+4}$.

Solution. It is easy to see that $\lim\limits_{n\to\infty} \dfrac{3n+8}{5n+4} = \dfrac{3}{5}$. But the value of $(-1)^n$ is $-1, 1, -1, 1$, etc. This means that one term of $(-1)^n \dfrac{3n+8}{5n+4}$ is close to $3/5$, but the next term is close to $(-3/5)$. Since the terms are not getting close to a single number, it follows that $\lim\limits_{n\to\infty} (-1)^n \dfrac{3n+8}{5n+4}$ does not exist.

Example 4. Let $a_n = 3 + (-1)^n$, find $\lim\limits_{n\to\infty} a_n$.

Solution. Write out a few terms of this sequence.

$$\{2, 4, 2, 4, 2, 4, 2, 4, \ldots\}.$$

The terms of this sequence do not get closer and closer to any particular number. If we think that the terms are getting close to 2, then the very next term is 4 which is not close to 2. We sometimes say that a sequence like this one oscillates. In any case it does not have a limit. The limit does not exist.

Example 5. Find $\lim_{n\to\infty} \sin(n)$.

Solution. Even though we do not have a simple way to compute $\sin(n)$ for large values of n we know that $\sin(n)$ keeps changing value as n gets larger and larger. The value of $\sin(n)$ does not keep getting closer and closer to any particular number. Therefore, $\lim_{n\to\infty} \sin(n)$ does not exist.

Example 6. Find $\lim\limits_{n\to\infty} \dfrac{5n^3 + 10n + 9}{3n^3 + 8n^2 + 4}$.

In order to get a clear understanding of the value of this fraction we multiply both numerator and denominator by $1/n^3$. This is the same as multiplying the fraction by 1. We get

$$\frac{(5n^3 + 10n + 9)\,(1/n^3)}{(3n^3 + 8n^2 + 4)\,(1/n^3)} = \frac{5 + (10/n^2) + (9/n^3)}{3 + (8/n) + (4/n^3)}.$$

As n gets larger and larger the expressions $10/n^2, 9/n^3, 8/n$ and $4/n^3$ get closer and closer to 0. It follows that

$$\lim_{n\to\infty} \frac{5 + (10/n^2) + (9/n^3)}{3 + (8/n) + (4/n^3)} = \frac{5 + 0 + 0}{3 + 0 + 0} = \frac{5}{3}.$$

We use the same idea to find the limits of any rational expression. In order to find the limit of any rational expression we multiply both the numerator and denominator by 1 over the highest power of n in the numerator. It is then easy to determine $\lim_{n\to\infty}$.

However, it is not so easy to find the limit for some more complicated sequences. We indicate the value of a few limits here for future reference. Since the proofs are a little difficult we do not give the proofs here.

Theorem. If $|r| < 1$, then $\lim\limits_{n\to\infty} r^n = 0$. Also if $|r| > 1$, then $\lim\limits_{n\to\infty} r^n$ does not exist.

This theorem says for example that $\lim\limits_{n\to\infty} (2/3)^n = 0$. It says $\lim\limits_{n\to\infty} e^{-n} = \lim\limits_{n\to\infty} (1/e^n) = 0$. It also says $\lim_{n\to\infty} (3/2)^n$ does not exist.

It is also true that $\lim_{n\to\infty} ne^{-n} = 0$ and $\lim_{n\to\infty} n^2 e^{-n} = 0$.

Theorem. It is true that

$$\lim_{n\to\infty} \ln n = +\infty.$$

212

When we write $\lim_{n \to \infty} \ln(n) = +\infty$ we mean that as n gets larger and larger, the value of $\ln(n)$ gets larger and larger. In particular, $\lim_{n \to \infty} \ln(n)$ does not exist. We write $\lim_{n \to \infty} \ln(n) = +\infty$ to give the reader some idea of the manner in which $\lim_{n \to \infty} \ln(n)$ does not exist.

Exercises

1. Evaluate the following limits:

    a) $\lim_{n \to \infty} \dfrac{3n^2 + 11}{5n^2 + 2n}$

    b) $\lim_{n \to \infty} \dfrac{8n^3 + 11n + 10}{3n^3 + 9n^2 + 5}$

    c) $\lim_{n \to \infty} \dfrac{4n^3 + 9}{8n^2 + 7n}$

    d) $\lim_{n \to \infty} \dfrac{8n^2 + 7n}{4n^3 + 9}$

    e) $\lim_{n \to \infty} \dfrac{12n^3 + 5n^2 + 8}{n^3 + 4n + 3}$

2. Evaluate the following limits:

    a) $\lim_{n \to \infty} (5/8)^{n-1}$

    b) $\lim_{n \to \infty} (4/3)^n$

    c) $\lim_{n \to \infty} [5 + 2(-1)^n]$

    d) $\lim_{n \to \infty} \ln(n + 1)$

    e) $\lim_{n \to \infty} [12(3/5)^{n-2}]$

    f) $\lim_{n \to \infty} (-1)^n \dfrac{4n + 11}{7n + 5}$

    g) $\lim_{n \to \infty} (n + 1)e^{-n}$

6703 Finite Arithmetic and Geometric Series

Definition. An arithmetic sequence $\{a_k\}$ is a sequence that has a common difference between successive terms. Two successive terms are denoted by a_k and a_{k+1}. This says that $\{a_k\}$ is an arithmetic sequence if $a_{k+1} - a_k = d$ or $a_{k+1} = a_k + d$, where d is the common difference.

Example 1. The sequence $1, 8, 15, 22, 29, \ldots$ is an arithmetic sequence with common difference 7, that is, each term is the previous term plus 7.

Example 2. The sequence $-8, -3, 2, 7, 12, 17, \ldots$ is an arithmetic sequence with common difference 5, that is, each term is the previous term plus 5. The first term of this sequence is $a_1 = -8$ and the second term is $a_2 = -3$. Note that

$$a_2 = -3 = -8 + 5$$
$$a_3 = 2 = -3 + 5 = (-8 + 5) + 5 = -8 + (2)(5)$$
$$a_4 = 7 = 2 + 5 = (-8 + (2)(5)) + 5 = -8 + (3)(5)$$
$$a_5 = 12 = 7 + 5 = (-8 + (3)(5)) + 5 = -8 + (4)(5)$$

We see a pattern here. The formula for a_k, the kth term of this arithmetic sequence is $a_k = -8 + (k-1)(5)$.

We use the same reasoning to show that given the arithmetic sequence $\{a_k\}$ the formula for the general term a_k is

$$a_k = a_1 + (k-1)d, \tag{3.1}$$

where a_1 is the first term and d is the common difference.

Example 3. Find the 50th term of the sequence $-10, -6, -2, 2, 6, 10, \ldots$. The first term is $a_1 = -10$ and the common difference is 4. Using the above formula for the general term, the formula for a_k is

$$a_k = -10 + (k-1)(4).$$

Replace k by 50 and we find that the 50th term is

$$a_{50} = -10 + (49)(4) = 186.$$

214

Summation notation

Sometimes a special notation is used to indicate the sum of a certain number of terms of a sequence. The capital Greek letter sigma, Σ, is used as a summation symbol. For example, $\sum\limits_{k=1}^{4} b_k$ means the same thing as the sum $b_1 + b_2 + b_3 + b_4$. The letter k, called the summation index, takes on all integer values from the lower limit 1 to the upper limit 4 inclusive. Some examples which illustrate the meaning of summation notation are:

$$\sum_{k=1}^{6} a_k = a_1 + a_2 + a_3 + a_4 + a_5 + a_6.$$

$$\sum_{k=1}^{5} k^2 = 1^2 + 2^2 + 3^2 + 4^2 + 5^2.$$

$$\sum_{k=1}^{50} b_k = b_1 + b_2 + b_3 + \ldots + b_{50}.$$

The sum of an arithmetic sequence is called an arithmetic series. The name "series" is used for a sum.

Example 4. Express the following arithmetic series using summation notation:

$$-11 - 5 + 1 + 7 + 13 + \cdots + 601.$$

Solution. We need a general formula for the terms a_k such that $a_1 = -11$, $a_2 = -5$, $a_3 = 1$, and so forth. The general formula for the kth term of an arithmetic sequence given by (3.1) is

$$a_k = a_1 + (k-1)d.$$

For this series $a_1 = -11$ and $d = 6$. This gives

$$a_k = -11 + (k-1)(6) = 6k - 17.$$

We express the series using summation notation as

$$\sum_{k=1}(6k - 17).$$

However, we still have a problem. We get the first term a_1 as $a_1 = 6(1) - 17 = -11$, but what number do we substitute for k in order to get the last term 601. What is the value of k such that $a_k = 601$? We need to use (3.1) which is

$$a_k = a_1 + (k-1)d$$
$$601 = -11 + (k-1)(6)$$
$$612 = (k-1)(6)$$
$$k = 103.$$

The number 601 is the 103rd term. Expressed using sum notation the series is

$$\sum_{k=1}^{103} (6k - 17).$$

Given the arithmetic sequence $\{a_k\}$ let $S_n = a_1 + a_2 + \ldots + a_n$, that is, let S_n denote the sum of the first n terms. It is fairly easy to show that the sum of a finite arithmetic series is one half the number of terms times the sum of the first term and the last term. This is stated as the following theorem.

Theorem 1. Suppose $a_1 + a_2 + \cdots + a_n$ is a finite arithmetic series. Let $S_n = a_1 + a_2 + \cdots + a_n$, then

$$S_n = \left(\frac{n}{2}\right)(a_1 + a_n). \tag{3.2}$$

Example 5. Find the sum of the following arithmetic series:

$$-8 - 3 + 2 + 7 + 12 + \ldots + 487.$$

In order to find the value of the sum we must find the number of terms. Recall (3.1) is the formula $a_k = a_1 + (k-1)d$ for the kth term. For this sequence $a_1 = -8$ and $d = 5$. Thus $a_k = -8 + (k-1)(5)$ or $a_k = -13 + 5k$. What number term is the last term 487? That is, how many terms are we adding together? Substituting $a_k = 487$ into $a_k = -13 + 5k$, we get $487 = -13 + 5k$ or $500 = 5k$ or $k = 100$, that is, $a_{100} = 487$. The number 487 is the 100th term or $n = 100$. One half of 100 is 50. Substituting into the formula (3.2) in Theorem 1, the sum is

$$S_{100} = 50(-8 + 487) = 23,950.$$

216

We have an alternate form for the formula for S_n which is often more useful. Since $a_n = a_1 + (n-1)d$ we can rewrite the sum formula for S_n as

$$S_n = \frac{n}{2}[a_1 + a_1 + (n-1)d] = \frac{n}{2}[2a_1 + (n-1)d]. \qquad (3.3)$$

Example 6. Find the sum of the arithmetic series

$$\sum_{k=1}^{25}(4k-3).$$

Solution. Note that the kth term a_k is given by $a_k = 4k - 3$. Hence, the next term $a_{k+1} = 4(k+1) - 3 = 4k + 1$. Recall that the common difference $d = a_{k+1} - a_k$. Therefore, $d = a_{k+1} - a_k = (4k+1) - (4k-3) = 4$. Each term is the previous term plus 4. This shows that the sequence $\{4k - 3\}$ is an arithmetic sequence with common difference equal to 4. We can also see this by writing out the first few terms. The first term is $a_1 = 4(1) - 3 = 1$. This means $4k - 3 = 1 + 4(k-1)$ and the finite series may be written as

$$1 + 5 + 9 + \ldots + 97 = \sum_{k=1}^{25}[1 + 4(k-1)] = \sum_{k=1}^{25}[a_1 + (k-1)d].$$

The important thing is that we have found the values of a_1 and d for this problem. We also have $n = 25$ and $a_{25} = 97$. Substituting these numbers into the formula (3.2) for the sum of an arithmetic series, we get

$$S_{25} = \frac{25}{2}[1 + 97] = 1225$$

$$\sum_{k=1}^{25}(4k-3) = 1,225.$$

Example 7. Consider the arithmetic sequence

$$-8, -2, 4, 10, 16, 22, \ldots$$

Find a formula for a_k, the kth term, and a formula for the sum S_n of the first n terms.

Solution. Clearly the common difference is $d = 6$ and the first term is $a_1 = -8$. The general formula for the kth term is $a_k = a_1 + (k-1)d$ as given by (3.1). Substituting $a_1 = -8$ and $d = 6$, we get a formula

$$a_k = -8 + (k-1)6 = 6k - 14,$$

for the kth term of this sequence. This tells us that the 20th term is $a_{20} = 6(20) - 14 = 106$. We can now restate the last part of the problem as: find the sum of the arithmetic series

$$S_n = \sum_{k=1}^{n} (6k - 14).$$

In order to find S_n, the sum of the first n terms, we use the formula (3.3) for S_n given above. It is

$$S_n = \frac{n}{2}[2a_1 + (n-1)d]$$

We substitute $a_1 = -8$ and $d = 6$ into this formula to get

$$S_n = \frac{n}{2}[2(-8) + (n-1)6] = \frac{n}{2}[6n - 22] = n(3n - 11).$$

This formula is correct for any value of n. We find the sum of the first 40 terms by letting $n = 40$.

$$S_{40} = 40(3(40) - 11) = 4360.$$

We can also use (3.2) which is $S_n = \frac{n}{2}(a_1 + a_n)$ to find the sum:

$$S_n = \frac{n}{2}[-8 + (6n - 14)] = n(3n - 11).$$

Example 8. Show that $\sum_{k=1}^{25} (4k - 3)$ is the same as $\sum_{n=4}^{28} (4n - 15)$.

Solution. Start with the sum $\sum_{n=4}^{28} (4n - 15)$ and change the summation index by letting $k = n - 3$ or $n = k + 3$. The values of n are $n = 4, 5, 6, \ldots 28$. But since $k = n - 3$ it follows that the values of k are $k = 1, 2, 3, \ldots 25$.

$$\sum_{n=4}^{28} (4n - 15) = \sum_{k=1}^{25} 4(k + 3) - 15 = \sum_{k=1}^{25} (4k - 3).$$

218

Geometric Sequences

Definition. A geometric sequence is a sequence in which we can obtain each term by multiplying the preceding term by a fixed number, r, called the common ratio of the sequence. We can find the common ratio of a geometric sequence by dividing any term by the preceding term.

Example 9. The sequence $3, 6, 12, 24, 48, \ldots$ is a geometric sequence with common ratio 2, that is, each term is the previous term multiplied by 2, that is, $r = 2$.

Example 10. The sequence $18, 6, 2, 2/3, 2/9, 2/27,$ is a geometric sequence with common ratio $1/3$, that is, each term is the previous term multiplied by $1/3$. For this sequence we see that

$$a_1 = 18$$
$$a_2 = 6 = (1/3)(18)$$
$$a_3 = 2 = (1/3)(6) = (1/3)(1/3)18 = (1/3)^2 18$$
$$a_4 = 2/3 = (1/3)(2) = (1/3)(1/3)^2 18 = (1/3)^3 18$$
$$a_5 = 2/9 = (1/3)(2/3) = (1/3)(1/3)^3 18 = (1/3)^4 18$$

We see that there is a pattern here. Without doing any computations we can see that $a_6 = (1/3)^5(18)$. This pattern shows us that the formula for a_k, the kth term of this geometric sequence, is

$$a_k = 18(1/3)^{k-1}.$$

Given a general geometric sequence $\{a_k\}$ with common ratio r, we have

$$a_2 = a_1 r$$
$$a_3 = a_2 r = (a_1 r)r = a_1 r^2$$
$$a_4 = a_3 r = (a_1 r^2)r = a_1 r^3$$
$$a_5 = a_4 r = (a_1 r^3)r = a_1 r^4.$$

Again we see a pattern. Without any more work we conclude that $a_6 = a_1 r^5$ and $a_7 = a_1 r^6$. The formula for the general term of a geometric sequence with first term a_1 and common ratio r is

$$a_k = a_1 r^{k-1}. \tag{3.4}$$

Example 11. Find the formula for the kth term of the geometric sequence

$$\frac{27}{2}, 9, 6, 4, \frac{8}{3}, \frac{16}{9}, \dots$$

Solution. Dividing any term by the preceding term we see that the common ratio is $r = 2/3$. The first term is $a_1 = (27/2)$. Using (3.4) the formula for the kth term is

$$a_k = a_1 r^{k-1} = \left(\frac{27}{2}\right)\left(\frac{2}{3}\right)^{k-1}.$$

The 10th term is (using $a_{10} = a_1 r^9$)

$$a_{10} = \frac{27}{2}\left(\frac{2}{3}\right)^9 = \frac{256}{729}.$$

Suppose $\{a_k\}$ is a geometric sequence. Let $S_n = a_1 + a_2 + \dots + a_n$, then S_n denotes the sum of the first n terms of the geometric sequence $\{a_k\}$. The sum of a finite part of a sequence like this is known as a finite series. This is a finite geometric series. It is fairly easy to prove that the formula for the sum of n terms of a geometric sequence is given as follows.

Theorem 2. The formula for the sum of a finite geometric series is

$$S_n = \sum_{k=1}^{n} a_1 r^{k-1} = a_1 \frac{1 - r^n}{1 - r}. \tag{3.5}$$

Example 12. Find the sum of the following geometric series

$$\frac{27}{2} + 9 + 6 + 4 + \dots + \frac{1024}{6561}.$$

Solution. For the geometric series in this problem, we have $a_1 = 27/2$ and $r = 2/3$. Using (3.4) the kth term of this geometric sequence is $(27/2)(2/3)^{k-1}$. We need to know the value of n that is, how many terms in this finite sum. What number term is $(1024)/(6561)$? What is the value of n such that

$$\frac{27}{2} + 9 + 6 + \dots + \frac{1024}{6561} = \sum_{k=1}^{n} \left(\frac{27}{2}\right)\left(\frac{2}{3}\right)^{k-1}?$$

What is the value of n such that

$$\frac{1024}{6561} = \frac{27}{2}\left(\frac{2}{3}\right)^{n-1}?$$

Multiplying both sides of this last equation by $2/27$, we get $\left(\frac{2}{3}\right)^{n-1} = \frac{2048}{177147}$. Using the calculator we find that $\left(\frac{2}{3}\right)^{11} = \frac{2048}{177,147}$. This shows us that $(2/3)^{n-1} = (2/3)^{11}$. This says that $n - 1 = 11$ or $n = 12$. This says that $(1024)/(6561)$ is the 12th term. Substituting $a_1 = 27/2$, $r = 2/3$, and $n = 12$ into the sum formula (3.5) for a geometric series, we get

$$S_{12} = a_1 \frac{1 - r^{12}}{1 - r} = \left(\frac{27}{2}\right)\frac{1 - \left(\frac{2}{3}\right)^{12}}{1 - (2/3)}$$

$$= \frac{81}{2}\left[1 - \left(\frac{2}{3}\right)^{12}\right] = \frac{527,345}{13,122}.$$

We can use (3.5) to find S_n for any n. The sum of any number n of terms of this geometric sequence is given by

$$S_n = \frac{27}{2}\frac{1 - (2/3)^n}{1 - 2/3} = \frac{81}{2}\left[1 - \left(\frac{2}{3}\right)^n\right].$$

Example 13. What is the sum of the finite geometric series:

$$\sum_{k=1}^{n} 4\frac{3^k}{5^{k-1}} = 12 + \frac{36}{5} + \frac{108}{25} + \ldots + \frac{4(3^n)}{5^{n-1}}?$$

Solution. First, we observe that this is a geometric series with common ratio $3/5$. In order to more easily apply the sum formula (3.5) we should write any geometric series in the standard form for a geometric series which is $\sum_{k=1}^{n} a_1 r^{k-1}$.

$$\sum_{k=1}^{n} 4\frac{3^k}{5^{k-1}} = \sum_{k=1}^{n} 4(3)\frac{3^{k-1}}{5^{k-1}} = \sum_{k=1}^{n} 12\left(\frac{3}{5}\right)^{k-1}.$$

221

For this geometric series it is now easy to see that $a_1 = 12$ and $r = 3/5$. Substituting $a_1 = 12$ and $r = 3/5$ into the formula (3.5), the sum of the first n terms of this series is given by

$$\sum_{k=1}^{n} 12 \left(\frac{3}{5}\right)^{k-1} = 12\frac{1-(3/5)^n}{1-(3/5)} = 30\left[1 - \left(\frac{3}{5}\right)^n\right].$$

When $n = 10$ this formula says that

$$12 + \frac{36}{5} + \frac{108}{25} + \ldots + 12\left(\frac{3}{5}\right)^9 = 30\left[1 - \left(\frac{3}{5}\right)^{10}\right].$$

Example 14. Find the sum of the geometric series $\sum_{k=1}^{8} 200\frac{3^{k-1}}{4^{k-2}}$.

Solution. We need to write this geometric series in standard form as given in (3.5).

$$\sum_{k=1}^{8} 200\frac{3^{k-1}}{4^{k-2}} = \sum_{k=1}^{8} 800\left(\frac{3}{4}\right)^{k-1}.$$

It is now easy to see that $n = 8$, $a_1 = 800$, and $r = 3/4$. Putting these numbers into the formula (3.5) for the sum of a geometric series, we get

$$S_8 = 800\frac{1-(3/4)^8}{1-(3/4)} = 3200[1-(3/4)^8] = 2879.64.$$

Example 15. Find the sum $\sum_{k=1}^{n} 405(2/3)^{k+1}$.

Solution. Always start any problem where we must find the sum of a geometric series by writing the geometric series in standard form $\sum_{k=1}^{n} a_1 r^{k-1}$ found in (3.5). We see that $r = 2/3$, but it does not have the correct power in order to be in standard form. We need the power of $(2/3)$ to be $k-1$. We rewrite as follows:

$$405(2/3)^{k+1} = 405(2/3)^2(2/3)^{k-1} = 180(2/3)^{k-1}.$$

We rewrite the sum in the standard form as follows:

$$\sum_{k=1}^{n} 405 \left(\frac{2}{3}\right)^{k+1} = \sum_{k=1}^{n} 180 \left(\frac{2}{3}\right)^{k-1}.$$

We easily see that $a_1 = 180$ and $r = 2/3$. Substituting these numbers into the formula (3.5) for the sum of a geometric series, we get

$$S_n = 180 \frac{1 - (2/3)^n}{1 - (2/3)} = 540[1 - (2/3)^n].$$

This formula for the sum S_n is good for any value of n.

In order to solve problems involving arithmetic and geometric series we need to know the formulas (3.1), (3.2), (3.3), (3.4), and (3.5).

Problems

1. Find the 80th term in the arithmetic sequence

$$3, 10, 17, 24, 31, 38, \ldots.$$

2. First write the following sum using sum notation and then find the sum.

$$-11 - 5 + 1 + 7 + 13 + 19 + \ldots + 589.$$

3. Find a formula for the following sums. The formula will involve n.

a) $\displaystyle\sum_{k=1}^{n}(3k + 5)$

b) $\displaystyle\sum_{k=1}^{n}(4k - 9)$

4. What number term is $2048/6561$ in the geometric sequence

$$18, 12, 8, \frac{16}{3}, \frac{32}{9}, \ldots \frac{2048}{6561}, \ldots$$

5. Write the following geometric series in standard form using sum notation and then find its sum.

$$\frac{16}{243} + \frac{8}{81} + \frac{4}{27} + \frac{2}{9} + \cdots + \frac{2187}{256}.$$

223

6. Find the sum $\displaystyle\sum_{k=1}^{9} \frac{125}{3}\left(\frac{3}{5}\right)^{k-1}$.

7. Find the sum of the first nine terms of the geometric series

$$9, 12, 16, 64/3, \ldots$$

8. Find an expression for the value of each of the following sums. The formula that we find for each sum will involve n.

a) $\displaystyle\sum_{k=1}^{n} 12(3/4)^{k-1}$

b) $\displaystyle\sum_{k=1}^{n} 4(3/5)^{k+2}$

c) $\displaystyle\sum_{k=1}^{n} (5/8)^{k+1}$

d) $\displaystyle\sum_{k=1}^{n} \frac{4^{k+1}}{5^k}$.

9. Explain why

$$\sum_{k=1}^{25}(4k + 13) \text{ is the same as } \sum_{n=4}^{28}(4n + 1).$$

$$\sum_{k=1}^{30}(5k + 12) \text{ is the same as } \sum_{n=3}^{32}(5n + 2).$$

10. Thomas places \$1,000 in a savings account. The bank pays interest at the rate of 5% per year with the interest compounded every year. How much money does Thomas have in the bank at the end of 6 years? How long would Thomas have to keep his money in this account to have \$2,078.93?

6705 Definition of Infinite Series

A series is the indicated sum of a sequence. For a finite series this sum is a finite sum. We start with a finite list of numbers:

$$a_1, a_2, a_3 \ldots a_n.$$

We add these numbers together and we get a finite sum

$$a_1 + a_2 + a_3 + \ldots + a_n.$$

For example, the following list of numbers is a sequence:

$$\frac{4}{3}, \frac{8}{9}, \frac{16}{27}, \frac{32}{81}, \ldots, \frac{2^{n+1}}{3^n}.$$

The sum of this list of numbers is the following series.

$$\frac{4}{3} + \frac{8}{9} + \frac{16}{27} + \frac{32}{81} + \ldots + \frac{2^{n+1}}{3^n}.$$

We can denote this very same finite series using summation notation as follows:

$$\sum_{k=1}^{n} \frac{2^{k+1}}{3^k}.$$

This notation indicates that in order to obtain the sum we substitute for k the numbers 1,2,3,... n into the formula $\frac{2^{k+1}}{3^k}$ and add the resulting numbers together, that is,

$$\sum_{k=1}^{n} \frac{2^{k+1}}{3^k} = \frac{4}{3} + \frac{8}{9} + \frac{16}{27} + \frac{32}{81} + \ldots + \frac{2^{n+1}}{3^n}.$$

What happens to such a sum for very large values of n? How can we define such a sum if we want to take the limit as $n \to +\infty$? This is a finite geometric series and so we know how to compute its sum. We use the formula $\sum_{k=1}^{n} a_1 r^{k-1} = a_1 \frac{1 - r^n}{1 - r}$. The sum is

$$\sum_{k=1}^{n} \frac{2^{k+1}}{3^k} = 4 \left[1 - \left(\frac{2}{3} \right)^n \right].$$

We also use the symbol $\sum_{k=1}^{\infty} a_k$ to denote the sum of the infinite series. Is the sum actually a real number? Does $\lim_{n \to \infty} 4[1 - (\frac{2}{3})^n]$ exist? Recall that $\lim_{n \to \infty} (2/3)^n = 0$ since $0 < 2/3 < 1$. It follows that $\lim_{n \to \infty} 4[1 - (2/3)^n] = 4$. Using this fact, we get

$$\lim_{n \to \infty} \sum_{k=1}^{n} \frac{2^{k+1}}{3^k} = \lim_{n \to \infty} 4\left[1 - \left(\frac{2}{3}\right)^n\right] = 4.$$

This would seem to be a logical way to define the sum of all (an infinite number) of the terms of $\sum_{k=1}^{\infty} \frac{2^{k+1}}{3^k}$. Using this idea we define the sum of any infinite series.

Definition 1. An infinite series is the sum of all the terms in an infinite sequence. The sum of an infinite series is denoted by

$$\sum_{k=1}^{\infty} a_k.$$

Definition 2. The value of an infinite sum is defined as follows:

$$\sum_{k=1}^{\infty} a_k = \lim_{n \to \infty} \sum_{k=1}^{n} a_k.$$

Look carefully at this definition. It tells us that there are two steps to finding the sum of an infinite series. The first step is to find a formula for the finite sum $\sum_{k=1}^{n} a_k$. The second step is to take the limit as $n \to \infty$.

The finite sums $\sum_{k=1}^{n} a_k$ are sometimes referred to as subtotals. The symbol $\lim_{n \to \infty} \sum_{k=1}^{n} a_k$ indicates that we keep adding more and more terms of the sum together and hope that the subtotals keep getting closer and closer to some fixed number. Said another way, the sequence of subtotals approaches a limit.

Example 1. Find the sum of the infinite series

$$\sum_{k=1}^{\infty} 5(3/4)^{k-1}.$$

Solution. First, we need to find the formula for the general finite sum

$$\sum_{k=1}^{n} 5(3/4)^{k-1}$$

where n can be any positive integer. Fortunately, this is a finite geometric series with $a_1 = 5$ and $r = 3/4$. Recall that the sum formula for a finite geometric series is

$$S_n = \sum_{k=1}^{n} a_1 r^{k-1} = a_1 \frac{1 - r^n}{1 - r}. \tag{5.1}$$

Substituting $a_1 = 5$ and $r = 3/4$ into the formula (5.1), we get that the sum of the finite series is

$$\sum_{k=1}^{n} 5(3/4)^{k-1} = 5\frac{1 - (3/4)^n}{1 - 3/4} = 20[1 - (3/4)^n].$$

The subtotal for the sum of first n numbers is $20[1 - (3/4)^n]$. Definition 2 then says that

$$\sum_{k=1}^{\infty} 5(3/4)^{k-1} = \lim_{n \to \infty} \sum_{k=1}^{n} 5(3/4)^{k-1} = \lim_{n \to \infty} 20[1 - (3/4)^n].$$

We need to know the limit $\lim_{n \to \infty} (3/4)^n$. Since $0 < 3/4 < 1$, we know that $\lim_{n \to \infty} (3/4)^n = 0$. Therefore,

$$\sum_{k=1}^{\infty} 5(3/4)^{k-1} = 20[1 - 0] = 20.$$

When an infinite series has a sum as this one does, then we say that the series is **convergent**. We would also say that the sum of this infinite series is 20.

Example 2. Find the sum of the infinite series

$$\sum_{k=1}^{\infty} 8^{k+1} 5^{-k}.$$

Solution. Definition 2 of the sum of an infinite series says that

$$\sum_{k=1}^{\infty} 8^{k+1} 5^{-k} = \lim_{n \to \infty} \sum_{k=1}^{n} 8^{k+1} 5^{-k}.$$

First, we need to find a general formula for the finite sum which is correct for every value of n. In order to do this, we write the finite geometric series in the standard form $\sum a_1 r^{k-1}$.

$$\sum_{k=1}^{n} 8^{k+1} 5^{-k} = \sum_{k=1}^{n} 8^2 (8^{k-1}) 5^{-1} (5^{-k+1}) = \sum_{k=1}^{n} \frac{64}{5} \left(\frac{8}{5}\right)^{k-1}.$$

This series is a geometric series with $a_1 = 64/5$ and $r = 8/5$. Substituting these numbers into the sum formula (5.1) for a finite geometric series, we get

$$\sum_{k=1}^{n} \frac{64}{5} \left(\frac{8}{5}\right)^{k-1} = \frac{64}{5} \frac{1 - (8/5)^n}{1 - (8/5)} = \frac{64}{3} \left[\left(\frac{8}{5}\right)^n - 1\right].$$

This completes the first step in finding the infinite sum. Next, we must take the limit. Applying Definition 2 for the sum of an infinite series, we get

$$\sum_{k=1}^{\infty} \frac{64}{5} \left(\frac{8}{5}\right)^{k-1} = \lim_{n \to \infty} \sum_{k=1}^{n} \frac{64}{5} \left(\frac{8}{5}\right)^{k-1} = \lim_{n \to \infty} \frac{64}{3} \left[\left(\frac{8}{5}\right)^n - 1\right].$$

We need the value of $\lim_{n \to \infty} (8/5)^n$. Since $8/5 > 1$, we know that $\lim_{n \to \infty} (8/5)^n$ does not exist. Recall that $\lim_{n \to \infty} r^n = 0$ if $|r| < 1$ and $\lim_{n \to \infty} r^n$ does not exist if $|r| > 1$. It follows that

$$\lim_{n \to \infty} \sum_{k=1}^{n} \left(\frac{64}{5}\right) \left(\frac{8}{5}\right)^{k-1}$$

228

does not exist. In this case, we say that the series $\sum\limits_{k=1}^{\infty}\left(\dfrac{64}{5}\right)\left(\dfrac{8}{5}\right)^{k-1}$ is **divergent**.

Example 3. Does the following infinite series converge or diverge

$$\sum_{k=1}^{\infty}\frac{1}{(4k-1)(4k+3)}?$$

Solution. Definition 2 of the sum of the infinite series says that

$$\sum_{k=1}^{\infty}\frac{1}{(4k-1)(4k+3)}=\lim_{n\to\infty}\sum_{k=1}^{n}\frac{1}{(4k-1)(4k+3)}.$$

As this equation clearly indicates finding $\sum\limits_{k=1}^{\infty}$ involves two steps. The first step is to find the finite sum $\sum\limits_{k=1}^{n}$ and the second step is to take $\lim\limits_{n\to\infty}$.

We need a formula for the finite sum

$$\sum_{k=1}^{n}\frac{1}{(4k-1)(4k+3)}.$$

We can show using mathematical induction that a formula for this finite sum is

$$\sum_{k=1}^{n}\frac{1}{(4k-1)(4k+3)}=\frac{n}{3(4n+3)}.$$

The definition of infinite series tells us that

$$\sum_{k=1}^{\infty}\frac{1}{(4k-1)(4k+3)}=\lim_{n\to\infty}\frac{n}{3(4n+3)}=\lim_{n\to\infty}\frac{1}{3(4+3/n)}=\frac{1}{12}.$$

This infinite series converges and its sum is 1/12.

Note that the subtotals $\sum_{k=1}^{n} \dfrac{1}{(4k-1)(4k+3)} = \dfrac{n}{3(4n+3)}$ form a sequence, that is, $S_n = \dfrac{n}{3(4n+3)}$ is a sequence. The sequence $\left\{ \dfrac{n}{3(4n+3)} \right\}$ is called the sequence of partial sums of the series.

Example 4. Does the infinite series $\sum_{k=1}^{\infty}(3k-11)$ converge?

Solution. Definition 2 of the sum of an infinite series says that

$$\sum_{k=1}^{\infty}(3k-11) = \lim_{n \to \infty} \sum_{k=1}^{n}(3k-11).$$

We need a formula for the general finite sum $\sum_{k=1}^{n}(3k-11)$. Writing out the sum can sometimes help.

$$\sum_{k=1}^{n}(3k-11) = -8 - 5 - 2 + 1 + 4 + \ldots + (3n-11).$$

We see that this is an arithmetic series with first term $a_1 = -8$ and common difference $d = 3$. The formula for the sum of the first n terms of an arithmetic series is

$$S_n = \dfrac{n}{2}[a_1 + a_n].$$

Substituting $a_1 = -8$ and $a_n = 3n - 11$ into this formula we get

$$\sum_{k=1}^{n}(3k-11) = \dfrac{n}{2}[-8 + (3n-11)] = \dfrac{n}{2}(3n-19).$$

We have a formula for the sum of n terms. The subtotals $S_n = \frac{n}{2}(3n-19)$ forms the sequence of partial sums. This completes the first step. In order to find the infinite sum, we take the limit:

$$\lim_{n \to \infty} \sum_{k=1}^{n}(3k-11) = \lim_{n \to \infty} \dfrac{n}{2}(3n-19).$$

As n gets larger and larger the numbers $(n/2)(3n-19)$ get even larger and larger. The limit does not exist. The sequence of partial sums does not have a limit. The series

$$\sum_{k=1}^{\infty}(3k-11) \text{ is divergent.}$$

Example 5. Show that the series $\sum_{k=1}^{\infty}\dfrac{1}{k(k+1)}$ converges and find its value.

Solution. We can show using mathematical induction that

$$\sum_{k=1}^{n}\frac{1}{k(k+1)} = 1 - \frac{1}{n+1}.$$

for all values of n. Definition 2 says that

$$\sum_{k=1}^{\infty}\frac{1}{k(k+1)} = \lim_{n\to\infty}\sum_{k=1}^{n}\frac{1}{k(k+1)} = \lim_{n\to\infty}\left(1 - \frac{1}{n+1}\right) = 1.$$

The infinite series $\sum_{k=1}^{\infty}\dfrac{1}{k(k+1)}$ converges and its value is 1.

Example 6. Find the repeating decimal $8.\overline{45}$ as a fraction. Recall that $8 + 0.\overline{45} = 8 + \frac{45}{100} + \frac{45}{(100)^2} + \frac{45}{(100)^3} + \cdots$.

Solution. The repeating decimal $0.\overline{45}$ is defined as

$$\overline{.45} = \frac{45}{100} + \frac{45}{10,000} + \frac{45}{1,000,000} + \cdots$$

$$= \sum_{k=1}^{\infty} 45\left(\frac{1}{100}\right)^k = \sum_{k=1}^{\infty}\frac{45}{100}\left(\frac{1}{100}\right)^{k-1}.$$

The finite sum $\sum_{k=1}^{n}\dfrac{45}{100}\left(\dfrac{1}{100}\right)^{k-1}$ is a geometric series with $a_1 = 45/100$ and $r = 1/100$. The formula (5.1) for the sum of a finite geometric series

231

tells us that

$$\sum_{k=1}^{n} \frac{45}{100} \left(\frac{1}{100}\right)^{k-1} = \frac{45}{100} \frac{1-(1/100)^n}{1-(1/100)}$$

$$= \frac{45}{99} \left[1 - \left(\frac{1}{100}\right)^n\right].$$

Definition 2 of the sum of an infinite series says

$$\sum_{k=1}^{\infty} \frac{45}{100} \left(\frac{1}{100}\right)^{k-1} = \lim_{n\to\infty} \frac{45}{99} \left[1 - \left(\frac{1}{100}\right)^n\right] = \frac{45}{99} = \frac{5}{11}.$$

Recall that $\lim_{n\to\infty} (1/100)^n = 0$. The limit exists and so the series is convergent to the value $5/11$. It follows that

$$8.\overline{45} = 8 + .\overline{45} = 8 + (5/11) = (93/11).$$

Theorem 0. If the infinite series $\sum_{k=1}^{\infty} a_k$ is convergent, then $\lim_{k\to\infty} a_k = 0$.

Proof. Let $S_n = \sum_{k=1}^{n} a_k = a_1 + a_2 + \ldots + a_n$. It follows that $S_n - S_{n-1} = a_n$. When we say $\sum_{k=1}^{\infty} a_k$ is convergent, then following Definition 2 we are saying that there exists a limit S such that $\lim_{n\to\infty} S_n = S$. Also it follows that $\lim_{n\to\infty} S_{n-1} = S$. Therefore,

$$\lim_{n\to\infty} a_n = \lim_{n\to\infty} (S_n - S_{n-1}) = \lim_{n\to\infty} S_n - \lim_{n\to\infty} S_{n-1} = S - S = 0.$$

Thus $\lim_{k\to\infty} a_k = 0$. Theorem 0 does <u>not</u> say: If $\lim_{k\to\infty} a_k = 0$, then $\sum_{k=1}^{\infty} a_k$ is convergent. In fact this last statement is false. Theorem 0 may be restated as follows: If $\lim_{k\to\infty} a_k$ does not exist or $\lim_{k\to\infty} a_k \neq 0$, then the series $\sum_{k=1}^{\infty} a_k$ is divergent (not convergent). This is really two theorems.

Theorem 1. If $\lim\limits_{k\to\infty} a_k$ does not exist, then the series $\sum\limits_{k=1}^{\infty} a_k$ is divergent.

Theorem 2. If $\lim\limits_{k\to\infty} a_k = L$ and $L \neq 0$, then $\sum\limits_{k=1}^{\infty} a_k$ is divergent.

Example 7. Show that the series $\sum\limits_{k=1}^{\infty} \dfrac{2k+1}{8k+9}$ is divergent.

Solution. Replace a_k by $\dfrac{2k+1}{8k+9}$ in Theorem 2 and we get the true statement.

(a) If $\lim\limits_{k\to\infty} \dfrac{2k+1}{8k+9} = \dfrac{1}{4}$ and $\dfrac{1}{4} \neq 0$, then $\sum\limits_{k=1}^{\infty} \dfrac{2k+1}{8k+9}$ is divergent.

Clearly $\lim\limits_{k\to\infty} \dfrac{2k+1}{8k+9} = \lim\limits_{k\to\infty} \dfrac{2+1/k}{8+9/k} = \dfrac{2}{8} = \dfrac{1}{4}$. Also $\dfrac{1}{4} \neq 0$. Since the hypotheses of the statement (a) is true we are able to conclude that the conclusion is true. Therefore,

$$\sum_{k=1}^{\infty} \frac{2k+1}{8k+9} \text{ is divergent.}$$

Note that $\sum_{k=1}^{\infty} \frac{2k+1}{8k+9}$ is a series while $\frac{2k+1}{8k+9}$ are the terms of a sequence.

Example 8. Show that the series $\sum\limits_{k=1}^{\infty} (-1)^k \dfrac{3k+1}{7k+4}$ is divergent.

Solution. Let us find the limit $\lim\limits_{k\to\infty} (-1)^k \dfrac{3k+1}{7k+4}$. First, note that
$\lim\limits_{k\to\infty} \dfrac{3k+1}{7k+4} = \lim\limits_{k\to\infty} \dfrac{3+1/k}{7+4/k} = \dfrac{3}{7}$. But multiplying by $(-1)^k$ causes the signs to alternate. The sequence tries to get close to $3/7$ and then to $-3/7$ as k gets large. It follows that

a) $\lim\limits_{k\to\infty} (-1)^k \dfrac{3k+1}{7k+4}$ does not exist.

233

Replace a_k with $(-1)^k \dfrac{3k+1}{7k+4}$ in Theorem 1. The result is

b) If $\lim\limits_{k \to \infty} (-1)^k \dfrac{3k+1}{7k+4}$ does not exist, then $\sum\limits_{k=1}^{\infty} (-1) \dfrac{3k+1}{7k+4}$ diverges. From the true statements (a) and (b) we conclude:

The series $\sum\limits_{k=1}^{\infty} (-1)^k \dfrac{3k+1}{7k+4}$ diverges.

Exercises

1. a) Find a formula for the sum S_n of n terms of the geometric series

$$S_n = \sum_{k=1}^{n} 2\frac{3^{k+1}}{5^{k-1}}.$$

b) Find $\lim\limits_{n \to \infty} S_n$.

c) Explain why the infinite series $\sum\limits_{k=1}^{\infty} 2\dfrac{3^{k+1}}{5^{k-1}}$ is convergent. What is its sum?

2. a) Find a formula for the sum S_n of n terms of the geometric series

$$\sum_{k=1}^{n} \frac{4^{k-1}}{3^{k+1}}.$$

b) Find $\lim\limits_{n \to \infty} S_n$.

c) Explain why the infinite series $\sum\limits_{k=1}^{\infty} \dfrac{4^{k-1}}{3^{k+1}}$ is divergent.

3. a) Find the formula for S_n the sum of the first n terms of the arithmetic series

$$\sum_{k=1}^{n} (2k-7).$$

b) Find $\lim\limits_{n \to \infty} S_n$.

234

c) Explain why the infinite series $\sum_{k=1}^{\infty}(2k - 7)$ is divergent.

4. Given that $\sum_{k=1}^{n} \dfrac{1}{(5k-3)(5k+2)} = \dfrac{n}{2(5n+2)}$.

Explain why $\sum_{k=1}^{\infty} \dfrac{1}{(5k-3)(5k+2)} = \dfrac{1}{10}$.

5. Show that the following infinite geometric series is convergent. Be sure to clearly include both steps.

$$\sum_{k=1}^{\infty} 15(5/8)^{k-1}.$$

6. Given that $\sum_{k=1}^{n} \dfrac{1}{(4k+5)(4k+9)} = \dfrac{n}{9(4n+9)}$. Is the infinite series

$\sum_{k=1}^{\infty} \dfrac{1}{(4k+5)(4k+9)}$ convergent or divergent?

7. Show that the series $\sum_{k=1}^{\infty} \dfrac{k+1}{8k+5}$ is divergent using Theorem 2.

8. Show that the series $\sum_{k=1}^{\infty} \dfrac{(-1)^{k+1}(k+2)}{10k+9}$ is divergent using Theorem 1.

9. Find the repeating decimal $6.\overline{4} = 6.444$ as a fraction.

6707 The Comparison Test

We are going to introduce a bit of mathematical notation. We do not absolutely require this notation, but it shortens the way we write sentences and everyone likes to write less. Consider the sentences:

(1) If "statement A" is true, then "statement B" is true.

The converse of (1) is:

(2) If "statement B" is true, then "statement A" is true.

Alternate way to phrase sentence (2).

(3) "Statement B" is true if "statement A" is true.

We can combine the two sentences (1) and (2) together as:

(4) "Statement A" is true if and only if "statement B" is true.

We then use "iff" to replace "if and only if" when we write sentences. Note that sentence (3) says exactly the same thing as does sentence (2). Sentence (1) and sentence (2) do not say the same thing. Consider the sentence: "If $(n+2) = 10$, then $(n+2)^2 = 10$, where n is a positive integer." This is true, we need $n = 8$. Consider the sentence: "If $(n+2)^2 = 100$, then $(n+2) = 10$." This is not true. It is true when $n = 8$, but is false when $n = -12$. The sentence is false. If a sentence is false for some (even one) value of the variable, then we say that it is false. When we say a sentence is true, we mean that it is true for all values of the variable (100% of the time).

If $\frac{5n+3}{2n^3+8} < \frac{3}{n^2}$, then $5n^3 + 3n^2 < 6n^3 + 24$. This is true because we can multiply both sides of less than by the same positive number namely $n^2(2n^3 + 8)$.

If $5n^3 + 3n^2 < 6n^3 + 24$, then $\frac{5n+3}{2n^3+8} < \frac{3}{n^2}$. This is true because we can divide both sides of less than by the same positive number. Combining these two sentences, we have

$$\frac{5n + 3}{2n^3 + 8} < \frac{3}{n^2} \text{ iff } 5n^3 + 3n^2 < 6n^3 + 24$$

If $5n^3 + 3n^2 < 6n^3 + 24$, then $3n^2 < n^3 + 24$. This is true because we can add the same positive number to both sides of less than. If $3n^2 < n^3 + 24$, then $5n^3 + 3n^2 < 6n^3 + 24$. Combining these two sentences

$$5n^3 + 3n^2 < 6n^3 + 24 \text{ iff } 3n^2 < n^3 + 24.$$

It is clearly true that $3n^2 < n^3 + 24$ for all positive integers n.

When we are working with only positive numbers we are allowed to "add the same number to both sides of the inequality" and "to multiply (divide) both sides of an inequality by the same number" just as we can with equations and "iff" is true. We use steps like this all the time when solving equations except we usually do not bother to write "iff". We assume that everyone knows that part. In our discussion of sequences and series we shall write "iff" when that is what we mean.

Example 1. Show that $\dfrac{5n + 3}{2n^3 + 8} < \dfrac{3}{n^2}$.

Solution. We will not use so many words as we did above but write out the solution as we usually would.

$$\frac{5n + 3}{2n^3 + 8} < \frac{3}{n^2} \text{ iff } 5n^3 + 3n^2 < 6n^3 + 24$$

$$\text{iff } 3n^2 < n^3 + 24.$$

We can often determine if a given series of positive terms is convergent or divergent by comparing it with another simpler series. The idea is to compare a series whose convergence or divergence is in question with a series that is known to be convergent or that is known to be divergent. We use the following theorem.

Theorem 1. The Comparison Test. Suppose that $\displaystyle\sum_{n=1}^{\infty} a_n$ and $\displaystyle\sum_{n=1}^{\infty} b_n$ are series of positive terms, that is, $a_n > 0$ and $b_n > 0$.

1. If $\displaystyle\sum_{n=1}^{\infty} b_n$ is convergent and if $a_n < b_n$ for all n, then $\displaystyle\sum_{n=1}^{\infty} a_n$ is also convergent.

2. If $\displaystyle\sum_{n=1}^{\infty} b_n$ is divergent and if $a_n > b_n$ for all n, then $\displaystyle\sum_{n=1}^{\infty} a_n$ is also divergent.

The proof of this theorem involves the idea that a bounded increasing sequences will always have a limit. We have chosen to skip this idea in our study and as a consequence we are unable to give a proof of this theorem. Both parts of this theorem are also true if we replace "for all n" with "$n \geq$ some positive integer" such as "for $n \geq 5$".

When we say that part (1) or part (2) of the Comparison Test is true we are saying that we can replace a_n by a specific expression involving n and replace b_n with another specific expression involving n and the sentence that results will be a true statement. For example, if we replace a_n by $\frac{3n+5}{4n^3+7}$ and b_n by $1/n^2$ in part (1), we get the sentence

If $\sum_{n=1}^{\infty} \frac{1}{n^2}$ is convergent and if $\frac{3n+5}{4n^3+7} < \frac{1}{n^2}$ for all n, then $\sum_{n=1}^{\infty} \frac{3n+5}{4n^3+7}$ is convergent.

This is a true statement because it comes from a true theorem. Note that this statement taken alone does not say that $\sum_{n=1}^{\infty} \frac{3n+5}{4n^3+7}$ is convergent.

Another example. If we replace a_n with $\frac{2n+1}{5n^2+8}$ and b_n with $\frac{1}{3n}$ in part (2) of the Comparison Test, we get the statement:

If $\sum_{n=1}^{\infty} \frac{1}{3n}$ is divergent and if $\frac{2n+1}{5n^2+8} > \frac{1}{3n}$ for all n, then $\sum_{n=1}^{\infty} \frac{2n+1}{5n^2+8}$ is divergent. This is a true statement because it comes from a true theorem. This statement taken alone does not say that $\sum_{n=1}^{\infty} \frac{2n+1}{5n^2+8}$ is divergent.

Theorem 2. If $\sum a_n$ is known to be convergent (divergent) and c is a fixed constant, then $\sum ca_n$ is convergent (divergent) and $\sum ca_n = c \sum a_n$.

We will denote infinite series using both $\sum_{k=1}^{\infty} a_k$ and $\sum_{n=1}^{\infty} a_n$. These two notations have the same meaning. For example, $\sum_{k=1}^{\infty} \frac{2k+1}{5k^2+8}$ denotes exactly the same sum as does $\sum_{n=1}^{\infty} \dfrac{2n+1}{5n^2+8}$.

In the previous section on the definition of infinite series we found that the series $\sum_{n=1}^{\infty} \dfrac{1}{n(n+1)}$ is convergent and its sum is 1. Starting with this fact it follows from Theorem 2, that the series $\sum_{n=1}^{\infty} \dfrac{2}{n(n+1)}$ is convergent and its sum is 2.

Example 2. Show that $\sum_{n=1}^{\infty} \dfrac{1}{n^2}$ is convergent using the Comparison Test.

Solution. The Comparison Test contains two separate statements. In order to prove that a series converges we must start with the statement which is part (1) of the Comparison Test. We start by substituting for a_n and b_n. We are trying to prove that $\sum(1/n^2)$ converges. This means we must replace a_n with $(1/n^2)$. We need to figure out what to use to replace b_n. For this example we will just use the series discussed above.

Replacing a_n by $1/n^2$ and b_n by $2/n(n+1)$ is part (1) of the Comparison Test, we get the following true statement. If $\sum_{n=1}^{\infty} \dfrac{2}{n(n+1)}$ is convergent and if $\dfrac{1}{n^2} < \dfrac{2}{n(n+1)}$, then $\sum_{n=1}^{\infty} \dfrac{1}{n^2}$ is convergent. As stated above the series $\sum_{n=1}^{\infty} \dfrac{2}{n(n+1)}$ is convergent. Next,

$$\frac{1}{n^2} < \frac{2}{n(n+1)} \text{ iff } n(n+1) < 2n^2$$
$$\text{iff } n < n^2$$
$$\text{iff } 1 < n.$$

It follows that

$$\frac{1}{n^2} < \frac{2}{n(n+1)} \text{ for } n > 1.$$

Since both parts of the hypotheses of the statement obtained from the Comparison Test are true the conclusion is true. Therefore, we conclude: "$\sum_{n=1}^{\infty} \dfrac{1}{n^2}$ is convergent". Using Theorem 2 we are able to conclude:

Corollary. The series $\sum_{n=1}^{\infty} \dfrac{c}{n^2}$ is convergent for any fixed number c.

The reasoning that we did in Example 2 conforms to good logical thinking. We showed that $\dfrac{1}{n^2} < \dfrac{2}{n(n+1)}$. The individual terms in $\sum_{n=1}^{\infty} \dfrac{1}{n^2}$ are less than the individual terms in $\sum_{n=1}^{\infty} \dfrac{2}{n(n+1)}$. The subtotals for $\sum_{n=1}^{\infty} 1/n^2$ which are $\sum_{n=1}^{N} \dfrac{1}{n^2}$ are less than the subtotals $\sum_{n=1}^{N} \dfrac{2}{n(n+1)} < 2 - \dfrac{2}{N+1}$. Since $\sum_{n=1}^{N} \dfrac{1}{n^2} < \sum_{n=1}^{N} \dfrac{2}{n(n+1)}$ it follows that $\sum_{n=1}^{N} \dfrac{1}{n^2} < 2$ for all N. The infinite series $\sum_{n=1}^{\infty} \dfrac{1}{n^2}$ must have a sum and this sum must be less than 2. In fact $\sum_{n=1}^{\infty} n^{-p} < 2$ for $p \geq 2$.

The statement "If $\sum_{n=1}^{\infty} \dfrac{2}{n(n+1)}$ is convergent and $\dfrac{1}{n^2} < \dfrac{2}{n(n+1)}$, then $\sum_{n=1}^{\infty} \dfrac{1}{n^2}$ is convergent" is called an implication or conditional statement. The phrase after "if" and before "then" is called the hypothesis. The phrase after "then" is called the conclusion. For this implication the hypothesis is "$\sum_{n=1}^{\infty} \dfrac{2}{n(n+1)}$ is convergent and $\dfrac{1}{n^2} < \dfrac{2}{n(n+1)}$" the conclusion is "$\sum_{n=1}^{\infty} \dfrac{1}{n^2}$ is convergent".

An implication can be true but this does not mean that the conclusion is true. Suppose you make the following true statement to a classmate. If someone gives me $2,000 tonight, then I will fly to England for the weekend. You are not saying that you will be flying to England for the weekend.

For a series of positive terms the more terms you add together the larger the subtotals get. If a series of positive terms is divergent (not convergent), then its sum must be plus infinity. For a divergent series of positive terms the subtotals approach infinity.

Theorem 3. The harmonic series $\sum_{n=1}^{\infty} \dfrac{1}{n}$ is divergent.

We can show that the harmonic series $\sum_{k=1}^{\infty} n^{-1}$ is divergent by direct calculations. However, this is a long calculation. We usually prove Theorem 3 using the inequality

$$\sum_{k=1}^{n} \frac{1}{k} > \int_{1}^{n} \frac{dx}{x} = \ln(n+1),$$

and the fact that $\lim_{u \to \infty} \ln n = +\infty$. This inequality is a special case of the more general statement: If $f(x)$ is a continuous monotone decreasing function, then

$$\int_{1}^{n+1} f(x)dx \le \sum_{k=1}^{n} f(k) < f(1) + \int_{1}^{n} f(x)dx.$$

Taking $\lim_{n \to \infty}$ we get what is usually called the integral test: If $f(x)$ is a continuous monotone decreasing function, then

$$\sum_{k=1}^{\infty} f(k) \text{ and } \int_{1}^{\infty} f(x)dx$$

are either both convergent or both divergent, the integral test is of limited use and will not be one of our official tests for convergence of a series. However, we can use it to show that $\sum_{n=1}^{\infty} n^{-3/2}$ is convergent, or to prove the p series theorem.

The fact that the harmonic series $\sum \frac{1}{n}$ is divergent is an interesting result. It may even be surprising. Note that the limit as n approaches infinity

241

of $1/n$ is zero. The harmonic series is divergent even though the terms being added together get infinitely small as n gets large. The harmonic series is an example and there are others where the limit as n approaches infinity of the terms of the series is zero, but still the series is divergent, that is, $\lim_{n\to\infty} 1/n = 0$ but $\sum_{n=1}^{\infty} 1/n$ is divergent.

Example 3. Show that $\displaystyle\sum_{n=1}^{\infty} \frac{1}{\sqrt{2n+1}}$ is divergent by comparing it to the harmonic series. This means use the Comparison Test.

Solution. In order to prove that a series diverges using the Comparison Test we must substitute into part 2. In the end we want to say that $\sum_{n=1}^{\infty} \frac{1}{\sqrt{2n+1}}$ is divergent. This means we must use $a_n = \frac{1}{\sqrt{2n+1}}$. We are comparing with the harmonic series $\displaystyle\sum_{n=1}^{\infty} \frac{1}{n}$. This means that we must use $b_n = \frac{1}{n}$. Replacing a_n by $1/\sqrt{2n+1}$ and b_n by $1/n$ in part (2) of the Comparison Test Theorem, we get the following statement (a) is a true statement:

Statement (a):

If $\displaystyle\sum_{n=1}^{\infty} \frac{1}{n}$ is divergent and if $\dfrac{1}{\sqrt{2n+1}} > \dfrac{1}{n}$, then $\displaystyle\sum_{n=1}^{\infty} \frac{1}{\sqrt{2n+1}}$ is divergent.

The fact that part (2) of the Comparison Test is a true theorem, means that the above statement which was obtained by substituting into the Comparison Test part (2) results is a true statement. The hypothesis of this statement is: "$\displaystyle\sum_{n=1}^{\infty} \frac{1}{n}$ is divergent and $\dfrac{1}{\sqrt{2n+1}} > \dfrac{1}{n}$". The conclusion is "$\displaystyle\sum_{n=1}^{\infty} \frac{1}{\sqrt{2n+1}}$ is divergent". We need to show that the hypothesis is true.

Statement (b): Theorem 3 tells us that $\displaystyle\sum_{n=1}^{\infty} \frac{1}{n}$ is divergent.

The inequality $\dfrac{1}{\sqrt{2n+1}} > \dfrac{1}{n}$ iff $n > \sqrt{2n+1}$ iff $n^2 > 2n+1$. Now $n^2 >$

242

$2n + 1$ is true for all n except $n = 1$ and $n = 2$. It follows that the following is true:

Statement (c): $\dfrac{1}{\sqrt{2n+1}} > \dfrac{1}{n}$ iff $n^2 > 2n + 1$ which is true for $n > 2$.

Since both parts in the hypothesis of the statement (a) are true, it follows that the conclusion is true. That is, the true statements (a), (b) and (c) taken together enables us to conclude that the following statement is true.

$$\sum_{n=1}^{\infty} \frac{1}{\sqrt{2n+1}} \text{ is divergent.}$$

Recall that we can still apply the Comparison Theorem even though the condition $n^2 > 2n + 1$ is not satisfied for $n = 1$ and $n = 2$ but is satisfied for $n > 2$.

Theorem 4. The p Series Theorem. The p series $\displaystyle\sum_{n=1}^{\infty} \frac{c}{n^p}$ is convergent if $p > 1$ and divergent if $p \leq 1$, where c is a constant.

This p Series Theorem says for example that $\sum 1/n^3$ and $\sum 4/n^{5/2}$ are convergent and that $\sum 1/\sqrt{n}$ and $\sum 5n^{-2/3}$ are divergent. It is easy to prove that p Series Theorem is true if $0 < p < 1$ or if $p > 2$, but is somewhat harder if $1 < p < 2$.

Example 4. Show that the following series is convergent by comparing it with the series $\displaystyle\sum_{n=1}^{\infty} \frac{6}{n^3}$:

$$\sum_{n=1}^{\infty} \frac{5n+3}{n^2(n^2+2)}.$$

Solution. We are trying to conclude that a series is convergent using the Comparison Test. In order to do this we must substitute into part (1). The statement of the problem makes it clear what expressions must be substituted for a_n and b_n. Replacing a_n by $\dfrac{5n+3}{n^2(n^2+2)}$ and b_n by $\dfrac{6}{n^3}$ in part (1) of the Comparison Test we obtain the true statement.

(a) If $\displaystyle\sum_{n=1}^{\infty} \frac{6}{n^3}$ is convergent and if $\dfrac{5n+3}{n^2(n^2+2)} < \dfrac{6}{n^3}$, then $\displaystyle\sum_{n=1}^{\infty} \frac{5n+3}{n^2(n^2+2)}$ is convergent.

(b) The series $\displaystyle\sum_{n=1}^{\infty} \frac{6}{n^3}$ is convergent because it is a p series with $p = 3$ and $c = 6$.

The inequality $\dfrac{5n+3}{n^2(n^2+2)} < \dfrac{6}{n^3}$ iff $5n^2+3n < 6n^2+12$ iff $3n < n^2+12$ which is clearly true for all n. Therefore,

(c) $$\frac{5n+3}{n^2(n^2+2)} < \frac{6}{n^3} \text{ for all } n.$$

The statements (a),(b), and (c) taken together say that

$$\sum_{n=1}^{\infty} \frac{5n+3}{n^2(n^2+2)} \text{ is convergent.}$$

Example 5. Is the following series convergent or divergent:

$$\sum_{n=1}^{\infty} \frac{3n+1}{n(n^2+4)}?$$

Solution. The first step in the solution of this problem is to decide the question: do we think that the series is convergent or do we think that it is divergent? There are two parts or statements to the Comparison Theorem. Based on the answer to this question we decide which of these two parts we want to substitute into. In order to decide this we first get an estimate of the size of the terms in the series. We find this estimate as follows:

$$\frac{3n+1}{n(n^2+4)} = \frac{3+1/n}{n^2+4} \approx \frac{3}{n^2}.$$

Since $\sum 3/n^2$ is convergent, our guess is that $\displaystyle\sum_{n=1}^{\infty} \frac{3n+1}{n(n^2+4)}$ is convergent.

We now try to prove that this guess is correct using the Comparison Theorem. We replace a_n by $\frac{3n+1}{n(n^2+4)}$ and b_n by $\frac{4}{n^2}$ (a little larger than $3/n^2$) in the Comparison Theorem and we get the following true statement.

244

(a) If $\displaystyle\sum_{n=1}^{\infty}\frac{4}{n^2}$ is convergent and if $\dfrac{3n+1}{n(n^2+4)}<\dfrac{4}{n^2}$, then $\displaystyle\sum_{n=1}^{\infty}\frac{3n+1}{n(n^2+4)}$ is convergent.

(b) The series $\displaystyle\sum_{n=1}^{\infty}\frac{4}{n^2}$ is convergent because it is a p-series with $p=2$ and $c=4$.

Cross multiplying we see that $\dfrac{3n+1}{n(n^2+4)}<\dfrac{4}{n^2}$ iff $(3n+1)n<4(n^2+4)$ iff $3n^2+n<4n^2+16$ iff $n<n^2+16$. Since $n<n^2+16$ is clearly true, it follows that

(c) $\qquad\qquad\qquad \dfrac{3n+1}{n(n^2+4)}<\dfrac{4}{n^2}$ is also true for all n.

The statements (b) and (c) say that the hypothesis of the implication (a) are true. The statements (a),(b), and (c) taken together say that

$$\sum_{n=1}^{\infty}\frac{3n+1}{n(n^2+4)}\text{ is convergent.}$$

Example 6. Does the series $\displaystyle\sum_{n=1}^{\infty}\frac{2n+1}{5n^2+4}$ converge or diverge? Justify your answer.

Solution. First, we must decide, do we think that the series converges or do we think that the series diverges? We want to use the Comparison Theorem. In order to decide this question we estimate the size of the terms of the series as follows:

$$\frac{2n+1}{5n^2+4}=\frac{2+1/n}{5n+4/n}\approx\frac{2}{5n}.$$

Since $\sum 2/5n$ diverges, we guess that $\displaystyle\sum_{n=1}^{\infty}\frac{2n+1}{5n^2+4}$ also diverges.

We now set out to prove that this guess is really correct. We replace a_n by $\dfrac{2n+1}{5n^2+4}$ and b_n by $\dfrac{1}{3n}$ in Part (2) of the Comparison Theorem. Note that $(1/3n)<(2/5n)$. This gives the following true statement:

(a) If $\displaystyle\sum_{n=1}^{\infty} \frac{1/3}{n}$ is divergent and $\displaystyle\frac{2n+1}{5n^2+4} > \frac{1}{3n}$, then $\displaystyle\sum_{n=1}^{\infty} \frac{2n+1}{5n^2+4}$ is divergent.

(b) The series $\displaystyle\sum \frac{1/3}{n}$ is divergent because it is a p series with $p = 1$ and $c = 1/3$.

Now $\displaystyle\frac{1}{3n} < \frac{2n+1}{5n^2+4}$ iff $5n^2 + 4 < 6n^2 + 3n$ iff $4 \le n^2 + 3n$. It is clearly true that $4 \le n^2 + 3n$. It follows that

(c) $$\frac{1}{3n} < \frac{2n+1}{5n^2+4} \text{ for all } n.$$

Both parts of the hypothesis of the above statement are true. The true statements (a), (b), and (c) taken together cause us to conclude that

$$\sum_{n=1}^{\infty} \frac{2n+1}{5n^2+4} \text{ is divergent.}$$

Problems

1. Show that $\displaystyle\frac{4n^2+3}{(n+5)(n^2+4)} < \frac{4}{n}$.

2. Given that the following three statements are true. The three statements (a),(b), and (c) taken together enable us to conclude that what other statement is true.

(a) If $\displaystyle\sum_{n=1}^{\infty} \frac{8}{n^2}$ converges and $\displaystyle\frac{7n^2+5}{n^4+8n^2+3} < \frac{8}{n^2}$ then $\displaystyle\sum_{n=1}^{\infty} \frac{7n^2+5}{n^4+8n^2+3}$ converges.

(b) The series $\displaystyle\sum_{n=1}^{\infty} \frac{8}{n^2}$ converges.

(c) The inequality $\displaystyle\frac{7n^2+5}{n^4+8n^2+3} < \frac{8}{n^2}$ is true for all n.

3. Given that each of the following three statements is true, what other statement are you able to conclude is also true?

246

(a) If $\displaystyle\sum_{n=1}^{\infty} \frac{1}{5n^{1/2}}$ diverges and $\displaystyle\frac{\sqrt{4n+5}}{9n+4} > \frac{1}{5n^{1/2}}$, then $\displaystyle\sum_{n=1}^{\infty} \frac{\sqrt{4n+5}}{9n+4}$ diverges.

(b) The series $\displaystyle\sum_{n=1}^{\infty} \frac{1}{5n^{1/2}}$ diverges.

(c) The inequality $\displaystyle\frac{\sqrt{4n+5}}{9n+4} > \frac{1}{5n^{1/2}}$ is true for all n.

4. Replace a_n by $\displaystyle\frac{5n+9}{(n+1)^2(n+2)^2}$ and b_n by $\displaystyle\frac{5}{n^3}$ in Part 1 of the Comparison Theorem.

   (a) What is the hypothesis of the statement that results? What is the conclusion?

   (b) Is the resulting statement true? Why?

5. Replace a_n by $\displaystyle\frac{n+5}{(n+1)(3n+2)}$ and b_n by $\displaystyle\frac{1}{3n}$ in part 2 of the Comparison Theorem.

   (a) What is the hypothesis of the statement that results?

   (b) Does this statement say that "$\displaystyle\sum_{n=1}^{\infty} \frac{(n+5)}{(n+1)(3n+2)}$ is divergent"?

6. For what values of n is $\displaystyle\frac{1}{\sqrt{2n+5}} > \frac{1}{n}$.

7. Show that the series $\displaystyle\sum_{n=1}^{\infty} \frac{7n+4}{n(n^2+5)}$ is convergent by using the Comparison Test and comparing it with $\displaystyle\sum_{n=1}^{\infty} \frac{8}{n^2}$.

8. Show that the series $\displaystyle\sum_{n=1}^{\infty} \frac{1}{\sqrt{4n^2+5}}$ is divergent by comparing it with the

harmonic series $\sum\limits_{n=1}^{\infty} \dfrac{c}{n}$ for an appropriate value of c.

9. Estimate the size of $\dfrac{3n^2 + 1}{(n+4)(n^2+5)}$ for large n.

10. Does the series $\sum\limits_{n=1}^{\infty} \dfrac{5n+8}{n(2n^2+3)}$ converge or diverge? Justify your answer using the Comparison Theorem. First estimate the size of $\dfrac{5n+8}{n(2n^2+3)}$.

11. Does the series $\sum\limits_{n=1}^{\infty} \dfrac{\sqrt{4n+3}}{5n+8}$ converge or diverge. Justify your answer using the Comparison Theorem.

12. a) For what values of n is $\dfrac{n+1}{(n+2)^{3/2}} > \dfrac{1}{\sqrt{2n}}$?

b) Use the comparison test to show that the following series is divergent:

$$\sum_{n=1}^{\infty} \dfrac{n+1}{(n+2)^{3/2}}.$$

6709 Alternating Series

Definition. A infinite series of the form $\sum_{n=1}^{\infty}(-1)^{n-1}b_n$ with $b_n > 0$ is called an alternating series.

The series $\sum_{n=1}^{\infty}(-1)^{n-1}\dfrac{n}{2n+1} = \dfrac{1}{3} - \dfrac{2}{5} + \dfrac{3}{7} - \dfrac{4}{9} + \ldots$ is an example of an alternating series. The word alternating refers to the fact that the signs of the terms alternate between $+$ and $-$. We have the following theorem which can be used to show that an alternating series is convergent.

The Alternating Series Theorem. Given the alternating series $\sum_{n=1}^{\infty}(-1)^{n-1}b_n$ with $b_n > 0$.

If (1) $\lim_{n\to\infty} b_n = 0$ and (2) $b_{n+1} < b_n$,

then $\sum_{n=1}^{\infty}(-1)^{n-1}b_n$ converges.

Note that the hypothesis, the phase which follows the word "if" in the statement of the theorem, contains two parts. This means that we will always have to show two things to be true in order to conclude that an alternating series is convergent. This theorem can not be used to show that an alternating series diverges. The proof of this theorem is fairly easy, but long. Note that this is the first theorem we are discussing in detail where the terms of the series can be negative as well as positive.

Example 1. Show that the following alternating series converges:

$$\sum_{n=1}^{\infty}(-1)^{n-1}\frac{n+2}{n^2+4}.$$

Solution. The clue that this is an alternating series is the $(-1)^{n-1}$. We can also write out a few terms of this series

$$(3/5) - (4/8) + (5/13) - (6/20) + (7/29) - (8/40) + \ldots$$

Note that the $+$ and $-$ signs alternate in this series. This is the feature that indicates that it is an alternating series. As soon as we are certain that this is an alternating series we try to show that it converges using the Alternating Series Theorem. In order to do this we replace b_n by $\frac{n+2}{n^2+4}$ in the Alternating Series Theorem. We get the following true statement:

(a) If $\displaystyle\lim_{n\to\infty} \frac{n+2}{n^2+4} = 0$ and $\displaystyle\frac{(n+1)+2}{(n+1)^2+4} < \frac{n+2}{n^2+4}$,

then $\displaystyle\sum_{n=1}^{\infty}(-1)^{n-1}\frac{n+2}{n^2+4}$ converges.

The hypothesis of this true statement is the clause that comes after the word "if" and before the comma. The conclusion is the clause that comes after the word "then". The hypothesis of this statement is:

$$\lim_{n\to\infty} \frac{n+2}{n^2+4} = 0 \text{ and } \frac{(n+1)+2}{(n+1)^2+4} < \frac{n+2}{n^2+4}.$$

The conclusion is:

$$\sum_{n=1}^{\infty}(-1)^{n-1}\frac{n+2}{n^2+4} \text{ converges.}$$

In order to conclude that the conclusion is true we must first show that the hypothesis is true.

Note that there are two parts to the hypothesis of this statement and that the two parts are connected by the word "and". Because the parts are connected by the word "and", we must show that both parts of the hypothesis are true in order to say that the hypothesis is true. First part:

$$(b) \qquad\qquad \lim_{n\to\infty} \frac{n+2}{n^2+4} = \lim_{n\to\infty} \frac{1+2/n}{n+4/n} = 0.$$

For this series $b_n = \frac{n+2}{n^2+4}$. We also need b_{n+1}. We get b_{n+1} by replacing n with $n+1$ in the formula for b_n. We must show $b_{n+1} < b_n$. We have

$$\frac{n+3}{n^2+2n+5} < \frac{n+2}{n^2+4} \text{ iff } (n+3)(n^2+4) < (n+2)(n^2+2n+5).$$

250

Multiplying this is the same as

$$n^3 + 3n^2 + 4n + 12 < n^3 + 4n^2 + 9n + 10$$

iff $2 < n^2 + 5n$ which is clearly true. Therefore, it is true for all n that

(c)
$$\frac{n+3}{n^2 + 2n + 5} < \frac{n+2}{n^2 + 4}.$$

Since both parts of the hypothesis have been shown to be true it follows that the conclusion is true. In other words true the statements (a),(b), and (c) all taken together cause us to conclude that

$$\sum_{n=1}^{\infty} (-1)^{n-1} \frac{n+2}{n^2 + 4} \quad \text{converges}$$

is a true statement.

We have shown that this series converges, but we have no idea what the sum of the series is. The Alternating Series Theorem give us no help in finding the sum of the series.

Example 2. Show that the following alternating series converges:

$$\sum_{k=1}^{\infty} (-1)^{k-1} \frac{1}{\sqrt{2k+5}}.$$

Solution. This is an alternating series. We can replace b_k in the Alternating Series Theorem by any numbers and we will always get a true statement. First, since $b_k = \frac{1}{\sqrt{2k+5}}$, then $b_{k+1} = \frac{1}{\sqrt{2k+7}}$. Replace b_k by $\frac{1}{\sqrt{2k+5}}$ in the Alternating Series Theorem and we get the following true statement.

(a) If $\lim\limits_{k \to \infty} \dfrac{1}{\sqrt{2k+5}} = 0$ and $\dfrac{1}{\sqrt{2k+7}} < \dfrac{1}{\sqrt{2k+5}}$, then $\sum\limits_{k=1}^{\infty} \dfrac{(-1)^{k-1}}{\sqrt{2k+5}}$ converges. The hypothesis of this implication is:

$$\lim_{k \to \infty} \frac{1}{\sqrt{2k+5}} = 0 \text{ and } \frac{1}{\sqrt{2k+7}} < \frac{1}{\sqrt{2k+5}}.$$

251

The conclusion is

$$\sum_{k=1}^{\infty} \frac{(-1)^{k-1}}{\sqrt{2k+5}} \text{ converges.}$$

We must show that the hypothesis is true. Clearly

(b)
$$\lim_{k\to\infty} \frac{1}{\sqrt{2k+5}} = 0.$$

Using properties of inequalities $\frac{1}{\sqrt{2k+7}} < \frac{1}{\sqrt{2k+5}}$ iff $\sqrt{2k+5} < \sqrt{2k+7}$ iff $2k+5 < 2k+7$ iff $5 < 7$. Therefore,

(c)
$$\frac{1}{\sqrt{2k+7}} < \frac{1}{\sqrt{2k+5}}.$$

Since both parts of the hypothesis are true it follows that the conclusion is true. What we are saying is that the true statements (a),(b), and (c) taken together causes us to conclude that

$$\sum_{k=1}^{\infty} \frac{(-1)^{k-1}}{\sqrt{2k+1}} \text{ converges.}$$

We know that this series converges, but we do not know its sum. Because we do not know what the exact sum of this series is we often try to find an approximate value for the exact sum.

Let us consider the problem of getting an approximate value for the sum of an alternating series. Consider the simple alternating series

$$\sum_{k=1}^{\infty} \frac{(-1)^{k+1}}{k^2}.$$

It is easy to show that this alternating series converges using the Alternating Series Theorem. Let us assume we have done that part. Once we know that the series is convergent we might ask: What is its exact sum? We can always approximate the exact sum by adding a few terms together. For example, the sum of 8 terms is

$$1 - (1/4) + (1/9) - (1/16) + (1/25) - (1/36) + (1/49) - (1/64) = 0.8156.$$

We feel fairly certain that if we were to add more terms we would get a more accurate approximation. For example, the sum of 16 terms is

$$1 - \frac{1}{4} + \frac{1}{9} - \frac{1}{16} + \ldots - \frac{1}{256} = 0.8206.$$

The question is: how accurate is this approximation? How accurate is the sum of the first 100 terms as an approximation? In the general discussion of this question we use the following notation. We can write any series as the sum of two parts as follows:

$$\sum_{k=1}^{\infty} (-1)^{k-1} b_k = \sum_{k=1}^{N} (-1)^{k-1} b_k + \sum_{k=N+1}^{\infty} (-1)^{k-1} b_k.$$

Here we are thinking of the entire series as expressed as a sum of the first N terms of the series plus the sum of the rest of the terms. We can use any counting number for N. When $N = 20$ this would be

$$\sum_{k=1}^{\infty} (-1)^{k-1} b_k = \sum_{k=1}^{20} (-1)^{k-1} b_k + \sum_{k=21}^{\infty} (-1)^{k-1} b_k.$$

Here we have expressed the series as the sum of the first 20 terms plus the sum of all the other terms starting with the 21st term. We have already used the notation

$$S_N = \sum_{k=1}^{N} (-1)^{k-1} b_k.$$

We will also use the notation

$$R_N = \sum_{k=N+1}^{\infty} (-1)^{k-1} b_k.$$

Using this notation

$$\sum_{k=1}^{\infty} (-1)^{k-1} b_k = S_N + R_N.$$

We have broken the sum of the entire series into two parts. The first sum S_N is a finite sum and so can be calculated. The second sum R_N can be thought of as the remainder of the series. Also since

$$\left[\sum_{k=1}^{\infty}(-1)^{k-1}b_k\right] - S_N = R_N$$

we see that R_N is the error we make when we approximate the whole sum $\sum_{k=1}^{\infty}(-1)^{k-1}b_k$ with just the finite sum $S_N = \sum_{k=1}^{N}(-1)^{k-1}b_k$. The following theorem gives us a maximum value or bound for the error term R_N.

Alternating Series Estimation Theorem. Given the alternating series $\sum_{k=1}^{\infty}(-1)^{k-1}b_k$ such that

(1) $\lim_{k\to\infty} b_k = 0$ (2) $b_{k+1} < b_k$,

then

$$|R_N| = \left|\sum_{k=N+1}^{\infty}(-1)^{k-1}b_k\right| < b_{N+1}.$$

Note that the theorem says $|R_N| < b_{N+1}$ not that $|R_N| = b_{N+1}$. The number R_N is the actual error. The number b_{N+1} is an error bound. The inequality $|R_N| < b_{N+1}$ says that the absolute value of the actual error is less than the error bound. Recall that in order to have an alternating series we must have $b_k > 0$. Let us use this theorem to estimate the error resulting when the finite sum

$$1 - (1/4) + (1/9) - (1/16) + \ldots - (1/256) = 0.8206$$

is used to estimate the whole sum $\sum_{k=1}^{\infty}\frac{(-1)^{k-1}}{k^2}$. Breaking up the series as was done in general earlier, we get

$$\sum_{k=1}^{\infty}\frac{(-1)^{k-1}}{k^2} = \sum_{k=1}^{16}\frac{(-1)^{k-1}}{k^2} + \sum_{k=17}^{\infty}\frac{(-1)^{k-1}}{k^2}$$

The difference between the exact value $\sum\limits_{k=1}^{\infty} \dfrac{(-1)^{k-1}}{k^2}$ and the approximate

value $\sum\limits_{k=1}^{16} \dfrac{(-1)^{k-1}}{k^2}$ is $\sum\limits_{k=17}^{\infty} \dfrac{(-1)^{k-1}}{k^2}$. Substituting $N = 16$ and $b_k = 1/k^2$

into $\left| \sum\limits_{k=N+1}^{\infty} (-1)^{k-1} b_k \right| < b_{N+1}$, which is the formula of the Alternating

Series Estimation Theorem, we get

$$\left| \sum_{N=17}^{\infty} (-1)^{k-1} \frac{1}{k^2} \right| < \frac{1}{(17)^2} = \frac{1}{289}.$$

The error is less than $1/289 = 0.0035$. This is a bound for the error and tells us that the actual value is between $0.8206 - 0.0035 = 0.8171$ and $0.8206 + 0.0035 = 0.8241$. The actual sum of the infinite series is 0.8235. The actual error is 0.0019. The actual error 0.0019 is less than the error bound of 0.0035 found using the estimation theorem. The error bound is almost always quite a bit larger than the actual error. However the error bound does enable us to say that although the actual error is unknown we can be sure that the actual error is less than the error bound.

Example 3. Assume that $\lim\limits_{k \to \infty} \dfrac{3k+2}{2k^3+5} = 0$ and $\dfrac{3(k+1)+2}{2(k+1)^3+5} < \dfrac{3k+2}{2k^3+5}$ are both true. Use the Alternating Series Estimation Theorem to find an upper bound for

$$\left| \sum_{k=15}^{\infty} (-1)^k \frac{3k+2}{2k^3+5} \right| \text{ and } \left| \sum_{k=50}^{\infty} (-1)^k \frac{3k+2}{2k^3+5} \right|.$$

Solution. Substituting $b_k = \dfrac{3k+2}{2k^3+5}$ in the Alternating Series Estimation Theorem, the hypothesis is:

$$\lim_{k \to \infty} \frac{3k+2}{2k^3+5} = 0 \text{ and } \frac{3(k+1)+2}{2(k+1)^3+5} < \frac{3k+2}{2k^3+5}.$$

We are given that this is true. This means that the Alternating Series Estimation Theorem applies to the series. Therefore, we are able to conclude

255

using first $N = 14$ and then $N = 49$ that

$$\left| \sum_{k=15}^{\infty} (-1)^k \frac{3k+2}{2k^3+5} \right| < \frac{3(15)+2}{2(15)^3+5} = \frac{47}{6755} < .007.$$

$$\left| \sum_{k=50}^{\infty} (-1)^k \frac{3k+2}{2k^3+5} \right| < \frac{3(50)+2}{2(50)^3+5} = \frac{152}{250005} < .00061$$

Example 4. Given $\sum_{k=1}^{\infty} (-1)^{k-1} \frac{k+2}{k^4+8}$. Suppose we approximate the exact sum using the finite sum $\sum_{k=1}^{20} (-1)^{k-1} \frac{k+2}{k^4+8}$, what is the error bound according to the Alternating Series Estimation Theorem?

Solution. We first show that both parts of the hypothesis of the Alternating Series Estimation Theorem are true by showing that

(1) $\displaystyle \lim_{k \to \infty} \frac{k+2}{k^4+8} = 0$

(2) $\displaystyle \frac{k+3}{(k+1)^4+8} < \frac{k+2}{k^4+8}.$

Let us assume that this has been done. Since both parts of the hypothesis are true, we can apply the Alternating Series Estimation Theorem to find a bound on the error.

$$\sum_{k=1}^{\infty} (-1)^{k-1} \frac{k+2}{k^4+8} = \sum_{k=1}^{20} (-1)^{k-1} \frac{k+2}{k^4+8} + \sum_{k=21}^{\infty} (-1)^{k-1} \frac{k+2}{k^4+8}.$$

The remainder or error term is

$$\sum_{k=21}^{\infty} (-1)^{k-1} \frac{k+2}{k^4+8}.$$

This sum gives the actual error. Replacing N by 20 in the Alternating Series Estimation Theorem, we have

$$\left| \sum_{k=21}^{\infty} (-1)^{k-1} \frac{k+2}{k^4+8} \right| < \frac{21+2}{(21)^4+8} = \frac{23}{194,489} = .000118$$

The error is bound 0.00012. Note that $N = 20$ is $N + 1 = 21$. We know that the actual error is less than the error bound.

Example 5. Given the alternating series $\sum\limits_{k=1}^{\infty} (-1)^{k-1} \dfrac{1}{k^3}$. Suppose we want to approximate the exact sum of the whole series using a finite sum. How many terms do we need in the finite sum in order to be sure that the actual error in the approximation is less than 10^{-4}?

Solution. First, we really need to make sure that the alternating series $\sum\limits_{k=1}^{\infty} \dfrac{(-1)^{k-1}}{k^3}$ converges. This is the same as showing that the hypothesis of the Alternating Series Estimation Theorem is true. The hypothesis of the Alternating Series Estimation Theorem is

$$(1) \ \lim_{k \to \infty} \frac{1}{k^3} = 0 \text{ and } (2) \ \frac{1}{(k+1)^3} < \frac{1}{k^3}.$$

Clearly $\lim\limits_{k \to \infty} k^{-3} = 0$. We have $(k+1)^{-3} < k^{-3}$ iff $k^3 < (k+1)^3$ iff $k < k + 1$. The hypothesis is true. We conclude that the conclusion of the estimation theorem is true. Using the general value N the Alternating Series Estimation Theorem now allows us to conclude that the following is true:

$$\left| \sum_{k=1}^{\infty} \frac{(-1)^{k-1}}{k^3} - \sum_{k=1}^{N} \frac{(-1)^{k-1}}{k^3} \right| = \left| \sum_{k=N+1}^{\infty} \frac{(-1)^{k-1}}{k^3} \right| < \frac{1}{(N+1)^3}.$$

The error bound is $1/(N+1)^3$. This inequality says that for any value of N the actual error is less than the error bound $1/(N+1)^3$. For this problem we need to choose a value of N such that

$$\frac{1}{(N+1)^3} < 10^{-4}.$$

This is the same as

$$(N+1)^3 > 10^4$$
$$N + 1 > 10^{4/3}$$
$$N + 1 > 21.54$$

We make the fraction $1/(N+1)$ smaller by making N larger. Also N must be a whole number. The smallest value of N we can choose in order to make the error bound less than 10^{-4} is $N = 21$. The error in approximating $\sum_{k=1}^{\infty} \frac{(-1)^{k-1}}{k^3}$ with the finite sum $\sum_{k=1}^{21} \frac{(-1)^{k-1}}{k^3}$ is less than 10^{-4} or is correct to 4 decimal places. Note that we usually consider the phrase "correct to 4 decimal places" to mean "$|R_N| < 10^{-4}$".

We now discuss a new theorem about alternating series. We easily see that $\lim_{n \to \infty} \frac{3n+2}{5n+7} = \frac{3}{5}$. Also $(-1)^{n-1}$ alternately takes the values $1, -1, 1, -1, \ldots$. It follows that for large n the fraction $(-1)^n \frac{3n+2}{5n+7}$ alternately takes the values close to $3/5$ and $-3/5$. Therefore

$$\lim_{n \to \infty} (-1)^n \frac{3n+2}{5n+7} \text{ does not exist.}$$

Using this same reasoning we easily see that the following lemma is true.

Lemma. If $\lim_{n \to \infty} b_n = L$ and $L \neq 0$, then $\lim_{n \to \infty} (-1)^n b_n$ does not exist.

Recall the following theorem from the section on definition of infinite series.

Theorem. If $\lim_{n \to \infty} (-1)^n b_n$ does not exist, then $\sum_{n=1}^{\infty} (-1)^{n-1} b_n$ is divergent.

Combining this theorem and the lemma we get the following theorem.

Divergence Theorem for Alternating Series. If $\lim_{n \to \infty} b_n = L$ and $L \neq 0$, then $\sum_{n=1}^{\infty} (-1)^{n-1} b_n$ is divergent.

Example 6. Show that the following alternating series is divergent using the Divergence Theorem for Alternating Series:

$$\sum_{n=1}^{\infty} (-1)^{n-1} \frac{n+3}{5n+8}.$$

Solution. Replacing b_n in The Divergence Theorem for Alternating Series with $b_n = \dfrac{n+3}{5n+8}$, we get the following true statement:

(a) If $\displaystyle\lim_{n\to\infty} \dfrac{n+3}{5n+8} = \dfrac{1}{5}$ and $1/5 \neq 0$, then $\displaystyle\sum_{n=1}^{\infty} (-1)^{n-1} \dfrac{n+3}{5n+8}$ is divergent.

Certainly (b) $1/5 \neq 0$. Also (c) $\displaystyle\lim_{n\to\infty} \dfrac{n+3}{5n+8} = \lim_{n\to\infty} \dfrac{1+3/n}{5+8/n} = \dfrac{1}{5}$. Since the hypothesis of the statement (a) is true it follows that the conclusion is true. The three true statements (a),(b), and (c) taken together cause us to conclude that

$$\sum_{n=1}^{\infty} (-1)^{n-1} \dfrac{n+3}{5n+8} \text{ is divergent.}$$

We only have two theorems about the convergence of alternating series. One is the Alternating Series Theorem which we use to show that an alternating series is convergent. The other is the Divergence Theorem for Alternating Series which we use to show that an alternating series is divergent as we did in the Example 6.

Example 7. Is the alternating series $\displaystyle\sum_{n=1}^{\infty} (-1)^{n-1} \dfrac{2n+3}{n^2+1}$ convergent or divergent?

Solution. We have only two theorems about alternating series. In order to apply either theorem we must find $\displaystyle\lim_{n\to\infty} \dfrac{2n+3}{n^2+1}$ since the value of this limit is part of the hypothesis of both theorems. Start by finding this limit.

$$\lim_{n\to\infty} \dfrac{2n+3}{n^2+1} = \lim_{n\to\infty} \dfrac{2+3/n}{n+1/n} = 0.$$

A limit of zero is part of the hypothesis of the Alternating Series Theorem. This means that we want to substitute into the Alternating Series Theorem. Let $b_n = \dfrac{2n+3}{n^2+1}$ in the Alternating Series Theorem and we get the following true statement:

(a) If $\lim\limits_{n\to\infty}\dfrac{2n+3}{n^2+1}=0$ and $\dfrac{2n+5}{n^2+2n+2}<\dfrac{2n+3}{n^2+1}$, then $\sum\limits_{n=1}^{\infty}(-1)^{n-1}\dfrac{2n+3}{n^2+1}$ is convergent.

(b) $\lim\limits_{n\to\infty}\dfrac{2n+3}{n^2+1}=\lim\limits_{n\to\infty}\dfrac{2/n+3/n^2}{1+1/n^2}=\dfrac{0}{1}=0$ was shown above.

Now $\dfrac{2n+5}{n^2+2n+2}<\dfrac{2n+3}{n^2+1}$ iff

$$2n^3+5n^2+2n+5<2n^3+7n^2+10n+6$$

iff $0<2n^2+8n+1$ which is clearly true. Therefore,

(c) $\qquad\qquad\dfrac{2n+5}{n^2+2n+5}<\dfrac{2n+3}{n^2+1}$ for all n.

The true statements (a),(b), and (c) taken together cause us to conclude that $\sum\limits_{n=1}^{\infty}(-1)^{n-1}\dfrac{2n+3}{n^2+1}$ is convergent.

Exercises

1. Given that the following statements are all true, what other statement are you able to conclude is also true?

(a) If $\dfrac{5n+13}{n^2+2n+5}<\dfrac{5n+8}{n^2+4}$ and $\lim\limits_{n\to\infty}\dfrac{5n+8}{n^2+4}=0$, then $\sum\limits_{n=1}^{\infty}(-1)^{n-1}\dfrac{5n+8}{n^2+4}$ converges.

(b) $\lim\limits_{n\to\infty}\dfrac{5n+8}{n^2+4}=0$

(c) $\dfrac{5n+13}{n^2+2n+5}<\dfrac{5n+8}{n^2+4}$ is true for all n.

2. Given that the following two statements are both true, what other statement are we able to conclude is also true?

(a) If $\lim\limits_{n\to\infty}\dfrac{2n+3}{5n+8}=\dfrac{2}{5}$ and $(2/5)\neq 0$, then $\sum\limits_{n=1}^{\infty}(-1)^{n-1}\dfrac{2n+3}{5n+8}$ is divergent.

260

(b) $\displaystyle\lim_{n\to\infty}\frac{2n+3}{5n+8}=\frac{2}{5}$.

3. a) Replace b_n by $\dfrac{2n+1}{n(n+1)}$ in the Alternating Series Theorem.

b) Show that $\displaystyle\lim_{n\to\infty}\frac{2n+1}{n(n+1)}=0$

c) Show that $\dfrac{2n+3}{(n+1)(n+2)}<\dfrac{2n+1}{n(n+1)}$.

d) Considering the true statements in parts a,b, and c taken together what can you conclude?

4. a) Replace b_n by $\dfrac{1}{\sqrt{4n^2+3}}$ in the Alternating Series Theorem.

b) Look at your answer in part (a). Show that the hypothesis of the implication which you wrote down in (a) is true.

c) Considering the true statements in parts (a) and (b) what are you able to conclude?

5. Show that the Alternating Series $\displaystyle\sum_{k=1}^{\infty}\frac{(-1)^{k-1}}{2k^2+5}$ is convergent.

6.a) Replace b_n by $\dfrac{3n+2}{5n+4}$ in the Alternating Series Divergence Theorem.

b) Show that $\displaystyle\lim_{n\to\infty}\frac{3n+2}{5n+4}=\frac{3}{5}$.

c) You have two true statements here. What conclusion can you reach?

7. Is the following alternating series convergent or divergent? Justify your answer by applying a theorem.

$$\sum_{k=1}^{\infty}(-1)^{k-1}\frac{5k+3}{k^2+4}$$

8. Is the following alternating series convergent or divergent? Justify your answer by applying a theorem.

$$\sum_{k=1}^{\infty}(-1)^{k+1}\frac{5k+3}{9k+4}.$$

9. Show that $\dfrac{2n+5}{5n+11} < \dfrac{2n+3}{5n+6}$. Is the alternating series $\displaystyle\sum_{n=1}^{\infty}(-1)^n\frac{2n+3}{5n+6}$ convergent or divergent?

10. Given that $\displaystyle\lim_{k\to\infty}\frac{3k+5}{7k^3+8}=0$ and that $\dfrac{3(k+1)+5}{7(k+1)^3+8} < \dfrac{3k+5}{7k^3+8}$, find the error bound of each of the following using the Alternating Series Estimation Theorem.

a) $\left|\displaystyle\sum_{k=25}^{\infty}(-1)^{k-1}\frac{3k+5}{7k^3+8}\right|$

b) $\left|\displaystyle\sum_{k=100}^{\infty}(-1)^{k-1}\frac{3k+5}{7k^3+8}\right|$

c) $\left|\displaystyle\sum_{k=1}^{\infty}(-1)^{k-1}\frac{3k+5}{7k^3+8}\right|$

d) $\left|\displaystyle\sum_{k=1}^{\infty}(-1)^{k-1}\frac{3k+5}{7k^3+8} - \sum_{k=1}^{42}(-1)^{k-1}\frac{3k+5}{7k^3+8}\right|$

e) $\left|\displaystyle\sum_{k=1}^{\infty}(-1)^{k-1}\frac{3k+5}{7k^3+8} - \sum_{k=1}^{100}(-1)^{k-1}\frac{3k+5}{7k^3+8}\right|$

11. a) Show that $\displaystyle\lim_{k\to\infty} k^{-4}=0$.

b) Show that $(k+1)^{-4} < k^{-4}$ for all k.

c) Find a bound for each of the following using the Alternating Series Estimation Theorem:

i) $\left| \displaystyle\sum_{k=30}^{\infty} \dfrac{(-1)^{k-1}}{k^4} \right|$

ii) $\left| \displaystyle\sum_{k=63}^{\infty} \dfrac{(-1)^{k-1}}{k^4} \right|$

iii) $\left| \displaystyle\sum_{k=1}^{\infty} \dfrac{(-1)^{k-1}}{k^4} - \sum_{k=1}^{30} \dfrac{(-1)^{k-1}}{k^4} \right|$

$\left| \displaystyle\sum_{k=1}^{\infty} \dfrac{(-1)^{k-1}}{k^4} - \sum_{k=1}^{59} \dfrac{(-1)^{k-1}}{k^4} \right|$

d) Find the smallest value of N such that we can be sure that

i) $\left| \displaystyle\sum_{k=N+1}^{\infty} \dfrac{(-1)^{k-1}}{k^4} \right| < 10^{-5}$

ii) $\left| \displaystyle\sum_{k=N+1}^{\infty} \dfrac{(-1)^{k-1}}{k^4} \right| < 10^{-7}$

12. Consider the alternating series $\displaystyle\sum_{k=1}^{\infty}(-1)^{k-1}\dfrac{k+4}{5k^3+8}$. Given that the hypothesis of the Alternating Series Estimation Theorem are satisfied.

a) Find a bound for $\left| \displaystyle\sum_{k=50}^{\infty}(-1)^{k-1}\dfrac{k+4}{5k^3+8} \right|$.

b) Suppose we approximate the infinite sum $\displaystyle\sum_{k=1}^{\infty}(-1)^{k-1}\dfrac{k+4}{5k^3+8}$ using the finite sum $\displaystyle\sum_{k=1}^{N}(-1)^{k-1}\dfrac{k+4}{5k^3+8}$, what is the error bound according to the Alternating Series Estimation Theorem?

c) Find a bound for the value of the following using the Alternating Series Estimation Theorem?

$$\left| \displaystyle\sum_{k=1}^{\infty}(-1)^{k-1}\dfrac{k+4}{5k^3+8} - \sum_{k=1}^{39}(-1)^{k-1}\dfrac{k+4}{5k^3+8} \right|.$$

13. Given that the hypothesis of the Alternating Series Estimation Theorem

are satisfied for the series $\displaystyle\sum_{k=1}^{\infty} \frac{(-1)^{k-1}}{5k^3 + 8}$ find the smallest value of N such that:

a) $\left| \displaystyle\sum_{k=N+1}^{\infty} \frac{(-1)^{k-1}}{5k^3 + 8} \right| < 10^{-4}.$

b) $\left| \displaystyle\sum_{k=N+1}^{\infty} \frac{(-1)^{k-1}}{5k^3 + 8} \right| < 10^{-8}.$

Remark. The following are the sums of some well known series:

(a) $\displaystyle\sum_{n=1}^{\infty} \frac{1}{n^2} = \frac{\pi^2}{6}$

(b) $\displaystyle\sum_{n=1}^{\infty} \frac{1}{n^4} = \frac{\pi^4}{90}$

(c) $\displaystyle\sum_{n=1}^{\infty} \frac{1}{(2n-1)^2} = \frac{\pi^2}{8}$

(d) $\displaystyle\sum_{n=1}^{\infty} \frac{(-1)^{n-1}}{n^2} = \frac{\pi^2}{12}$

(e) $\displaystyle\sum_{n=1}^{\infty} \frac{(-1)^{n-1}}{2n-1} = \frac{\pi}{4}$

(f) $\displaystyle\sum_{n=1}^{\infty} \frac{(-1)^{n-1}}{(2n-1)^3} = \frac{\pi^3}{32}$

6711 Ratio Test

One of the easiest ways to tell if an infinite series is convergent is to use the ratio test. The down side of the ratio test is that it often fails to tell us if the series is convergent or if it is divergent. In many problems we are unable to reach any conclusion after applying the ratio test.

Theorem. The Ratio Test. Given the infinite series $\sum_{n=1}^{\infty} a_n$, let L denote the limit

$$L = \lim_{n \to \infty} \left| \frac{a_{n+1}}{a_n} \right|.$$

1. If $0 \leq L < 1$, then the series $\sum_{n=1}^{\infty} a_n$ is convergent.

2. If $L > 1$ or $L = \infty$, then the series $\sum_{n=1}^{\infty} a_n$ is divergent.

3. If $L = 1$, the test gives no information about convergence. Test fails.

We could now prove that the ratio test theorem is a true statement. We would only need the idea of absolute convergence and theorems which we have already stated. However, the proof is long and so we choose to omit it. Note that the numbers a_n can be either positive or negative in the ratio test.

Example 1. Is the series $\sum_{n=1}^{\infty} \frac{n(2n+5)}{2^n}$ convergent or divergent? Justify your answer.

Solution. Apply the ratio test. The first step is to find the ratio a_{n+1}/a_n. In order to find the ratio we need a_n and a_{n+1}. We obtain a_{n+1} by replacing n with $n+1$ in the expression for a_n. For this problem

$$a_n = \frac{n(2n+5)}{2^n} \quad \text{and} \quad a_{n+1} = \frac{(n+1)(2n+7)}{2^{n+1}}.$$

The ratio is found by dividing a_{n+1} by a_n. In order to divide the fraction a_{n+1} by the fraction a_n we invert the fraction a_n and multiply. The ratio is

$$\frac{a_{n+1}}{a_n} = \frac{(n+1)(2n+7)}{2^{n+1}} \div \frac{n(2n+5)}{2^n}$$

$$= \frac{(n+1)(2n+7)}{2^{n+1}} \cdot \frac{2^n}{n(2n+5)} = \frac{(n+1)(2n+7)}{2n(2n+5)}.$$

When we simplified we used the fact that $2^n/2^{n+1} = 1/2$. We did this by subtracting powers. Next find L which is the limit of the ratio. Since all these numbers are positive we do not need to worry about taking the absolute value.

$$L = \lim_{n \to \infty} \frac{(n+1)(2n+7)}{2n(2n+5)} = \lim_{n \to \infty} \frac{(n+1)(2n+7)(1/n^2)}{(2n)(2n+5)(1/n^2)}$$

$$= \lim_{n \to \infty} \frac{(1+1/n)(2+7/n)}{2(2+5/n)} = \frac{1}{2}.$$

Note that we multiplied the numerator and denominator both by the same number $(1/n^2)$. This did not change the value of the fraction. Replacing L by $1/2$ and a_n by $\dfrac{n(2n+5)}{2^n}$ in Part 1 of the Ratio Test, we get the following true statement:

If $1/2 < 1$, then $\displaystyle\sum_{n=1}^{\infty} \frac{n(2n+5)}{2^n}$ converges.

Clearly $1/2 < 1$ is true. Therefore, we are able to conclude that

"$\displaystyle\sum_{n=1}^{\infty} \frac{n(2n+5)}{2^n}$ converges" is true.

Example 2. Is the series $\displaystyle\sum_{n=1}^{\infty} \frac{(-1)^{n-1}3^n}{2n^2+5}$ convergent or divergent?

Solution. We want to use the ratio test. First, find the ratio which is a_{n+1}/a_n. Note that

$$a_n = \frac{(-1)^{n-1}3^n}{2n^2+5} \text{ and } a_{n+1} = \frac{(-1)^n 3^{n+1}}{2(n+1)^2+5}.$$

Remember that in order to divide by the fraction a_n we invert and multiply. Inverting a_n and multiplying we get the ratio

$$\frac{a_{n+1}}{a_n} = \frac{(-1)^n 3^{n+1}}{2n^2 + 4n + 7} \cdot \frac{2n^2 + 5}{(-1)^{n-1} 3^n} = -\frac{3(2n^2 + 5)}{2n^2 + 4n + 7}.$$

When simplifying this fraction we used the fact that $(-1)^n / (-1)^{n-1} = -1$ and $3^{n+1}/3^n = 3$ which we get by subtracting powers. Next we take the absolute value and then find the limit. We get

$$L = \lim_{n \to \infty} \left| \frac{a_{n+1}}{a_n} \right| = \lim_{n \to \infty} \frac{3(2n^2 + 5)}{2n^2 + 4n + 7}$$

$$= \lim_{n \to \infty} \frac{6 + 15/n^2}{2 + 4/n + 7/n^2} = \frac{6}{2} = 3.$$

Replacing L by 3 and a_n by $\dfrac{(-1)^{n-1} 3^n}{2n^2 + 5}$ in the second part of the Ratio Test Theorem, we get the following true statement:

If $3 > 1$, then $\displaystyle\sum_{n=1}^{\infty} \frac{(-1)^{n-1} 3^n}{2n^2 + 5}$ is divergent.

Clearly $3 > 1$. Therefore, we are able to conclude that $\displaystyle\sum_{n=1}^{\infty} \frac{(-1)^{n-1} 3^n}{2n^2 + 5}$ is divergent.

Example 3. Apply the ratio test to the series $\displaystyle\sum_{n=1}^{\infty} (-1)^{n-1} \frac{3n + 5}{7n^2 + 10}$.

Solution. First, we have

$$a_n = \frac{(-1)^{n-1}(3n + 5)}{7n^2 + 10} \quad \text{and} \quad a_{n+1} = \frac{(-1)^n(3n + 8)}{7n^2 + 14n + 17}.$$

The ratio is a_{n+1} divided by a_n. In order to divide we invert and multiply. After inverting the fraction a_n we compute the ratio as follows:

$$\frac{a_{n+1}}{a_n} = \frac{(-1)^n(3n + 8)}{7n^2 + 14n + 17} \cdot \frac{7n^2 + 10}{(-1)^{n-1}(3n + 5)} = \frac{-(3n + 8)(7n^2 + 10)}{(7n^2 + 14n + 17)(3n + 5)}.$$

267

Take the absolute value and then multiply the numerator and denominator by $1/n^3$. This does not change the value of the fraction. Taking the limit, we get

$$\lim_{n\to\infty} \left| \frac{a_{n+1}}{a_n} \right| = \lim_{n\to\infty} \frac{(3+8/n)(7+10/n^2)}{(7+14/n+17/n^2)(3+5/n)} = 1.$$

We get $L = 1$. When $L = 1$, the ratio test gives no information. We must use some other theorem to determine if this series converges. For this series we could use the Alternating Series Theorem to show that it converges.

Exercises

1. a) Consider the infinite series $\displaystyle\sum_{n=1}^{\infty} \frac{2n+5}{(n+1)2^n}$. Find a_n and a_{n+1} for this series.

b) Find $\displaystyle\lim_{n\to\infty} \left| \frac{a_{n+1}}{a_n} \right|$.

c) Substitute into the appropriate part of the Ratio Test Theorem. What conclusions do you reach?

2. a) Consider the series $\displaystyle\sum_{n=1}^{\infty} \frac{(-1)^{n-1}3^n}{2n^2+5n+4}$. Find a_n and a_{n+1} for this series.

b) Find the limit of the absolute value of the ratio.

c) Substitute into the appropriate part of the Ratio Test Theorem. What are you able to conclude?

3. a) Consider the series $\displaystyle\sum_{n=1}^{\infty} \frac{(-1)^{n-1}}{3n^2+5}$. Find a_n and a_{n+1} for this series.

b) Find the limit of the absolute value of the ratio.

c) Substitute into the appropriate part of the Ratio Test Theorem. What are you able to conclude?

4. Consider the series $\displaystyle\sum_{n=1}^{\infty} \frac{(2n+1)2^n}{(5n+3)3^n}$. Use the Ratio Test Theorem to determine if this series is convergent or divergent. This means substitute into the appropriate part of the ratio test. Show all steps.

6713 Power Series

An infinite series of the form $\sum_{n=0}^{\infty} b_n x^n$ is called a power series. The numbers b_n are constants depending on n whereas x is an independent variable. The variable x can be replaced with any real number. Consider the power series

$$\sum_{n=0}^{\infty} \frac{nx^n}{n^2 + 4}$$

where x is a variable. We get somewhat different series by replacing x with different numbers. If we replace x by 2 we get the series $\sum_{n=0}^{\infty} \frac{n2^n}{n^2 + 4}$. If we replace x by 3/4, we get the series $\sum_{n=0}^{\infty} \frac{n(3/4)^n}{n^2 + 4}$. Since we are allowed to replace x by different numbers the series may converge when we replace x by certain numbers and diverge when we replace x by other numbers. This means that when given a power series we ask ourselves the question: for exactly which values of x does this series converge and for exactly which values of x does it diverge?

A remark on the notation for power series. Most series considered up to this point have been written using $\sum_{n=1}^{\infty}$. The first term considered is $n = 1$. On the other hand, power series are usually written with $\sum_{n=0}^{\infty}$. For power series the first term considered is the $n = 0$ term. This is the usual way to write power series since this causes the powers of x to be of the form x^n. When testing a series for convergence, it makes no difference whether we start n with $n = 0$ or $n = 1$. The tests for convergence involve $\lim_{n \to \infty}$ and do not depend on what happens with the small values of n.

Example 1. For what values of x does the power series $\sum_{n=0}^{\infty} \frac{2^n x^n}{n^2 + 1}$ converge?

Solution. In order to determine the values of x for which a power series converges we always apply the Ratio Test. The first step in the ratio test is to find the ratio a_{n+1}/a_n. For this power series

$$a_n = \frac{2^n x^n}{n^2 + 1} \text{ and } a_{n+1} = \frac{2^{n+1} x^{n+1}}{n^2 + 2n + 2}.$$

Recall that in order to divide we invert and multiply. The ratio is

$$\frac{a_{n+1}}{a_n} = \frac{2^{n+1} x^{n+1}}{n^2 + 2n + 2} \cdot \frac{n^2 + 1}{2^n x^n} = \frac{2x(n^2 + 1)}{n^2 + 2n + 2}.$$

We simplified using the fact that $x^{n+1}/x^n = x$ and $2^{n+1}/2^n = 2$. Recall that we must take the absolute value of the ratio when we find L. The number L in the Ratio Test is given by

$$L = \lim_{n \to \infty} \left| \frac{a_{n+1}}{a_n} \right| = \lim_{n \to \infty} \left| \frac{2x(n^2 + 1)}{n^2 + 2n + 2} \right|$$

We can factor $|x|$ out in front of the limit sign since $|x|$ does not depend on n.

$$L = 2|x| \lim_{n \to \infty} \frac{n^2 + 1}{n^2 + 2n + 2} = 2|x|.$$

Since $L = 2|x|$ we replace L with $2|x|$ and a_n with $\dfrac{2^n x^n}{n^2 + 1}$ in the Ratio Test. This gives us the following two true statements:

1) If $2|x| < 1$, then $\displaystyle\sum_{n=1}^{\infty} \frac{2^n x^n}{n^2 + 1}$ converges.

2) If $2|x| > 1$, then $\displaystyle\sum_{n=1}^{\infty} \frac{2^n x^n}{n^2 + 1}$ diverges.

Note that "$2|x| < 1$" is the same as "$x > -1/2$ and also that $x < 1/2$". We often write this as $-1/2 < x < 1/2$. The statement "$2|x| > 1$" is the same as "either $x < -1/2$ or $x > 1/2$". Recall that when $L = 1$ there is no conclusion in the Ratio Test. For this series $L = 1$ is $2|x| = 1$, that is, $x = -1/2$ or $x = 1/2$. We do not know from the Ratio Test if this

271

series converges or if it diverges when $x = -1/2$ and $x = 1/2$. The values $x = -1/2$ and $x = 1/2$ are known as *end points* for the power series. We will not try to determine if a power series converges or if it diverges at its end points. We could use other theorems and show that this particular power series converges for $x = -1/2$ and $x = 1/2$. We will not do this. Given a power series we will only apply the ratio test in order to find for what values of x it converges. For a power series suppose $L = |x|/R$, then R is called the radius of convergence of the power series. For this series we say that the radius of convergence is $R = 1/2$.

Example 2. For what values of x does the following power series converge?

$$\sum_{n=0}^{\infty} \frac{(n+1)^2 x^n}{5^n}$$

Solution. We always apply the Ratio Test and only the Ratio Test to determine the values of x for which a power series converges. For this power series

$$a_n = \frac{(n+1)^2 x^n}{5^n} \quad \text{and} \quad a_{n+1} = \frac{(n+2)^2 x^{n+1}}{5^{n+1}}$$

We need to find a_{n+1} divided by a_n. Recall that to divide we invert and multiply. After inverting the fraction a_n, the ratio is

$$\frac{a_{n+1}}{a_n} = \frac{(n+2)^2 x^{n+1}}{5^{n+1}} \cdot \frac{5^n}{(n+1)^2 x^n} = \frac{x (n+2)^2}{5 (n+1)^2}.$$

We used the fact that $x^{n+1}/x^n = x$ and $5^n/5^{n+1} = 1/5$. Do not forget to take absolute value when finding L.

$$L = \lim_{n\to\infty} \left| \frac{x (n+2)^2}{5 (n+1)^2} \right| = \frac{|x|}{5} \lim_{n\to\infty} \frac{(1+2/n)^2}{(1+1/n)^2} = \frac{|x|}{5}.$$

We can factor $|x|/5$ out in front of the limit sign since $|x|/5$ does not depend on n. Replacing L with $|x|/5$ and a_n with $\dfrac{(n+1)^2 x^n}{5^n}$ in the ratio test, we get the following true statements

1. If $\dfrac{|x|}{5} < 1$, then $\displaystyle\sum_{n=0}^{\infty} \dfrac{(n+1)^2 x^n}{5^n}$ converges.

2. If $\dfrac{|x|}{5} > 1$, then $\displaystyle\sum_{n=0}^{\infty} \dfrac{(n+1)^2 x^n}{5^n}$ diverges.

Note that $\dfrac{|x|}{5} < 1$ is the same as $-5 < x < 5$, which is the same as $-5 < x$ and at the same time $x < 5$. Also note that $\dfrac{|x|}{5} > 1$ is the same as "$x < -5$ or $x > 5$". The end points for this power series are $x = -5$ and $x = 5$. We do not try to decide if the power series is convergent for these two values of x. Using other theorems we could show that this series is divergent when $x = -5$ and when $x = 5$. The radius of convergence for this power series is 5. The interval of convergence is $-5 < x < 5$.

Example 3. For what values of x does the power series $\displaystyle\sum_{n=0}^{\infty} \dfrac{x^n}{n!}$ converge?

Solution. For this power series

$$a_n = \frac{x^n}{n!} \text{ and } a_{n+1} = \frac{x^{n+1}}{(n+1)!}$$

The ratio is

$$\frac{a_{n+1}}{a_n} = \frac{x^{n+1}}{(n+1)!} \cdot \frac{n!}{x^n} = \frac{x(n!)}{(n+1)!}$$

The definition of factorial says that $4! = 1 \cdot 2 \cdot 3 \cdot 4$ and $6! = 1 \cdot 2 \cdot 3 \cdot 4 \cdot 5 \cdot 6$. This means that $6! = 6(5!)$ and $10! = 10(9!)$. In general $(n+1)! = (n+1)(n!)$. This can be rewritten as

$$\frac{1}{n+1} = \frac{n!}{(n+1)!}$$

Using this we reduce the fraction as follows:

$$\frac{a_{n+1}}{a_n} = \frac{x(n!)}{(n+1)!} = \frac{x}{n+1}.$$

273

Taking the limit of this ratio we find L:

$$L = \lim_{n \to \infty} \frac{|a_{n+1}|}{|a_n|} = \lim_{n \to \infty} \frac{|x|}{n+1} = 0.$$

This limit is zero for all x. Replacing L by 0 and a_n by $x^n/n!$ in the first part of the ratio test, we get the true statement.

If $0 < 1$, then $\displaystyle\sum_{n=0}^{\infty} \frac{x^n}{n!}$ converges.

From this we conclude that $\displaystyle\sum_{n=0}^{\infty} \frac{x^n}{n!}$ converges for all values of x.

Exercises

1. a) Consider the power series $\displaystyle\sum_{n=0}^{\infty} \frac{(2n+3)x^n}{(n^2+1)2^n}$. Find a_n and a_{n+1} for this series.

b) Find $\displaystyle\lim_{n \to \infty} \left| \frac{a_{n+1}}{a_n} \right|$.

c) Substitute into the Ratio Test Theorem. What conclusions do you reach?

2. (a) Consider the power series $\displaystyle\sum_{n=0}^{\infty} (-1)^n \frac{(n^2+4)x^n}{3n+5}$. Find a_n and a_{n+1} for this series.

b) Next find the ratio a_{n+1}/a_n and then find the limit of the absolute value of the ratio, that is, find L.

c) Substitute into the Ratio Test Theorem. What conclusions do you reach?

3. a) Consider the power series $\displaystyle\sum_{n=0}^{\infty} (-1)^n \frac{x^{2n}}{(2n)!}$. Find a_n and a_{n+1} for this series.

b) Find the ratio $\dfrac{a_{n+1}}{a_n}$.

274

c) Find the limit of the absolute value of the ratio.

d) Substitute into the Ratio Test Theorem. What conclusions do you reach?

4.(a) Consider the power series $\sum_{n=0}^{\infty}(-1)^n(n+1)3^nx^n$. Find a_n and a_{n+1} for this series.

b) Find the limit of the absolute value of the ratio.

c) Substitute into the Ratio Test Theorem. What is your conclusion?

6715 Taylor's Series and Polynomials

We are now going to discuss: given a function $f(x)$ find the Taylor's series for $f(x)$. In order to find the Taylor's series for $f(x)$, the function $f(x)$ must have an infinite number of derivatives.

Definition. Taylor's Series. The formula for the Taylor's series for a function $f(x)$ about $a = 0$ is given by

$$f(x) = \sum_{n=0}^{\infty} \frac{f^{(n)}(0)}{n!} x^n$$

$$= f(0) + \frac{f'(0)}{1}x + \frac{f''(0)}{2!}x^2 + \frac{f'''(0)}{3!}x^3 + \frac{f^{(4)}(0)}{4!}x^4 + \ldots$$

Example 1. Find the Taylor's series about $a = 0$ (also called the Maclaurian series) for the function $f(x) = e^{5x}$.

Solution. We need to find $f^{(n)}(0)$ in order to substitute for it in the formula for Taylor's series. This means we need to find all the derivatives of the function $f(x)$. The first few derivatives are:

$$f'(x) = 5e^{5x} \quad f''(x) = 5^2 e^{5x} \quad f'''(x) = 5^3 e^{5x}$$
$$f^{(4)}(x) = 5^4 e^{5x} \quad f^{(5)}(x) = 5^5 e^{5x} \quad f^{(6)}(x) = 5^6 e^{5x}.$$

We need an expression for $f^{(n)}(x)$ which is good for any positive integer n. The hardest part of finding the Taylors series for a function $f(x)$ is to find a general formula for the general nth derivative of $f(x)$. We do this by looking at the pattern of the first half dozen or so derivatives. For this example note that the 4th derivative has 5^4 as a factor, the 5th derivative has 5^5, the 6th derivative has 5^6. From this pattern we conclude that for any nth derivative we would have 5 to nth power or 5^n as the coefficient of e^{5x}. Therefore,

$$f^{(n)}(x) = 5^n e^{5x}.$$

This says that the 15th derivative is given by $f^{(15)}(x) = 5^{15} e^{5x}$. We need the nth derivative evaluated when $x = 0$.

$$f^{(n)}(0) = 5^n e^0 = 5^n.$$
$$\frac{f^{(n)}(0)}{n!} = \frac{5^n}{n!}$$

We substitute this into $\sum_{n=0}^{\infty} \dfrac{f^{(n)}(0)}{n!} x^n$ and find that the Taylor's series for e^{5x} is

$$e^{5x} = \sum_{n=0}^{\infty} \frac{5^n}{n!} x^n.$$

Note that this Taylor's Series is a power series. We also have the question: for what values of x does this Taylor's series for e^{5x} converge? Since it is a power series we determine for what values of x this series converges by using the Ratio test. Indeed any Taylor's Series is a power series. This means that we can determine for what values of x any Taylor's series converges by applying the ratio test. However, in a complete discussion of Taylor's series we do not need to apply the ratio test to determine for what values of x the Taylor's series converges. In fact as part of a complete discussion of Taylor's series we find out not only for what values of x the series converges but also the sum of the series. The sum of the Taylor's series for the function $f(x)$ is the function $f(x)$ for all values of x for which the series converges. Starting with a function $f(x)$ we can find the Taylor's series for $f(x)$. We then use the ratio test to determine for what values of x the series converges. The sum of the Taylor's series is equal to $f(x)$ for whatever values of x the series converges. The series $\sum_{n=0}^{\infty} \dfrac{5^n x^n}{n!}$ converges for all real members x. We find this fact by applying the ratio test. This means that the sum of the series is e^{5x} for any real number x. Knowing the sum of the series can be very helpful.

In general finding the Taylor's series for a given function is a difficult task. The difficult part is to find a formula for $f^{(n)}(x)$. For this reason, we will not find many Taylor's series by directly using the formulas given above. However, we do need to know a few basic Taylor's series. As part of the work in finding these Taylor's series we could also find the sum of the series and the values of x for which the series converge. The Taylor's series for a few common functions are given below.

$$\frac{1}{1-x} = \sum_{n=0}^{\infty} x^n = 1 + x + x^2 + x^3 + \dots \quad \text{for } |x| < 1.$$

$$e^x = \sum_{n=0}^{\infty} \frac{x^n}{n!} = 1 + x + \frac{x^2}{2} + \frac{x^3}{6} + \dots, \text{ for all } x$$

$$\sin x = \sum_{n=0}^{\infty} (-1)^n \frac{x^{2n+1}}{(2n+1)!} = x - \frac{x^3}{3!} + \frac{x^5}{5!} - \frac{x^7}{7!} + \dots \text{ for all } x$$

$$\cos x = \sum_{n=0}^{\infty} (-1)^n \frac{x^{2n}}{(2n)!} = 1 - \frac{x^2}{2} + \frac{x^4}{4!} - \frac{x^6}{6!} + \dots \text{ for all } x$$

$$\arctan x = \sum_{n=0}^{\infty} (-1)^n \frac{x^{2n+1}}{2n+1} = x - \frac{x^3}{3} + \frac{x^5}{5} - \frac{x^7}{7} + \dots \text{ for } |x| < 1.$$

The values of x for what these series converge are indicated. We find these values using the ratio test. The sum of the series is equal to the function for these values of x.

A series of the form $\sum_{n=0}^{\infty} a_n x^n$ is called a power series. Note that these Taylor's series are power series. When we find a Taylor's series, we also call it a power series. In some more advanced discussions we sometimes talk about power series that are not Taylor's series. In these advanced discussions the term "Taylor's series" is used to indicate that the series was originally found using the derivative formulas.

Definition. The Taylor polynomial $T_n(x)$ of order n of the function $f(x)$ about the value $a = 0$ is given by

$$T_n(x) = f(0) + f'(0)x + \frac{f''(0)}{2!}x^2 + \frac{f'''(0)}{3!}x^3 + \dots + \frac{f^{(n)}(0)}{n!}x^n.$$

The Taylor polynomial $T_n(x)$ of order n for $f(x)$ is the terms up to x^n of the Taylor's series for $f(x)$. The Taylor polynomial $T_3(x)$ of order 3 for $f(x)$ is the terms up to x^3 of the Taylor's series for $f(x)$. The general formula for $T_3(x)$ is

$$T_3(x) = f(0) + f'(0)x + \frac{f''(0)}{2!}x^2 + \frac{f'''(0)}{3!}x^3.$$

The Taylor polynomial $T_5(x)$ of order 5 for $f(x)$ is the terms up to x^5 of the Taylor's series for $f(x)$. The general formula for $T_5(x)$ is

$$T_5(x) = f(0) + f'(0)x + \frac{f''(0)}{2!}x^2 + \frac{f'''(0)}{3!}x^3 + \frac{f^{(4)}(0)}{4!}x^4 + \frac{f^{(5)}(0)}{5!}x^5.$$

Example 2. The Taylor polynomial $T_4(x)$ of degree 4 for $f(x) = e^x$ is

$$T_4(x) = 1 + x + \frac{x^2}{2!} + \frac{x^3}{3!} + \frac{x^4}{4!}$$

We obtain this polynomial by looking at the Taylor's series for e^x and copying down the terms up to x^4.

Example 3. The Taylor polynomial $T_6(x)$ of degree 6 for $f(x) = \cos x$ is

$$T_6(x) = 1 - \frac{x^2}{2} + \frac{x^4}{4!} - \frac{x^6}{6!}$$

We get the polynomial $T_6(x)$ for $\cos x$ by looking at the Taylor's series for $\cos x$ and copying down the terms up to x^6.

Example 4. Find $T_4(x)$, the Taylor polynomial of degree 4 about $x = 0$, for $f(x) = \sin 5x + \cos 4x$.

Solution. We are not given the Taylor's series for this function. In order to find this Taylor polynomial we must use the derivative formulas. We start by finding the first four derivatives. They are

$$f(x) = \sin 5x + \cos 4x$$
$$f'(x) = 5\cos 5x - 4\sin 4x$$
$$f''(x) = -25\sin 5x - 16\cos 4x$$
$$f'''(x) = -125\cos 5x + 64\sin 4x$$
$$f^{(4)}(x) = 625\sin 5x + 256\cos 4x$$

Next, we evaluate these derivatives when $x = 0$.

$$f(0) = 1 \qquad f'(0) = 5 \qquad f''(0) = -16$$
$$f'''(0) = -125 \qquad f^{(4)}(0) = 256$$

Next divide each derivative by the appropriate factorial:

$$\frac{f''(0)}{2} = -8, \quad \frac{f'''(0)}{3!} = -\frac{125}{6} \quad \text{and} \quad \frac{f^{(4)}(0)}{4!} = \frac{256}{24} = \frac{32}{3}.$$

279

Substituting into the formula for the Taylor polynomial, we get the polynomial for $f(x) = \sin 5x + \cos 4x$.

$$T_4(x) = 1 + 5x - 8x^2 - \frac{125}{6}x^3 + \frac{32}{3}x^4.$$

Example 5. Find $T_5(x)$, the Taylor polynomial of degree 5, for $f(x) = (9 + 4x)^{1/2}$.

Solution. Since we do not know the Taylor's series for $(9 + 4x)^{1/2}$ we must use the derivative formulas to find $T_5(x)$. In order to substitute into the formula for $T_5(x)$ we must find the first five derivatives of $f(x) = (9+4x)^{1/2}$.

$$f(x) = (9 + 4x)^{1/2}$$
$$f'(x) = 2(9 + 4x)^{-1/2}$$
$$f''(x) = -4(9 + 4x)^{-3/2}$$
$$f'''(x) = 24(9 + 4x)^{-5/2}$$
$$f^{(4)}(x) = -240(9 + 4x)^{-7/2}$$
$$f^{(5)}(x) = 3360(9 + 4x)^{-9/2}$$

Next we evaluate these derivatives when $x = 0$.

$$f(0) = 3 \qquad f'(0) = 2/3 \qquad f''(0) = -4/27$$
$$f'''(0) = 24/243 \quad f^{(4)}(0) = -240/2187 \quad f^{(5)}(0) = 3360/19683$$

Next divide each derivative by appropriate factorial.

$$\frac{f''(0)}{2} = -\frac{2}{27} \qquad\qquad \frac{f'''(0)}{3!} = \frac{4}{243}$$

$$\frac{f^{(4)}(0)}{4!} = -\frac{240}{24(2187)} = -\frac{10}{2187} \qquad \frac{f^{(5)}(0)}{5!} = \frac{28}{19683}$$

Substituting into the formula for the Taylor polynomial, we get

$$T_5(x) = 3 + \frac{2x}{3} - \frac{2x^2}{27} + \frac{4x^3}{243} - \frac{10x^4}{2187} + \frac{28x^5}{19683}.$$

280

So far we have discussed two methods for finding the Taylor polynomial of a function. First method, if we are real lucky the polynomial we want to find is just the first few terms of a known Taylor's series. Second method, we can take the derivatives of the function and substitute into the general formula for the Taylor polynomial. It is always possible to use this second method but it can be long.

There are also other methods for finding Taylor polynomials which we are now going to discuss. Our third method is to find the Taylor polynomial of a function by manipulating the polynomial of a related function.

Example 6. Find the Taylor polynomial of degree 5 for $\sin 6x$ and for $\sin(x/3)$. Start with the known Taylor polynomial for $\sin x$.

Solution. The Taylor polynomial of degree 5 for $\sin x$ is

$$\sin x \approx x - \frac{x^3}{6} + \frac{x^5}{120}.$$

We find this polynomial by copying the terms up to x^5 in the power series for $\sin x$. Replacing x with $6x$ we obtain the polynomial of degree 5 for $\sin 6x$:

$$\sin 6x \approx 6x - \frac{(6x)^3}{6} + \frac{(6x)^5}{120}.$$

The Taylor polynomial $T_5(x)$ for $\sin 6x$ is

$$T_5(x) = 6x - \frac{216x^3}{6} + \frac{7776x^5}{120} = 6x - 36x^3 + \frac{324}{5}x^5.$$

Replacing x with $x/3$ we obtain the polynomial of degree 5 for $\sin(x/3)$.

$$\sin\frac{x}{3} \approx \frac{x}{3} - \frac{(x/3)^3}{6} + \frac{(x/3)^5}{120}$$

$$\approx \frac{x}{3} - \frac{x^3}{162} + \frac{x^5}{29160}.$$

Example 7. Find the Taylor polynomial of degree 4 for $f(x) = (4+x)^{-1}$.

Solution. We start by finding the Taylor polynomial of degree 4 for $(1-x)^{-1}$. It is the terms up to x^4 of power series for $(1-x)^{-1}$ which is given above. The Taylor polynomial for $(1-x)^{-1}$ is

$$(1-x)^{-1} \approx 1 + x + x^2 + x^3 + x^4.$$

281

Replacing x with $(-x/4)$ we get the polynomial of degree 4 for $[1+(x/4)]^{-1}$.

$$\frac{1}{1+(x/4)} \approx 1 - \frac{x}{4} + \frac{x^2}{4^2} - \frac{x^3}{4^3} + \frac{x^4}{4^4}$$

$$\frac{4}{4+x} \approx 1 - \frac{x}{4} + \frac{x^2}{16} - \frac{x^3}{64} + \frac{x^4}{256}$$

Dividing both sides by 4 we get

$$\frac{1}{4+x} \approx \frac{1}{4} - \frac{x}{16} + \frac{x^2}{64} - \frac{x^3}{256} + \frac{x^4}{1024}.$$

We could also find this exact same polynomial by substituting $f(x) = (4 + x)^{-1}$ into the derivative formulas.

Fourth method. We are now going to discuss a fourth method for finding a Taylor polynomial. This method involves differentiating the function and differentiating the corresponding polynomial to obtain a new function and a new polynomial. We use the following theorem which is easy to prove.

Theorem 1. If $P_n(x)$ is the Taylor polynomial of order n of $f(x)$, then $P_n'(x)$ is the Taylor polynomial of order $(n-1)$ of $f'(x)$.

Example 8. Find the Taylor polynomial for $(4 + x)^{-2}$.

Solution. Note that $\frac{d}{dx}(4+x)^{-1} = (-1)(4+x)^{-2}$. The Taylor polynomial for $(4+x)^{-1}$ is given above. We differentiate the function and we differentiate the polynomial. Differentiating both sides gives

$$-\frac{1}{(4+x)^2} \approx -\frac{1}{16} + \frac{x}{32} - \frac{3x^2}{256} + \frac{x^3}{256}$$

$$\frac{1}{(4+x)^2} \approx \frac{1}{16} - \frac{x}{32} + \frac{3x^2}{256} - \frac{x^3}{256}$$

Note that this is $T_3(x)$ the Taylor polynomial for $(4 + x)^{-2}$ of degree 3.

Fifth method. Finally, we are going to discuss a fifth method for finding a Taylor polynomial. This method involves integrating the function and integrating the corresponding polynomial to obtain a new function and a new polynomial. We use the following theorem which is easy to prove.

Theorem 2. If $P_n(x)$ is the Taylor polynomial of order n of $f(x)$, then $\int P_n(x)dx$ is the Taylor polynomial of order $(n+1)$ of $\int f(x)dx$. The constant of integration must be assigned the appropriate value.

Example 9. Find the Taylor polynomial $T_5(x)$ for $\ln(4+x)$.

Solution. Note that $\int(4+x)^{-1}dx = \ln(4+x) + C$. We start with

$$(4+x)^{-1} \approx \frac{1}{4} - \frac{x}{16} + \frac{x^2}{64} - \frac{x^3}{256} + \frac{x^4}{1024},$$

which was found above. First,

$$\int \left[\frac{1}{4} - \frac{x}{16} + \frac{x^2}{64} - \frac{x^3}{256} + \frac{x^4}{1024}\right] dx$$

$$= C + \frac{x}{4} - \frac{x^2}{32} + \frac{x^3}{3(64)} - \frac{x^4}{4(256)} + \frac{x^5}{5(1024)}.$$

This tells us that

$$\ln(4+x) \approx C + \frac{x}{4} - \frac{x^2}{32} + \frac{x^3}{3(64)} - \frac{x^4}{4(256)} + \frac{x^5}{5(1024)}.$$

In order to find the constant of integration C let us substitute $x = 0$ into both sides. We get

$$\ln 4 = C$$

The value of the constant of integration is $C = \ln 4$.

$$\ln(4+x) \approx \ln 4 + \frac{x}{4} - \frac{x^2}{32} + \frac{x^3}{192} - \frac{x^4}{1024} + \frac{x^5}{5120}.$$

We can also find the value of C as follows. The expression

$$C + \frac{x}{4} - \frac{x^2}{32} + \frac{x^3}{3(64)} - \frac{x^4}{4(256)} + \frac{x^5}{5(1024)}$$

is the Taylor polynomial $T_5(x)$ for $\ln(4 + x)$. We do not know the constant term C in this polynomial. However, the constant term in the Taylor polynomial for $f(x)$ is always given by $f(0)$. It follows that $C = f(0)$ or $C = \ln(4 + 0) = \ln 4$.

$$T_5(x) = \ln 4 + \frac{x}{4} - \frac{x^2}{32} + \frac{x^3}{192} - \frac{x^4}{1024} + \frac{x^5}{5120}.$$

Essentially all the Taylor polynomials we have discussed so far have been in powers of $(x - 0)^n$. It is possible to use a center $b \neq 0$ for a Taylor polynomial. The Taylor polynomial $T_4(x)$ about the general number b rather than $b = 0$ is

$$T_4(x) = f(b) + f'(b)(x - b) + \frac{f''(b)}{2!}(x - b)^2 + \frac{f^{(3)}(b)}{3!}(x - b)^3 + \frac{f^{(4)}(b)}{4!}(x - b)^4$$

Example 10. Find the Taylor polynomial $T_4(x)$ for $f(x) = \sqrt{x}$ about $a = 4$.

Solution. First, note that $b \neq 0$. The general formula for the Taylor polynomial of order 4 when $b \neq 0$ is given above. We need to find the first 4 derivatives of $f(x) = \sqrt{x}$.

$$f'(x) = \frac{1}{2}x^{-1/2} \qquad\qquad f''(x) = \frac{1}{2}(-\frac{1}{2})x^{-3/2}$$

$$f'''(x) = \frac{1}{2}(-\frac{1}{2})(-\frac{3}{2})x^{-5/2} \qquad f^{(4)}(x) = \frac{1}{2}(-\frac{1}{2})(-\frac{3}{2})(-\frac{5}{2})x^{-7/2}$$

We need to evaluate the given function and these derivatives for $a = 4$.

$$f(4) = \sqrt{4} = 2$$
$$f'(4) = (1/2)(4)^{-1/2} = 1/4$$
$$f''(4) = -(1/4)(4)^{-3/2} = -1/32$$
$$f'''(4) = (3/8)(4)^{-5/2} = 3/256$$
$$f^{(4)}(4) = -(15/16)(4)^{-7/2} = -(15/2048).$$

Dividing by the appropriate factorial, we get

$$\frac{f^{(2)}(4)}{2} = -\frac{1}{64} \qquad\qquad \frac{f'''(4)}{3!} = \frac{1}{512}$$

$$\frac{f^{(4)}(4)}{4!} = -\frac{5}{16,384}.$$

Substituting into the formula we get the Taylor polynomial

$$T_4(x) = 2 + \frac{1}{4}(x-4) - \frac{1}{64}(x-4)^2 + \frac{1}{512}(x-4)^3 - \frac{5}{16384}(x-4)^4.$$

We can also find the Taylor polynomial for $\sqrt{x}$ about $b = 4$ by making a substitution. Let $y = x - 4$ or $x = y + 4$. Now find the Taylor polynomial of degree 4 about $y = 0$ for $\sqrt{y+4}$. After some work we find the polynomial for $\sqrt{y+4}$. This work is almost exactly the same as the work done above to find the polynomial for $\sqrt{x}$ in powers of $(x-4)$.

$$\sqrt{y+4} \approx 2 + \frac{y}{4} - \frac{1}{64}y^2 + \frac{1}{512}y^3 - \frac{5}{16384}y^4.$$

Replace y with $x - 4$ and we get

$$\sqrt{x} \sim 2 + \frac{1}{4}(x-4) - \frac{1}{64}(x-4)^2 + \frac{1}{512}(x-4)^3 - \frac{5}{16384}(x-4)^4.$$

This method of finding the Taylor polynomial for $f(x)$ about a number b where $b \neq 0$ can be used for any function. This makes it clear why we only practice finding polynomials where the center is $b = 0$.

We have a special name which we discussed earlier for the Taylor polynomial of order one. The general Taylor polynomial of order 1 for $f(x)$ about $x = b$ is

$$T_1(x) = f(b) + f'(b)(x - b).$$

This polynomial is also known as the linearization of $f(x)$ at $x = b$.

Example 11. Let $f(x) = (4x+9)^{3/2}$. Find the Taylor polynomial of order 1 for $f(x)$ about $x = 10$. This is called the linearization of $(4x+9)^{3/2}$ about $x = 10$.

Solution. We need to find $f'(x)$.

$$f'(x) = (3/2)(4x+9)^{1/2}(4) = 6(4x+9)^{1/2}.$$

$$f(10) = (49)^{3/2} = 343, \text{ and } f'(10) = 6(49)^{1/2} = 42$$

The Taylor polynomial is

$$T_1(x) = 343 + 42(x - 10).$$

The Taylor polynomial of order one of $f(x)$ about $x = b$ is also called the linearization of $f(x)$ at $x = b$. The linearization of $f(x) = (4x + 9)^{3/2}$ at $x = 10$ is

$$L(x) = 343 + 42(x - 10).$$

Let us review our more difficult methods for finding Taylor polynomials. The general Taylor polynomial for $f(x)$ about $b = 0$ is called $T_n(x)$. In particular the polynomial of degree 4 is denoted by $T_4(x)$. The general formula for $T_4(x)$ is

$$T_4(x) = f(0) + f'(0)x + \frac{f''(0)}{2!}x^2 + \ldots + \frac{f^{(4)}(0)}{4!}x^4.$$

Example 12. The infinite series for $(1 + x)^{1/2}$ is

$$(1 + x)^{1/2} = 1 + \frac{1}{2}x + \sum_{n=2}^{\infty} \frac{(-1)^{n-1}}{2^n(n!)}[1 \cdots 3 \cdots 5 \cdots (2n - 3)]x^n.$$

We find the Taylor polynomial $T_4(x)$ for $(1 + x)^{1/2}$ by using the general formula for $T_4(x)$ given above or by copying the first 5 terms of this series. The Taylor polynomial $T_4(x)$ for $f(x) = (1 + x)^{1/2}$ is

$$(1 + x)^{1/2} \approx T_4(x) = 1 + \left(\frac{1}{2}\right)x - \left(\frac{1}{8}\right)x^2 + \left(\frac{1}{16}\right)x^3 - \left(\frac{5}{128}\right)x^4.$$

Find the polynomial $T_4(x)$ for $(9 + 4x)^{1/2}$.

Solution. We use the third method. The third method is to start with the polynomial of a related function and then manipulate both the function and the polynomial. Note that $(9 + 4x)^{1/2} = 3[1 + (4x/9)]^{1/2}$. Replacing x in the given relationship with $4x/9$, we get

$$(1 + 4x/9)^{1/2} \approx 1 + \left(\frac{1}{2}\right)\left(\frac{4x}{9}\right) - \left(\frac{1}{8}\right)\left(\frac{4x}{9}\right)^2 + \left(\frac{1}{16}\right)\left(\frac{4x}{9}\right)^3 - \frac{5}{128}\left(\frac{4x}{9}\right)^4.$$

Multiplying both sides by 3 and simplifying, we get

$$[9 + 4x]^{1/2} \approx 3 + \frac{2x}{3} - \frac{2x^2}{27} + \frac{4x^3}{243} - \frac{10x^4}{2187}. \tag{7.1}$$

286

This is the Taylor polynomial of degree 4 for $(9 + 4x)^{1/2}$.

Example 13. Starting with the polynomial $T_4(x)$ for $(9 + 4x)^{1/2}$ found in Example 12 find the Taylor polynomial for $(9 + 4x)^{-1/2}$.

Solution. Note that $\dfrac{d}{dx}(9 + 4x)^{1/2} = (2)(9 + 4x)^{-1/2}$. We use the fourth method which is to differentiate the function and the polynomial. We then apply Theorem 1. The Taylor polynomial for $(9 + 4x)^{1/2}$ was found above as (7.1). Differentiating both sides of (7.1), we get

$$(2)(9 + 4x)^{-1/2} \approx \frac{2}{3} - \frac{4x}{27} + \frac{4x^2}{81} - \frac{40x^3}{2187}.$$

Dividing both sides by 2 gives

$$(9 + 4x)^{-1/2} \approx \frac{1}{3} - \frac{2x}{27} + \frac{2x^2}{81} - \frac{20x^3}{2187}.$$

This is the polynomial of degree 3 for $(9 + 4x)^{-1/2}$.

Example 14. Starting with the Taylor polynomial of degree 4 for $(9 + 4x)^{1/2}$, find a Taylor polynomial for $(9 + 4x)^{3/2}$.

Solution. We use the fifth method which is to integrate the function and the polynomial. We then apply Theorem 2. Since $\int (9 + 4x)^{1/2} dx = (1/6)(9 + 4x)^{3/2}$, we can find a polynomial for $(9 + 4x)^{3/2}$ by integrating the polynomial for $(9 + 4x)^{1/2}$. Integrating both sides of the relationship (7.1), we get

$$\frac{1}{6}(9 + 4x)^{3/2} \approx C + 3x + \frac{x^2}{3} - \frac{2x^3}{81} + \frac{x^4}{243} - \frac{2x^5}{2187}.$$

Multiplying both sides by 6, we get

$$(9 + 4x)^{3/2} \approx 6C + 18x + 2x^2 - \frac{4x^3}{27} + \frac{2x^4}{81} - \frac{4x^5}{729}.$$

We evaluate the constant of integration by substituting $x = 0$ into both sides. This gives $27 \approx 6C$. Substituting this value for $6C$, we get

$$(9 + 4x)^{3/2} \approx 27 + 18x + 2x^2 - \frac{4x^3}{27} + \frac{2x^4}{81} - \frac{4x^5}{729}.$$

This is the Taylor polynomial for $(9 + 4x)^{3/2}$ of order 5. If we are ever in doubt about the value of the constant term in the Taylor polynomial for a function $f(x)$, we can just recall that the constant term is always $f(0)$. For this function $f(0) = (9 + 0)^{3/2} = 27$.

Exercises

1. Use the derivative formulas for finding a Taylor polynomial to find the Taylor polynomial $T_4(x)$ for each of the following functions:

   a) $f(x) = \sin 3x + \cos 5x$ \qquad b) $f(x) = \sqrt{25 + 4x}$

2. By looking at the appropriate power series find the Taylor polynomial $T_4(x)$ for each of the following functions:

   a) $\cos x$ \qquad b) $\arctan x$ \qquad c) $\sin x$

3. Start with an appropriate Taylor polynomial found by looking at a known Taylor's series, and by substituting into it find the Taylor polynomial $T_5(x)$ for each of the following functions:

   a) $\dfrac{1}{1 - x^2}$ \qquad b) $\cos(5x)$ \qquad c) e^{-x^2}

4. Start with the Taylor polynomial of degree 5 for $(1 - x)^{-1}$ and find the Taylor polynomial of degree 5 for $f(x) = (5 + x)^{-1}$.

5. Start with the Taylor polynomial of degree 6 for $\arctan x$ and find the Taylor polynomial of degree 5 for $f(x) = (1 + x^2)^{-1}$ using differentiation.

6. The Taylor polynomial of degree 4 for $(1 + x)^{3/2}$ about $x = 0$ is

$$(1 + x)^{3/2} \approx 1 + \frac{3}{2}x + \frac{3}{8}x^2 - \frac{1}{16}x^3 + \frac{3}{128}x^4.$$

Find the Taylor polynomial of degree 4 for $(9 + 2x)^{3/2}$.
Find the Taylor polynomial of degree 3 for $(9 + 2x)^{1/2}$.
Find the Taylor polynomial of degree 4 for $(9 + 2x)^{5/2}$.

7. The Taylor polynomial of degree 4 for $(8 + x)^{2/3}$ about $x = 0$ is

$$4 + \frac{x}{3} - \frac{x^2}{144} + \frac{x^3}{2592} - \frac{7x^4}{248832}.$$

Find the Taylor polynomial of degree 3 about $x = 0$ for $(8 + x)^{-1/3}$. Find the Taylor polynomial $T_4(x)$ of degree 4 about $x = 0$ for $(8 + x)^{5/3}$.

8. Let $f(x) = x^{3/2}$. Find the linearization of $f(x)$ at $x = 64$.

9. Let $f(x) = \frac{x}{\sqrt{5+x}}$. Find the linearization of $f(x)$ at $x = 4$.

10. The Taylor polynomial of 4th degree for $(8 + 3x)^{1/3}$ is

$$(8 + 3x)^{1/3} \approx 2 + \frac{x}{4} - \frac{x^2}{32} + \frac{5x^3}{768} - \frac{5x^4}{3072}.$$

a) Use this to find $T_3(x)$ for $(8 + 3x)^{-2/3}$.

b) Use this to find $T_5(x)$ for $(8 + 3x)^{4/3}$.

6717 Applications of Taylor Polynomials

The reason we find the Taylor polynomial $T_n(x)$ of a function $f(x)$ is that we want to use the polynomial $T_n(x)$ to approximate the function $f(x)$. This is an important idea for us. For example, we found the Taylor polynomial $T_5(x)$ for $f(x) = (9 + 4x)^{3/2}$. This polynomial $T_5(x)$ is

$$(9 + 4x)^{3/2} \sim 27 + 18x + 2x^2 - \frac{4x^3}{27} + \frac{2x^4}{81} - \frac{4x^5}{729}.$$

Suppose we want to calculate a value for $(9 + 1.2)^{3/2}$. Note that this is $x = 0.3$. Forgetting about our calculator we could use the polynomial $T_5(x)$ and calculate $T_5(0.3)$ to approximate $(9 + 1.2)^{3/2}$. Substituting $x = 0.3$ into the polynomial, we get

$$(9 + 1.2)^{3/2} \approx 27 + 18(0.3) + 2(0.3)^2 - \frac{4(0.3)^3}{27} + \frac{2(0.3)^4}{81} - \frac{4(0.3)^5}{729}$$

$$= 32.5761867.$$

This is an approximate value of $(10.2)^{3/2}$. How close is this approximate value to the actual value? Without a calculator how do we answer this question? As it turns out we can answer this question using the Alternating Series Estimation Theorem. We need to look at exactly what the Alternating Series Estimation Theorem tells us when we apply the estimation theorem to power series. Recall the theorem

Alternating Series Estimation Theorem. If $\lim_{k \to \infty} b_k = 0$ and $b_{k+1} < b_k$, then

$$\left| \sum_{k=N+1}^{\infty} (-1)^k b_k \right| < b_{N+1}.$$

We did not compute the 6th term of the power series for $(9 + 4x)^{3/2}$. Also we did not show that the power series for $(9 + 4x)^{3/2}$ is an alternating series. In order to apply the Alternating Series Theorem we need to know that we have a convergent alternating series and we need to know the formula for the terms. Actually the series for $(9 + 4x)^{3/2}$ is alternating only if $x > 0$. But we still need to do more work in order to be able to apply the theorem to the above calculation. Let us begin with an example for which it is easy to apply the theorem. We need to consider an example where we know the whole power series. Consider the power series

$$\sum_{k=0}^{\infty}(-1)^k\frac{x^{2k}}{(2k+1)^3}.$$

This series converges for $|x| \le 1$. This is clearly an alternating series with $b_k = \frac{x^{2k}}{(2k+1)^3}$. We can easily show that this series satisfies the hypothesis of the Alternating Series Estimation Theorem. Since $\lim\limits_{k\to+\infty} x^{2k} = 0$ if $|x| < 1$, it follows that $\lim\limits_{k\to\infty} \frac{x^{2k}}{(2k+1)^3} = 0$ if $|x| \le 1$. Clearly $(2k+1)^3 < (2k+3)^3$. This implies that $1 < \left(\frac{2k+3}{2k+1}\right)^3$. This in turn implies that $x^2 < \left(\frac{2k+3}{2k+1}\right)^3$ for $|x| \le 1$. This last inequality is the same as

$$\frac{x^{2k+2}}{(2k+3)^3} < \frac{x^{2k}}{(2k+1)^3} \text{ for } x \le 1.$$

Thus we are able to apply the Alternating Series Estimation Theorem to this series when $|x| \le 1$.

Example 1. Suppose we require that $|x| \le 0.6$ and use $\sum_{k=0}^{6}(-1)^k\frac{x^{2k}}{(2k+1)^3}$ to estimate the value of $\sum_{k=0}^{\infty}(-1)^k\frac{x^{2k}}{(2k+1)^3}$. What is the maximum possible error? What is the error bound given by the Alternating Series Estimation Theorem?

Solution. Using $N = 6$ the Alternating Series Estimation Theorem says

$$\left|\sum_{k=7}^{\infty}(-1)^k\frac{x^{2k}}{(2k+1)^3}\right| \le \frac{|x|^{14}}{15^3} \text{ for } |x| \le 1. \tag{7.2}$$

If we restrict x to $|x| \le 0.6$, then this inequality is

$$\left|\sum_{k=7}^{\infty}(-1)^k\frac{x^{2k}}{(2k+1)^3}\right| \le \frac{(0.6)^{14}}{15^3} < .0000003 = 3(10^{-7}).$$

How do we use this in calculations? First,

$$\sum_{k=0}^{\infty}(-1)^k \frac{x^{2k}}{(2k+1)^3} = \sum_{k=0}^{6}(-1)^k \frac{x^{2k}}{(2k+1)^3} + \sum_{k=7}^{\infty}(-1)^k \frac{x^{2k}}{(2k+1)^3}$$

$$\sum_{k=0}^{\infty}(-1)^k \frac{x^{2k}}{(2k+1)^3} - \sum_{k=0}^{6}(-1)^k \frac{x^{2k}}{(2k+1)^3} = \sum_{k=7}^{\infty}(-1)^k \frac{x^{2k}}{(2k+1)^3} \qquad (7.3)$$

$$\left|\sum_{k=0}^{\infty}(-1)^k \frac{x^{2k}}{(2k+1)^3} - \sum_{k=0}^{6}(-1)^k \frac{x^{2k}}{(2k+1)^3}\right| < 0.0000003 \text{ if } |x| \leq 0.6.$$

The difference between the full sum $\sum_{k=0}^{\infty}(-1)^k \frac{x^{2k}}{(2k+1)^3}$ and the polynomial $\sum_{k=0}^{6}(-1)^k \frac{x^{2k}}{(2k+1)^3}$ is less than 0.0000003 when we restrict x to be such that $|x| \leq 0.6$. If we want to find the whole sum $\sum_{k=0}^{\infty}(-1)^k \frac{x^{2k}}{(2k+1)^3}$ we cannot add an infinite sum in any easy way. However, if $|x| \leq 0.6$ we can for sure get the value of the infinite sum correct to six decimal places by substituting into the finite sum $\sum_{k=0}^{6}(-1)^k \frac{x^{2k}}{(2k+1)^3}$. We know for sure that the actual error is less than the error bound 0.0000003.

Example 2. Suppose we are required to find the value of the infinite sum $\sum_{k=1}^{\infty}(-1)^k \frac{x^{2k}}{(2k+1)^3}$ and the error must be less than 10^{-8}. If we use $\sum_{k=0}^{6}(-1)^k \frac{x^{2k}}{(2k+1)^3}$ to do our calculations what values of x are we allowed to use?

Solution. Note that when $k = 6$ the highest power on x is 12. This means that the value of k is not the order of the polynomial. The order of this polynomial is 12. We recopy (7.3) from above

$$\left|\sum_{k=0}^{\infty}(-1)^k \frac{x^{2k}}{(2k+1)^3} - \sum_{k=0}^{6}(-1)^k \frac{x^{2k}}{(2k+1)^3}\right| = \left|\sum_{k=7}^{\infty}(-1)^k \frac{x^{2k}}{(2k+1)^3}\right|$$

Also from above we recopy the inequality (7.2)

$$\left|\sum_{k=7}^{\infty}(-1)^k \frac{x^{2k}}{(2k+1)^3}\right| \leq \frac{|x|^{14}}{3375} \text{ for } |x| \leq 1.$$

The difference between using the polynomial $\sum_{k=0}^{6}(-1)^k\frac{x^{2k}}{(2k+1)^3}$ and using the complete series $\sum_{k=0}^{\infty}(-1)^k\frac{x^{2k}}{(2k+1)^3}$ to do the calculations is less than $\frac{|x|^{14}}{3375}$. The error made by using the polynomial is less than $\frac{|x|^{14}}{3375}$. If we select values of x such that $\frac{|x|^{14}}{3375} < 10^{-8}$, then for these values of x the error will be less than 10^{-8}. Let us solve

$$|x|^{14} < 3375(10^{-8})$$
$$|x| < 0.4792.$$

If $|x| < 0.4792$, then the error term satisfies

$$\left|\sum_{k=7}^{\infty}(-1)^k\frac{x^{2k}}{(2k+1)^3}\right| < 10^{-8}.$$

There is a small point we should discuss about the statement "the bound is less than 10^{-8} for $|x| < 0.4792$". When computing the 14th root of $3375(10^{-8})$ the calculator says 0.47928. We can <u>not</u> round this up to 0.4793. The statement "the error bound is less than 10^{-8} for $x = 0.4793$" is not correct because this number is larger than 0.47928. We cannot "round up" this answer. In fact, when $x = 0.4793$ the error bound is

$$\frac{(0.4793)^{14}}{3375} = 1.00053 \times 10^{-8} > 10^{-8}.$$

Example 3. Suppose we use the polynomial

$$\sum_{k=0}^{4}(-1)^k\frac{x^{2k}}{(2k+1)^3} = 1 - \frac{x^2}{3^3} + \frac{x^4}{5^3} - \frac{x^6}{7^3} + \frac{x^8}{9^3}$$

to approximate the whole series $\sum_{k=0}^{\infty}(-1)^k\frac{x^{2k}}{(2k+1)^3}$. Suppose we require that the error in this calculation be less than 10^{-6}. What values of x may we use in our calculations?

Solution. When we say that the maximum error is 10^{-6} we are saying the error bound is 10^{-6}. This is the same as saying that we require

$$\left|\sum_{k=5}^{\infty}(-1)^k\frac{x^{2k}}{(2k+1)^3}\right| < 10^{-6}.$$

The Alternating Series Estimation Theorem says that

$$\left| \sum_{k=5}^{\infty} (-1)^k \frac{x^{2k}}{(2k+1)^3} \right| < \frac{x^{10}}{11^3} \text{ for } |x| \leq 1.$$

The error bound is $|x|^{10}(11^{-3})$. For what values of x is $\frac{|x|^{10}}{11^3} < 10^{-6}$? These are the values of x such that

$$|x|^{10} < (11)^3 10^{-6} = 0.001331$$
$$|x| < 0.5157$$

If $|x| < 0.5157$, then

$$\left| \sum_{k=0}^{\infty} (-1)^k \frac{x^{2k}}{(2k+1)^3} - \sum_{k=0}^{4} (-1)^k \frac{x^{2k}}{(2k+1)^3} \right| < 10^{-6}.$$

If $|x| < 0.5157$, we can calculate the value of the sum of the whole series by just calculating the value of the polynomial and the error bound will be less than 10^{-6}. Since the error bound is less than 10^{-6} we can be sure that the actual error made is less than 10^{-6}.

Example 4. Suppose we are trying to find the sum of the infinite series $\sum_{k=0}^{\infty} (-1)^k \frac{x^{2k}}{(2k+1)^3}$ using the polynomial $\sum_{k=0}^{N} (-1)^k \frac{x^{2k}}{(2k+1)^3}$ and using values of x such that $|x| \leq 0.5$. How many terms must the polynomial have in order to be sure that the error in the computations is less than 10^{-7}? What is the smallest value of N such that the error bound is less than 10^{-7}? Recall that this is a convergent alternating series for $|x| \leq 1$.

Solution. The general polynomial of degree $2N$ is written as

$$\sum_{k=0}^{N} (-1)^k \frac{x^{2k}}{(2k+1)^3}.$$

If $|x| \leq 0.5$, what is the smallest value of N that we can use and still be sure that the error is less than 10^{-7}? Without giving a value to either x or

N, the Alternating Series Estimation Theorem says

$$\left| \sum_{k=0}^{\infty} (-1)^k \frac{x^{2k}}{(2k+1)^3} - \sum_{k=0}^{N} (-1)^k \frac{x^{2k}}{(2k+1)^3} \right| = \left| \sum_{k=N+1}^{\infty} (-1)^k \frac{x^{2k}}{(2k+1)^3} \right|$$

$$< \frac{|x|^{2N+2}}{[2(N+1)+1]^3} \text{ for } |x| \le 1.$$

Note that when $k = N+1$ the power on x is $2N+2$. For this problem, we have $|x| \le 0.5$. This says $|x|^{2N+2} \le (0.5)^{2N+2}$. For any N if $|x| \le 0.5$ the error is less than

$$\frac{(0.5)^{2N+2}}{(2N+3)^3}.$$

We are required to choose a value of N such that the error is less than 10^{-7}. We need to choose a value of N such that

$$\frac{(0.5)^{2N+2}}{(2N+3)^3} < 10^{-7}.$$

The only way to solve this inequality is by trial and error (guess and check). We substitute some values for N in the inequality and see what is the smallest number that satisfies the inequality. Try $N = 5$. When $N = 5$, the inequality is

$$\frac{(0.5)^{12}}{(13)^3} = 1.11(10^{-7}) > 10^{-7}.$$

The value $N = 5$ is not large enough. For $N = 6$ the inequality is

$$\frac{(0.5)^{14}}{(15)^3} = 1.8 \times 10^{-8} < 10^{-7}.$$

The value $N = 6$ is the smallest value of N that gives a quotient less than 10^{-7}. Thus we see that

$$\left| \sum_{k=0}^{\infty} (-1)^k \frac{x^{2k}}{(2k+1)^3} - \sum_{k=0}^{6} (-1)^k \frac{x^{2k}}{(2k+1)^3} \right| < 10^{-7}$$

whenever $|x| \le 0.5$. We can use the polynomial $T_{12}(x) = \sum_{k=0}^{6} (-1)^k \frac{x^{2k}}{(2k+1)^3}$ to do calculations and when $|x| \le 0.5$ the error bound will be less than 10^{-7}. The actual error is less than the error bound. Note that $N = 6$ gives the polynomial $T_{12}(x)$.

Example 5. The infinite series for $\arctan x$ is

$$\arctan x = \sum_{k=0}^{\infty}(-1)^k \frac{x^{2k+1}}{2k+1} \qquad |x| < 1.$$

This is a convergent alternating series. We can write this infinite series using two sums as

$$\sum_{k=0}^{N}(-1)^k \frac{x^{2k+1}}{2k+1} + \sum_{k=N+1}^{\infty}(-1)^k \frac{x^{2k+1}}{2k+1} = T_{2N+1}(x) + R_{2N+1}(x).$$

Suppose we want to use the Taylor polynomial $T_{2N+1}(x)$ to approximate the value of $\arctan x$. Without a calculator we need to know the answer to the question: How close is the value of the polynomial $T_{2N+1}(x)$ to the value of the function $\arctan x$?

Solution. We can answer this question using the Alternating Series Estimation Theorem. If $|x| < 1$ the series for $\arctan x$ is a convergent alternating series and we can apply the Alternating Series Estimation Theorem. This theorem says for all $|x| < 1$ that

$$|R_{2N+1}(x)| = \left| \sum_{k=N+1}^{\infty}(-1)^k \frac{x^{2k+1}}{2k+1} \right| < \frac{|x|^{2N+3}}{2N+3}.$$

If $|x| \leq (1/4)$, then the maximum value for $|R_{2N+1}(x)|$ is

$$|R_{2N+1}(x)| \leq \frac{(1/4)^{2N+3}}{2N+3} = \frac{1}{(2N+3)4^{2N+3}},$$

which is true for all N. Let us substitute a value for N. If $|x| \leq 1/4$ and $N = 5$, the inequality for $|R_{11}(x)|$ is

$$|R_{11}(x)| = \left| \sum_{k=6}^{\infty}(-1)^k \frac{x^{2k+1}}{2k+1} \right| < \frac{|x|^{13}}{13} \leq \frac{(1/4)^{13}}{13} \leq .0000000012$$

This says that if $|x| < 1/4$, then

$$\left| \arctan x - \sum_{k=0}^{5} (-1)^k \frac{x^{2k+1}}{2k+1} \right| < .0000000012 = 1.2(10^{-9})$$

If $|x| \leq (1/2)$, then $|x|^{2N+3} \leq (1/2)^{2N+3}$. If $|x| < 1/2$, the above inequality becomes

$$|R_{2N+1}| \leq \frac{(1/2)^{2N+3}}{2N+3},$$

which is true for all N. Let us substitute a value for N. If $|x| < 1/2$ and $N = 6$, then

$$\left| \arctan x - \sum_{k=0}^{6} (-1)^k \frac{x^{2k+1}}{2k+1} \right| = \left| \sum_{k=7}^{\infty} (-1)^k \frac{x^{2k+1}}{2k+1} \right| < \frac{(1/2)^{15}}{15} < 0.0000021.$$

Example 6. If $|x| \leq 0.6$, what is the smallest value of N such that

$$| \arctan x - T_{2N+1}(x)| < 10^{-5}$$

is true? What is the smallest value of N such that the error bound is less than 10^{-5}?

Solution. For $|x| < 1$ and all N it is true that

$$\arctan x - T_{2N+1}(x) = \sum_{k=0}^{\infty} (-1)^k \frac{x^{2k+1}}{2k+1} - \sum_{k=0}^{N} (-1)^k \frac{x^{2k+1}}{2k+1}$$

$$= \sum_{k=N+1}^{\infty} (-1)^k \frac{x^{2k+1}}{2k+1} = R_{2N+1}(x).$$

Since $|x| \leq 0.6$, we can use the fact that $|x|^{2N+3} \leq (0.6)^{2N+3}$ in the error term.

$$| \arctan x - T_{2N+1}(x)| \leq \frac{|x|^{2N+3}}{2N+3} \leq \frac{(0.6)^{2N+3}}{2N+3}$$

We need to find N such that

$$\frac{(0.6)^{2N+3}}{2N+3} < 10^{-5}.$$

We do this using the calculator to guess and check. Substituting $N = 6$ we find that
$$\frac{(0.6)^{15}}{15} = 3.13(10^{-5}) > 10^{-5}.$$

Substituting $N = 7$ we find that

$$\frac{(0.6)^{17}}{17} = 9.96(10^{-6}) < 10^{-5}.$$

Note that $N = 7$ gives an error bound less than 10^{-5} while $N = 6$ gives an error bound greater than 10^{-5}. This shows that $N = 7$ is the smallest integer such that
$$\frac{(0.6)^{2N+3}}{2N+3} < 10^{-5}.$$

We have shown that if $|x| \leq 0.6$, then

$$\left| \arctan x - \sum_{k=0}^{7} (-1)^k \frac{x^{2k+1}}{2k+1} \right| < 10^{-5}.$$

We have looked at several examples where we determined the error bound for the error made when using a Taylor polynomial to approximate an entire Taylor's Series. Let us actually do a few calculations where we compare the value obtained using a Taylor polynomial with the sum of the entire Taylor's Series.

The Taylor polynomial $T_9(x)$ for $\arctan x$ is

$$T_9(x) = x - \frac{x^3}{3} + \frac{x^5}{5} - \frac{x^7}{7} + \frac{x^9}{9}.$$

When $x = 0.4$ the value of the polynomial is

$$T_9(0.4) = (0.4) - \frac{(0.4)^3}{3} + \frac{(0.4)^5}{5} - \frac{(0.4)^7}{7} + \frac{(0.4)^9}{9} = 0.380509736.$$

The actual value of $\arctan(0.4)$ is

$$\arctan(0.4) = 0.380506377.$$

Note that these values agree for five decimal places. Using $x = 0.5$ gives

$$T_9(0.5) = 0.463684275 \text{ and } \arctan(0.5) = 0.463647609.$$

Note that these values agree for four decimal places. Using $x = 0.2$ gives

$$T_9(0.2) = 0.197395561 \text{ and } \arctan(0.2) = 0.197395559.$$

These values agree for eight decimal places. The smaller the value of x the smaller the difference between $T_9(x)$ and $\arctan(x)$.

The power series for e^{-x} is $e^{-x} = \sum_{n=0}^{\infty} \frac{(-1)^n x^n}{n!}$. This is an alternating series for $x > 0$. For $x > 0$ this series satisfies the hypothesis that allow us to apply the Alternating Series Estimation Theorem. The polynomial $T_5(x)$ for $e^{-x} = \exp(-x)$ is

$$T_5(x) = 1 - x + \frac{x^2}{2} - \frac{x^3}{6} + \frac{x^4}{24} - \frac{x^5}{120}.$$

Substituting $x = 0.2$, we get

$$T_5(0.2) = 1 - (0.2) + \frac{(0.2)^2}{2} - \frac{(0.2)^3}{6} + \frac{(0.2)^4}{24} - \frac{(0.2)^5}{120}$$
$$= 0.818730666.$$

On the other hand
$$\exp(-0.2) = 0.818730753.$$

The actual error when using $T_5(0, 2)$ to approximate $e^{-0.2}$ is $e^{-0.2} - T_5(0.2) = 8.7(10^{-8})$. The error bound $|R_5(0.2)| = \frac{(0.2)^6}{6!} = 8.9(10^{-8})$. Note that the error bound is larger than the actual error.

In passing we should note that $\sum a_n x^n$ for $a_n > 0$ is an alternating series for $x < 0$.

The Taylor polynomial $T_6(x)$ for $\cos x$ is

$$T_6(x) = 1 - \frac{x^2}{2} + \frac{x^4}{24} - \frac{x^6}{720}.$$

After doing some computations we find

$$\cos(0.3) - T_6(0.3) = 0.955336489 - 0.955336487$$
$$= 0.000000002$$
$$\cos(0.5) - T_6(0.5) = 0.877582561 - 0.877582465$$
$$= 0.000000096$$
$$\cos(0.7) - T_6(0.7) = 0.764842187 - 0.764840765$$
$$= 0.000001422.$$

The smaller the value of x the smaller the actual error made when using $T_6(x)$ to approximate $\cos x$.

Example 7. The Taylor polynomial for $\sin x$ of degree 5 is

$$\sin x \approx x - \frac{x^3}{6} + \frac{x^5}{120}.$$

The Taylor polynomial for $\sin x^2$ of degree 10 is obtained by replacing x by x^2 in this polynomial.

$$\sin(x^2) \approx x^2 - \frac{x^6}{6} + \frac{x^{10}}{120}.$$

We can find an approximate value for the integral $\int_0^{1/2} \sin(x^2)dx$ by integrating

$$\int_0^{1/2} \sin(x^2)dx \approx \int_0^{1/2} \left[x^2 - \frac{x^6}{6} + \frac{x^{10}}{120} \right] dx$$

$$= \frac{x^3}{3} - \frac{x^7}{42} + \frac{x^{11}}{1320} \Big|_0^{1/2}$$

$$= \frac{(0.5)^3}{3} - \frac{(0.5)^7}{42} + \frac{(0.5)^{11}}{1320} = 0.041481024.$$

The Alternating Series Estimation Theorem says for $|x| \leq 1$ that

$$\left| \sin x^2 - \left[x^2 - \frac{x^6}{6} + \frac{x^{10}}{120} \right] \right| < \frac{x^{14}}{7!}.$$

300

Next, $\int_0^{1/2} \dfrac{x^{14}}{7!} dx = \dfrac{(0.5)^{15}}{75600} = 4.037 \times 10^{-10}$. It follows that

$$\left| \int_0^{1/2} \sin(x^2)dx - 0.041481024 \right| < 4.037 \times 10^{-10}.$$

The value of the integral $\int_0^{1/2} \sin(x^2)dx$ which we found using the polynomial to be 0.041481024 is correct to nine decimal places.

Whenever we have an alternating series we always use the Alternating Series Estimation Theorem to find a bound for the error made when using a finite sum to estimate the infinite sum. However, there are other methods (theorems, formula) that may be used to find an error bound. In particular these are methods that may be used when the series is not an alternating series. These methods (formulas) are more complicated. Another real problem is that these methods are miscellaneous. In some sense every type of series has its own method. The most well known of these is usually called Taylor's inequality. The Taylor's inequality gives an error bound which is greater than the bound given by the Alternating Series Estimation Theorem. It is also more complicated to find.

By way of illustration we are going to look at estimating the error for a series of positive terms. Some notation. Given a series of positive terms Σa_n we will define a function $f(x)$ such that $f(n) = a_n$. Also we will always assume that $f(x)$ is a monotone decreasing function. We now state a formula similar to the one we gave when discussing the divergence of the harmonic series $\Sigma 1/n$.

THEOREM 1. Suppose $a_k > 0$, then

$$\sum_{k=N+1}^{\infty} a_k < \int_N^{\infty} f(x)dx.$$

Example 8. Find a bound for using the finite sum $\Sigma_{k=1}^N 1/k^2$ to estimate the infinite sum $\sum_{k=1}^{\infty} 1/k^2$, that is find a bound for

$$\sum_{k=N+1}^{\infty} \frac{1}{k^2} = \sum_{k=1}^{\infty} \frac{1}{k^2} - \sum_{k=1}^N \frac{1}{k^2}.$$

Solution. Applying the formula from Theorem 1

$$\sum_{k=N+1}^{\infty} \frac{1}{k^2} < \int_{N}^{\infty} \frac{1}{x^2} dx = -\frac{1}{x} \Big|_{N}^{\infty} = \frac{1}{N}.$$

Example 9. Find a bound for the sum

$$\sum_{k=N+1}^{\infty} \frac{1}{(k+2)(k+5)}.$$

Solution. Applying Theorem 1 we conclude

$$\sum_{k=N+1}^{\infty} \frac{1}{(k+2)(k+5)} < \int_{N}^{\infty} \frac{dx}{(x+2)(x+5)}.$$

Now for $b > N$

$$\int_{N}^{b} \frac{dx}{(x+2)(x+5)} = \frac{1}{3} \int_{N}^{b} \left(\frac{1}{x+2} \right) - \left(\frac{1}{x+5} \right) dx$$

$$= \frac{1}{3} \ln \left[\frac{x+2}{x+5} \right]_{N}^{b} = \frac{1}{3} \ln \frac{b+2}{b+5} - \frac{1}{3} \ln \frac{N+2}{N+5}.$$

$$\lim_{b \to \infty} \left[\ln \frac{b+2}{b+5} - \ln \frac{N+2}{N+5} \right] = -\ln \left[\frac{N+2}{N+5} \right] = \ln \left[\frac{N+5}{N+2} \right].$$

$$\frac{1}{3} \ln \left[\frac{N+5}{N+2} \right] = \frac{1}{3} \ln \left[1 + \frac{3}{N+2} \right].$$

Recall that $\ln(1+x) < x$ for $0 < x < 1$. Thus

$$\frac{1}{3} \left[1 + \frac{3}{N+2} \right] < \frac{1}{N+2}.$$

It follows that

$$\sum_{k=N+1}^{\infty} \frac{1}{(k+2)(k+5)} < \frac{1}{N+2}.$$

302

Alternately, since $\dfrac{1}{(k+2)(k+5)} < \dfrac{1}{k^2}$

$$\sum_{k=N+1}^{\infty} \frac{1}{(k+2)(k+5)} < \sum_{k=N+1}^{\infty} \frac{1}{k^2} < \frac{1}{N}.$$

Example 10. Find an upper bound for

$$\sum_{k=N+1}^{\infty} \frac{1}{k^2+9}.$$

Solution. Using Theorem 1

$$\sum_{k=N+1}^{\infty} \frac{1}{k^2+9} < \int_{N}^{\infty} \frac{dx}{x^2+9}.$$

$$\int_{N}^{\infty} \frac{dx}{x^2+9} = -\int_{1/N}^{0} \frac{1/y^2\ dy}{1/y^2+9} = \int_{0}^{1/N} \frac{dy}{9y^2+1}$$

$$= \frac{1}{3} \arctan 3y \ \big|_{0}^{1/N} = \frac{1}{3} \arctan \frac{3}{N} < \frac{1}{N}.$$

Hence,

$$\sum_{k=N+1}^{\infty} \frac{1}{k^2+9} < \frac{1}{N}.$$

Alternately, since $k^2 + 9 > k^2$,

$$\sum_{k=N+1}^{\infty} \frac{1}{k^2+9} < \sum_{k=N+1}^{\infty} \frac{1}{k^2} < \frac{1}{N}.$$

Exercises

1. Use the Alternating Series Estimation Theorem to find a bound for each of the following expressions. Assume that these alternating series converge.

a) $\left| \sum_{k=10}^{\infty} \frac{(-1)^k k}{(k+2)^3} \right|$

b) $\left| \sum_{k=20}^{\infty} \frac{(-1)^k (k+1)}{(k^2+1)^2} \right|$

303

c) $\left| \sum_{k=12}^{\infty} \frac{(-1)^k x^{2k}}{(k+2)^4} \right|$ for $|x| \leq 1$

d) $\left| \sum_{k=0}^{\infty} \frac{(-1)^k (k+1)}{k^5+10} - \sum_{k=0}^{11} \frac{(-1)^k (k+1)}{k^5+10} \right|$

2. The hypothesis of the Alternating Series Estimation Theorem is satisfied by the series $\sum_{k=0}^{\infty} \frac{(-1)^k x^{2k}}{(k^2+3)^2}$ for $|x| \leq 1$. In (a) and (b) using the Alternating Series Estimation Theorem find the bound for the given expression.

a) $\left| \sum_{k=0}^{\infty} \frac{(-1)^k x^{2k}}{(k^2+3)^2} - \sum_{k=0}^{9} \frac{(-1)^k x^{2k}}{(k^2+3)^2} \right|$ for $|x| \leq 0.6$.

b) $\left| \sum_{k=15}^{\infty} \frac{(-1)^k x^{2k}}{(k^2+3)^2} \right|$ for $|x| \leq 0.8$.

c) For what values of x is $\left| \sum_{k=9}^{\infty} \frac{(-1)^k x^{2k}}{(k^2+3)^2} \right| < 10^{-6}$?

d) For what values of x is $\left| \sum_{k=6}^{\infty} \frac{(-1)^k x^{2k}}{(k^2+3)^2} \right| < 10^{-8}$?

3. Find the Taylor polynomial $T_5(x)$ for $\sin x$. Find the actual value of each of the following. Then compute the error bound given by the Alternating Series Estimation Theorem.

a) $\sin(0.3) - T_5(0.3)$.

b) $\sin(0.5) - T_5(0.5)$.

c) $\sin(0.7) - T_5(0.7)$.

4. Consider the series $\sum_{k=20}^{\infty} (-1)^{k-1} \frac{(k+1)x^{2k}}{5k^3+8}$. According to the Alternating Series Estimation Theorem if $|x| \leq 0.8$ what is an error bound for this series? You may assume that this series satisfies the hypothesis of the Alternating Series Estimation Theorem for $|x| \leq 1$.

5. Suppose we wish to approximate the infinite sum $\displaystyle\sum_{k=0}^{\infty}\frac{(-1)^k}{(k+2)^3}$ by the finite sum $\displaystyle\sum_{k=0}^{N}\frac{(-1)^k}{(k+2)^3}$. What is the smallest value of N we can use in order to be sure that the error in the approximation is less than 10^{-5}?

6. For what values of x can we be sure that $\left|\cos x - [1 - \dfrac{x^2}{2} + \dfrac{x^4}{4!}]\right| < 10^{-4}$?

7. For what values of x can we be sure that

$$\left|\arctan x - \sum_{k=0}^{8}(-1)^k\frac{x^{2k+1}}{2k+1}\right| < 10^{-6}?$$

8. If $|x| \leq 0.5$, what is the minimum value of N such that we can be sure according to the Alternating Series Estimation Theorem that

$$\left|\arctan x - \sum_{k=0}^{N}(-1)^{-k}\frac{x^{2k+1}}{2k+1}\right| < 10^{-7}?$$

You may assume that the series for $\arctan(x)$ satisfies the hypothesis of the Alternating Series Estimation Theorem.

9. If $|x| \leq 0.6$ what is the minimum value of N such that we can be sure that

$$\left|\sum_{k=0}^{\infty}\frac{(-1)^k x^{2k}}{(3k+4)^2} - \sum_{k=0}^{N}\frac{(-1)^k x^{2k}}{(3k+4)^2}\right| < 10^{-6}?$$

You may assume that this infinite series satisfies the hypothesis of the Alternating Series Estimation Theorem for $|x| \leq 1$.

10. If $|x| \leq 0.6$ what is the minimum value of N such that we can be sure that

$$\left|\sum_{k=0}^{\infty}\frac{(-1)^k x^{2k}}{(3k+4)^2} - \sum_{k=0}^{N}\frac{(-1)^k x^{2k}}{(3k+4)^2}\right| < 10^{-6}?$$

You may assume that this infinite series satisfies the hypothesis of the Alternating Series Estimation Theorem for $|x| \leq 1$.

11. Use Theorem 1 to find a bound for

$$\sum_{k=N+1}^{\infty} \frac{1}{k^4}.$$

12. Find a bound for

(a) $\displaystyle\sum_{k=N+1}^{\infty} \frac{1}{(k+5)(k+10)}$

(b) $\displaystyle\sum_{k=N+1}^{\infty} \frac{k+4}{(k^2+4)(k+5)}.$

6719 Manipulating Power Series

The standard notation for a power series is to write the power series so that it has the form

$$\sum_{n=0}^{\infty} a_n x^n,$$

where a_n depends on n but not on x. The essential feature for a power series to be in standard notation is that the power on x is n. However, for various reasons we sometimes find ourselves writing a power series in non standard form. When this happens, we should always convert to the standard form when writing the final answer to the problem. Recall that if there are any negative powers on x, then the series is not called a power series. If any of the powers on x are fractions or are negative we do not call the series a power series.

Example 1. Consider the series $\sum_{n=3}^{\infty} (2n + 3) x^{n-3}$. Write this power series in standard notation

Solution. This is a power series but it is not written in standard notation. Note that the power on x is never negative. In order to be in standard notation, the power on x must be n. We need to rewrite this series using a different summation. Since we are using n as the summation index in the given problem, let us use k as our new summation index. We want to write the series in the form $\sum (\ ) x^k$. This would mean that the power series is in standard notation. Clearly, we need to make a substitution using $k = n - 3$, that is, we want to write k where we are now writing $n - 3$ in the given series. This is $k = n - 3$ or $n = k + 3$. Using this substitution the coefficient is

$$2n + 3 = 2(k + 3) + 3 = 2k + 9.$$

We are going to rewrite the series as

$$\sum_{k=}^{\infty} (2k + 9) x^k.$$

An important question is: What are the values of k in order for this new

307

sum to give the same terms as $\sum_{n=3}^{\infty}(2n+3)x^{n-3}$. In the given sum

$$n = 3, 4, 5, 6, \ldots$$

This means that

$$k = n - 3 = 0, 1, 2, 3, \ldots$$

Therefore, the sum $\sum_{k=0}^{\infty}(2k+9)x^k$ is the same as $\sum_{n=3}^{\infty}(2n+3)x^{n-3}$. If we write out the first few terms of either sum, we get

$$9 + 11x + 13x^2 + 15x^3 + 17x^4 + \ldots.$$

Example 2. Consider the power series $\sum_{n=0}^{\infty}(3n+1)x^{n+2}$. Rewrite this power series in standard form.

Solution. Since the given power on x is $n+2$, we would like to rewrite the sum so that the power on x is k. Clearly, this means we must use the substitution $k = n+2$. This is $n = k-2$. Therefore, $3n+1 = 3(k-2)+1 = 3k - 5$. This tells us that the sum has the form

$$\sum_{k=}^{\infty}(3k-5)x^k.$$

But how does the summation index k count the terms? Does it start with k equal to zero or with k equal to some other value? The values of n are

$$n = 0, 1, 2, 3, 4, \ldots$$

This means that the values of k are

$$k = n + 2 = 2, 3, 4, 5, 6, \ldots$$

Thus we see that the sum $\sum_{k=2}^{\infty}(3k-5)x^k$ is the same as $\sum_{n=0}^{\infty}(3n+1)x^{n+2}$.

It is also clear that we can write the final answer using n instead of k if we wish, that is $\sum_{n=2}^{\infty} (3n-5)x^n$ has the same meaning as $\sum_{k=2}^{\infty} (3k-5)x^k$.

Example 3. Fill in the missing value of n and the missing coefficient of x^k in

$$\sum_{n=}^{\infty} (2n+5)x^{n-3} = \sum_{k=0}^{\infty} (\quad)x^k.$$

Solution. Comparing the powers of x we see that $n-3=k$ or $n=k+3$. This means $2n+5 = 2(k+3)+5 = 2k+11$. We have $k = 0,1,2,3,\dots$. This means $n = k+3 = 3,4,5,6,\dots$. Filling in the blanks, we have

$$\sum_{n=3}^{\infty} (2n+5)x^{n-3} = \sum_{k=0}^{\infty} (2k+11)x^k.$$

We have spent a lot of time starting with the Taylor polynomial for one function, then manipulating the given polynomial in some way to obtain the Taylor polynomial of another function. We can do this same thing with power series. We start with one power series and by manipulating it in some way we can obtain another power series. We shall start looking at such manipulation problems by just looking at how to manipulate the power series without thinking about what function the power series represents.

Given the Power Series $\sum_{n=0}^{\infty} (n^2+3)x^n$ which converges for $|x| < 1$, we can get a new power series by replacing x with multiples of x. If we replace x in this series with $-x/4$, we get the power series

$$\sum_{n=0}^{\infty} (n^2+3)\left(-\frac{x}{4}\right)^n = \sum_{n=0}^{\infty} (-1)^n \frac{(n^2+3)x^n}{4^n}.$$

This series converges for $|x| < 4$. We can get a new power series by replacing x with some power of x. We can determine the radius of convergence using the ratio test. If we replace x by x^2, we get the new power series

$$\sum_{n=0}^{\infty} (n^2+3)(x^2)^n = \sum_{n=0}^{\infty} (n^2+3)x^{2n}.$$

This power series converges for $|x| < 1$.

We can obtain new Taylor's series by adding together known Taylor's Series

Example 4. Find the Taylor's Series (power series) for

$$f(x) = \frac{1}{x^2 + 2x - 15} = \frac{1}{(x-3)(x+5)}.$$

Solution. Note that

$$\frac{1}{(x-3)(x+5)} = -\frac{1}{8}\left[\frac{1}{3-x} + \frac{1}{5+x}\right].$$

We have previously found the power series

$$\frac{1}{3-x} = \sum_{n=0}^{\infty} \frac{x^n}{3^{n+1}} \text{ and } \frac{1}{5+x} = \sum_{n=0}^{\infty} \frac{(-1)^n x^n}{5^{n+1}}.$$

It follows that

$$\frac{1}{3-x} + \frac{1}{5+x} = \sum_{n=0}^{\infty} \frac{x^n}{3^{n+1}} + \sum_{n=0}^{\infty} \frac{(-1)^n x^n}{5^{n+1}}.$$

$$\frac{1}{(x-3)(x+5)} = \frac{1}{8}\sum_{n=0}^{\infty}\left[\frac{(-1)^{n+1}}{5^{n+1}} - \frac{1}{3^{n+1}}\right]x^n.$$

We can get a new power series by differentiating the given power series. We can differentiate a power series by differentiating it term by term. The derivative of the power series $\sum_{n=0}^{\infty}(n^2 + 3)x^n$ is

$$\frac{d}{dx}\left[\sum_{n=0}^{\infty}(n^2 + 3)x^n\right] = \sum_{n=0}^{\infty}(n^2 + 3)\frac{d}{dx}(x^n) = \sum_{n=0}^{\infty}(n^2 + 3)nx^{n-1}.$$

310

The resulting series also converges for $|x| < 1$. When we differentiate the constant term in this series we got zero. In order to see this more clearly let us write out the first few terms of this series.

$$\frac{d}{dx}(3 + 4x + 7x^2 + 12x^3 + 19x^4 + \ldots) = 4 + 14x + 36x^2 + 76x^3 + \ldots.$$

Since the term $(n^2 + 3)nx^{n-1}$ is equal to zero when $n = 0$, we can write the sum starting with $n = 1$. This says that

$$\sum_{n=0}^{\infty}(n^2 + 3)nx^{n-1} = \sum_{n=1}^{\infty}(n^2 + 3)nx^{n-1}.$$

This power series is not in standard form. In the standard notation for power series the power of x is n not $n - 1$. We need to change notation so that we have x^k. We make a change in summation index. We change from n to k using $k = n - 1$ or $n = k + 1$. Since $k = n - 1$ and $n = 1, 2, 3, 4 \ldots$, it follows that $k = 0, 1, 2, 3, \ldots$. This means that

$$\sum_{n=1}^{\infty}(n^2 + 3)nx^{n-1} = \sum_{k=0}^{\infty}[(k + 1)^2 + 3](k + 1)x^k$$

$$= \sum_{k=0}^{\infty}(k^2 + 2k + 4)(k + 1)x^k.$$

This last power series is the derivative of the given series and is in standard notation.

Finally, just as with polynomials we can integrate power series. We can integrate a power series by integrating it term by term. Let us integrate the given series:

$$\int\left[\sum_{n=0}^{\infty}(n^2 + 3)x^n\right]dx = \sum_{n=0}^{\infty}(n^2 + 3)\int x^n dx$$

$$= C + \sum_{n=0}^{\infty}(n^2 + 3)\frac{x^{n+1}}{n + 1}.$$

311

The standard notation for power series is to have x^n not x^{n+1}. Since this series has x^{n+1} we change the summation index. We change summation index from n to k using the relation $n + 1 = k$. Since $k = n + 1$ and $n = 0, 1, 2, \ldots$ it follows that $k = 1, 2, 3, \ldots$.

$$C + \sum_{n=0}^{\infty} (n^2 + 3) \frac{x^{n+1}}{n+1} = C + \sum_{k=1}^{\infty} [(k-1)^2 + 3] \frac{x^k}{k}$$

$$= C + \sum_{k=1}^{\infty} (k^2 - 2k + 4) \frac{x^k}{k}.$$

This power series is the indefinite integral of the given series. Note that it contains a constant of integration.

Let us state formally the facts that we just used.

Theorem 1. Suppose $f(x) = \sum_{n=0}^{\infty} a_n x^n$ for $|x| < R$, that is, the power series converges for $|x| < R$, then the following are true

$$\frac{d}{dx}[f(x)] = \frac{d}{dx} \sum_{n=0}^{\infty} a_n x^n = \sum_{n=0}^{\infty} a_n \frac{d}{dx}[x^n] = \sum_{n=0}^{\infty} n a_n x^{n-1} \text{ for } |x| < R.$$

$$\int [f(x)]dx = \int \left[\sum_{n=0}^{\infty} a_n x^n \right] dx = \sum_{n=0}^{\infty} a_n \int x^n dx$$

$$= C + \sum_{n=0}^{\infty} a_n \frac{x^{n+1}}{n+1} \text{ for } |x| < R.$$

The proof of this theorem involves ideas that are more advanced than we have discussed. We are unable to prove it unless we go back and discuss these ideas.

Example 5. Let us apply these ideas to a known power series. Start with the geometric series

$$\frac{1}{1-x} = \sum_{n=0}^{\infty} x^n = 1 + x + x^2 + x^3 + x^4 + \ldots.$$

This series converges for $|x| < 1$. We can obtain the power series for other functions by manipulating this power series. When we replace x by $5x$, we get

$$\frac{1}{1-5x} = \sum_{n=0}^{\infty} (5x)^n = 1 + 5x + 25x^2 + 125x^3 + \ldots$$

Thus, we obtain the power series for $f(x) = (1-5x)^{-1}$. This series converges for $5|x| < 1$.

We can replace x by $x/5$ in the power series for $(1-x)^{-1}$ to get

$$\frac{1}{1-(x/5)} = \sum_{n=0}^{\infty} (x/5)^n = 1 + \frac{x}{5} + \frac{x^2}{25} + \frac{x^3}{125} + \frac{x^4}{625} + \ldots$$

This series converges for $|x|/5 < 1$ or $|x| < 5$. Note that $\dfrac{1}{1-(x/5)} = \dfrac{5}{5-x}$.

Dividing both sides by 5, we get the power series for $(5-x)^{-1}$. It is

$$\frac{1}{5-x} = \sum_{n=0}^{\infty} \frac{x^n}{5^{n+1}} = \frac{1}{5} + \frac{x}{25} + \frac{x^2}{125} + \frac{x^3}{625} + \ldots \text{ for } |x| < 5.$$

This is the same series we would get if we found the Taylor's Series for $(5-x)^{-1}$ using the formula for finding Taylor's Series.

Example 6. Given the power series for $(5-x)^{-1}$, find the power series for $(5-x)^{-2}$.

Solution. Since $\dfrac{d}{dx}(5-x)^{-1} = (5-x)^{-2}$, we use differentiation. Applying Theorem 1 to the series for $(5-x)^{-1}$, we get

$$\frac{d}{dx}\left[\sum_{n=0}^{\infty} \frac{x^n}{5^{n+1}}\right] = \sum_{n=0}^{\infty} \frac{1}{5^{n+1}} \frac{d}{dx}(x^n) = \sum_{n=0}^{\infty} \frac{nx^{n-1}}{5^{n+1}}.$$

Note that the value of $\dfrac{nx^{n-1}}{5^{n+1}}$ is zero when $n = 0$. With this in mind it follows that

$$\sum_{n=0}^{\infty} \frac{nx^{n-1}}{5^{n+1}} = \sum_{n=1}^{\infty} \frac{nx^{n-1}}{5^{n+1}}.$$

313

Let us convert this power series to standard notation. We change the summation index from n to k using $n = k+1$. Since $k = n-1$ and $n = 1, 2, 3, \ldots$ it follows that $k = 0, 1, 2, \ldots$. Also $n - 1 = k$ and $n + 1 = k + 2$. Therefore,

$$\sum_{n=1}^{\infty} \frac{nx^{n-1}}{5^{n+1}} = \sum_{k=0}^{\infty} \frac{(k+1)x^k}{5^{k+2}}.$$

Since $\dfrac{d}{dx}(5 - x)^{-1} = \dfrac{d}{dx}\left[\displaystyle\sum_{n=0}^{\infty} \frac{x^n}{5^{n+1}}\right]$ for $|x| < 5$, and since

$\dfrac{d}{dx}(5 - x)^{-1} = (5 - x)^{-2}$, it follows that

$$\frac{1}{(5 - x)^2} = \sum_{n=0}^{\infty} \frac{(n + 1)x^n}{5^{n+2}}, \text{ for } |x| < 5.$$

This is the power series we would get if we found the Taylor's series for $(5 - x)^{-2}$ using the formula for finding a Taylor's series.

Example 7. Starting with the power series for $(5 - x)^{-1}$ find the power series for $\ln(5 + x)$.

Solution. Since $\int (5 + x)^{-1} dx = \ln(5 + x)$, we can also use integration to find the new series.

Replacing x with $-x$ in $\dfrac{1}{5 - x} = \displaystyle\sum_{n=0}^{\infty} \frac{x^n}{5^{n+1}}$ gives

$$\frac{1}{5 - (-x)} = \sum_{n=0}^{\infty} \frac{(-x)^n}{5^{n+1}} \text{ or } \frac{1}{5 + x} = \sum_{n=0}^{\infty} \frac{(-1)^n}{5^{n+1}} x^n.$$

Applying the second part of Theorem 1 to find the integral of the infinite series, we get

$$\int \frac{dx}{5 + x} = \sum_{n=0}^{\infty} \frac{(-1)^n}{5^{n+1}} \int x^n dx.$$

$$\ln(5 + x) = C + \sum_{n=0}^{\infty} \frac{(-1)^n}{5^{n+1}} \frac{x^{n+1}}{n + 1}.$$

314

We would like to know the value of the constant of integration, C. Note that all the terms of the series involving x^{n+1} are equal to zero when $x = 0$. We evaluate the constant of integration C by substituting $x = 0$ into the equality. This gives

$$\ln 5 = C + 0.$$

We can also find C by recalling that the constant term in a Taylor's series is given by $f(0)$. Now that we have the power series we should write it in standard form. In order to do this we change the summation index by replacing n by $k - 1$. This gives us

$$\sum_{n=0}^{\infty} \frac{(-1)^n x^{n+1}}{5^{n+1}(n+1)} = \sum_{k=1}^{\infty} \frac{(-1)^{k-1} x^k}{5^k (k)}.$$

This says that the power series for $\ln(5 + x)$ is

$$\ln(5 + x) = \ln 5 + \sum_{k=1}^{\infty} \frac{(-1)^{k-1}}{5^k} \frac{x^k}{k} \text{ for } |x| < 5.$$

Let us review our more difficult methods for finding Taylor polynomials After working with infinite series we often start to think that an infinite sum is exactly the same as a finite sum. But finding an infinite sum, the sum of an infinite series, is a two step process. This process does not obey all the rules that a finite sum obeys. In a finite sum we can add the numbers up in any order and the sum is always the same. This is not always true for an infinite sum. We can show that

$$\ln 2 = 1 - \frac{1}{2} + \frac{1}{3} - \frac{1}{4} + \frac{1}{5} - \frac{1}{6} + \frac{1}{7} - \frac{1}{8} + \frac{1}{9} - \frac{1}{10} + \cdots$$

Suppose we rearrange the order in which we add the numbers.

$$\left(1 + \frac{1}{3}\right) - \frac{1}{2} + \left(\frac{1}{5} + \frac{1}{7}\right) - \frac{1}{4} + \left(\frac{1}{9} + \frac{1}{11}\right) - \frac{1}{6} + \cdots$$

We add two of the plus numbers then one of the minus numbers. We can show that the sum then is $(3/2) \ln 2$. By changing the order of addition we changed the sum.

315

When we introduced Taylor's Series we defined the Taylor's Series for a function $f(x)$ about the center $x = 0$ as

$$\sum_{n=0}^{\infty} \frac{f^{(n)}(0)}{n!} x^n.$$

We can generalize this idea and discuss the Taylor's Series for a function $f(x)$ about the general point $x = b$. It is defined as follows

$$\sum_{n=0}^{\infty} \frac{f^{(n)}(b)}{n!} (x - b)^n.$$

We can, of course, find such a Taylor's Series by finding $f^{(n)}(b)$ and substituting into the formula. Doing this is a small variation in finding the series about $x = 0$. It is usually easier to find the series about $x = b$ by first making a translation of independent variable using $g(y) = f(y+b)$ and then finding the series for $g(y)$ about $y = 0$. This means that every question about the Taylor's Series for $f(x)$ about $x = b$ is answered by referring to a series about $y = 0$.

Example 8. Find the Taylor's Series for $f(x) = \sin x$ with center at $x = \pi/4$.

Solution. We want to find a_n such that the series has the form

$$\sin x = \sum_{n=0}^{\infty} a_n (x - \pi/4)^n.$$

Let $y = x - \pi/4$, then

$$\sin(y + \pi/4) = \sin y \cos(\pi/4) + \cos y(\sin \pi/4)$$
$$= (\sqrt{2}/2)[\cos y + \sin y]$$

Recall the Taylor's Series (power series)

$$\cos y = \sum_{n=0}^{\infty} \frac{(-1)^n y^{2n}}{(2n)!} \quad \text{and} \quad \sin y = \sum_{n=0}^{\infty} \frac{(-1)^n y^{2n+1}}{(2n + 1)!}$$

for all y. Thus we can write

$$\sin(y + \pi/4) = (\sqrt{2}/2) \left[\sum_{n=0}^{\infty} \frac{(-1)^n y^{2n}}{(2n)!} + \sum_{n=0}^{\infty} \frac{(-1)^n y^{2n+1}}{(2n + 1)!} \right].$$

Replacing y with $x - (\pi/4)$, we get

$$\sin x = \frac{\sqrt{2}}{2} \left[\sum_{n=0}^{\infty} \frac{(-1)^n (x - \pi/4)^{2n}}{(2n)!} + \sum_{n=0}^{\infty} \frac{(-1)^n (x - \pi/4)^{2n+1}}{(2n+1)!} \right]$$

$$= \frac{\sqrt{2}}{2} \left[1 + (x - \pi/4) - \frac{(x - \pi/4)^2}{2!} - \frac{(x - \pi/4)^3}{3!} + \frac{(x - \pi/4)^4}{4!} \right.$$

$$\left. + \frac{(x - \pi/4)^5}{5!} - \frac{(x - \pi/4)^6}{6!} - \cdots \right].$$

Exercises

1. Explain why

$$\sum_{n=0}^{\infty} \frac{n x^{n-1}}{n^2 + 2} = \sum_{n=1}^{\infty} \frac{n x^{n-1}}{n^2 + 2}.$$

2. Write the following power series in standard notation:

(a) $\displaystyle\sum_{n=1}^{\infty} \frac{n x^{n-1}}{n^2 + 2}$

(b) $\displaystyle\sum_{n=1}^{\infty} \frac{(2n + 3) x^{n+2}}{n^2 + 2}$

(c) $\displaystyle\sum_{n=3}^{\infty} \frac{(n - 1) x^{n-2}}{n^2 + 1}$

3. Explain why

$$\sum_{n=4}^{28} \frac{n - 3}{2n + 3} = \sum_{k=1}^{25} \frac{k}{2k + 9}.$$

4. Fill in the missing value of n and the missing coefficient of x^k in

$$\sum_{n=}^{\infty} \frac{3n + 2}{n + 1} x^{n-4} = \sum_{k=0}^{\infty} (\quad) x^k.$$

5. Given the geometric series $\displaystyle\frac{1}{1 - x} = \sum_{n=0}^{\infty} x^n$ find the power series for $\displaystyle\frac{1}{4 + x}$.

317

6. Given that the power series for $(8 + x)^{-1}$ is $\displaystyle\sum_{n=0}^{\infty} \frac{(-1)^n x^n}{8^{n+1}}$ find the power series for $(8 + x)^{-2}$.

7. Given that $(3 + x)^{-1} = \displaystyle\sum_{n=0}^{\infty} \frac{(-1)^n}{3^{n+1}} x^n$ for $|x| < 1$, find the power series for $\ln(3 + x)$.

8. Write the following power series in standard form

$$\sum_{n=0}^{\infty} n(n - 1)(2n + 3)^2 x^{n-2}$$

9. Starting with the power series for $\sinh(x) = \displaystyle\sum_{n=0}^{\infty} \frac{x^{2n+1}}{(2n + 1)!}$ use integration to find the power series for $\cosh(x)$.

10. Find the Taylor's Series (power series) for

$$f(x) = \frac{1}{x^2 + 2x - 24} = \frac{1}{(x - 4)(x + 6)}.$$

11. Find the Taylor's Series for $\cos x$ about $x = (\pi/4)$, that is, find a_n such that

$$\cos x = \sum_{n=0}^{\infty} a_n (x - \pi/4)^n.$$

6771 Vectors in Three Space

We locate numbers in a two dimensional coordinate plane using an x coordinate and a y coordinate. Recall that two intersecting lines determine a unique point. We use this fact to locate points in a coordinate plane. The x coordinate of a point in the coordinate plane is $x = a$ if the point lies on the line which passes through the point $x = a$ on the x axis and is perpendicular to the x axis. The y coordinate of a point in the coordinate plane is $y = b$ if the point lies on the line which passes through the point $y = b$ on the y axis and is perpendicular to the y axis.

We are now going to locate points in a coordinate space (three dimensional). For three dimensional space we use three reference lines which are all perpendicular to each other and all pass through the same point, the origin. We label points on these lines using the same distance as the unit distance on each line. The lines are labeled the x-axis, the y axis, and the z axis. Since we have already discussed the xy coordinate plane the easy way to think of this is that we start with an xy plane and then draw the z axis perpendicular to the xy plane and through the origin. We can also describe the origin as the intersection of 3 coordinate planes. One plane is the plane containing the x and y axis. The second plane is the one containing the x and z axis, and the third plane is the plane containing the y and z axis. These coordinate planes are called the xy plane, the xz plane, and the yz plane. Recall that two intersecting lines determine a plane. We usually think of the xy plane as horizontal and the z axis as vertical.

A point in three dimensional space has x coordinate $x = a$ if the point is in the plane which passes through the point $x = a$ on the x axis and is perpendicular to the x axis (parallel to the yz plane). A point has y coordinate $y = b$ if the point is in the plane which passes through the point $y = b$ on the y axis and is perpendicular to the y axis (parallel to the xz plane). A point has z coordinate $z = c$ if the point is in the plane which passes through the point $z = c$ on the z axis and is perpendicular to the z axis (parallel to the yz plane). Thus the location of a point is uniquely determined. The three coordinate planes are the xy plane, the xz plane, and the yz plane. These planes divide space into eight parts called octants. The first octant is the one eight of space in which all the coordinates of all the points are positive.

Example 1. Locate the point with coordinates $(3, 4, 5)$.

The first number 3 is the x coordinate, the second number 4 is the y coordinate, and the third number 5, is the z coordinate.

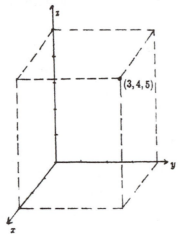

We often think of the point with coordinates $(3, 4, 5)$ as 3 units from the yz plane ($x = 3$), 4 units from the xz plane ($y = 4$) and 5 units from the xy plane ($z = 5$). A point with x coordinate $x = -3$ is three units on the other side of the yz plane.

All of our drawings of objects in three space must be done in a plane. This presents visualization problems. Different people have different ideas about what makes a good visualization.

Distance Formula in Three Dimensions. Consider the line segment in space connecting the point with coordinates (x_1, y_1, z_1) and the point with coordinates (x_2, y_2, z_2). The length of this line segment is equal to

$$\sqrt{(x_2 - x_1)^2 + (y_2 - y_1)^2 + (z_2 - z_1)^2}.$$

We prove that this formula is correct using the pythagoren theorem. This formula might even be called the pythagoren theorem in three dimensions.

Example 2. Find the length of the line segment from $(5, -3, 8)$ to $(12, -7, -9)$.

Solution.

$$\text{Length} = \sqrt{(12 - 5)^2 + (-7 + 3)^2 + (-9 - 8)^2} = \sqrt{354}.$$

Vectors

Definition. The term vector is used to indicate a quantity which has both magnitude and direction. Vectors are denoted by placing an arrow over the letter as in $\vec{v}$.

Vectors are used to indicate quantities such as displacement, velocity, or force that have both magnitude and direction. A vector can be represented by an arrow or a directed line segment. The length of the arrow or line segment represents the magnitude of the vector and the arrow points in the direction of the vector. This means that vectors have an initial point and a terminal point. Vectors that have the same magnitude and direction are equal or equivalent. They need not have the same initial point in order to be equal. We can move a vector about in space as long as we do not change its magnitude or direction. However, we do have a standard position for a vector. The standard position of a vector is when its initial point is the origin $(0,0,0)$. The vector with initial point $(0,0,0)$ and terminal point (a, b, c) is denoted by

$$a\vec{i} + b\vec{j} + z\vec{k}.$$

The vector $\vec{i}$ is a vector one unit long with initial point $(0,0,0)$ and terminal point $(1,0,0)$. The vector $\vec{j}$ is vector with initial point $(0,0,0)$ and terminal point $(0,1,0)$. The vector $\vec{k}$ is a vector with initial point at $(0,0,0)$ and terminal point $(0,0,1)$. Consider the vector $a\vec{i} + b\vec{j} + c\vec{k}$. The number a is called the component of the vector in the $\vec{i}$ direction. The number b is called the component in the $\vec{j}$ direction. The number c is called the component in the $\vec{k}$ direction. When discussing vectors in space, we always assume that we have a given coordinate system. This coordinate system must be a right handed coordinate system.

Definition. Two vectors are equal if and only if all three corresponding components are equal.

Example 3. When we say that $a\vec{i} + b\vec{j} + c\vec{k} = 3\vec{i} + 4\vec{j} + 5\vec{k}$ this is the same as saying $a = 3$, $b = 4$, and $c = 5$. One vector equation is equivalent to three scalar equations.

Example 4. Find the vector $\vec{v}$ if the initial point of the vector is $(4, -5, 2)$ and the terminal point is $(9, 4, -6)$.

Solution. The change in the x coordinate is from 4 to 9. Since $9 - 4 = 5$ the change in x coordinate is 5. The change in the y coordinate is from -5 to 4. Since $4 - (-5) = 9$, the change in y coordinate is 9. The change in z coordinate is from 2 to -6. Since $-6 - 2 = -8$, the change in the z coordinate is -8. The vector is $\vec{v} = 5\vec{i} + 9\vec{j} - 8\vec{k}$.

$$(4, -5, 2) \xrightarrow{5\vec{i}+9\vec{j}-8\vec{k}} (9, 4, -6)$$

Suppose the vector $\vec{v} = a\vec{i} + b\vec{j} + c\vec{k}$ is placed in standard position with initial point at $(0, 0, 0)$ and terminal point at (a, b, c). We can use the formula for the length of a line segment to find the length of the vector.

Theorem. Let $\vec{v} = a\vec{i} + b\vec{j} + c\vec{k}$, then the length of $\vec{v}$ is

$$\|\vec{v}\| = \sqrt{a^2 + b^2 + c^2}.$$

Definition. A vector $\vec{u}$ is called a unit vector if $\|\vec{u}\| = 1$.

Example 5. Show that the vector

$$\vec{u} = (2/7)\vec{i} - (6/7)\vec{j} + (3/7)\vec{k}$$

is a unit vector.

Solution. $\|\vec{u}\|^2 = (2/7)^2 + (-6/7)^2 + (3/7)^2 = 1$.

Geometric Definition of Sum of Two Vectors. Given the two vectors $\vec{v}$ and $\vec{w}$ the sum of these two vectors $\vec{v} + \vec{w}$ is found as follows. The vectors are positioned such that the initial point of $\vec{w}$ is at the terminal point of $\vec{v}$, then the vector which is the sum $\vec{v} + \vec{w}$ is the vector from the initial point of $\vec{v}$ to the terminal point of $\vec{w}$.

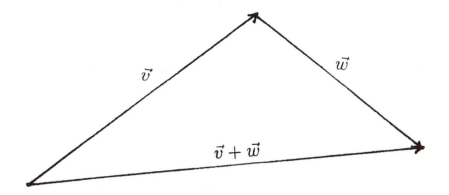

We are able to draw this picture in a plane. It is the plane through the following three points: the initial point of $\vec{v}$, the terminal point of $\vec{v}$, and the terminal point of $\vec{w}$.

Algebraic Definition of Addition of Vectors. The sum of two vectors is obtained by adding corresponding components, that is, if $\vec{v} = a\vec{i} + b\vec{j} + c\vec{k}$ and $\vec{w} = d\vec{i} + e\vec{j} + f\vec{k}$, then

$$\vec{v} + \vec{w} = (a + d)\vec{i} + (b + e)\vec{j} + (c + f)\vec{k}.$$

We should now prove that these two definitions are the same. We should prove that if we add two vectors using one of the definitions then the sum is the same as if we added the same two vectors using the other definition. We shall assume that this is true, that is, we assume that addition using either definition gives the same result.

Example 6. If $\vec{v} = 12\vec{i} - 8\vec{j} + 14\vec{k}$ and $\vec{w} = -5\vec{i} + 18\vec{j} + 6\vec{k}$, then

$$\vec{v} + \vec{w} = (12 - 5)\vec{i} + (-8 + 18)\vec{j} + (6 + 14)\vec{k} = 7\vec{i} + 10\vec{j} + 20\vec{k}.$$

Geometric Definition of Multiplication by a Scalar. If s is a scalar and $\vec{v}$ is a vector, then the product $s\vec{v}$ is a vector of length $|s|\|\vec{v}\|$. If $s > 0$, then the direction of $s\vec{v}$ is the same as the direction of $\vec{v}$. If $s < 0$, then the direction of vector $s\vec{v}$ is in the opposite direction of $\vec{v}$.

Algebraic Definition of Multiplication by a Scalar. If s is a scalar and $\vec{v} = a\vec{i} + b\vec{j} + c\vec{k}$ is a vector, then the scalar s times the vector $\vec{v}$ is the vector $s\vec{v} = (sa)\vec{i} + (sb)\vec{j} + (sc)\vec{k}$. Note that the vectors $\vec{v}$ and $s\vec{v}$ are parallel. Also if any two vectors are parallel, then one is a scalar multiple of the other.

Example 7. If $\vec{v} = 8\vec{i} - 4\vec{j} + 7\vec{k}$ and $\vec{w} = 3\vec{i} + 9\vec{j} - 6\vec{k}$, find $5\vec{v} - 8\vec{w}$.

Solution.

$$\begin{aligned} 5\vec{v} - 8\vec{w} &= 5(8\vec{i} - 4\vec{j} + 7\vec{k}) - 8(3\vec{i} + 9\vec{j} - 6\vec{k}) \\ &= (40\vec{i} - 20\vec{j} + 35\vec{k}) + (-24\vec{i} - 72\vec{j} + 48\vec{k}) \\ &= 16\vec{i} - 92\vec{j} + 83\vec{k} \end{aligned}$$

Exercises

1. Locate the point with coordinates $(5, 3, 4)$ in an xyz coordinate systems.

2. Find the vector with initial point $(8, -3, 5)$ and terminal point $(-3, 9, 20)$. Find the length of this vector.

3. If $\vec{v} = 11\vec{i} - 5\vec{j} + 9\vec{k}$ and $\vec{w} = 4\vec{i} + 6\vec{j} - 7\vec{k}$, find $6\vec{v} - 4\vec{w}$.

4. Find a unit vector $\vec{u}$ parallel to the vector $\vec{v} = 6\vec{i} - 4\vec{j} + 12\vec{k}$.

6775 Dot Product in Space

We also define the dot product of two vectors in three dimensions. Before we can state the geometric definition we need to define the angle between two vectors. If $\vec{v}$ and $\vec{w}$ are two vectors, we find the angle between these two vectors as follows. First, place both vectors in standard position with their initial points at the origin, $(0,0,0)$. Next consider the plane determined by the origin, the terminal point of $\vec{v}$, and the terminal point of $\vec{w}$. Recall that three points determine a plane. Note that both the vectors $\vec{v}$ and $\vec{w}$ are line segments in this plane. We define the angle θ between the two vectors $\vec{v}$ and $\vec{w}$ to be the angle satisfying $0 \le \theta \le \pi$ which is measured in this plane.

Geometric Definition of Dot Product. If $\vec{v}$ and $\vec{w}$ are two vectors, then the dot product $\vec{v} \cdot \vec{w}$ is the scalar given by

$$\vec{v} \cdot \vec{w} = \|\vec{v}\| \, \|\vec{w}\| \cos\theta,$$

where θ is the angle between $\vec{v}$ and $\vec{w}$.

Algebraic Definition of Dot Product. Given the vectors $\vec{v} = a\vec{i} + b\vec{j} + c\vec{k}$ and $\vec{w} = d\vec{i} + e\vec{j} + f\vec{k}$, the dot product is

$$\vec{v} \cdot \vec{w} = ad + be + cf.$$

We should prove that these two definitions of dot product are equivalent, that is, we should prove that

$$\|\vec{v}\| \, \|\vec{w}\| \cos\theta = ad + be + cf.$$

We will not give the proof here although the proof only involves some complicated trigonometry. We see that it is much easier to find the dot product of two vectors using the algebraic definition than it is using the geometric definition. However, the geometric definition does have one interesting application. We can use the geometric definition to find the angle between two vectors. In order to do this we rewrite the formula as

$$\cos\theta = \frac{\vec{v} \cdot \vec{w}}{\|\vec{v}\| \, \|\vec{w}\|}.$$

Example 1. Find the angle in radians between the vector $\vec{v} = 5\vec{i} - 4\vec{j} + 7\vec{k}$ and the vector $\vec{w} = 6\vec{i} - 3\vec{j} + 2\vec{k}$.

Solution. We have $\vec{v} \cdot \vec{w} = 30 + 12 + 14 = 56$, $\|\vec{v}\|^2 = 25 + 16 + 49 = 90$ and $\|\vec{w}\|^2 = 49$.

$$\cos\theta = \frac{56}{7\sqrt{90}} = 0.8433.$$

The angle between the two vectors is

$$\theta = 0.5675.$$

Theorem. Suppose $\vec{v} \neq \vec{0}$ and $\vec{w} \neq \vec{0}$. If $\vec{v} \cdot \vec{w} = 0$, then the vectors $\vec{v}$ and $\vec{w}$ are perpendicular. Alternately we say that the vectors are orthogonal.

Proof. Note that $\cos\theta = 0$ and so $\theta = \pi/2$.

Example 2. Show that $\vec{v} = 5\vec{i} + 3\vec{j} - 6\vec{k}$ and $\vec{w} = -3\vec{i} + 7\vec{j} + \vec{k}$ are perpendicular (orthogonal). Let $\vec{z} = 5\vec{i} + 4\vec{j} - 13\vec{k}$. Show that $\vec{w} \cdot \vec{z} = 0$.

Solution. We have $\vec{v} \cdot \vec{w} = -15 + 21 - 6 = 0$. This means $\cos\theta = 0$ or $\theta = \pi/2$. The angle between the two vectors is $\pi/2$. Therefore, the vectors are perpendicular to each other. Also $\vec{w} \cdot \vec{z} = -15 + 28 - 13 = 0$. There are lots of vectors perpendicular to $\vec{w}$.

Definition of Projection. Given the two vectors $\vec{v}$ and $\vec{w}$ we define the projection of $\vec{v}$ onto $\vec{w}$ as the vector

$$\text{Proj}_{\vec{w}}\vec{v} = \left(\frac{\vec{v} \cdot \vec{w}}{\|\vec{w}\|}\right)\frac{\vec{w}}{\|\vec{w}\|} = \left[\frac{\vec{v} \cdot \vec{w}}{\|\vec{w}\|^2}\right]\vec{w}.$$

Note that the vector $\text{Proj}_{\vec{w}}\vec{v}$ is a scalar multiple of $\vec{w}$. The scalar $\frac{\vec{v} \cdot \vec{w}}{\|\vec{w}\|}$ is the component of the vector $\text{Proj}_{\vec{w}}\vec{v}$ in the direction of $\vec{w}$. This scalar is denoted by $\text{Comp}_{\vec{w}}\vec{v}$. This says that $\text{Proj}_{\vec{w}}\vec{v} = (\text{Comp}_{\vec{w}}\vec{v})\frac{\vec{w}}{\|\vec{w}\|}$. This means that $\text{Proj}_{\vec{w}}\vec{v}$ and $\vec{w}$ have the same direction or exactly opposite directions. Note that $\frac{\vec{w}}{\|\vec{w}\|}$ is a vector one unit long and that the length of the vector $\text{Proj}_{\vec{w}}\vec{v}$ is $\frac{|\vec{v} \cdot \vec{w}|}{\|\vec{w}\|}$. We can picture the relationship between the vectors $\vec{v}$, $\vec{w}$, and $\text{Proj}_{\vec{w}}\vec{v}$. First place $\vec{v}$ and $\vec{w}$ in the standard position with their initial point at the origin. Consider the plane determined by the three points: the origin, the terminal point of $\vec{v}$, and the terminal point of $\vec{w}$. The following diagram is drawn in this plane.

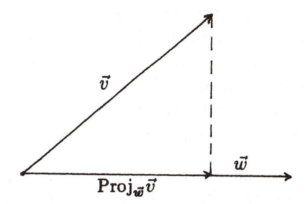

The dotted line is the line perpendicular to the line containing the vector $\vec{w}$ and through the tip of $\vec{v}$. The initial point of $\text{Proj}_{\vec{w}}\vec{v}$ is $(0,0,0)$ and the terminal point is where the dotted line through the tip of $\vec{v}$ intersects the line through the vector $\vec{w}$.

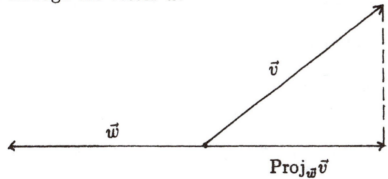

The vector $\text{Proj}_{\vec{w}}\vec{v}$ is also called the vector component of $\vec{v}$ in the direction of $\vec{w}$ or the vector component of $\vec{v}$ along the vector $\vec{w}$. The vector component is also given by $[\text{Comp}_{\vec{w}}\vec{v}]\frac{\vec{w}}{\|\vec{w}\|}$. The vector component and the scalar component are used in the physical applications of vectors.

Example 3. Given the vectors $\vec{v} = 3\vec{i} - 5\vec{j} + 7\vec{k}$ and $\vec{w} = 6\vec{i} + 2\vec{j} - 4\vec{k}$, find the vector $\text{Proj}_{\vec{w}}\vec{v}$.

Solution. We have $\vec{v} \cdot \vec{w} = 18 - 10 - 28 = -20$ and $\|\vec{w}\|^2 = 36 + 4 + 16 = 56$. It follows that

$$\text{Proj}_{\vec{w}}\vec{v} = \frac{-20}{56}\left[6\vec{i} + 2\vec{j} - 4\vec{k}\right] = -\frac{5}{7}\left[3\vec{i} + \vec{j} - 2\vec{k}\right].$$

Suppose an object is moved from (a, b, c) to the point (d, e, f) along the line segment connecting these two points, then the displacement vector for this object is

$$\vec{D} = (d - a)\vec{i} + (e - b)\vec{j} + (f - c)\vec{k}.$$

If during the motion the object is acted on by a constant force $\vec{F}$, then the work done by the force in moving the object is given by $\vec{F} \cdot \vec{D}$. Note that the dot product of the vectors $\text{Proj}_{\vec{D}}\vec{F}$ and $\vec{D}$ is the same as $\vec{F} \cdot \vec{D}$. This says that the only part of the force $\vec{F}$ that counts in doing the work is the part given by $\text{Proj}_{\vec{D}}\vec{F}$. Note that work is a scalar. We often use the fact that $\text{Comp}_{\vec{D}}\vec{F} = \|\vec{F}\| \cos\theta$ in solving work problems.

Example 4. An object is moved from $(3, 4, 5)$ to $(2, -3, -7)$ along a line segment connecting these two points with distance measured in feet. The object is acted on by the constant force $\vec{F} = 5\vec{i} - 4\vec{j} + 7\vec{k}$ pounds during the move. Strictly speaking pounds is a scalar unit. We should say that $\|\vec{F}\|$ is measured in pounds. Find the work done by this force.

Solution. The displacement vector is

$$\vec{D} = (2-3)\vec{i} + (-3-4)\vec{j} + (-7-5)\vec{k} = -\vec{i} - 7\vec{j} - 12\vec{k}.$$

We often say that the units of $\vec{D}$ are feet, but strictly speaking we should say that $\|\vec{D}\|$ is measured in feet. The work done is

$$\text{work} = (5\vec{i} - 4\vec{j} + 7\vec{k}) \cdot (-\vec{i} - 7\vec{j} - 12\vec{k}) = -5 + 28 - 84$$
$$= -61 \text{ ft lbs}$$

Exercises

1. Find the angle in radians between the vector $\vec{v} = 5\vec{i} + 4\vec{j} - 8\vec{k}$ and the vector $\vec{w} = 6\vec{i} - 3\vec{j} + 7\vec{k}$.

2. Given the vectors $\vec{v} = 3\vec{i} - 2\vec{j} + 5\vec{k}$ and $\vec{w} = 4\vec{i} + 6\vec{j} + 7\vec{k}$ find $\text{Proj}_{\vec{w}}\vec{v}$ and $\text{Proj}_{\vec{v}}\vec{w}$. Find $\text{Comp}_{\vec{w}}\vec{v}$ and $\|\vec{v}\| \cos\theta$.

3. Given the vectors $\vec{v} = 4\vec{i} - 3\vec{j} + 5\vec{k}$ and $\vec{w} = 6\vec{i} + 5\vec{j} - 8\vec{k}$ find $\text{Proj}_{\vec{w}}\vec{v}$ and $\text{Proj}_{\vec{v}}\vec{w}$. Find the angle between $\vec{v}$ and $\vec{w}$.

4. An object is moved from $(-2, 5, 11)$ to $(3, 8, -1)$ along a line segment connecting these two points with distance measured in feet. The object is acted on by the constant force $\vec{F} = 4\vec{i} - 3\vec{j} - 5\vec{k}$ pounds during the move. Find the work done by this force. Find $\text{Comp}_{\vec{D}}\vec{F}$.

328

5. An object is moved from $(3, -8, 5)$ to $(19, -14, 25)$ along a line segment connecting these two points with distance measured in meters. The object is acted on by the constant force $\vec{F} = 5\vec{i} + 3\vec{j} - 4\vec{k}$ Newtons during the move. Find the work done by this force.

6. Let $\vec{v} = 4\vec{i} - 3\vec{j} + 2\vec{k}$, $\vec{w} = 7\vec{i} + 6\vec{j} - 5\vec{k}$, and $\vec{z} = 5\vec{i} - 4\vec{j} + 16\vec{k}$. Show that $\vec{v}$ and $\vec{w}$ are perpendicular. Find the angle between $\vec{v}$ and $\vec{z}$. Find the angle between $\vec{w}$ and $\vec{z}$.

6778 Cross Product

The cross product of two vectors $\vec{v}$ and $\vec{w}$ is another vector denoted by $\vec{v} \times \vec{w}$. Since the dot product $\vec{v} \cdot \vec{w}$ is a scalar we often call the cross product the vector product. The cross product $\vec{v} \times \vec{w}$ is only defined when $\vec{v}$ and $\vec{w}$ are three dimensional vectors.

Geometric Definition of Cross Product. Given two non zero vectors $\vec{v}$ and $\vec{w}$ the cross product $\vec{v} \times \vec{w}$ is the unique vector with the following three properties.

(1) If θ is the angle between $\vec{v}$ and $\vec{w}$, then the length of the cross product is $\|\vec{v} \times \vec{w}\| = \|\vec{v}\| \, \|\vec{w}\| \sin \theta$.

(2) The vector $\vec{v} \times \vec{w}$ is perpendicular to both the vectors $\vec{v}$ and $\vec{w}$. This says that $(\vec{v} \times \vec{w}) \cdot \vec{v} = 0$ and $(\vec{v} \times \vec{w}) \cdot \vec{w} = 0$.

(3) Of the two possible directions the direction of $\vec{v} \times \vec{w}$ is determined by the right hand rule. The vectors $\vec{v} \times \vec{w}$ and $\vec{w} \times \vec{v}$ are in opposite directions, that is, $\vec{w} \times \vec{v} = -(\vec{v} \times \vec{w})$.

Algebraic Definition of Cross Product. Given the vectors $\vec{v} = a_1 \vec{i} + b_1 \vec{j} + c_1 \vec{k}$ and $\vec{w} = a_2 \vec{i} + b_2 \vec{j} + c_2 \vec{k}$, the cross product is given by

$$\begin{vmatrix} \vec{i} & \vec{j} & \vec{k} \\ a_1 & b_1 & c_1 \\ a_2 & b_2 & c_2 \end{vmatrix} = \vec{i} \begin{vmatrix} b_1 & c_1 \\ b_2 & c_2 \end{vmatrix} - \vec{j} \begin{vmatrix} a_1 & c_1 \\ a_2 & c_2 \end{vmatrix} + \vec{k} \begin{vmatrix} a_1 & b_1 \\ a_2 & b_2 \end{vmatrix}.$$

$$= (b_1 c_2 - b_2 c_1)\vec{i} - (a_1 c_2 - a_2 c_1)\vec{j} + (a_1 b_2 - a_2 b_1)\vec{k}.$$

This formula is correct only if we are using a right handed coordinate system.

Theorem. Given two vectors $\vec{v}$ and $\vec{w}$ the cross product of these two vectors computed using the geometric definition of cross product is the same vector as the vector found by computing the cross product using the algebraic definition of cross product.

Example 1. If $\vec{v} = 3\vec{i} + 4\vec{j} - \vec{k}$ and $\vec{w} = -5\vec{i} + 2\vec{j} + 7\vec{k}$, find $\vec{v} \times \vec{w}$.

Solution.

$$\begin{vmatrix} \vec{i} & \vec{j} & \vec{k} \\ 3 & 4 & -1 \\ -5 & 2 & 7 \end{vmatrix} = \vec{i}\begin{vmatrix} 4 & -1 \\ 2 & 7 \end{vmatrix} - \vec{j}\begin{vmatrix} 3 & -1 \\ -5 & 7 \end{vmatrix} + \vec{k}\begin{vmatrix} 3 & 4 \\ -5 & 2 \end{vmatrix}$$

$$= \vec{i}(28 + 2) - \vec{j}(21 - 5) + \vec{k}(6 + 20)$$

$$= 30\vec{i} - 16\vec{j} + 26\vec{k}.$$

Example 2. Let $\vec{v}$ denote the vector whose initial point is $(3, -5, 1)$ and whose terminal point is $(7, 1, -10)$. Let $\vec{w}$ denote the vector whose initial point is $(3, -5, 1)$ and whose terminal point is $(11, -7, 6)$. Find a vector which is perpendicular to both $\vec{v}$ and $\vec{w}$.

Solution. The point $(3, -5, 1)$ is located by the vector $3\vec{i} - 5\vec{j} + \vec{k}$ and the point $(7, 1, -10)$ is located by the vector $7\vec{i} + \vec{j} - 10\vec{k}$.

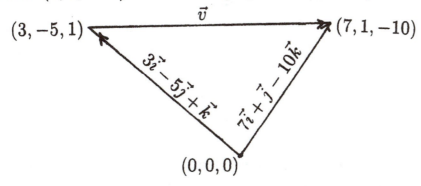

It follows that $3\vec{i} - 5\vec{j} + \vec{k} + \vec{v} = 7\vec{i} + \vec{j} - 10\vec{k}$. Therefore, $\vec{v} = 4\vec{i} + 6\vec{j} - 11\vec{k}$. In a similar way $3\vec{i} - 5\vec{j} + \vec{k} + \vec{w} = 11\vec{i} - 7\vec{j} + 6\vec{k}$. Therefore, $\vec{w} = 8\vec{i} - 2\vec{j} + 5\vec{k}$.

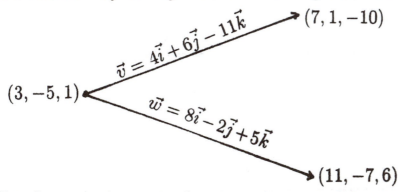

This figure is drawn in the plane determined by the three points $(3, -5, 1)$, $(7, 1, -10)$, and $(11, -7, 6)$. According to the geometric definition of cross

331

product a vector perpendicular to both these vectors is given by the cross product of the two vectors. The cross product of $\vec{v}$ and $\vec{w}$ is

$$\begin{vmatrix} \vec{i} & \vec{j} & \vec{k} \\ 4 & 6 & -11 \\ 8 & -2 & 5 \end{vmatrix} = \vec{i}(30-22) - \vec{j}(20+88) + \vec{k}(-8-48)$$

$$= 8\vec{i} - 108\vec{j} - 56\vec{k}.$$

Note that this vector is perpendicular to the plane determined by the three points $(3,-5,1)$, $(7,1,-10)$, and $(11,-7,6)$. Also the vector $\vec{w} \times \vec{v} = -(\vec{v} \times \vec{w}) = -8\vec{i} + 108\vec{j} + 56\vec{k}$ is perpendicular to this plane. It is just in the opposite direction.

Suppose that the vector $\vec{v}$ and $\vec{w}$ are placed in standard position with their initial points at the origin. In this situation these vectors determine a parallelogram.

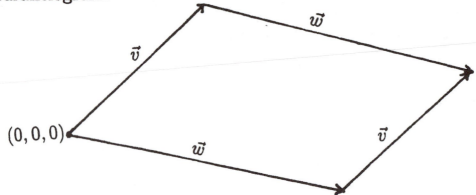

Theorem 2. The area of the parallelogram determined by the vectors $\vec{v}$ and $\vec{w}$ is given by $\|\vec{v} \times \vec{w}\| = \|\vec{v}\| \, \|\vec{w}\| \sin\theta$.

Proof. Redraw the parallelogram and extend the vector $\vec{w}$. Label the vertices of the parallelogram.

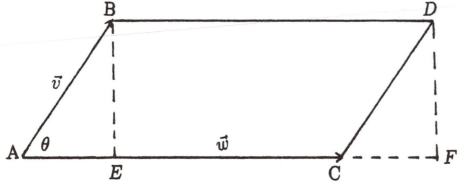

Draw a line segment perpendicular to AC and through the point B. Label the intersection of this line segment with the line AC as E. Draw a line segment perpendicular to AC and through the point D. Label the intersection of this line segment with the line AC extended as F. The triangles AEB and CFD are congruent. This means that the area of $\triangle AEB$ equals the area of $\triangle CFD$. It follows that:

Area of parallelogram $ABCD$ = Area of rectangle $EBDF$.

Area of $EBDF$ = (length of BD)times(length BE).

Let θ denote the angle between $\vec{v}$ and $\vec{w}$.

Length of BE = (length of AB)$\sin\theta = \|\vec{v}\| \sin\theta$.

Also length of $BD = \|\vec{w}\|$. It follows that

Area of parallelogram $ABCD = \|\vec{v}\| \, \|\vec{w}\| \sin\theta$.

Since $\|\vec{v} \times \vec{w}\| = \|\vec{v}\| \, \|\vec{w}\| \sin\theta$. It follows that the area of the parallelogram determined by the vector $\vec{v}$ and $\vec{w}$ is given by $\|\vec{v} \times \vec{w}\|$.

Example 3. Given the parallelogram with the four vertices $(2,0,3)$, $(8,-2,8)$, $(6,3,-7)$, and $(12,1,-2)$, find the area of this parallelogram.

Solution. We need to be sure that this is really a parallelogram.

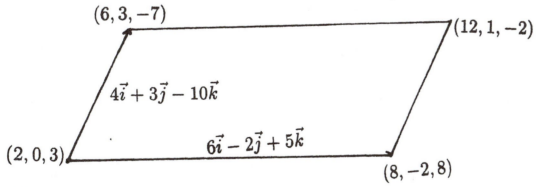

Note that

$$6\vec{i} + 3\vec{j} - 7\vec{k} - (2\vec{i} + 3\vec{k}) = 4\vec{i} + 3\vec{j} - 10\vec{k}$$

$$12\vec{i} + \vec{j} - 2\vec{k} - (8\vec{i} - 2\vec{j} + 8\vec{k}) = 4\vec{i} + 3\vec{j} - 10\vec{k}$$

$$8\vec{i} - 2\vec{j} + 8\vec{k} - (2\vec{i} + 3\vec{k}) = 6\vec{i} - 2\vec{j} + 5\vec{k}$$

$$12\vec{i} + \vec{j} - 2\vec{k} - (6\vec{i} + 3\vec{j} - 7\vec{k}) = 6\vec{i} - 2\vec{j} + 5\vec{k}$$

Since the opposite sides are the same vector this is a parallelogram. This is a parallelogram determined by the vectors

$$\vec{v} = 4\vec{i} + 3\vec{j} - 10\vec{k} \text{ and } \vec{w} = 6\vec{i} - 2\vec{j} + 5\vec{k}.$$

The cross product $\vec{v} \times \vec{w}$ is given by

$$\begin{vmatrix} \vec{i} & \vec{j} & \vec{k} \\ 4 & 3 & -10 \\ 6 & -2 & 5 \end{vmatrix} = \vec{i}(15 - 20) - \vec{j}(20 + 60) + \vec{k}(-8 - 18)$$

$$= -5\vec{i} - 80\vec{j} - 26\vec{k}.$$

$$\|\vec{v} \times \vec{w}\|^2 = \| - 5i - 80j - 26k\|^2 = (-5)^2 + (-80)^2 + (-26)^2 = 7101.$$

The area of the parallelogram is $\sqrt{7101} = 84.27$ square units.

The most common use of the cross product is to find torque. Torque is a vector $\vec{T}$ and $\|\vec{T}\|$ is measured in ft-lbs or newton-meters. These units are often called lbs-ft and meters-newton. We should not say that $\|\vec{T}\|$ is measured in Joules because Joules is an energy unit.

Suppose we have a line L in space. Let $\vec{r}$ denote a vector from some point on this line to a point P in space. The vector $\vec{r}$ need not be perpendicular to the line L. Note that a vector from some point on the line L to a given fixed point in space is uniquely determined. Think of the line L as an axis and $\vec{r}$ as a steel rod connecting point P to some point on line L. The rod $\vec{r}$ is the force arm. Suppose a vector force $\vec{F}$ is acting at the point P, then the torque exerted by the force $\vec{F}$ about the line L is given by the cross product

$$\vec{T} = \vec{r} \times \vec{F}.$$

Let $\vec{u}$ be a unit vector in the direction of the line L. Note that the vector $\vec{u}$ is not directly involved in the computation of the torque vector $\vec{T}$. However, in applications we are usually interested in the effect of torque about the line L. We would usually want to compute $\text{Proj}_{\vec{u}}\vec{T}$ and $\vec{T} - \text{Proj}_{\vec{u}}\vec{T}$ and look at these vectors (or their magnitudes). In first year physics courses most of the discussion of torque and angular motion is limited to the cases where the vectors $\vec{r}$ and $\vec{F}$ are in plane perpendicular to the line L. When this is the case we can avoid vectors because $\vec{T}$ is parallel to the line L that is, $\vec{u}$ is parallel to $\vec{T}$. This means that $\|\vec{T}\| = \|\text{Proj}_{\vec{u}}\vec{T}\|$.

Suppose we have a wheel in the xy plane which can rotate about the line which is the z axis. Let us apply a constant force at a fixed point on the edge of the wheel. After the wheel has turned through an angle θ the position of this point is given by

$$\vec{r}(\theta) = \|\vec{r}\|[(\cos\theta)\vec{i} + (\sin\theta)\vec{j}].$$

Consider

$$\vec{R}(\theta) = (\cos\theta)\vec{i} + (\sin\theta)\vec{j}$$
$$\vec{R}'(\theta) = (-\sin\theta)\vec{i} + (\cos\theta)\vec{j}.$$

This means that $\vec{R}'(\theta) = (-\sin\theta)\vec{i} + (\cos\theta)\vec{j}$ is a tangent vector to the circle which is the graph of $\vec{R}(\theta) = (\cos\theta)\vec{i} + (\sin\theta)\vec{j}$. Let us apply a constant force to the point on the wheel located by

$$\|\vec{r}\|[(\cos\theta)\vec{i} + (\sin\theta)\vec{j}]$$

of magnitude

$$\vec{F} = \|\vec{F}\|[(-\sin\theta)\vec{i} + (\cos\theta)\vec{j}].$$

This is a force tangent to the wheel of magnitude $\|\vec{F}\|$. The torque of this force about the line L which is the axis of the wheel is

$$\vec{T} = [\|\vec{r}\|[(\cos\theta)\vec{i} + (\sin\theta)\vec{j}]] \times [\|F\|[(-\sin\theta)\vec{i} + (\cos\theta)\vec{j}]]$$

$$= \|\vec{r}\| \, | \, \|\vec{F}\| \begin{vmatrix} \vec{i} & \vec{j} & \vec{k} \\ \cos\theta & \sin\theta & 0 \\ -\sin\theta & \cos\theta & 0 \end{vmatrix}.$$

$$= \|\vec{r}\|\|\vec{F}\|\vec{k}.$$

The torque is a vector of magnitude $\|\vec{r}\|\|\vec{F}\|$ along the axis of the wheel. Note that $\vec{T}$ is perpendicular to both $\vec{r}$ and $\vec{F}$.

335

This example shows that we are able to discuss torque without using vectors if we only discuss motion about a line which is perpendicular to the plane which contains both the displacement vector and the force vector. In this case, the magnitude of torque is just the product of distance and the magnitude of the force. In these cases the word torque is sometimes used to indicate the magnitude of the torque vector.

In angular motion torque plays the roll of force in that torque equals the second moment or moment of inertia times angular acceleration. The power output of an engine equals the product of the torque and the angular velocity.

Torque also plays a roll in fixed situations. Note that if we take two vectors $\vec{v}$ and $\vec{w}$ of constant length and vary the angle between the vectors then $\|\vec{v} \times \vec{w}\|$ has a maximum value when $\sin \theta = 1$ or $\theta = \pi/2$. We need to remember this where trying to turn a bolt with a wrench. Let QR denote a line through the axis of the bolt. Suppose $\vec{r}$ is the vector from the head of the bolt the point P at the end of the wrench. Suppose a force $\vec{F}$ is applied at point P at the end of the wrench. The torque about the line QR is $\vec{r} \times \vec{F}$. Torque is what turns the bolt. If we want the maximum torque about the line QR, the axis of the bolt, then we need to have $\vec{r} \times \vec{F}$ to be parallel to QR. This means that the plane determined by $\vec{r}$ and $\vec{F}$ has QR as a normal. Also the angle between $\vec{r}$ and $\vec{F}$ should be $\pi/2$ in order to maximize $\|\vec{r} \times \vec{F}\|$. If $\vec{r} \times \vec{F}$ is not parallel to QR then we can snap the head off the bolt.

Exercises

1. Given $\vec{v} = 3\vec{i} - 4\vec{j} + 6\vec{k}$ and $\vec{w} = 7\vec{i} + 5\vec{j} - 2\vec{k}$, find the vector $\vec{v} \times \vec{w}$.

2. Let $\vec{v}$ denote the vector with initial point $(-3, 2, 5)$ and terminal point $(4, 6, -1)$. Let $\vec{w}$ denote the vector with initial point $(5, -2, 7)$ and terminal point $(8, 4, -1)$. Find the vector $\vec{v} \times \vec{w}$.

3. Consider the plane determined by the three points $(3, 0, -2)$, $(11, -5, 2)$, and $(3, 7, 4)$. Find two vectors in this plane. Find a vector which is perpendicular to these two vectors.

4. Consider the parallelogram determined by the two vectors $5\vec{i} - 8\vec{j} + 3\vec{k}$

and $4\vec{i} + 7\vec{j} - 6\vec{k}$. Find the area of this parallelogram.

5. Consider the parallelogram determined by the four points $(3, 0, -2)$, $(3, 7, 4)$, $(11, -5, 2)$, and $(11, 2, 8)$. Find the area of this parallelogram.

6. The crank arm of a bicycle is 3/8 ft long. A woman who weighs 120 lbs puts her weight on one petal. Use the following right handed coordinate system. The x axis is perpendicular bicycle and points to the right when facing forward. The y axis and points toward the front. The z axis is perpendicular to the ground. Suppose the crank arm has rotated through an angle θ in the clockwise direction with vertical $\theta = 0$, $0 < \theta < \pi$. Find the displacement $\vec{r}$ and the force vector $\vec{F}$. Find the torque vector generated by the woman.

6782 Three Dimensional Position Vector

Suppose we want to graph a three dimensional vector function $\vec{r}(t)$ of one independent variable. This is done in three dimensional space using the same idea that we used to graph vector functions in two dimensional space. For a given value of t say $t = t_0$ the vector $\vec{r}(t_0)$ locates a point on the graph. Points are located as before. The vector $2\vec{i} - 3\vec{j} + 4\vec{k}$ locates the point $(2, -3, 4)$ in space. The vector $7\vec{i} + 12\vec{j} - 11\vec{k}$ locates the point $(7, 12, -11)$ in space. In general, if we place the vector $a\vec{i} + b\vec{j} + c\vec{k}$ in standard position with its initial point at $(0,0,0)$, then its terminal point is at (a, b, c). The vector $a\vec{i} + b\vec{j} + c\vec{k}$ locates the point (a, b, c). Graphing the vector function $\vec{r}(t)$ means that we plot all the points located by all the vectors $\vec{r}(t)$ as t takes on all its allowed values.

Example 1. Plot the graph of the vector function $\vec{r}(t) = (3t - 2)\vec{i} + (-2t + 5)\vec{j} + (t + 3)\vec{k}$.

Solution. The vector $\vec{r}(-1) = -5\vec{i} + 7\vec{j} + 2\vec{k}$ locates the point $(-5, 7, 2)$. The vector $\vec{r}(0) = -2\vec{i} + 5\vec{j} + 3\vec{k}$ locates the point $(-2, 5, 3)$. The vector $\vec{r}(2) = 4\vec{i} + \vec{j} + 5\vec{k}$ locates the point $(4, 1, 5)$. We plot the points $(-5, 7, 2)$, $(-2, 5, 3)$ and $(4, 1, 5)$ in space. We also find and plot other points as needed. We then connect these points with a smooth curve. This is the graph of the vector function $\vec{r}(t)$. The graph is a one dimensional figure in three dimensional space. The particular curve in this example is actually a straight line. A one dimensional figure is somewhat hard to visualize in three dimensional space. We usually do not try to draw a visualization of the graph of the vector function $\vec{r}(t)$ in space except for some special cases.

Example 2. Plot the graph of the vector function $\vec{r}(t) = (2\cos t)\vec{i} + (2\sin t)\vec{j} + t\vec{k}$.

Solution. We can easily find the coordinates of a few points on this curve. We have $\vec{r}(0) = 2\vec{i}$. The vector $2\vec{i}$ locates the point $(2, 0, 0)$ which is on the curve. We have $\vec{r}(\pi/2) = 2\vec{j} + (\pi/2)\vec{k}$. The vector $2\vec{i} + (\pi/2)\vec{k}$ locates the point $(0, 2, \pi/2)$ which is on the curve. However, instead of trying to draw this graph by finding a lot of points and connecting them with a smooth curve, let us try to describe the graph. Note that the x and y

coordinates are always given by $x = 2\cos t$ and $y = 2\sin t$. It follows that $x^2 + y^2 = 4\cos^2 t + 4\sin^2 t = 4$. The equation $x^2 + y^2 = 4$ is the equation of a circle of radius 2. This means that as t increases the x and y coordinates of the points on the graph are always on a circle of radius 2. As t increases the points go around and around this circle. The z coordinate is $z = t$. As t increases the z coordinate increases. Thus, when $t = 0$ the curve is at $(2, 0, 0)$. **As t increases the curve goes around the circle of radius 2 and up in a spiral.**

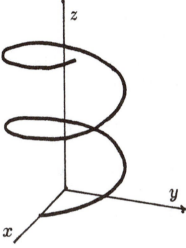

Example 3. Find the vector function whose graph is the line through the point $(2, 3, 4)$ and parallel to the vector $\vec{v} = 5\vec{i} - 4\vec{j} + 7\vec{k}$.

Solution. We can not draw a good visualization of the three dimensional situation. However, all the important items in this example are in the same plane. We can draw a visualization in this plane.

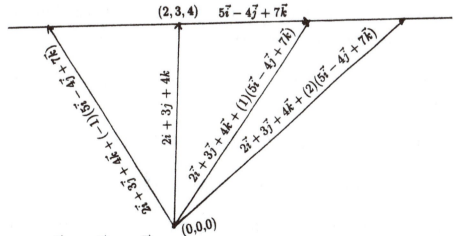

The vector $2\vec{i} + 3\vec{j} + 4\vec{k}$ locates the point $(2, 3, 4)$. Using the geometric definition of vector addition we see that the vector

$$2\vec{i} + 3\vec{j} + 4\vec{k} + (1)(5\vec{i} - 4\vec{j} + 7\vec{k})$$

locates a point on this line. Also using the geometric definition of vector addition we see that the following vectors also locate points on the line.

$$2\vec{i} + 3\vec{j} + 4\vec{k} + (2)(5\vec{i} - 4\vec{j} + 7\vec{k})$$
$$2\vec{i} + 3\vec{j} + 4\vec{k} + (-1)(5\vec{i} - 4\vec{j} + 7\vec{k}).$$

The vector $2\vec{i} + 3\vec{j} + 4\vec{k} + (t)(5\vec{i} - 4\vec{j} + 7\vec{k})$ locates a point on the line for any value of the scalar t. All the points on this line are given by

$$\vec{r}(t) = 2\vec{i} + 3\vec{j} + 4\vec{k} + t(5\vec{i} - 4\vec{j} + 7\vec{k})$$

when we use all values of t. The line is the graph of the vector function

$$\vec{r}(t) = 2\vec{i} + 3\vec{j} + 4\vec{k} + t(5\vec{i} - 4\vec{j} + 7\vec{k}).$$

The vector $5\vec{i} - 4\vec{j} + 7\vec{k}$ is called the direction vector of the line. Any line in space is determined by one point on the line and the direction vector of the line. The graph of the vector function

$$\vec{r}(t) = (\text{vector locating one point on the line}) + t(\text{direction vector})$$

is the line.

Example 4. Find the vector function $\vec{r}(t)$ whose graph is the line through the point $(-7, 2, 5)$ and parallel to the vector $8\vec{i} - 9\vec{j} + 11\vec{k}$.

Solution. The vector from $(0, 0, 0)$ to $(-7, 2, 5)$ is $-7\vec{i} + 2\vec{j} + 5\vec{k}$. The graph of the vector function

$$\vec{r}(t) = (-7\vec{i} + 2\vec{j} + 5\vec{k}) + t(8\vec{i} - 9\vec{j} + 11\vec{k})$$

is the line through the point $(-7, 2, 5)$ parallel to the vector $8\vec{i} - 9\vec{j} + 11\vec{k}$.

Example 5. Find the vector function whose graph is the line through the two points $(-3, 5, 8)$ and $(2, -4, 6)$.

Solution. We have a point on the line, in fact two points. We need to find a direction vector. Let $\vec{v}$ denote the vector with initial point $(-3, 5, 8)$ and terminal point $(2, -4, 6)$, then $\vec{v}$ is a direction vector for the line. The vector $-3\vec{i} + 5\vec{j} + 8\vec{k}$ locates the point $(-3, 5, 8)$ and the vector $2\vec{i} - 4\vec{j} + 6\vec{k}$ locates the point $(2, -4, 6)$.

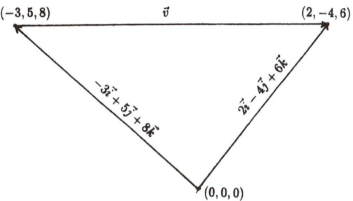

The geometric definition of vector addition tells us that

$$-3\vec{i} + 5\vec{j} + 8\vec{k} + \vec{v} = 2\vec{i} - 4\vec{j} + 6\vec{k}$$

$$\vec{v} = 5\vec{i} - 9\vec{j} - 2\vec{k}.$$

The line through $(-3, 5, 8)$ and $(2, -4, 6)$ is parallel to $5\vec{i} - 9\vec{j} - 2\vec{k}$. The vector $5\vec{i} - 9\vec{j} - 2\vec{k}$ is a direction vector for the line. The graph of the vector function

$$\vec{r}(t) = (-3\vec{i} + 5\vec{j} + 8\vec{k}) + t(5\vec{i} - 9\vec{j} - 2\vec{k})$$
$$= (5t - 3)\vec{i} + (-9t + 5)\vec{j} + (-2t + 8)\vec{k}$$

is the line through the two points $(-3, 5, 8)$ and $(2, -4, 6)$.

Example 6. At what point do the following two lines intersect.

$$\vec{r}(t) = (4t - 17)\vec{i} + (7t - 13)\vec{j} + (-3t + 19)\vec{k}$$
$$\vec{R}(s) = (3s + 1)\vec{i} + (-4s)\vec{j} + (7s + 24)\vec{k}$$

Solution. In order for the lines to intersect the same point must be on both lines. In order for the lines to intersect there must be a value of t say t_0 and a value of s say s_0 such that $\vec{r}(t_0) = \vec{R}(s_0)$. We need to find a value of t and a value of s such that

$$(4t - 17)\vec{i} + (7t - 13)\vec{j} + (-3t + 19)\vec{k} = (3s + 1)\vec{i} + (-4s)\vec{j} + (7s + 24)\vec{k}.$$

341

In order for this vector equality to be true the corresponding components must be equal.

$$4t - 17 = 3s + 1 \qquad\qquad 4t - 3s = 18$$
$$7t - 13 = -4s \qquad \text{or} \qquad 7t + 4s = 13$$
$$-3t + 19 = 7s + 24. \qquad -3t - 7s = 5$$

We have three equations in two unknowns. Such a system may not have a solution. Solve the top equation for s and we get

$$s = (4/3)t - 6.$$

Substitute this for s in the other two equations and we get

$$
\begin{aligned}
7t + 4[(4/3)t - 6] &= 13 \\
7t + (16/3)t - 24 &= 13 \\
(37/3)t &= 37 \\
t &= 3
\end{aligned}
\qquad \text{and} \qquad
\begin{aligned}
-3t - 7[(4/3)t - 6] &= 5 \\
-3t - (28/3)t + 42 &= 5 \\
-(37/3)t &= -37 \\
t &= 3
\end{aligned}
$$

Using $t = 3$, we have $s = (4/3)(3) - 6 = -2$. The solution is $t = 3$ and $s = -2$. We have

$$\vec{r}(3) = -5\vec{i} + 8\vec{j} + 10\vec{k}$$
$$\vec{R}(-2) = -5\vec{i} + 8\vec{j} + 10\vec{k}.$$

The point of intersection is $(-5, 8, 10)$.

Example 7. Find the intersection point of the two lines that are the graphs of the following vector functions.

$$\vec{r}(t) = (2t + 3)\vec{i} - 3t\vec{j} + (4t - 5)\vec{k}$$
$$\vec{R}(s) = (4s - 5)\vec{i} + (-s + 6)\vec{j} + (5s)\vec{k}.$$

Solution. We need to find a point that is on both curves. We need to find a value of t and a value of s such that

$$(2t + 3)\vec{i} - 3t\vec{j} + (4t - 5)\vec{k} = (4s - 5)\vec{i} + (-s + 6)\vec{j} + (5s)\vec{k}.$$

In order for this vector equation to be true the corresponding components must be equal. We need to find a value of t and a value of s such that

$$2t + 3 = 4s - 5$$
$$-3t = -s + 6$$
$$4t - 5 = 5s$$

These equations simplify to

$$2t - 4s = -8$$
$$-3t + s = 6$$
$$4t - 5s = 5.$$

Solving the second equation for s we get

$$s = 3t + 6$$

Substitute this into the first equation

$$2t - 4(3t + 6) = -8$$
$$2t - 12t - 24 = -8$$
$$-10t = 16$$
$$t = -8/5.$$

Replacing t with $-8/5$ in the first equation, we get

$$2(-8/5) + 3 = 4s - 5$$
$$\frac{24}{5} = 4s$$
$$6/5 = s$$

Substituting $t = -8/5$ and $s = 6/5$ into the third equation, we have

$$4(-8/5) - 5(6/5) = 5$$
$$-62 = 25.$$

The statement $-62 = 25$ is not true. The values $t = -8/5$ and $s = 6/5$ are solutions of the first two equations but are not a solution of the third equation. There is no values of s and t that is a solution of all three equations. This system of three equations in two unknowns has no solution. These two lines do not have a point in common. These two lines in space do not intersect. Note that these lines are not parallel since their direction vectors are not parallel. Such lines are called skew lines.

A fact that causes a lot of confusion when graphing vector functions is that a linear shift in the t variable in a given vector function $\vec{r}(t)$ results in a new vector function $\vec{R}(s)$ which graphs as the same curve as the original function $\vec{r}(t)$. The two different functions have the same graph. Consider the vector function

$$\vec{r}(t) = (3t - 5)\vec{i} + (-4t + 7)\vec{j} + (5t + 8)\vec{k}.$$

This function graphs as a line through $(-5, 7, 8)$ with direction vector $3\vec{i} - 4\vec{j} + 5\vec{k}$. Define a new vector function by

$$\vec{R}(s) = \vec{r}(s + 3) = [3(s + 3) - 5]\vec{i} + [-4(s + 3) + 7]\vec{j} + [5(s + 3) + 8]\vec{k}$$
$$= (3s + 4)\vec{i} + (-4s - 5)\vec{j} + (5s + 23)\vec{k}.$$

The function $\vec{R}(s)$ graphs as a line through the point $(4, -5, 23)$ with direction vector $3\vec{i} - 4\vec{j} + 5\vec{k}$. But $\vec{r}(3) = 4\vec{i} - 5\vec{j} + 23\vec{k}$ and thus the point $(4, -5, 23)$ is a point on the graph of $\vec{r}(t)$. Note both $\vec{r}(t)$ and $\vec{R}(s)$ have $3\vec{i} - 4\vec{j} + 5\vec{k}$ as direction vector. Thus, both $\vec{r}(t)$ and $\vec{R}(s)$ graph as the same line.

Example 8. The graph of the vector function $\vec{r}(t) = (3t - 7)\vec{i} + (5t + 2)\vec{j} + (-4t + 1)\vec{k}$ is a line. A point on this line is located by the vector $\vec{r}(3) = 2\vec{i} + 17\vec{j} - 11\vec{k}$. Find a vector function $\vec{R}(s)$ whose graph is this same line and is such that $\vec{R}(0) = 2\vec{i} + 17\vec{j} - 11\vec{k}$.

Solution. There are a couple of easy ways to find $\vec{R}(s)$. One way is to make $\vec{R}(s)$ a translation of $\vec{r}(t)$. This means that $\vec{R}(s)$ is found by translating the variable t in $\vec{r}(t)$. Since $\vec{r}(3) = 2\vec{i} + 17\vec{j} - 11\vec{k}$ we let $s = t - 3$ or $t = s + 3$. This means that $\vec{R}(s) = \vec{r}(s + 3)$ and so $\vec{R}(0) = \vec{r}(3)$. Using this we get

$$\vec{R}(s) = \vec{r}(s + 3) = [3(s + 3) - 7]\vec{i} + [5(s + 3) + 2]\vec{j} + [-4(s + 3) + 1]\vec{k}.$$

This gives

$$\vec{R}(s) = (3s + 2)\vec{i} + (5s + 17)\vec{j} + (-4s - 11)\vec{k}.$$

Note that $\vec{R}(0) = 2\vec{i} + 17\vec{j} - 11\vec{k}$ and the direction vector in $\vec{R}(s)$ is $3\vec{i} + 5\vec{j} - 4\vec{k}$ the same as the direction vector for $\vec{r}(t)$. The point $(2, 17, -11)$ is on both the graph of $\vec{r}(t)$ and $\vec{R}(s)$. The graph of both $\vec{r}(t)$ and $\vec{R}(s)$ have the same direction vector. The lines must be the same line.

Alternately we can look at this problem as we are given a point $(2, 17, -11)$ on the line located by the vector $2\vec{i} + 17\vec{j} - 11\vec{k}$. Also by looking at the given vector function for the line we see that its direction vector is $3\vec{i} + 5\vec{j} - 4\vec{k}$. Therefore, a vector function whose graph is the line in question is $\vec{R}(s) = 2\vec{i} + 17\vec{j} - 11\vec{k} + s(3\vec{i} + 5\vec{j} - 4\vec{k})$.

Exercises

1. Find the vector function whose graph is the line through the point $(3, -2, 5)$ and parallel to the vector $\vec{v} = 6\vec{i} + 7\vec{j} - 4\vec{k}$.

2. Find the vector function whose graph is the line through the two points $(5, 0, -4)$ and $(4, -5, -4)$.

3. The graphs of the following two vector functions are lines. Find the point where these lines intersect.

$$\vec{r}(t) = (3t - 1)\vec{i} + (5t - 15)\vec{j} + (-4t + 7)\vec{k}$$
$$\vec{R}(s) = (-2s + 4)\vec{i} + (6s + 12)\vec{j} + (-s - 7)\vec{k}.$$

4. The graphs of the following two vector functions are lines. Find the point where these lines intersect, if they do intersect.

$$\vec{r}(t) = (5t + 3)\vec{i} + (-2t + 7)\vec{j} + (4t + 5)\vec{k}$$
$$\vec{R}(s) = (8)\vec{i} + (6s + 1)\vec{j} + (3s + 2)\vec{k}.$$

5. The graph of the following two vector functions are lines. Find the point where these lines intersect. Show that the lines are perpendicular.

$$\vec{r}(t) = (4t - 5)\vec{i} + (-5t + 11)\vec{j} + (3t)\vec{k}$$
$$\vec{R}(s) = (7s)\vec{i} + (2s - 6)\vec{j} + (-6s + 15)\vec{k}.$$

6. The graph of each of the following vector functions is a line. Show that the graphs are actually the same line.

$$\vec{r}(t) = (-6t + 6)\vec{i} + (4t - 7)\vec{j} + (-10t + 11)\vec{k}$$
$$\vec{R}(s) = (9s + 18)\vec{i} + (-6s - 15)\vec{j} + (15s + 31)\vec{k}$$

7. The graph of the vector function $\vec{r}(t) = (5t - 8)\vec{i} + (-3t + 11)\vec{j} + (6t + 7)\vec{k}$ is a line. A point on this line is located by the vector $\vec{r}(2) = 2\vec{i} + 5\vec{j} + 19\vec{k}$. Find a vector function $\vec{R}(s)$ by translating $\vec{r}(t)$ whose graph is the same line and is such that $\vec{R}(0) = 2\vec{i} + 5\vec{j} + 19\vec{k}$.

8. The graph of the vector function

$$\vec{r}(t) = (3t + 7)\vec{i} + (-4t + 8)\vec{j} + (5t + 9)\vec{k}$$

is a line. The point $(16, -4, 24)$ on this line is located by the vector $\vec{r}(3) = 16\vec{i} - 4\vec{j} + 24\vec{k}$. Find a vector function $\vec{R}(s)$ whose graph is the same line but is such that $\vec{R}(0) = 16\vec{i} - 4\vec{j} + 24\vec{k}$.

6786 Vector Calculus in Three Space

Given the vector valued function

$$\vec{r}(t) = f(t)\vec{i} + g(t)\vec{j} + h(t)\vec{k}$$

the real functions $f(t), g(t)$, and $h(t)$ are called the component functions of $\vec{r}(t)$. The derivative of the vector function $\vec{r}(t)$ is defined for any value $t = b$ as

$$\vec{r}\,'(b) = \lim_{t \to b} \frac{\vec{r}(b) - \vec{r}(t)}{b - t}.$$

We can easily show from this that

$$\vec{r}\,'(t) = f'(t)\vec{i} + g'(t)\vec{j} + h'(t)\vec{k}.$$

Example 1. Given that $\vec{r}(t) = (\sin 2t)\vec{i} + (t^3 - 8t)\vec{j} + (t^{5/3} - 7)\vec{k}$, then

$$\vec{r}\,'(t) = (2\cos 2t)\vec{i} + (3t^2 - 8)\vec{j} + [(5/3)t^{2/3}]\vec{k}$$

A visualization of the vector $\vec{r}(b) - \vec{r}(t)$ is

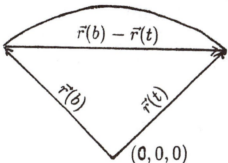

The expression $\frac{1}{b-t}$ is a scalar which is larger than 1 when $0 < b - t < 1$.

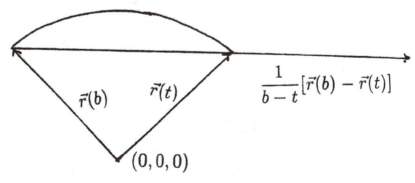

Clearly when we take the limit as $t \to b$ the point located by $\vec{r}(t)$ approaches the point located by $\vec{r}(b)$. Also the vector

$$\frac{1}{b-t}[\vec{r}(b) - \vec{r}(t)]$$

approaches a tangent vector to the curve $\vec{r}(t)$ at the point located by $\vec{r}(b)$.

Theorem. Consider the curve which is the graph of the vector function $\vec{r}(t)$. The vector $\vec{r}\,'(b)$ is a tangent vector to this curve at the point located by $\vec{r}(b)$.

Example 2. Consider the curve in space which is the graph of $\vec{r}(t) = (t^3 + 8t)\vec{i} + (3t^2 - 5t)\vec{j} + t^4\vec{k}$. Find a tangent vector to this curve at the points on the curve where $t = -1, t = 2$, and $t = 3$.

Solution. The derivative of $\vec{r}(t)$ is given by $\vec{r}\,'(t) = (3t^2 + 8)\vec{i} + (6t - 5)\vec{j} + (4t^3)\vec{k}$. When $t = -1$ we have $\vec{r}(-1) = -9\vec{i} + 8\vec{j} + \vec{k}$ and $\vec{r}\,'(-1) = 11\vec{i} - 11\vec{j} - 4\vec{k}$. The vector $\vec{r}\,'(-1) = 11\vec{i} - 11\vec{j} - 4\vec{k}$ is a tangent vector to the space curve which is the graph of $\vec{r}(t)$ at the point $(-9, 8, 1)$.

When $t = 2$, we have $\vec{r}(2) = 24\vec{i} + 2\vec{j} + 16\vec{k}$ and $\vec{r}\,'(2) = 20\vec{i} + 7\vec{j} + 32\vec{k}$. The vector $\vec{r}\,'(2) = 20\vec{i} + 7\vec{j} + 32\vec{k}$ is a tangent vector to the curve which is the graph of $\vec{r}(t)$ at the point $(24, 2, 16)$.

When $t = 3$, we have $\vec{r}(3) = 51\vec{i} + 12\vec{j} + 81\vec{k}$ and $\vec{r}\,'(3) = 35\vec{i} + 13\vec{j} + 108\vec{k}$. The vector $\vec{r}\,'(3) = 35\vec{i} + 13\vec{j} + 108\vec{k}$ is a tangent vector to the curve at the point $(51, 12, 81)$.

Example 3. Consider the curve in space which is the graph of the vector function

$$\vec{r}(t) = (t^3 - 6t)\vec{i} + (10 - 2t^2)\vec{j} + (t^2 + 5t + 3)\vec{k}.$$

Find the point on this curve corresponding to $t = 2$. Find the vector function whose graph is the tangent line to this curve at the point where $t = 2$.

Solution. We need to find the point on the curve located by the vector $\vec{r}(2)$. Since $\vec{r}(2) = -4\vec{i} + 2\vec{j} + 17\vec{k}$, the point on the curve has coordinates $(-4, 2, 17)$. The tangent line to this curve at this point is a line in space. A line is determined by one point on the line and the direction vector of the line. The tangent line to the curve also goes through the point on the curve. One point on the tangent line is $(-4, 2, 17)$. The tangent vector to the curve at this point would also be the direction vector of the tangent line.

$$\vec{r}\,'(t) = (3t^2 - 6)\vec{i} + (-4t)\vec{j} + (2t + 5)\vec{k}$$
$$\vec{r}\,'(2) = 6\vec{i} - 8\vec{j} + 9\vec{k}.$$

The vector $6\vec{i} - 8\vec{j} + 9\vec{k}$ is a tangent vector to the curve at $(-4, 2, 17)$ and so is also a direction vector for the tangent line. Therefore, the graph of the vector function

$$\vec{R}(s) = (-4\vec{i} + 2\vec{j} + 17\vec{k}) + s(6\vec{i} - 8\vec{j} + 9\vec{k})$$

is a tangent line to the curve which is the graph of the vector function $\vec{r}(t)$ at the point $(-4, 2, 17)$. Note that $\vec{R}(2) = 8\vec{i} - 14\vec{j} + 35\vec{k}$. The point $(8, -14, 35)$ is on the tangent line. The graph of the vector function $\vec{T}(s) = (8\vec{i} - 14\vec{j} + 35\vec{k}) + s(6\vec{i} - 8\vec{j} + 9\vec{k})$ is also this same tangent line.

Example 4. Show that the curve which is the graph of the vector function $\vec{r}(t) = (t^2 + 5t)\vec{i} + (2t + 6)\vec{j} + (t^3 + 2)\vec{k}$ passes through the point $(24, 12, 29)$.

Solution. We need to show that there is a value of t such that

$$(t^2 + 5t)\vec{i} + (2t + 6)\vec{j} + (t^3 + 2)\vec{k} = 24\vec{i} + 12\vec{j} + 29\vec{k}.$$

In order for this vector equation to be true the corresponding components must be equal. We must find a value of t such that

$$t^2 + 5t = 24$$
$$2t + 6 = 12$$
$$t^3 + 2 = 29$$

Start with the easy middle equation which is $2t + 6 = 12$. This gives $t = 3$. Does this value of t work for the other two equations?

$$(3)^2 + 5(3) = 24$$
$$(3)^3 + 2 = 29.$$

Yes, it does work. We have $\vec{r}(3) = 24\vec{i} + 12\vec{j} + 29\vec{k}$. Thus, the graph of the vector function $\vec{r}(t)$ does pass through the point $(24, 12, 29)$.

Suppose $\vec{R}(t)$ is a vector function in two dimensional space. We showed in our previous work that the arc length of the graph of $\vec{R}(t)$ from the point $\vec{R}(a)$ to the point $\vec{R}(b)$ is given by

$$\int_a^b \|\vec{R}\,'(t)\|dt.$$

Suppose we have a vector function $\vec{r}(t) = f(t)\vec{i} + g(t)\vec{j} + h(t)\vec{k}$ in three dimensional space. Consider the arc in space which is the graph of this vector function from the point located by $\vec{r}(a)$ to the point located by $\vec{r}(b)$. In a similar way we can show that the arc length of this arc is given by

$$\text{Arc length} = \int_a^b \sqrt{[f'(t)]^2 + [g'(t)]^2 + [h'(t)]^2}\, dt.$$

We know that $\vec{r}\,'(t) = [f'(t)]\vec{i} + [g'(t)]\vec{j} + [h'(t)]\vec{k}$, and so

$$\|r'(t)\| = \sqrt{[f'(t)]^2 + [g'(t)]^2 + [h'(t)]^2}$$

It follows that we can write the formula for arc length as

$$\text{Arc length} = \int_a^b \|\vec{r}\,'(t)\|dt.$$

Example 5. Consider the graph of the vector function

$$\vec{r}(t) = (18t)\vec{i} + (4\sqrt{6}t^{3/2})\vec{j} + (3t^2)\vec{k}.$$

Find the arc length of the section of curve from $(18, 4\sqrt{6}, 3)$ to $(54, 36\sqrt{2}, 27)$.

Solution. We need to find $\|\vec{r}\,'(t)\|$.

$$\vec{r}\,'(t) = (18)\vec{i} + (6\sqrt{6}t^{1/2})\vec{j} + (6t)\vec{k}$$
$$\|\vec{r}\,'(t)\|^2 = (18)^2 + (6\sqrt{6}t^{1/2})^2 + (6t)^2$$
$$= 324 + 216t + 36t^2$$
$$= 36(9 + 6t + t^2) = 36(3 + t)^2$$

350

Taking the square root, we get

$$\|\vec{r}\,'(t)\| = 6(3 + t).$$

Note that $\vec{r}(1) = (18)\vec{i} + (4\sqrt{6})\vec{j} + 3\vec{k}$ and $\vec{r}(3) = 54\vec{i} + (36\sqrt{2})\vec{j} + 27\vec{k}$. The section of curve in question is given by $1 \le t \le 3$. The arc length is

$$\int_1^3 \|\vec{r}\,'(t)\| dt = \int_1^3 6(3 + t) dt = 60.$$

Definition. Suppose $\vec{r}(t) = [f(t)]\vec{i} + [g(t)]\vec{j} + [h(t)]\vec{k}$, then the indefinite integral of $\vec{r}(t)$ is given by

$$\int \vec{r}(t) dt = \left[\int f(t) dt\right]\vec{i} + \left[\int g(t) dt\right]\vec{j} + \left[\int h(t) dt\right]\vec{k}.$$

The definite integral of $\vec{r}(t)$ over the interval $a \le t \le b$ is defined as

$$\int_a^b \vec{r}(t) dt = \left[\int_a^b f(t) dt\right]\vec{i} + \left[\int_a^b g(t) dt\right]\vec{j} + \left[\int_a^b h(t) dt\right]\vec{k}.$$

Example 6. Given that $\vec{r}(t) = (\cos 3t)\vec{i} + (4t^3 + 5)\vec{j} + (6t^2 + 8t + 3)\vec{k}$, find $\int \vec{r}(t) dt$.

Solution. We simply find the indefinite integral of each component function

$$\int \vec{r}(t) dt = \left[\int (\cos 3t) dt\right]\vec{i} + \left[\int (4t^3 + 5) dt\right]\vec{j} + \left[\int (6t^2 + 8t + 3) dt\right]\vec{k}$$

$$= [(1/3)\sin(3t)]\vec{i} + [t^4 + 5t]\vec{j} + [2t^3 + 4t^2 + 3t]\vec{k} + a\vec{i} + b\vec{j} + c\vec{k},$$

where $a\vec{i} + b\vec{j} + c\vec{k}$ is a constant vector.

Example 7. Given that $\vec{r}(t) = (4t+5)\vec{i} + (3t^2)\vec{j} + (4t^3 + 6t)\vec{k}$, find $\int_1^3 \vec{r}(t) dt$.

Solution. We simply find the definite integral of each component function.

$$\int_1^3 \vec{r}(t) dt = \left[\int_1^3 (4t + 5) dt\right]\vec{i} + \left[\int_1^3 (3t^2) dt\right]\vec{j} + \left[\int_1^3 (4t^3 + 6t) dt\right]\vec{k}$$

$$= [(2t^2 + 5t)\,|_1^3]\vec{i} + [t^3\,|_1^3]\vec{j} + [(t^4 + 3t^2)\,|_1^3]\vec{k}$$

$$= 26\vec{i} + 26\vec{j} + 104\vec{k}.$$

There is another common integral and for it we integrate along a curve in 3 space. It is usually called a line integral. The most common way to motivate this integral is to use the example of work. Let us recall a simple work problem.

Example 8. Suppose a particle is moved along a straight line from $(-4, 9, 3)$ to $(8, -2, 11)$. Suppose this particle is acted on during the move by the constant force $\vec{F} = 5\vec{i} + 7\vec{j} - 3\vec{k}$. Find the work done on the particle during the move by the force $\vec{F}$.

Solution. The displacement vector for this particle is

$$\vec{D} = (8\vec{i} - 2\vec{j} + 11\vec{k}) - (-4\vec{i} + 9\vec{j} + 3\vec{k}) = 12\vec{i} - 11\vec{j} + 8\vec{k}.$$

Recall that work equals force vector dotted with displacement vector.

$$\vec{F} \cdot \vec{D} = (5\vec{i} + 7\vec{j} - 3\vec{k}) \cdot (12\vec{i} - 11\vec{j} + 8\vec{k}) = -41.$$

These units would be ft-lbs or Newton meters.

The problem of computing work gets a lot more complicated if the path is not a line and the force is not constant. Suppose we have a variable force $\vec{F}(x, y, z)$ and also that the particle is moving along a curved path in space. Suppose the path is the graph of the vector function $\vec{r}(t) = x(t)\vec{i} + y(t)\vec{j} + z(t)\vec{k}$ for $a < t \leq b$. We can find the force at any point on the curve by substituting. Force is given by $\vec{F}(t) = \vec{F}(x(t), y(t), z(t))$.

Let us suppose we want to find the work done by the force $\vec{F}(x, y, z)$ in moving the object along the curve c given by $\vec{r}(t)$ from the point located by $\vec{r}(a)$ to the point located by $\vec{r}(b)$. Since we know how to find work when force is constant and curve is a line segment we proceed as follows. We start by dividing the t axis $a \leq t \leq b$ into subintervals of length $\Delta t = \frac{b-a}{n}$. Consider the typical interval

$$\Delta t = t_k - t_{k-1}, \ k = 1, 2, \cdots n.$$

$$t_{k-1}^* = (1/2)(t_k + t_{k-1}).$$

Note that t_k^* denotes the midpoint of the interval $t_{k-1} \leq t \leq t_k$. The force on the object at $t = t_k^*$ is given by

$$\vec{F}(t_k^*) = \vec{F}(x(t_k^*), y(t_k^*), z(t_k^*)).$$

Let us approximate the force everywhere on the entire small section of curve from $\vec{r}(t_{k-1})$ to $\vec{r}(t_k)$ by $\vec{F}(t_k^*)$. Let us approximate this short section of curve using the secant line from $\vec{r}(t_{k-1})$ to $\vec{r}(t_k)$. This secant line is given by the vector $\vec{r}(t_k) - \vec{r}(t_{k-1})$. The work done by the force $\vec{F}(t)$ in moving the object from the point located by $\vec{r}(t_{k-1})$ to the point located by $\vec{r}(t_k)$ is approximately equal to

$$\vec{F}(x(t_k^*), y(t_k^*), z(t_k^*)) \cdot [\vec{r}(t_k) - \vec{r}(t_{k-1})].$$

In more compact notation

$$\vec{F}(t_k^*) \cdot [\vec{r}(t_k) - \vec{r}(t_{k-1})].$$

Divide and multiply by the scalar $t_k - t_{k-1} = \Delta t$ and the result is

$$\vec{F}(t_k^*) \cdot \frac{\vec{r}(t_k) - \vec{r}(t_{k-1})}{t_k - t_{k-1}} (t_k - t_{k-1})$$

This is a real valued function defined on the t number line for $a \leq t \leq b$. We take the limit as $n \to \infty$ and we get

$$\int_a^b \vec{F}(t) \cdot \vec{r}\,'(t) dt.$$

Note that

$$\lim_{n \to \infty} \frac{\vec{r}(t_k) - \vec{r}(t_{k-1})}{t_k - t_{k-1}} = \vec{r}\,'(t).$$

Therefore, we define the work done by a variable force $\vec{F}(t)$ in moving a particle from point $\vec{r}(a)$ to the point $\vec{r}(b)$ along the curve $\vec{r}(t)$, $a \leq t \leq b$, as

$$\text{work} = \int_a^b \vec{F}(t) \cdot \vec{r}\,'(t) dt.$$

This is known as a line integral from $\vec{r}(a)$ to $\vec{r}(b)$ along the curve $\vec{r}(t)$. We also use the notation

$$\int_a^b \vec{F}(t) \cdot \vec{r}\,'(t)dt = \int_c \vec{F}(t) \cdot \vec{r}\,'(t)dt = \int_c x(t)dx + y(t)dy + z(t)dz.$$

Example 9. Let Let $\vec{F}(x, y, z) = (x+y)\vec{i} + 3y^2\vec{j} + 4(x+z)\vec{k}$. Let the curve c be given by $\vec{r}(t) = t^2\vec{i} + (5t + 7)\vec{j} + 3t^2\vec{k}$ for $0 \leq t \leq 2$. Evaluate the line integral

$$\int_c \vec{F}(t) \cdot \vec{r}\,'(t)dt = \int_c (x+y)dx + 3y^2 dy + 4(x+z)dz.$$

Solution. We have

$$\vec{F}(t) = (x+y)\vec{i} + 3y^2\vec{j} + 4(x+z)\vec{k}$$
$$= (t^2 + 5t + 7)\vec{i} + 3(5t + 7)^2\vec{j} + 4(4t^2)\vec{k}$$
$$\vec{r}\,'(t) = 2t\vec{i} + 5\vec{j} + 6t\vec{k}$$
$$\vec{F}(t) \cdot \vec{r}\,'(t) = 2t(t^2 + 5t + 7) + 5(75t^2 + 210t + 147) + 6(16t^3).$$
$$\vec{F}(t) \cdot \vec{r}\,'(t) = 98t^3 + 385t^2 + 1064t + 735.$$
$$\int_0^2 \vec{F}(t) \cdot \vec{r}\,'(t)dt = \int_0^2 (98t^3 + 385t^2 + 1064t + 735)dt$$
$$= \frac{98}{4}t^4 + \frac{385}{3}t^3 + 532t^2 + 735t \Big|_0^2$$
$$= \frac{15050}{3}.$$

Example 10. Suppose the force on a particle is given by $\vec{F}(t) = (t+3)\vec{i} + (t^2)\vec{j} + (4t - 2)\vec{k}$ as it moves along the curve $\vec{r}(t) = (t^2 + 5)\vec{i} + (4t + 7)\vec{j} + t^3\vec{k}$ for $0 \leq t \leq 3$. Find the work done by the force.

Solution. Note that

$$\vec{r}\,'(t) = (2t)\vec{i} + (4)\vec{j} + (3t^2)\vec{k}.$$

We have

$$\vec{F}(t) \cdot \vec{r}\,'(t) = 2t^2 + 6t + 4t^2 + 12t^3 - 6t^2$$

$$= 12t^3 + 6t$$

$$\text{work} = \int_0^3 (12t^3 + 6t)dt = 3t^4 + 3t^2 \mid_0^3 = 270.$$

Units could be ft-lbs. This would assume force given in lbs and lengths in ft.

Exercises

1. Consider the curve in space which is the graph of the vector function

$$\vec{r}(t) = (3t^2 + 5t)\vec{i} + (2t^3 - 8)\vec{j} + (t^3 + 2t^2 + 5)\vec{k}.$$

Find a tangent vector to this curve at the points on the curve where $t = -1$, $t = 1$, and $t = 2$.

2. Consider the curve in space which is the graph of the vector function

$$\vec{r}(t) = (t^2 + 6t)\vec{i} + (4t + 3)\vec{j} + (t^3 + 3)\vec{k}.$$

Find a tangent vector to this curve at the point $(-5, -1, 2)$ and at the point $(16, 11, 11)$.

3. Consider the curve in space which is the graph of the vector function

$$\vec{r}(t) = (3t^2)\vec{i} + (4t^2 + 3)\vec{j} + (2/3)t^3\vec{k}.$$

Find the arc length of the section of this curve corresponding to $0 \leq t \leq 12$.

4. Consider the curve in space which is the graph of the vector function

$$\vec{r}(t) = (3\sqrt{2}t^2)\vec{i} + (6t + 5)\vec{j} + (2t^3)\vec{k}.$$

Find the arc length of the section of this curve which begins at $(3\sqrt{2}, 11, 2)$ and ends at $(27\sqrt{2}, 23, 54)$.

5. Let $\vec{r}(t) = (6t^2 + 8t)\vec{i} + (2\sin 4t)\vec{j} + [(1+t^2)^{-1}]\vec{k}$. Find $\int \vec{r}(t)dt$.

6. Let $\vec{r}(t) = (6t^2 + 8t)\vec{i} + (10 - 3t^2)\vec{j} + (4t^3 - 8t^2)\vec{k}$. Evaluate the definite integral

$$\int_1^3 \vec{r}(t)dt.$$

7. Consider the curve which is the graph of the vector function

$$\vec{r}(t) = (t^2 + 4t - 7)\vec{i} + (5t^3 + 8t)\vec{j} + (t^4)\vec{k}.$$

Find a vector function $\vec{R}(s)$ whose graph is the tangent line to this curve at the point where $t = -1$. Find a vector function whose graph is the tangent line to this curve at the point where $t = 2$.

8. Consider the curve which is the graph of the vector function

$$\vec{r}(t) = (t^3 - 4t^2 + 7t)\vec{i} + (t^2 - 4t)\vec{j} + (3t^2 - 8)\vec{k}.$$

Find a vector function $\vec{R}(s)$ whose graph is a tangent line to this curve at the point $(4, -3, -5)$. Find a vector function whose graph is the tangent line to this curve at the point $(12, -3, 19)$.

9. Suppose $\vec{F}(x, y, z) = (2x + z)\vec{i} + (3z)\vec{j} + (y^2)\vec{k}$ and

$$\vec{r}(t) = (t^2 + 5)\vec{i} + (4t - 3)\vec{j} + (t^2 + t)\vec{k} \text{ for } 0 \le t \le 3,$$

then find the line integral

$$\int_c \vec{F}(x, y, z) \cdot \vec{r}\,'(t)dt.$$

10. Suppose a particle is moving along the curve $\vec{r}(t)$ for $0 \le t \le 3$ such that the force on the particle at any time t is given by

$$\vec{F}(t) = (2t^2 + 3t)\vec{i} + (5t + 8)\vec{j} + (t^3)\vec{k}.$$

Also

$$\vec{r}(t) = (5t + 8)\vec{i} + (t^2 + 4)\vec{j} + (5t^2)\vec{k}.$$

Find the work done by the force $\vec{F}$ in moving the particle along the curve.

6790 Velocity in Three Dimensions

Suppose that the position of an object moving in space is given by $\vec{r}(t)$, where $\vec{r}(t)$ is a three dimensional vector function and t is time. For example, suppose $\vec{r}(t) = (t^3 + 2t)\vec{i} + (t^3 + 4)\vec{j} + (2t^2 + 6t)\vec{k}$. When $t = 1$, we have $\vec{r}(1) = 3\vec{i} + 5\vec{j} + 8\vec{k}$. The vector $3\vec{i} + 5\vec{j} + 8\vec{k}$ locates the point $(3, 5, 8)$ in space. This says that when $t = 1$ the object is at the point in space with coordinates $(3, 5, 8)$. When $t = 3$, we have $\vec{r}(3) = 33\vec{i} + 31\vec{j} + 36\vec{k}$. The vector $33\vec{i} + 31\vec{j} + 36\vec{k}$ locates the point $(33, 31, 36)$. This means that when $t = 3$ the object is at the point in space with coordinates $(33, 31, 36)$. When using this method to locate an object in space at time t, we say that the position of an object at time t is given by $\vec{r}(t)$. The units of length along the axis can be feet, meters, or any length unit. We can measure time in seconds, minutes, or any other time unit.

If $h > 0$, then $\vec{r}(b+h) - \vec{r}(b)$ is the vector from the position of the object at time $t = b$ to the position of the object at the later time $t = b + h$. The vector $\vec{r}(b + h) - \vec{r}(b)$ gives the change in position. It is the displacement vector for the object during the time interval $b \le t \le b + h$. Change in position divided by charge in time is average velocity.

$$\text{Average velocity} = \frac{\vec{r}(b + h) - \vec{r}(b)}{h}.$$

The instantaneous velocity is the limit of the average velocity as the length of the change in time interval goes to zero. Recall the definition of derivative

$$\vec{r}\,'(b) = \lim_{h \to 0} \frac{\vec{r}(b + h) - \vec{r}(b)}{h}.$$

Theorem 1. If the position of an object at time t is given by $\vec{r}(t)$, then the instantaneous velocity of the object at time t is given by the derivative $\vec{r}\,'(t)$.

In a similar way, we can conclude that the following theorem is true.

Theorem 2. If the position of an object at time t is given by $\vec{r}(t)$, then the instantaneous acceleration of the object at time t is given by $\vec{r}\,''(t)$.

Example 1. Suppose that an object is moving in space such that at time t its position is given by

$$\vec{r}(t) = (t^3 - 4t)\vec{i} + (3t^2 + 8t + 2)\vec{j} + (2t^3 + t^2)\vec{k}.$$

Find its velocity and acceleration when $t = 1$ and when $t = 2$.

Solution. We have

$$\vec{r}\,'(t) = (3t^2 - 4)\vec{i} + (6t + 8)\vec{j} + (6t^2 + 2t)\vec{k}$$
$$\vec{r}\,''(t) = (6t)\vec{i} + (6)\vec{j} + (12t + 2)\vec{k}$$
$$\vec{r}(1) = -3\vec{i} + 13\vec{j} + 3\vec{k}$$
$$\vec{r}\,'(1) = -\vec{i} + 14\vec{j} + 8\vec{k}$$
$$\vec{r}\,''(1) = 6\vec{i} + 6\vec{j} + 14\vec{k}$$

At time $t = 1$ the particle is at the point $(-3, 13, 3)$ and its velocity is given by the vector $-\vec{i} + 14\vec{j} + 8\vec{k}$ and its acceleration is given by the vector $6\vec{i} + 6\vec{j} + 14\vec{k}$.

When $t = 2$, we have

$$\vec{r}(2) = 0\vec{i} + 30\vec{j} + 20\vec{k}$$
$$\vec{r}\,'(2) = 8\vec{i} + 20\vec{j} + 28\vec{k}$$
$$\vec{r}\,''(2) = 12\vec{i} + 6\vec{j} + 26\vec{k}$$

At time $t = 2$ the particle is at the point $(0, 30, 20)$. The velocity of the particle is given by the vector $8\vec{i} + 20\vec{j} + 28\vec{k}$ and its acceleration is given by the vector $12\vec{i} + 6\vec{j} + 26\vec{k}$.

Example 2. Suppose that the acceleration of a certain particle moving in space is given at any time t by

$$\vec{a}(t) = (12t^2)\vec{i} + (6t)\vec{j} + 8\vec{k}.$$

Also suppose that the velocity at time $t = 0$ given by the vector $5\vec{i} - 7\vec{j} + 10\vec{k}$, and the position at time $t = 0$ is given by the vector $8\vec{i} + 11\vec{j} - 9\vec{k}$. This says $\vec{v}(0) = 5\vec{i} - 7\vec{j} + 10\vec{k}$ and $\vec{r}(0) = 8\vec{i} + 11\vec{j} - 9\vec{k}$. Find a vector function $\vec{r}(t)$ which gives the position of the particle in space at any time t.

Solution. If we suppose we have a particle of mass m, then the force required to produce this acceleration is $m\vec{a}(t) = m[(12t^2)\vec{i} + (6t)\vec{j} + 8\vec{k}]$. Let $\vec{r}(t)$

denote the vector function such that the position of the particle at any time t is located by the vector $\vec{r}(t)$. Theorems 1 and 2 tell us that, the velocity of this particle is $\vec{r}\,'(t)$ and the acceleration is $\vec{r}\,''(t)$. It follows that

$$\vec{r}\,''(t) = 12t^2\vec{i} + 6t\vec{j} + 8\vec{k}.$$

Taking the indefinite integral, we get

$$\vec{r}\,'(t) = 4t^3\vec{i} + 3t^2\vec{j} + 8t\vec{k} + a\vec{i} + b\vec{j} + c\vec{k}.$$

We are given that $\vec{r}\,'(0) = 5\vec{i} - 7\vec{j} + 10\vec{k}$. It follows that the constant vector is

$$a\vec{i} + b\vec{j} + c\vec{k} = 5\vec{i} - 7\vec{j} + 10\vec{k}.$$

$$\vec{r}\,'(t) = 4t^3\vec{i} + 3t^2\vec{j} + 8t\vec{k} + 5\vec{i} - 7\vec{j} + 10\vec{k}$$
$$= (4t^3 + 5)\vec{i} + (3t^2 - 7)\vec{j} + (8t + 10)\vec{k}.$$

Finding the indefinite integral of $\vec{r}\,'(t)$, we get

$$\vec{r}(t) = (t^4 + 5t)\vec{i} + (t^3 - 7t)\vec{j} + (4t^2 + 10t)\vec{k} + a\vec{i} + b\vec{j} + c\vec{k}.$$

We are given that $\vec{r}(0) = 8\vec{i} + 11\vec{j} - 9\vec{k}$. It follows that $a\vec{i} + b\vec{j} + c\vec{k} = 8\vec{i} + 11\vec{j} - 9\vec{k}$. Therefore,

$$\vec{r}(t) = (t^4 + 5t)\vec{i} + (t^3 - 7t)\vec{j} + (4t^2 + 10t)\vec{k} + 8\vec{i} + 11\vec{j} - 9\vec{k}.$$
$$= (t^4 + 5t + 8)\vec{i} + (t^3 - 7t + 11)\vec{j} + (4t^2 + 10t - 9)\vec{k}.$$

Example 3. A magical force is acting on an object such that its position at time t with t measured in seconds is given by $\vec{r}(t) = (t^2 - 5t)\vec{i} + (4t + 3)\vec{j} + t^3\vec{k}$. At time $t = 2\,\mathrm{sec}$ the force is turned off so that the object then moves in space free of all forces. Find a function $\vec{Q}(t)$ which gives the position of the particle for $t \geq 2$ and t is the same t as the t in $\vec{r}(t)$.

Solution. Since $\vec{r}(2) = -6\vec{i} + 11\vec{j} + 8\vec{k}$, the position of the object at time $t = 2$ is $(-6, 11, 8)$. The velocity of the object when $0 \leq t \leq 2$ is given by $\vec{r}\,'(t) = (2t - 5)\vec{i} + 4\vec{j} + 3t^2\vec{k}$. When $t > 2$ the position of the object is no longer given by $\vec{r}(t)$. For $t > 2$, there is no force on the object and so it will move in a straight line. Since force is turned off at $t = 2$ the object will move along the tangent line to the curve $\vec{r}(t)$. This would be the tangent line to the curve at the point located by $\vec{r}(2)$. As the object moves along the tangent line its velocity would not change. The velocity would always be given by the constant vector $\vec{r}\,'(2)$. The vector $\vec{r}\,'(2)$ not only gives the velocity of the object, but it is also a tangent vector to the curve at $\vec{r}(2)$. We have

$$\vec{r}\,'(2) = -\vec{i} + 4\vec{j} + 12\vec{k}.$$

Usually we use the vector function

$$\vec{R}(s) = (-6\vec{i} + 11\vec{j} + 8\vec{k}) + s(-\vec{i} + 4\vec{j} + 12\vec{k})$$

and say that the graph of this vector function is the tangent line to the curve $\vec{r}(t)$ at the point $\vec{r}(2)$. As usual we use s as the parameter in the tangent function instead of t. The s in

$$(-6\vec{i} + 11\vec{j} + 8\vec{k}) + s(-\vec{i} + 4\vec{j} + 12\vec{k})$$

is not time as given in the problem. For this reason we do not wish to use the function $\vec{R}(s)$ to represent the position of the object at time $= s$.

We do not wish to use this function to represent the position of the object because when we substitute $s = 2$ in this formula we get $-8\vec{i} + 19\vec{j} + 32\vec{k}$ and do not get $-6\vec{i} + 11\vec{j} + 8\vec{k}$. The object is at the point $(-6, 11, 8)$ at time$= 2$. We need a vector function $\vec{Q}(t)$ whose graph is the tangent line but for which $\vec{Q}(2) = -6\vec{i} + 11\vec{j} + 8\vec{k}$. Let us recall that there are other vector functions whose graph is the tangent line. We can get the desired function by making a shift in the s variable. The s variable is not time as we want to measure it. When $s = 0$ we want time $= 2$. This means $s = t - 2$ where t is time. Let $\vec{Q}(t)$ denote the above vector function for the tangent line but with s replaced with $t - 2$, that is, $\vec{Q}(t) = \vec{R}(t - 2)$. This makes $\vec{Q}(2) = \vec{R}(0)$.

$$\vec{Q}(t) = (-6\vec{i} + 11\vec{j} + 8\vec{k}) + (t - 2)(-\vec{i} + 4\vec{j} + 12\vec{k})$$
$$= (-4i + 3j - 16k) + t(-\vec{i} + 4\vec{j} + 12\vec{k}).$$

The graph of $\vec{Q}(t)$ is a tangent line to the curve which is the graph of $\vec{r}(t)$ at the point located by $\vec{r}(2)$. The velocity of a particle whose position is given by $\vec{Q}(t)$ is $-\vec{i} + 4\vec{j} + 12\vec{k}$ which is the desired value of velocity. Also $\vec{Q}(2) = -6\vec{i} + 11\vec{j} + 8\vec{k}$ which is the desired position vector when $t = 2$. Thus, the position of the object for $t \geq 2$ is given by $\vec{Q}(t)$. When $t = 4$, we have

$$\vec{Q}(4) = -8\vec{i} + 19\vec{j} + 32\vec{k}.$$

At time $t = 4\,\mathrm{sec}$ the object is at the point $(-8, 19, 32)$.

Consider the curve which is the graph of the vector function $\vec{r}(t)$, $-\infty < t < \infty$. Suppose we make a shift in the independent variable t. Let $s = t - b$. The graph of $\vec{R}(s) = \vec{r}(s+b)$, $-\infty < s < \infty$ is the same as the graph of $\vec{r}(t)$, $-\infty < t < +\infty$. A shift in the independent variable in a vector function does not change the curve which is the graph of the vector function.

Example 4. Consider the curve which is the graph of the vector function

$$\vec{r}(t) = (t^3 - 9t)\vec{i} + (t^2 - 4)\vec{j}, \quad -\infty < t < +\infty,$$

in the xy plane. The graph of

$$\begin{aligned}\vec{R}(s) = \vec{r}(s + 2) &= [(s + 2)^3 - 9(s + 2)]\vec{i} + [(s + 2)^2 - 4]\vec{j} \\ &= [s^3 + 6s^2 + 3s - 10]\vec{i} + [s^2 + 4s]\vec{j} \quad -\infty < s < +\infty\end{aligned}$$

is the same curve. Graphing $\vec{r}(t)$ for $-4 \leq t \leq 4$ shows that this curve contains a loop. The graph of $\vec{r}(t)$ for $-3 \leq t \leq 3$ is this loop in the xy plane. Note that $-3 \leq s + 2 \leq 3$ is the same as $-5 \leq s \leq 1$. The graph of $\vec{R}(s) = (s^3 + 6s^2 + 3s - 10)\vec{i} + (s^2 + 4)\vec{j}$ for $-5 \leq s \leq 1$ is exactly this same loop. Note that $\vec{r}(-3) = 5\vec{j}$ and $\vec{r}(3) = 5\vec{j}$ while $\vec{R}(-5) = 5\vec{j}$ and $\vec{R}(1) = 5\vec{j}$.

Exercises

1. Suppose that an object is moving in space such that at time t the position of the object is given by

$$\vec{r}(t) = (t^4 - 3t^2)\vec{i} + (2t^3 + 7t)\vec{j} + (t^2 + 8t + 6)\vec{k}.$$

Find the position, the velocity, and the acceleration of this object at time $t = 1$ and at time $t = 3$.

2. Suppose that an object is moving in space such that at time t the acceleration at time t of the object is given by

$$\vec{a}(t) = (4t + 5)\vec{i} + (12t + 3)\vec{j} + 6t\vec{k}.$$

Also suppose that the velocity at time $t = 0$ is $7\vec{i} - 5\vec{j} + 12\vec{k}$ and that the position of the object at time $t = 0$ is the point located by $9\vec{i} + 11\vec{j} + 8\vec{k}$. This says $\vec{v}(0) = 7\vec{i} - 5\vec{j} + 12\vec{k}$ and $\vec{r}(0) = 9\vec{i} + 11\vec{j} + 8\vec{k}$. Find the vector function $\vec{r}(t)$ which gives the position of the particle in space at any time t.

3. A magical force is acting on an object such that its position at time t, with t measured in seconds, is given by

$$\vec{r}(t) = (2t^2 - 3t)\vec{i} + (t^3 - 1)\vec{j} + (5t + 3)\vec{k}$$

At time $t = 2\,\text{sec}$ the force is turned off so that the object then moves in space free of all forces. Find a vector function $\vec{Q}(t)$ such that the position of the particle on the tangent line is given by $\vec{Q}(t)$ when $t \geq 2$. The t in $\vec{r}(t)$ and $\vec{Q}(t)$ should both represent the same time.

4. A magical force is acting on an object such that its position at time t is given by

$$\vec{r}(t) = (t^3 + 2t + 1)\vec{i} + (4t - 3)\vec{j} + (3t - t^2)\vec{k}.$$

At time $t = 3\,\text{sec}$ the force is turned off so that the object then moves in space free of all forces. Find a vector function $\vec{Q}(t)$ which will give the position of the object in space for all t such that $t \geq 3$. An expression for the magical force is $\vec{F}(t) = m[(6t)\vec{i} - 2\vec{k}]$.

5. An object is moving in space such that its velocity at any time t is given by

$$\vec{v}(t) = (t^2 + 3t)\vec{i} + (4t - 5)\vec{j} + (t^3 + 2)\vec{k}.$$

Find the length of the line segment connecting the position of the object at time $t = 0$ and the position of the object at time $t = 2$.

6. An object is moving in space such that its position at any time t is given by

$$\vec{r}(t) = (4t + 3)\vec{i} + (8/3)t^{3/2}\vec{j} + (t^2 + 2)\vec{k}.$$

Find the distance along the curved path that the object travels during the time interval $0 \leq t \leq 2$.

CONTENTS

CHAPTER 1

AN OVERVIEW OF TELEVISION PRODUCTION

"I believe that good television can make our world a better place."

—Christiane Amanpour, CNN Reporter

"Any program, at any time, on any device, at any location."

—Frank Beacham, Director and Producer

TERMS

DVE: Digital video effect equipment, working with the switcher, is used to create special effects between video images. A DVE could also refer to the actual effect.

DSLR: A still camera (digital single lens reflex camera) that shoots video, allowing the photographer to see the image through the lens that will capture the image.

Linear editing: The copying, or dubbing, of segments from the master tape to another tape in sequential order.

New media: Any media that is made available for on-demand delivery. For instance, programming available online would be considered "new media."

Nonlinear editing: The process in which the recorded video is transferred onto a computer. Then the footage can be arranged and rearranged, special effects and graphics can be added, and the audio can be adjusted using editing software.

Prosumer equipment: Prosumer equipment, sometimes known as *industrial equipment*, is a little heavier-duty and sometimes employs a few professional features (such as interchangeable lenses on a camera), but may still have many of the automatic features that are included on the consumer equipment.

Second screen: Today's television audience is not just watching television; they are also on a computer of some type, often called the "second screen."

Smartphones: Smartphones have become the video camera of choice for amateurs. As the image quality has grown, smartphones have been adopted by news stations, documentary producers, and other professionals as a backup camera.

Switcher (vision mixer): Used to switch between video inputs (cameras, graphics, video players, etc.).

"The definition of television is changing. Let's take an iPod, which was an MP3 device a short time ago. When you put video on that device, it becomes a television. So I make the case that television is actually growing to other devices. It's because of the programming, it's the quality, it's the story line. It's all those things we associate with television programming. 'Television' was never the box—it was the programming that was on the box."

—Chris Pizzuro, Vice President of Digital News Media, Turner Entertainment

WHAT IS TELEVISION?

"In the next decade, 75 percent of all channels will be born on the Internet. The Web is poised to become the premium channel for entertainment distribution within the next decade."

—Robert Kyncl, Head of Global Partnerships, YouTube

Defining television can be quite difficult. It used to be easy; television directors and producers knew that their final program would be generally viewed on a 19-inch television set located in a home. As you know, that is no longer the case. Television's definition now embraces new technologies such as large-screen televisions, computers, the Internet, tablets, and smartphones. These changes have brought many new players into the television industry. Online networks now create programming that is as high quality as what we have seen in the past on networks. With their lower cost overhead, they increasingly have the money and audience to pull in some of the best program creators.

Today's viewing audience lives in a hyperconnected world. They do not distinguish between programming as being on television or online; they are looking for quality content that is accessible wherever they want it. Bottom line, television has to create the best possible experience for the greatest number of consumers in the widest viewing environment—in the kitchen, bedroom, living room, on laptops, and mobile devices, as well as in the home theater. What a challenge!

The Second Screen

Today's television audience is not just watching television; they are also on a computer or smartphone of some type, often called the "second screen." One of Nielsen's studies shows that more than 85 percent of mobile and computer users access the Web while watching TV. However, only 24 percent were actually looking at content related to the TV program, while others used it to text family and friends (56 percent), visit social networks (40 percent), and browse unrelated content (37 percent).

Those statistics are of great interest to television executives who are looking for ways to integrate social media with television programming. For example, in New Zealand, TVNZ has launched a youth channel that has created an interactive entertainment show that features chat and commentary driven by Facebook. Viewers can give their opinions using polling, write comments, and even include their profile photos, which are shown on the program.

Robin Sloan, from Twitter's media partnership team, thinks there is definitely an appetite for more integration. "People like to talk about the programming as it is happening," he said. "At this stage, they [TV executives] are primarily using Twitter to engage their existing audience and give them something to talk about. Our goal is to get Twitter integrated into TV shows. It means that people think about Twitter as a source of really great content, and frankly it means that Twitter gets in front of a really big audience."

"Social TV is a modern version of the old days of gathering round the TV to watch a variety show on a Saturday night," said Reggie James, managing director of marketing agency Digital Clarity.

TELEVISION PRODUCTION

Although the television medium has experienced transforming technical changes in the past decade, it is important to keep in mind that the key to great television is still storytelling. As equipment has evolved and become increasingly affordable and adaptable, production techniques have also evolved in order to take advantage of these new opportunities.

Equipment Has Become Simpler to Use

You've probably already discovered how even inexpensive consumer camcorders can produce extremely detailed images under a wide range of conditions (Figure 1.1). Camera circuitry automatically adjusts and compensates to give you a good picture. A photographer needs to do little more than point the camera, follow the subject, and zoom in and out. To pick up audio, we can simply clip a small lavaliere microphone onto a person's jacket, give him or her a handheld microphone, or just use the microphone attached to the camera. As for lighting, today's cameras are so sensitive that

FIGURE 1.1 *Consumer camcorders provide an incredible amount of quality, which is now available to the amateur.*
Source: Photo courtesy of JVC.

they work in daylight or whatever artificial light happens to be around. So where's the mystery? Why do we need to study television/video techniques? Today, anyone can get results.

The Illusion of Reality

"You must use the camera and microphone to produce what the brain perceives, not merely what the eye sees. Only then can you create the illusion of reality."
—Roone Arledge, Former Producer, ABC Television

One of the basic truths about photography, television, and film is that the camera always lies. On the face of it, it's reasonable to assume that if you simply point your camera and microphone at the scene, you will convey an accurate record of the action to your audience. But, if we are honest, the camera and microphone inherently transform "reality."

There can be considerable differences between what is actually happening, what your viewers are seeing, and what they think they are seeing. How the audience interprets space, dimension, atmosphere, time, and so on will depend on a number of factors, such as the camera's position, the lens angle, lighting, editing, the accompanying sound, and, of course, their own personal experience.

We can use this gap between the actual and the apparent to our advantage. It allows us to deliberately select and arrange each shot to affect an audience in a specific way. It gives us the opportunity to devise different types of persuasive and economical production techniques.

If a scene looks "real," the audience will invariably accept it as such. When watching a film, the audience will still respond by sitting on the edge of their seats to dramatic situations. Even though they know that the character hanging from the cliff is really safe and is accompanied by a nearby production crew, it does not override their suspended disbelief.

Even if you put together a disjointed series of totally unrelated shots, your audience will still attempt to rationalize and interpret what they are seeing (music videos and experimental films rely heavily on this fact to sustain interest). If you use a camera casually, the images will still unpredictably influence your audience. Generally speaking, careless or inappropriate production techniques will usually confuse, puzzle, and bore your audience. The show will lack a logical and consistent form. Systematic techniques are a must if you want to catch and hold audience attention and interest.

It's All About How You Do It

At first thought, learning about television production would seem to be just a matter of mastering the equipment. But let's think for a moment. How often have you heard two people play the same piece of music yet achieve entirely different results? The first instrumentalist may hit all the right notes but the performance may sound dull and uninteresting. The second musician's more sensitive approach stirs our emotions with memorable sound.

Of course, we could simply assume that the second musician had greater talent. But this "talent" generally comes from painstaking study and effective techniques. Experience alone is not enough—especially if it perpetuates incorrect methods. Even quite subtle differences can influence the quality and impact of a performance. You'll find parallel situations in television production practices.

Techniques Will Tell

It's common for three directors to shoot the same action, and yet produce quite diverse results:

* In a "shooting by numbers" approach, the first director may show us everything that's going on, but follow a dull routine: the same old wide shot to begin with, followed by close-up shots of whoever is speaking, with intercut

"reaction" shots of the listeners.

- The second director may worry so much about getting "unusual" shots that he or she actually ends up distracting us from the subject itself.
- The third director's smooth sequence of shots somehow manages to create an interesting, attention-grabbing story. The audience feels involved in what is going on.

Clearly, it's not simply a matter of pointing the camera and staying in focus.

Similarly, two different people can light the same setting. The first person illuminates the scene clearly enough, but the second somehow manages to build a persuasive atmospheric effect that enhances the show's appeal. These are the kinds of subtleties you will learn about as we explore techniques.

Having the Edge

Working conditions have changed considerably over the years. Earlier equipment often required the user to have deep technical understanding to operate it effectively and keep it working. Some of the jobs on the production crew, such as camera, audio, lighting, videotape operation, and editing, were all handled by engineers who specialized in that specific area.

In today's highly competitive industry, in which equipment is increasingly reliable and operation is simplified, there is a growing use of multitasking. Individuals need to acquire a variety of skills, rather than specialize in one specific skill or craft. Also, instead of permanent in-house production crews, the trend is to use freelance personnel on short-term contracts for maximum economy and flexibility. Today,

FIGURE 1.2 *Laptop computers, or even mobile phones, allow an editor to create programming anywhere.*

companies often send a single person out on location with a lightweight camera to record the images and sound, use a laptop computer system to edit the results, and return with a complete program ready to put on the air (Figure 1.2).

The person with greater know-how and adaptability has an edge. Job opportunities vary considerably. The person who specializes in a single craft can develop specific aptitudes in that field. However, the person who can operate a camera today, light a set tomorrow, and subsequently handle the sound has more opportunities in today's market.

Although a single person can accomplish many roles, television still relies on teamwork. Results depend not only on each person knowing his or her own job, but also on his or her understanding of what others are trying to accomplish.

Studying this book will give you a number of major advantages:

- By taking the trouble to understand the fundamentals of the equipment that you are using, you'll be able to rapidly assimilate and adapt when new gear comes along. After that, it's just a matter of discovering any operational differences and different features.
- It will help you to anticipate problems and avoid problems before they happen.
- When unexpected difficulties arise, as they inevitably will at some point, you will recognize them and quickly compensate. For example, when the talent has a weak voice, you may be able to tighten the shot a little to allow the sound boom to come a little closer without getting into the shot.

OVERVIEW

Before we begin our journey, let's take an overview of the terrain we will be covering. This will help to familiarize you with the areas that you are going to have to deal with and give a general idea of how they interrelate.

Organization

Although organizational basics follow a recognizable pattern for all types of television production, the actual format the director uses will always be influenced by such factors as the following:

- Whether the production is taking place in a studio or on location.
- Whether it is to be transmitted live or recorded for transmission later.

- Whether the action can be repeated (to correct errors, adjust shots) or is a one-time opportunity that has to be captured the first time around.
- Any restrictions due to limitations in time, equipment, and space.
- Whether there is an audience.
  In some situations, a multi-camera setup is the best

solution for shooting an event (this is when the cameras are controlled by a production team in a control room). At other times, the director may choose to stand beside a single camera, guiding each shot from a nearby video monitor (Figure 1.3).

HISTORY OF TELEVISION

1923: Russian immigrant Vladimir Kosma Zworykin patents the iconoscope, the first television transmission tube. He patents the first color tube in 1925.

1926: John Logie Baird, credited with inventing mechanical television, is the first to transmit a television image using a mechanical television.

1927: Philo T. Farnsworth transmits the first all-electronic television image.

1928: The first television is sold—a Daven for US$75.

June 2, 1931: The earliest true broadcast (available to the public) of an outdoor sporting event is the BBC's coverage of The Derby (horse race) at Epsom in Great Britain. The production's mechanical television equipment utilizes one camera.

1936: The British Broadcasting Corporation (BBC) debuts the world's first television service with three hours of programming a day.

1939: First major league baseball and football (American style college and NFL) games are telecast.

1952: First videotape used.

1955: The Helivision anti-vibration helicopter camera mount is invented by French director/cinematographer Albert Lamorisse.

April 1960: Ampex introduces the Intersync accessory, which makes it possible to cut to or from videotape without rolls or discontinuity and to do dissolves and some special effects.

1962: The first transatlantic television transmission occurs via the Telestar satellite, making worldwide television and cable networks a reality.

December 7, 1963: CBS airs the first instant replay during a football game in Philadelphia.

April 1967: Ampex introduces the first battery-powered portable high-band color tape recorder.

Weighing 35 pounds, it can record for 20 minutes. The accompanying camera weighs 13 pounds.

1968: NHK in Japan begins work on high-definition television (HDTV).

1969: First handheld video camera is developed.

1969: First exhibition of HDTV (Japan).

June 3, 1989: Japan begins regular high-definition television transmissions by satellite.

November 2003: MobiTV becomes the first streaming television content service that delivers live television programming to mobile phones.

January 1, 2010: Sky launches the world's first all-3D channel, Sky 3D.

2010: NHK shoots the Tokyo Marathon in ultra-high definition (UHD), previously called super hi-vision, which includes 22.2 multichannel sound.

February 2012: First Super Bowl streamed becomes the largest social media event for TV in history. There are over 2 million unique users with 4.5 million live streams online, a milestone for Internet live distribution of video. Over 5.4 million people post 12.2 million comments during the game.

June 2013: World's first 4K OB vehicle, Telegenic's T25 truck, uses seven 4K cameras to produce the Confederations Cup in Brazil.

May 2014: First 4K Ultra HD live streaming event is produced by French broadcaster TDF of the French Open 2014 tennis championship.

(Compiled with input from Iain Baird, earlytelevision.org, Ed Reitan's Color Television History, Philo T. Farnsworth Archives, Lytle Hoover, terramedia.co.uk, Alexander B. Magoun, David Sarnoff Library, sportandtechnology.com, tvhistory.tv, oldradio.com, Tom Genova, and the British Museum.)

FIGURE 1.3 *Directors using a single camera often guide the shot by viewing the camera image on a monitor.*

FIGURE 1.4 *Cameras with large lenses are used as "fixed" or "stationary" cameras.*

Source: Photo courtesy of Thomson/Grass Valley.

Planning and Performance

In order to create a smooth-flowing live television production, the director needs to understand the event; for example, what is going to happen next, where people are going to stand, what they are going to do, their moves, what they are going to say, and so on. Although there will be situations in which the director has no option but to extemporize and select shots spontaneously, quality results are more likely when action and camera treatment are planned in advance.

In more complex productions, it is usually necessary for performers and crews to work following a production schedule, which is based on the script. This serves as a regulatory framework throughout the show. Action and dialogue are rehearsed to allow the production team to check their camera shots, lighting, set sound levels, rehearse cues, and so on. These rehearsals give the crew a chance to see what the director is going to do. They also allow the director to see what does or does not work. In a drama production, actors have usually memorized all their dialogue (learned their lines), and every word and move is rehearsed before the actual shoot begins.

Shooting the Action

You can shoot action in several ways:

- As a continuous process (live-to-recorder or tape), recording everything that happens.
- Dividing the total action into a series of separately recorded sequences (scenes or acts).
- Analyzing each action sequence, putting them into a series of separately recorded shots with variations in viewpoints and/or subject sizes. Action may be repeated to facilitate later editing.

Later, in Chapters 9 and 10, we will look at the advantages and limitations of these various methods.

Cameras

Today's cameras range from large network cameras with huge lenses to lightweight designs that are adaptable to field and studio use (Figures 1.4–1.6). For documentaries and newsgathering, even smaller handheld units can play a valuable role.

Video Recording

For convenience and greater flexibility, most television programs are recorded. Historically, video was recorded on a tape. However, today the picture and sound are usually recorded on a hard drive, disk, or flash memory (Figure 1.7).

FIGURE 1.5 *Some small lightweight cameras are designed to be worn on a head.*

Source: Photo by Andrew Wingert.

FIGURE 1.6 *DSLRs (digital single-lens reflex) shoot still photos and high-quality video, all on the same camera.*

Source: Photo courtesy of Canon.

In some situations, sound may be recorded on a separate audio recorder, too. The video recorder may be:

- Often integrated into the actual camera unit.
- In a separate nearby portable unit, which is connected to the camera by cable.
- Housed in a central video recording area in a remote van or nearby room.

In a multi-camera production, the separate outputs of the cameras are to be switched or blended together. This task is usually carried out with a production switcher (Figure 1.8). The program is generally recorded on a central video recorder. Alternatively, each camera's output may be recorded separately on individual video recording decks (called an isolated camera or ISO camera) and their shots are edited together during an editing session.

Additional Image Sources

Additional image sources such as graphics, animations, still shots, digital video effects (DVEs), and other picture sources may be inserted into the program during production or added to the final project recording during the postproduction editing session.

Program Sound

Typically, a microphone is clipped to the speaker's clothing, handheld, or attached to a sound boom or other fitting. Music, sound effects, commentary, and the like can either be

FIGURE 1.7 *Hard drives are increasingly used to record video images.*

Source: Photo courtesy of Doremi Labs.

FIGURE 1.8 *Production switchers are used to switch between two or more live cameras during a project.*

FIGURE 1.9 *A wireless microphone is placed on the talent and lights are used to boost the natural light.*

FIGURE 1.10 *Set building requires skilled craftspeople.*

played into the program's soundtrack during the main taping session or added later during postproduction (Figure 1.9).

Lighting

Lighting can significantly contribute to the success of a presentation, whether it is augmenting the natural light or providing totally artificial illumination. Lighting techniques involve carefully blending the intensities and texture (hardness or diffusion) of the light, with selectively arranged light direction and coverage, to bring out specific features of the subject and/or scene (Figure 1.9).

Sets and Scenic Design

Scenic design, or providing appropriate surroundings for the action, creates a specific ambience for the program. The setting may include an existing location, sets that are built for the program, or virtual sets that can be used to simulate an environment (Figure 1.10).

Makeup and Costume (Wardrobe)

In larger productions, these areas are overseen by specialists. But in smaller productions, the responsibility for these areas may be given to someone else, such as a production assistant (Figure 1.11).

FIGURE 1.11 *The wardrobe department at a major studio.*

Editing

There are two forms of editing:

- *Live editing* occurs during the actual performance. A technical director, or vision mixer, cuts or dissolves between video sources (multiple cameras, graphics, etc.) using a production switcher (vision mixer) directly to air or to a record medium.
- *Postproduction editing* occurs after all of the program materials (video, audio, and graphics) have been compiled. The chosen shots or segments are then placed together in the appropriate order to create the final program.

Two basic systems are used in postproduction editing:

- In *linear tape editing*, specific segments from the original footage tape are selected and then copied from one tape deck to another tape deck to form a master tape. The content is placed on the master tape in a linear order. Significant changes to the edited master are difficult, as the program was assembled in a linear fashion on a tape. Linear systems usually require separate graphics and audio equipment (Figure 1.12). Linear editing equipment is rarely found today.
- In *nonlinear editing*, portions of the original footage are usually transferred onto a hard drive. Nonlinear editing systems allow random access to the individual video and audio clips and allow an unlimited number of changes to the program, as the clips can be easily reconfigured and manipulated on the hard drive. The final master is then output to a recording medium. Nonlinear systems usually include graphics and audio processing software (Figure 1.13). Nonlinear equipment is by far the most popular type of editing equipment.

Postproduction Audio (Program Sound)

In addition to the natural sound from the action, productions may include music, sound effects, and narration received from a variety of sources.

As with picture editing, the audio may be selected and mixed live during the actual production. Alternatively, the final soundtrack may be built during the postproduction session.

Distribution

The whole concept of television has changed. In the past, viewers were required to watch by appointment, a time that was designated by the networks or other gatekeepers. Today, viewers are able to watch almost any program they want, on any viewing device that may be nearby, at any time. Online, DVRs and mobile phones have provided the technology to allow the audience to view what they want, when they want (Figure 1.14).

Program creators also have the ability to distribute anything they create due to online channels and portals at no cost. A high school student can create a short film and get it seen by hundreds or thousands of people if it has a relevant subject and is created well. Even low-quality programs can go viral, seen by millions of people.

FIGURE 1.12 *Digital laptop linear editing system.*
Source: Photo by Austin Brooks.

FIGURE 1.13 *Nonlinear editor.*
Source: Photo by Austin Brooks.

FIGURE 1.14 *Mobile phones, tablets, and computers are changing the way the audience watches television.*

"Every time a new medium comes along in the television marketplace, the feeling is that it is going to knock out the old and that we should be threatened by this new technology. Instead, as content providers, we see great opportunity from this for ourselves and also for the advertisers that use our medium to communicate to their consumers."

—David Poltrack, President of CBS Vision

Audience Impact

It's pretty obvious that to achieve quality and professional results, you need to be able to competently use the equipment. However, it's all too easy to develop the mechanical know-how and create cool digital effects while overlooking the most important issue: how best to communicate to your audience.

Ideally, the techniques you choose should arise from the nature and purpose of the production. The camera techniques, lighting, audio, and set design should all be selected to communicate your story to a specific audience. For example, some excellent documentary programs have been ruined by obtrusive, repetitive, and irrelevant background music.

We will be exploring various ways in which techniques can influence your audience's response to what they are seeing. There is nothing mysterious about this idea: program making is a persuasive craft, the same as marketing, advertising, and all other presentational fields. Learning how to control and adjust production techniques to achieve the kind of impact you need is one of the director's primary skills.

Audiences don't usually give much thought to whether the program they are watching is actually film, television, or video, or a mixture of all these media. In fact, in today's multimedia world, it is becoming increasingly difficult to discern where one medium ends and another begins. The viewer is concerned only with the effectiveness of the material: whether it holds his or her attention; whether it is interesting, amusing, stimulating, gripping, intriguing, entertaining; and so on.

The convergence of media has blurred the demarcations between the various communication media. The same programming can be found on a television station, television network, in a theater, online, available on DVD, or accessible through a mobile phone.

"If you don't take advantage of interactivity, you're not maximizing the opportunity. We're asking advertisers to create something more interesting than the 30-second spot."

—Rick Mandler, Disney/ABC

TELEVISION, VIDEO, OR FILM?

Whether you are involved with television, video, or film production, you will find that the program-making principles and the equipment used can be virtually identical. The terms have blurred over recent years:

- *Television* usually refers to professional programming that is distributed via land transmitters, cables, satellites, or online to a general audience. This programming may be broadcast or distributed online.
- *Video* usually refers to productions that are created and distributed by individuals or corporations that are not meant for general consumption. Examples of videos could be home videos, employee training or in-store marketing productions. These would be non-broadcast television productions.
- Traditionally, *film* meant that the project was being shot on celluloid. Since most films are shot on video due to the financial savings, *film* is generally the term employed when referring to a narrative or documentary. When needed, video can be transferred to film for theater projection. However, many theaters now use video projectors. It is little wonder that people talk about "filming" or "photographing" with the video camera (Figure 1.15).

FIGURE 1.15 *A low-budget film shot using a video format.*
Source: Photo by Naomi Friedman.

Television Versus New Media

There is always a question about what "new media" is. It is a term we hear frequently. When the European Broadcast Union recognized that its members did not understand the definitions, it worked with those members to define the term. The organization's basic question was: how can we distinguish television from the new media? The organization adopted two basic terms: *linear service* and *nonlinear service*.

Television is considered a linear service—that is, the broadcasting of a program where the network or station decides when the program will be offered, no matter what distribution platform is used. Although there are many new distribution platforms (satellite, broadband Internet, iPod/PDA, and the cell phone), if television uses the platform, it is a linear service.

On the other hand, the nonlinear services equal the new media, which means making programs available for on-demand delivery. For instance, video-on-demand can use any platform. It is the *demand* that makes the difference. Most people's early experience of Web video was characterized by tiny standard-definition windows. However, now that HD can easily be streamed and 4K is appearing on some websites, major content owners such as networks recognize that they will have to step up to this mark to win and retain Web viewers.

The Equipment Is Always Changing

If you are uncomfortable with change, you may not want to work in television on the technical side. The technology keeps changing at an incredible pace. However, the good news is that the equipment also keeps increasing in quality and affordability. The bad news is that the viewers' expectations rise with the improvements in equipment. Mastering the equipment will leave you free to concentrate on the creative aspects of the job.

However, a word of warning. It is easy to want to try out the various "bells and whistles" available in new technology just because they are there. Directors often get tempted to introduce variety: adding a wipe here or some other exciting new effect there. Camera operators may be tempted to use their equipment with a certain "panache": an impressively rapid dolly move here, a fast zoom there. There's the temptation to dramatize the lighting treatment or add a little extra to the sound. However, keep in mind that appropriateness is the watchword. You don't want to create your program in a way that would distract the audience from the meaning and impact of the content.

Today's Equipment

Broadly speaking, the equipment used for television/video program making today tends to fall into the following three categories. However, the lines are rapidly blurring between them.

Consumer Equipment

Consumer equipment is intended for hobbyists and/or family use. This equipment tends to make almost everything automatic (auto focus, exposure, etc.); it usually has a lot of extra special effects (fades, masks); and it is usually very easy to use. The equipment is designed for occasional or light use. Of course, this equipment is the least expensive (Figure 1.16).

FIGURE 1.16 *Consumer high-quality video camera.*
Source: Photo courtesy of Panasonic.

FIGURE 1.17 *Prosumer video camera.*
Source: Photo courtesy of Canon.

Prosumer Equipment

Prosumer equipment, sometimes known as industrial equipment, is a little heavier-duty, and sometimes employs a few professional features (such as an interchangeable lens on a camera), but it may still have some of the automatic features that are included on the consumer equipment, which can also be manually adjusted. Because of its combination of portability and quality, this type of lower-cost equipment is often used by professionals. Prosumer equipment is generally medium-priced (Figure 1.17).

Professional Equipment

Professional equipment is usually designed for heavy-duty everyday use, usually has many more adjustable features than consumer/prosumer equipment, and includes the highest-quality components, such as high-quality lenses, on a camera (Figure 1.18). Professional equipment must:

- Function at the highest standards, providing high definition and color fidelity under a variety of conditions, low picture noise, and no visible defects or distortions (artifacts).
- Be extremely stable and reliable.
- Be adjusted and maintained to stringent standards, and provide consistent results (for instance, an image recorded on one machine must reproduce identically on another machine using the same standards).
- Withstand quite rough handling (such as vibration, bumps and jolts, dust, heat, and rain) when used under very demanding conditions.

FIGURE 1.18 *This professional 4K camera is used in both television and digital cinema productions.*
Source: Photo courtesy of Canon.

"It's funny to me that I'm pretty against auto-settings but I'm going to be working on a piece soon that's going to be 75 percent iPhone video. Videography seems to be going two ways; super-fast and easy (smartphones), and high-quality HD cinematography (4K cameras and DSLRs). The audience seems fine with both depending on the situation. Being able to adapt is key right now."

—Nathan White, News Photographer

THE PRODUCTION TEAM

In a well-coordinated production group, members continually interrelate. A good director will allow for the practical problems that a set designer has to face. The designer arranges a setting to help the lighting director achieve the most effective results. Similarly, the lighting director needs to rationalize treatment with the makeup artist. Of course, there are always individuals who concentrate on their own contribution to the exclusion of others. But when you work in a cooperative team in which each member appreciates the other person's aims and problems, difficulties are somehow minimized.

The Hidden Factors of Production

As you will discover, there are many hidden factors that directly affect how a television director works on a television production, such as:

- The program's budget.
- The amount of time allocated for rehearsal and recording.
- The available studio space.
- The type of equipment obtainable, and its flexibility.
- The size and the experience of the production team.
- Support/backup facilities.

All aspects of the production must be arranged and adjusted to suit these parameters.

The Daily Routine

In a surprisingly short time, many of the procedures and operations that mystified you not so long ago will quickly become second nature. Therein lies a trap for the unwary. It's all too easy to learn techniques by memory and go on to apply them as a comfortable routine. In a busy schedule, it is a temptation, of course, to apply regular solutions that have been successful before, rather than to work out innovative and creative approaches. However, if your goal is programming that will interest and engage your audience, you cannot settle for the routine.

Production Issues

Program making is an absorbing, extremely satisfying process, but there are always limitations. Aspirations are one thing, but the achievable can be quite another. We all live with limitations such as: How much will it cost? Do we have enough time? What happens if we go over the scheduled time? Is the right crew available? Do we have the equipment needed? Are there any obstacles that could get in the way?

Very few productions evolve without a hitch of some kind. There will always be things that just do not work, the last-minute interruption, the prop that breaks, the missed cue, and so on. Particularly when things go wrong, it's very easy to become more preoccupied with the mechanics of the situation; for example, how much recording time remains, camera moves, or microphone shadows, rather than aesthetic issues or potential audience responses. These can get pushed into the background.

It is not surprising that during the adrenaline rush of a live show, program makers can lose sight of the value and purpose of the end product their audience will see. The more fully you understand production principles and problems, the freer you are to think about the significance of what you are doing.

Keep the Audience in Mind

One of the greatest difficulties that everyone working on any production will have is how to assess how the audience will respond. The viewer is seeing it for the first time, and usually the only time. The production team has become overly familiar with all its aspects. Every person in the team is concentrating on his or her own specific contribution. While the director is worried about the talent's performance:

- The set designer has noticed where some scenic flats do not fit properly.
- The lighting director is worried about a boom shadow.
- The audio person finds the air conditioning noise to be obtrusive behind quiet speech.
- The makeup artist is disturbed by a perspiring forehead.
- The costume designer has noticed wrinkles in a shirt.
- The video operator is preoccupied with color correction.
- The producer is concerned with the costs of overrunning the scheduled time.

It is not easy to assess from the audience's viewpoint. And when you have put all that effort into a project, it's hard to accept that your audience may be watching the show while working on their computer and be totally unaware of any of these problems.

INTERVIEW WITH A PROFESSIONAL: VICKY COLLINS

Briefly define your job. I produce news, long-form stories, and nonfiction programming for network news, magazine and documentary shows, nonprofits, Olympics and multi-event sports, and the Web. I integrate social networking, blogging, and photography into my work to promote projects.

What do you like about your job? I love the variety of assignments, the travel, and the opportunity to work on projects that make a difference. I also enjoy the opportunity to put together teams and collaborate on projects.

FIGURE 1.19
Vicky Collins, producer.

What are the types of challenges that you face in your position? Mostly the feast or famine nature of freelancing. When you are deep into production, you don't have time for marketing, then when production ends you are looking for the next project again. Keeping work consistent is challenging. Also, budgets continue to tighten and clients often want more for less. Producers have to be resourceful and flexible to make money in this kind of economy. Skill sets need to be broadened to include shooting, editing, blogging, and social networking to stay relevant.

How do you prepare for a production? I do lots of pitching and research. Some people come to me with their projects (network news) and I go to others with ideas that turn into productions (long-form storytelling).

continued

Interview with a Professional: Vicky Collins—*continued*

What suggestions or advice do you have for someone interested in a position such as yours? Anyone can produce a YouTube video but good storytelling and journalism are still essential. Producers need to be able to shoot and edit these days. Also, they need to be expert social networkers. The route people took when I started (going through smaller markets to the network) is no longer necessary or recommended. If you really want to do this, do this. Produce something important and put it on the Web. Put yourself in the middle of the action and bring home a story that gets noticed.

Vicky Collins is a freelance television producer who has worked for NBC News, created long-form storytelling for HDNet's World Report, and worked on multiple Olympics and sports events.

REVIEW QUESTIONS

1. How do you define television?
2. Why do we need to study television production?
3. What are the advantages and disadvantages to being able to multitask in television production?
4. Compare the two different types of editing (live and postproduction).
5. What are the primary differences among video, television, and film?
6. What are the differences between linear and nonlinear editing?
7. What are some of the "hidden" factors that must be thought through on a production?

CHAPTER 2

THE PEOPLE WHO MAKE
IT HAPPEN

"The producer has the final word on all matters—but don't tell
that to the talent; most think they do."

—Joseph Maar, Director

TERMS

A-1: The senior audio person on a production.

A-2: The audio assistant, who reports to the A-1. See "audio assistant" within this chapter.

Director: Responsible for telling the story.

EIC (engineer in charge): Responsible for maintaining and troubleshooting all equipment on a truck or in a specific studio. See "studio engineer" in this chapter.

Field mixer: A portable audio mixer used by audio personnel to record a quality audio signal. This term could also refer to the person operating the field mixer.

Freelance: Independent contractors who provide services to multiple organizations, hiring out their skills on an as-needed basis.

Per diem: The per diem is a stipend often paid to freelancers that covers any incidental costs such as laundry and meals.

TD (technical director): Responsible for operating the television production switcher.

Visual storytelling: The use of images to convey a compelling story.

One of the reasons that this chapter is near the beginning of the book is to underscore the importance of people in production. A good crew is far more important than the latest equipment. The crew will make or break the production. It really does not matter what equipment you have if you don't have knowledgeable people running it. A great crew can make a boring event exciting. A mediocre crew can make an exciting event boring.

As you would expect, organizations differ greatly in how they describe each job. We will try to highlight the most popular descriptions of each job in this chapter. You will note that there are also overlaps between some of these positions. Each company decides how best to deal with those overlaps.

THE PRODUCTION CREW

Television production crews greatly differ in size. There is not a "perfect" size for all crews. The right size for a project depends on the type of project that you are working on and the preferences of the director.

At one end of the scale, there are directors who initiate the program idea, write the script, and even pre-design the settings. They cast and rehearse the performers, guide the production team, and, having recorded the show, control the postproduction. Smaller crews have become more prevalent with the introduction of more portable and affordable equipment. This type of production crew is very popular with documentaries and low-budget productions (Figure 2.1).

At the other end of the scale, there are big-budget productions with large crews, in which the director relies heavily on the production team to provide him or her with quality sets, lighting, sound, and camerawork. The director can then concentrate on directing talent and shot selection. Dramatic productions have traditionally used a larger crew (Figure 2.2).

Another type of crew includes the director in a "big picture" role. This type of project includes a number of separate stories or segments that are independently prepared by members of the production staff. The director is then responsible for visualizing the production treatment that will coordinate the various components. This type of situation occurs in many magazine programs, newscasts, and current affairs programs. Obviously, in this type of production, members of the crew other than the director have significant impact on the various components of the program.

There is a place for all levels of expertise in the wide spectrum of television production. Like other craftspeople, crew members become skilled in their specific field. Someone

FIGURE 2.1 *Small crews have become more prevalent with the introduction of more portable high-quality cameras.*

FIGURE 2.2 *Large productions may require an army of people.*

FIGURE 2.3 *A dramatic director gives guidance to an actor during a single-camera production.*

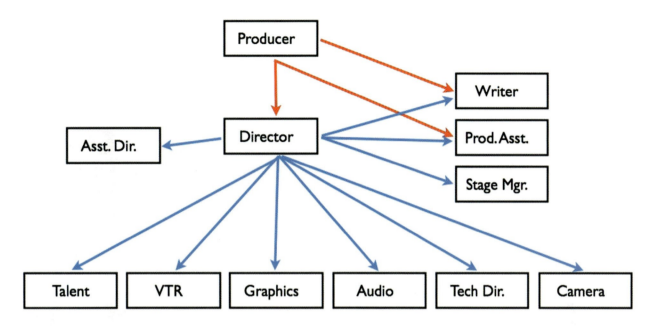

FIGURE 2.4 *A common diagram of how a television production crew works.*

whose talent lies in drama production would probably lack the edge-of-seat intuition of a good sports director: an almost clairvoyant ability to anticipate action and to take spontaneous shot opportunities. On the other hand, someone specializing in sports might be lost in the world of drama, in which the director, with painstaking shot-by-shot planning, guides performance and develops dramatic impact. And when presenting a symphony orchestra, one needs skills that are far removed from those involved in shooting a documentary on location (Figure 2.3).

For smaller productions, the director may combine the functions of both director and producer. Having been allocated a working budget, the person in this dual role is responsible for the entire business and artistic arrangements—origination, interpretation, casting, staging, and treatment—subsequently directing the studio operations and postproduction editing. Generally, the producer serves as the business head of a production and is thus responsible for organization, finance, and policy, serving as the artistic and business coordinator. The director is then free to concentrate on the program's interpretation, as well as the staging and direction of the subject being presented (Figure 2.4).

The common thread between all of these different production types is that the director is responsible for telling the story, whether it is a dramatic show or a sports production. Knowledge of visual storytelling is essential.

Members of the Production Crew

Although (as mentioned before) crew members' job descriptions may vary from company to company, this section provides some of the most common descriptions. Please note that some responsibilities differ between various types of programming, different companies, and different countries. For example, the producer's role in a dramatic production is different from the producer's role in sports.

Executive Producer

The executive producer is responsible for the overall organization and administration of the production group (e.g., a series of programs devoted to a specific field). He or she controls and coordinates the business management, including negotiating rights, program budget, and who gets hired, and sometimes may be involved in major creative decisions and/or concerned with wider issues such as funding, backing, and coproduction arrangements.

Producer

The producer is generally responsible for the management of a specific production. Usually, the producer is concerned with the choice of supervisory staff and crew, interdepartmental coordination, script acceptance, and production scheduling. The producer may select or initiate the program concepts and work with writers. He or she may assign the production's

THE CREW

Each studio facility is designed differently. The control room of the studio or the truck may be separate rooms or have everyone in one big room. Whatever the layout, everyone has his or her own area and everyone needs to be able to communicate clearly to everyone else. Communication may occur through the intercom or crew members may be close enough to each other that they can hear what is being said (Figure 2.5).

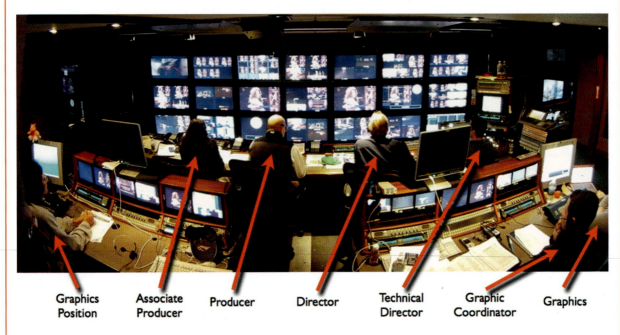

| Graphics Position | Associate Producer | Producer | Director | Technical Director | Graphic Coordinator | Graphics |

FIGURE 2.5 *While control room layouts differ from facility to facility, this is a very common layout for the crew of a large production.*

FIGURE 2.6 *Producers from different news shows confer on the handling of assignments.*

Source: Photo by Jon Greenhoe.

director, and is responsible for meeting deadlines, production planning, location projects, rehearsals, production treatment, and so on. Producers may also become involved in such specifics as craft or union problems, assessing postproduction treatment, and the final program format. The producer reports to the executive producer (Figure 2.6).

Usually the director is responsible for the visualization of the script or event and the producer stays out of the actual hands-on production. However, in sports and news, the producer generally gets more involved with working with the talent, determining replays, inserting pre-produced packages, the timing of the show, and guiding the general direction of the program during a live production.

Assistant Producer/Associate Producer

The assistant producer or associate producer (AP) is responsible for assisting the producer. His or her responsibilities, as assigned by the producer, may include coordinating appointments and production schedules, making sure that contracts are completed, booking guests, creating packages, and supervising postproduction. This person may be assigned some of the same responsibilities of an associate director. The AP reports to the producer.

Director

Ultimately, the director is the individual responsible for creatively visualizing the script or event. Directors must be able to effectively communicate their vision to the crew. They also have to be able to be team builders, moving the crew toward that vision. This role involves advising, guiding, and coordinating the various members on the production team (scenic, lighting, sound, cameras, costume, and so on) and approving their anticipated treatment. The director may choose and hire performers/talent/actors (casting), envision and plan the camera treatment (shots and camera movements) and editing, and direct/rehearse the performers during pre-rehearsals (Figure 2.7).

During studio rehearsals, the director guides and cues performance through the floor manager, and instructs the camera, sound crews, and technical director (vision mixer). He or she also evaluates the crew's contributions (sets, camerawork, lighting, sound, makeup, costume, graphics, and the like).

In some situations, the director may choose to operate the production switcher (e.g., a local news show) and guide and coordinate postproduction editing and audio sweetening.

As mentioned before, the director's job can range in practice from being the sole individual creating and coordinating the production to a person directing a camera and sound crew with material organized by others. The director reports to the producer.

FIGURE 2.7 *The director is the individual responsible for creatively visualizing the script or event.*

Assistant Director/Associate Director

The assistant director or associate director (AD) is responsible for assisting the director. Functions may include supervising pre-rehearsals, location organization, and similar events on the director's behalf. During rehearsal and the actual production, the AD is generally in the production control room and may be responsible for lining up shots, graphics, and tapes, so that they are ready for the director's cue. He or she may also be responsible for checking on special shots (such as chromakey), giving routine cues (tape inserts), and so on, while the director guides the actual performance and cameras. The AD also advises the director of upcoming cues and may assist in offline editing (timings and edit points). The AD may also check program timing and help the director with postproduction. This person may be assigned some of the same responsibilities of an associate producer. The AD reports to the director, or sometimes to the producer (Figure 2.8).

Production Assistant

The production assistant (PA) helps the director and/or producer with production needs. These may include supervising the production office (making copies, making coffee, and running errands), pre-rehearsals, and location organization. Their responsibilities may also include logging tapes, taking notes during production meetings, and similar tasks. During rehearsals and recording, this person may assist the producer/director with graphics or serve as a floor manager.

Producer's Assistant

The producer's assistant usually works with the director. This role may be very close to a PA's responsibilities. He or she may also check performance against the script and continuity. The producer's assistant may be assigned to line up shots,

FIGURE 2.8 *The assistant director, left in red, is helping the director with script cues for a dramatic production.*
Source: Photo by Tyler Hoff

prepare inserts, and so on, while the director guides the performance and camera crew. He or she may also ready the director, crew, and recorder/player regarding upcoming cues. The director may assign the producer's assistant to note the

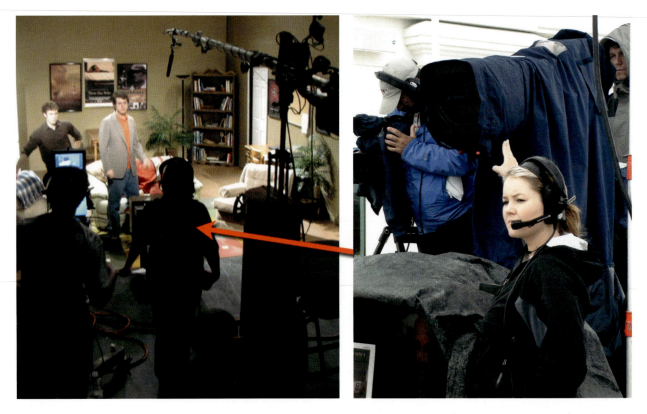

FIGURE 2.9 *The stage manager represents the director on the set or in the studio. She is shown giving the actors instructions from the director on the set of a sitcom. The second photo shows the floor manager letting the talent know which camera is currently on-air during an ESPN production.*
Source: Photos by Josh Taber.

durations/time code of each take and keep the director's notes regarding changes to be made and retakes.

Floor Manager/Stage Manager

The floor manager (FM)/stage manager (SM) is the director's primary representative and contact on the studio floor, in the broadcast booth or on the field of play (Figure 2.9). The FM may be used to cue performers and direct the floor crew. In the studio, the FM is responsible for general studio organization, safety, discipline (e.g., noise control), and security. An assistant floor manager may be used to ensure that the talent is present. The FM reports to the producer or director.

Production Manager/Line Producer

The production manager is responsible to the producer and director for maintaining the production within the allocated budget. He or she also may serve in administrative functions relative to the production.

Technical Director/Vision Mixer

The technical director (TD) generally sits next to the director in the control room and is responsible for operating the television production switcher (and perhaps for electronic effects). The TD may also serve as the crew chief. He or she reports to the director (Figure 2.10).

Makeup Artist/Makeup Supervisor

The makeup artist designs, prepares, and applies makeup to the talent, aided by makeup assistants and hair stylists. The supervisor is generally responsible for multiple makeup artists, to ensure consistency (Figure 2.11).

Costume Designer

The costume designer designs and selects performers' costumes (wardrobe). He or she may be assisted by dressers and wardrobe handlers (Figure 2.12).

Graphics Designer/Graphics Artist

The graphics designer is responsible for the design and preparation of graphics for a series of shows or just one individual show (Figure 2.13).

Graphics Operator

The graphics operator is the person who actually implements the graphics during the production. He or she is responsible for organizing and typing on-screen text and titles for a

FIGURE 2.10 *Technical directors "edit" a live program by utilizing a production switcher and multiple cameras.*

FIGURE 2.11 *A makeup artist prepares talent for a production.*
Source: Photo by Sarah Seaton.

Figure 2.12 *A costume designer designs costumes for a production.*

production, either doing so during the production or storing them for later use. Operators may also serve as designers/artists (Figure 2.13).

Lighting Director

The lighting director is responsible for designing, arranging, and controlling all lighting treatment, both technically and artistically. This responsibility may include indoor and/or outdoor lighting situations (Figure 2.14).

Video Operator/Shader/Vision Control

The video operator is responsible for controlling the picture quality by utilizing test equipment to adjust the video equipment. There are a variety of adjustments that can be made, including exposure, black level, and color balance. Operations are closely coordinated with the lighting director (Figure 2.15).

Camera Operator/Cameraperson/
Photographer/Cinematographer/Videographer

The camera operators are responsible for setting up the cameras (unless they are already set up in a studio situation) and then operating the cameras to capture the video images as requested by the director. On small productions, the videographer may have a lot of creative control over the image. In a multi-camera production, the director usually makes most of the final creative decisions about the shot. In a dramatic production, the camera videographer may be called a cinematographer (Figure 2.16).

Camera Assistant

The camera assistant is responsible for assisting the camera operator in setting up the camera. He or she is also responsible for making sure that the camera operator is safe (keeping him or her from tripping over something or falling), keeping people from walking in front of the camera when it is on, keeping the camera cable from getting tangled, and guiding the camera operator during moving shots. A camera assistant may also push a camera dolly if needed (Figure 2.17).

Focus Puller

The focus puller is responsible for adjusting the focus so that the camera operator can concentrate on composition. The focus puller uses a marked lens to establish focus. This position is rarely used except in dramatic shooting (Figure 2.18).

FIGURE 2.13 *Graphics operators insert data into predesigned graphics for a production.*

FIGURE 2.14 *Lighting director.*

FIGURE 2.15 *Video operators are responsible for matching the cameras and other input devices.*

FIGURE 2.16 *Camera operators are responsible for setting up the cameras, as well as operating them.*
Source: Photos by Josh Taber.

FIGURE 2.17 *Camera assistants are responsible for keeping the camera operator safe and the cable untangled. They also may be required to push the camera dolly.*

Audio Mixer/Audio Engineer/Sound Supervisor/Senior Audio Technician

The audio mixer (A-1) is responsible for the technical and artistic quality of the program sound. This job includes determining the number and placement of the microphones required for the production. He or she also makes sure that the audio cables are properly plugged into the audio mixer and is responsible for the final mix (audio levels, balance, and tonal quality) of the final production. Audio personnel are also generally responsible for the intercom system used by the crew (Figure 2.19).

FIGURE 2.18 *The person on the left is the focus puller. Having a focus puller allows the camera operator to concentrate on composition.*

FIGURE 2.19 *The field audio mixer is using a small portable mixer to adjust the levels on the talent's mic during the remote shooting of a segment for Access Hollywood. The photo on the right is an audio mixer, or A-1, mixing the television production live.*

Audio Assistant

Supervised by the A-1, the audio assistant (A-2) is responsible for positioning microphones, running audio cables, operating sound booms, troubleshooting audio problems, and operating field audio equipment (Figure 2.20).

FIGURE 2.20 *Audio assistants are often assigned to handhold microphones around the field of play.*

Set Designer/Scenic Designer

The set designer is responsible for conceiving, designing, and organizing the entire scenic treatment. His or her responsibilities may even include designing the on-screen graphics. The set designer is generally responsible for supervising the set crew during setting, dressing, and striking the sets.

Set Crew/Stage Crew/Grips/Floor Crew

The set crew are responsible for the sets or scenery such as erecting, setting, or resetting scenery, props, and action cues (such as rocking a vehicle). In some organizations, they may initially set up and dress settings.

Electricians

Electricians are responsible for rigging and setting lamps and electrical apparatuses, including electrical props.

Technical Manager/Studio Supervisor/Remote Supervisor/Technical Coordinator

The technical manager coordinates and is responsible for all technical operations of the production. He or she may book the facilities, check technical feasibility, make sure that everything is working correctly, and ensure safety.

Studio Engineer/Maintenance Engineer/Engineer-in-Charge

Engineers are responsible for maintaining and troubleshooting all camera and sound equipment in a production. Maintenance engineers usually are assigned to do regular maintenance on the equipment, the studio engineer is responsible for the studio, and an engineer-in-charge (EIC) is responsible for a production truck.

Utility

Utilities assist the engineering staff by helping carry gear and cables, setting up equipment, and laying cables (Figure 2.21).

VTR Operator/Tape Operator

The VTR (videotape recorder) operator is responsible for recording the program and playing pre-produced packages or replays that will be inserted into the program. Although still referred to as the VTR or tape operator, the operator may actually be recording and playing back programming from a memory card, hard disk, or hard drive (Figure 2.22).

Writer

The writer is responsible for writing the script. Occasionally, the producer or director will write material. At times, writers are assisted by a researcher, who obtains data, information, and references for the production writer.

Editor

The editor selects, compiles, and cuts video and audio to produce programs. He or she may assemble clips into segments and those segments into programs, or may just correct mistakes that occurred during the production process (Figure 2.23).

Talent

Talent is generally defined as people who are heard or appear on television. There are a number of different types of talent, including actors, anyone being interviewed, reporters, sportscasters, anchors, commentators, and so on.

In the news field, it is not uncommon for a person to serve as a one-person television production crew: interviewing, shooting, and appearing as talent (Figure 2.24).

Actors

Actors generally work from an established script. By playing a role, they create a character for the audience. Actors usually rehearse before the camera rehearsal (Figure 2.25).

FIGURE 2.21 *Utilities assist the engineering staff by helping carry gear and cables, setting up equipment, and laying cables.*

FIGURE 2.22 *A VTR operator is using a hard drive recorder to record and play back video.*

FIGURE 2.23 *The editor is responsible for assembling the shot footage into a useful finished production.*

FIGURE 2.24 *Many television stations, due to tight budgets, have moved to one-person news crews. In the above photo, the reporter asks the questions and holds the microphone, but is also holding the camera on her shoulder.*

FIGURE 2.25 *Actors play a role, creating a character for the audience.*

FIGURE 2.26 *People who appear on television as themselves are often called talent or performers.*

FIGURE 2.27 *Anchors usually provide a level of consistency on a show by being the person who opens the show, introduces others, and ends the show.*

Source: Photo by Jon Greenhoe.

Performers

Performer is usually a term used to describe people who are appearing in front of the camera as themselves. This group could include interviewers, announcers, newscasters, anchorpersons, hosts, and so on. Performers often address the audience directly through the camera using scripted, semi-scripted, or unscripted material. Sometimes the word "performer" refers to a person providing a specific act (such as a juggler) (Figure 2.26).

Anchor

The anchor generally sits at a desk or table of some type and provides the consistent talent that pulls the show together. Anchors generally open the show, introduce on-air talent such as reporters or other in-studio talent, and then end the show (Figure 2.27).

Guests

Guests are invited "personalities," specialists, or members of the public—usually people without experience in front of the camera. Guests could include interviewees, contestants, audience contributors, and so on.

There are other people who may be involved in a production as well, depending on the size of the staff. These people may include security, prop manager, stunt personnel, catering, and others.

THE FREELANCE CREW

Many companies today have moved from a full-time professional production staff to utilizing a freelance crew. Freelancers are independent contractors who work for multiple organizations, hiring out their production skills on an as-needed basis. There are freelancers available who can fill every one of the production positions. Freelancers are generally paid on an hourly or day-rate basis. If they are traveled far from their home base, freelancers usually get lodging and a per diem added to their contract. The per diem is a stipend that covers any incidental costs such as laundry and meals.

WHAT DO YOU WEAR?

What production crew members wear differs depending on your job, who will see you, if you are indoor or out, and the event.

If you are in the studio, the public generally does not see the crew. The most common clothes are as casual as jeans. However, studios can often be cold, which means it is good to have a couple of layers so that clothing can be removed or added, depending on the temperature. Setup days are always casual. Some inside events, such as major awards ceremonies, may dictate that the crew seen by the public actually dress well in order to fit into the overall environment during the shooting.

Outdoor work is much more unpredictable. In the morning, it may be sunny, and by afternoon it may be pouring rain and cold. That means that multiple layers and types of clothing should be carried. Depending on where you are, that may include a T-shirt and jeans if it is hot, a long-sleeved shirt and hat to project you against the sun, and a warm jacket to project you from the cold.

Whatever the situation, it is always important to look professional. You will get more respect from others, which will help you get your job done.

What Do You Bring With You?

Bringing the right supplies with you can make all the difference in the world on a shoot. Again, it depends where you are. It is always important to have something to write with and paper. Here are some of the other things that should be considered:

- Hat.
- Sunblock.
- Bug spray.
- Sunglasses.
- Snacks.
- Multitool (Figure 2.28).
- Simple first-aid supplies.
- Small flashlight.

FIGURE 2.28 *A variety of multitools are available.*

Source: Photo courtesy of Leatherman.

INTERVIEW WITH A PROFESSIONAL: TAYLOR VINSON

Briefly define your job. When acting as a multimedia journalist, my job is to take a television news story from conception to completion, all on my own. This means I handle making calls, setting up appointments, getting to and from the story's location(s), shooting, conducting interviews, gathering information, logging video, writing, editing, and presenting the story on-air.

What do you like about your job? The luxury of being a "one-man band" is that you do not spend time trying to communicate how you want your story to turn out to the other members of your team, since, in effect, you are the team. There is a great deal of control over how the video you shoot lines up the script that you write. Experienced one-man bands will often "write in their heads" as they are shooting a story out on location. Decision-making in the field is many times easier and faster for one-man bands. You do not have to be concerned about making sure another person on the news crew has his or her needs or wants met. Exceptional stories from one-man bands can sometimes have a kind of cohesiveness that often eludes storytelling done by two-person crews.

FIGURE 2.29
Taylor Vinson, news videographer/multimedia journalist.

What are the types of challenges that you face in your position? Here are a few of the challenges facing a multimedia journalist:

- It can be especially difficult to be a multimedia journalist at a large-scale breaking news scene. You cannot be shooting a scene efficiently *and* running around talking to witnesses and gathering the small tidbits of information that are the lifeblood of a reporter who is thrown into a rapidly developing story. In these situations, all you can do is your best.
- I cannot stress enough how helpful it can be for managers to give their one-man bands resources that allow them to complete and send back their stories, with video, remotely. There have been days when I have spent over five hours of my nine-hour work day just *driving*, because the story's location was far away, and to complete it I had to drive all the way back to the station. Ideally, one-man bands should have a laptop computer equipped with nonlinear editing software and mobile Internet connectivity, allowing the multimedia journalist to send his or her video back to the station from the story's physical location.
- Shooting video of yourself in the field while you are presenting as a reporter can be difficult and time-consuming. Pullout LCD screens on cameras can aid one-man bands in this process.

How do you prepare for a production? Before I head out, I always make sure I have charged batteries and all my equipment, including my camera, microphones, tripod, lighting equipment, and rain protection. If I am driving far away, I make sure I have some food and beverages in the car so I don't have to stop often on the way. If I have time, I make sure I've been briefed on the story by a manager and that I have all the phone numbers I will need to gather necessary information. If I am not currently dressed in a shirt and tie for on-camera presentation, I make sure I have those clothes in the car so I can change into them if I need to.

What suggestions or advice do you have for someone interested in a position such as yours? Make sure that you get to know all aspects of television news and how to execute different functions in a timesaving manner. Learn how to shoot efficiently. Learn how to write conversationally. Learn how to present enthusiastically yet naturally on camera. Learn how to edit quickly. Learn what your producers and managers want and *why* they want it.

Taylor Vinson is a news videographer, one-man band, and/or multimedia journalist for WFTS-TV in Tampa, Florida. His previous work has included working as a videographer/reporter for WLEX in Lexington, Kentucky, and as a reporter for NBC.com.

INTERVIEW WITH A PROFESSIONAL: PHIL BOWDLE

Briefly define your job. I oversee all aspects of our church communications, including video, design, print, Web, marketing, and public relations.

What do you like about your job? I love doing a job where I can see tangible results and I believe in the cause.

What are the types of challenges that you face in your position?

*   The industry changes so quickly that it is difficult to keep up to date.
*   Managing the demands and timelines of a busy organization.
*   Developing systems for teams (staff and volunteer) to accomplish big projects.

FIGURE 2.30
Phil Bowdle, director of communications.

How do you prepare for a production? As communications director, I map out each stage of the project into smaller lists of tasks in a project management system called Basecamp. It is there that I assign tasks and milestones to staff and volunteers to accomplish the project. In my role, I also am involved in revisions and approvals of all design/video elements.

What suggestions or advice do you have for someone interested in a position such as yours? You don't have to be an expert in any one thing. But it helps a ton to have a great working knowledge of how all the different pieces of communications work together to communicate effectively. Skill in design/video/writing/etc. is important, but learning how to effectively work and lead people is equally as important.

Phil Bowdle is the director of communications at a church in Georgia.

REVIEW QUESTIONS

1. What are the main differences between shooting a scripted drama and a sports event?
2. Why do some productions require large crews, while other productions can be shot with just a couple of people?
3. What is the difference between a director and a technical director?
4. What is the difference between an actor and a performer?
5. Ultimately, who is responsible for the final program?

CHAPTER 3

THE TELEVISION
PRODUCTION FACILITY

"Today you don't need to spend millions of dollars on equipment
to produce high-value broadcasts."
—Mark Scarpa, Producer and Director, *The X Factor Digital*

TERMS

Batten: The bar to which studio lights are connected.

Boom pole: A pole that is used to hold a microphone close to a subject.

Camera control unit (CCU): Equipment that controls the camera from a remote position. The CCU includes adjusting the camera, luminance, color correction, aperture, and so on.

Chromakey: Utilizing a production switcher, the director can replace a specific color (usually green or blue) with another image source (still image, live video, prerecorded material, and so on).

Control room: The television studio control room, sometimes known as a gallery, is where the director controls the production. Although the control room equipment may vary, they all include video and audio monitors, intercoms, and a switcher.

DVE: Digital video effects equipment, in combination with the switcher, is used to create special effect transitions between video images. A DVE could also refer to the actual effect instead of the equipment.

Flats: Freestanding background set panels.

Intercom: A wired or wireless communication link between members of the production crew.

Monitors: Monitors were designed to provide accurate, stable video image quality. They do not include tuners and may not include audio speakers.

OB van: See remote truck below.

Program monitor: Also known as the "on-air" monitor, the program monitor shows the actual program that is being broadcast or recorded.

Props: Items used to decorate the set or items used by the actors. Stage props could include items such as chairs, news desk and tables. Hand properties could include any items that are touched by the talent.

Remote truck (also known as an outside broadcasting or OB van): A mobile television control room that is used away from the studio.

Sitcom: A situation comedy television program.

Studio: An area designed to handle a variety of productions; a wide-open space equipped with lights, sound control, and protection from the impact of weather.

Switcher (vision mixer): A device used to transition between video inputs (cameras, graphics, video players, and so on).

Today, someone with a handheld camcorder and a very modest computer can produce results that not so long ago would have required the combined services of a large production team and a great deal of equipment.

Nowadays, the television camera is free to shoot virtually everywhere. The audience has come to accept and expect the camera's flexibility. Whether the pictures that they are watching are from a camera in the studio or from outer space, intense close-up shots from a microscope or from a "critter camera" attached to a swimming seal, or thrilling shots from a skydiver's helmet camera, these diverse channels are all grist for the endless mill of television programming, and they are accepted as "normal."

PRODUCTION METHODS

The way to develop a production depends on a number of factors:

- Whether the show is live (being seen by the viewers as it is happening), or whether you are recording the action for subsequent postproduction.

- Whether you shoot continuously, in the order in which the action will be seen, or selectively in an order arranged to suit the production's efficiency.

- Whether you are able to control and direct the action you are shooting, or whether you are required to grab shots wherever you are able.

As new facilities have been developed, established methods of creating programs have grown. Wherever possible, productions are recorded, and have come to rely more and more on postproduction editing techniques—using all kinds of digital effects to enhance audience appeal.

As you will see, although many situations involve little more than uncomplicated switching between shots, in others the editing process—which determines how shots are selected and arranged—becomes a subtle art.

All television productions have a number of common features—the specific skills required of the director and crew can vary considerably with the type of show that is being produced. The program material itself can determine how you present it. Some types of productions follow a prepared plan, and others have to rely heavily on spontaneous decisions. Let's look at some examples:

Interviews and talk shows: Approaches here are inevitably somewhat standardized, with shots concentrating on what people have to say and how they react (Figure 3.1).

Newscasts: Most news programs follow a similar format. "Live on air" in the studio, newscasters present the news seated behind a central desk, reading from teleprompters and introducing stories from various contributory sources—pre-produced packages, live on-site reporters, archive material, graphics, and so on. There also may be brief interviews, either in the studio or via a display screen. As newscasts are continuously reviewed and revised, there is a behind-the-scenes urgency, particularly for late breaking news stories. The closely coordinated team is continually assessing and editing incoming material, preparing commentary, and assembling illustrations and graphics (Figure 3.2).

FIGURE 3.1 *On-location interviews can be difficult, due to the lack of control of the surroundings.*
Source: Photo courtesy of Sodium.

FIGURE 3.2 *The newscast studio.*
Source: Photo by Jon Greenhoe.

FIGURE 3.3 *Television sports programs require the director to predict accurately where the action is going and they often require specialty camera gear, such as the jib shown.*

FIGURE 3.4 *Sitcoms are generally shot in a studio with a live audience.*
Source: Photo by Tyler Hoff.

FIGURE 3.5 *Concerts can be difficult to shoot if rehearsals are not available.*
Source: Photo by Paul Dupree.

Sports programs: Each type of sport or game poses its own specific problems for the director. Shooting conditions vary considerably. On one hand is the relatively localized action of the boxing ring. On the other is the fast ebb-and-flow action of the football field. When presenting the wide-ranging action of a golf tournament, marathon runners, a horse race, or a bicycle race, will the cameras follow along with the action or shoot from selected vantage points? Sometimes, several different events are taking place simultaneously. Can they all be covered effectively on-air or will selective segments be shown on a delayed basis? In each case, the director's aim is to always be "in the right place at the right time." While conveying a sense of continuity, the camera must not only capture the highlights, but be ready to record the unexpected. Slow-motion replays help the commentators and audience analyze the action (Figure 3.3).

Comedy: Most comedy shows follow the familiar "realistic" sitcom format. The studio production is staged to enable the studio audience (seen or only heard) to see and react to the action. They can watch any recording or pre-produced inserts on hanging video monitors. Additional "appropriate" laughter/applause or even "reaction shots" may be added

during postproduction editing (Figure 3.4). Some sitcoms have been moved out of the studio and are now being shot in the field.

Music and dance: Productions can range from straightforward performance to elaborate visual presentations in which images (and sound) involve considerable postproduction (e.g., creating montages, slow-motion sequences, color changes, or animation effects). Particularly where the sound arrangements are complex, the on-camera performance may be lip-synced to a previously recorded soundtrack (Figure 3.5).

FIGURE 3.6 *The first photo shows a dramatic production being shot with a smartphone. The second photo shows a program being shot with a DSLR.*

Drama: Drama productions usually follow a very carefully planned process in which the dialogue, action, camerawork, and sound and lighting treatments are fully scripted. The show may be recorded continuously or in segments. Most drama today relies heavily on postproduction (e.g., adding sound effects, music, post-synching). Figure 3.6 shows single-camera dramatic productions.

THE VENUE

Today, television productions are shot under a variety of conditions:

- In a fully equipped television/video studio.
- In an extemporized studio, set up just for the occasion.
- On location, in an existing interior (such as a public building).
- On location, in the open air.

Each locale has its specific advantages and limitations. Although a studio has all the facilities that we may need (cameras, lights, audio), we have to face the fact that there is nothing to shoot there (except the walls) until we create a set of some sort, which then needs to be decorated and lit appropriately. The running costs of providing these conditions can be considerable (Figure 3.7).

On location, you may have a ready-made environment in which to shoot, and perhaps daylight to provide the illumination. But there are various new problems, from variable weather conditions and background noises to traffic and

FIGURE 3.7 *Designing and building a set in a studio can be very expensive compared to going to a location where a scenic background already exists.*

bystanders. You are normally away from your base, with its backup services (e.g., spare equipment, maintenance, and so on).

The Television Studio

Studios are designed to handle a variety of productions with their wide-open spaces and their ability to hang and supply power to lights anywhere in the space. Studios are ideal, because they protect productions from the impact of weather such as snow and rain, they are independent from the time of day (productions can be lit as though it is daylight or nighttime), and they allow for sound control. They are used for many dramatic productions, news shows, and talk shows (Figures 3.8 and 3.9). Although in practice television studios vary from the modest to purpose-built giants, all seem somehow to share a certain indefinable atmosphere.

At first glance, a studio may have the feel of a deserted warehouse. It generally is well soundproofed in order to block unwanted sounds from being recorded on the program microphones. The acoustically treated walls are also designed to reduce echo. If you look up, you'll see that the studio has been designed with a framework of bars or battens suspended from the ceiling. These battens are designed to hold all of the required lights, as well as to support scenery. Once the scenery is brought in and the lights are hung, adjusted, and turned on, the atmosphere can become completely transformed.

The set may be permanent in studios, or a temporary structure may be built for a specific show. Furnishings and props are added to the set by the set designer/decorator. Once the basic set is in place, a lighting crew begins to position and adjust each lamp to a meticulously prepared lighting plot designed by the lighting director. The camera and sound crews move their equipment from storage into the opening positions, preparing for the upcoming rehearsals. Everywhere is urgent action: completing tasks, keeping the studio clean, and making last-minute changes.

To the first-time viewer, with the army of production personnel moving around the studio doing their own jobs, it can look like chaos. However, there really is a system in place that works. The various pieces of the production jigsaw combine to provide the show that has been planned and prepared. The cameras set up their shots and are seen on the monitor wall screens in the control room.

Within hours, the production will have completed rehearsals, the crew will have learned and practiced the director's treatment, performances hopefully will have reached

FIGURE 3.8 *The outside of a Hollywood television studio, or soundstage.*

FIGURE 3.9 *This set, inside a studio, allowed the producers to create a program that looked as though it was shot in the evening, even though it was daylight outside.*

their peak, and the program will have been recorded or transmitted. And then, if it is a temporary set, the whole set will be "struck" and tucked away in storage or discarded, equipment will be stored, and floors cleaned. And then it changes back to the original "warehouse."

The Television Studio in Action

During the rehearsal, the show is being monitored only internally. However, if it is a live transmission, at the scheduled time the studio video (pictures) and audio (sound) signals will be fed via the coordinating television control room to the recorders or immediately to the transmitters.

FIGURE 3.10 *Set pieces are put into place and each light is then meticulously placed for a specific purpose.*
Source: Second photo by Tyler Hoff.

Although sets can be electronically inserted, most are freestanding panels referred to as flats. Many studios have "permanently" built-in or arranged scenery ready for their regular newscasts, cooking programs, interviews, and the like. Sets may include three-walled rooms, a section of a street scene, or even a summer garden. Each set is lit with lights that may be suspended from the ceiling, clamped to sets, or even attached to floor stands. It is hard to appreciate how much care has gone into the fact that each light has been

placed and angled with precision, for a specific purpose (Figure 3.10).

Despite the number of people working around a set during the production, it is surprising how quiet the place usually is. Only the dialogue between the actors should be audible. Camera dollies should quietly move over the specially leveled floor, as the slightest bumps can shake the image. People and equipment move around silently, choreographed, systematically, and smoothly, to an unspoken plan.

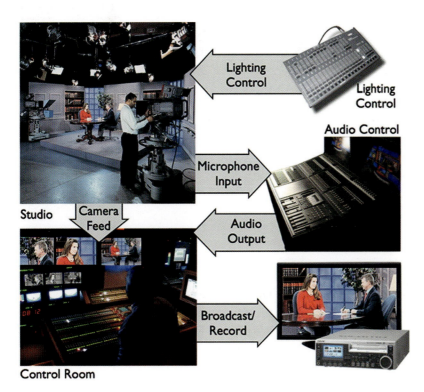

FIGURE 3.11

The studio production center. Lighting, audio, and camera signals are sent to the control room. The mixed signal is then recorded or transmitted live.

Source: Photos courtesy of Fischer, Sony, and Panasonic.

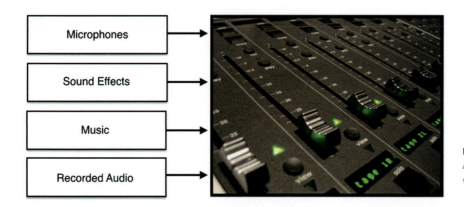

FIGURE 3.12
Audio mixers receive a wide variety of sources.

However, if you were to put on an intercom headset, you would enter into a different world! You would hear the continuous instructions from the unseen director in the production control room: guiding, assessing, querying, explaining, cueing, warning, correcting—coordinating the studio crew through their headsets. The director uses the intercom to guide the production crew. In the studio, the crews operating the cameras, microphones, lighting, set, and so on hear the intercom through their headsets—information that is unheard by the performers/talent or the studio microphones. The floor manager, the director's link with the studio floor, is responsible for diplomatically relaying the director's instructions and observations to the performers with hand signals (Figure 3.11).

Although a garden scene in the studio may not look real in person and up close, on camera the effect can be idyllic. A quiet birdsong can be used from a sound effects CD to complete the illusion.

Microphones, such as an attached mic or a boom mic, are used to capture all of the sounds. In a nearby room, an audio mixer is responsible for mixing and blending the various audio levels, from the actor's performance to the theme music. All of the audio sources flow into the audio mixer. Besides microphones, audio sources can include recorded voices, sound effects, and music from a variety of devices (Figure 3.12).

When shows are broadcast live, they must be shot continuously. However, during a rehearsal or a show that is going to be recorded (it can be completed in postproduction), the director has the ability to stop and start. If the director sees a bad shot, has a technical problem in the control room, notices a problem with the talent, or doesn't like something about the shot, he or she can tell the floor manager to stop the action while the problem is sorted out. Once the situation is remedied, the show can go on.

The Television Studio Control Room

The television studio control room, sometimes known as a gallery in Europe, is the nerve center where the director, accompanied by a support group, controls the production. Control rooms can be segmented into separate rooms or areas. However, there are smaller control rooms, or even one-piece switcher/monitor wall packages, that merge many of these operations into one area. A large control room has more room and flexibility, but requires more people. A one-piece system can be operated by one person but is limited in the number of cameras it can include (Figures 3.13–3.16).

The director can have many people trying to get his or her attention in the control room. Of course, there is another whole group of people in the studio. However, in the control room, the director needs to review graphics, listen to the assistant director, and direct and respond to the audio personnel, video shaders, playback, the technical director, and sometimes the producer (Figure 3.17).

The director usually sits in the television control room—although sometimes sitcom directors prefer to be out in the studio—watching a large group of video monitors called the monitor wall (Figure 3.17). The smaller monitors show the displays from each camera being used, plus a variety of image sources such as graphics, animations, and satellite feeds. There are usually two larger screens. One is generally the preview monitor, which is the director's "quality control" monitor, and which allows him or her to assess upcoming shots, video effects, combined sources, and the like. The second monitor is the "on-air" or "program" monitor, which shows what is actually being broadcast or recorded.

The director's attention is divided between the various input monitors, the selected output on the on-air monitor, and the program audio from a nearby loudspeaker.

As mentioned before, the director instructs the production team and floor crew through an intercom headset (earphone

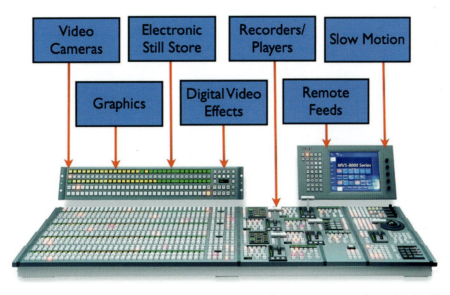

FIGURE 3.13 *Video switchers can have many sources sending signals to them. Keep in mind that each area listed in blue may represent one source or many sources, such as cameras.*

Source: Photo courtesy of Sony.

FIGURE 3.14 *This one-piece system includes audio mixing, graphics, video switching, and the monitor wall. The second photo shows the device in use by CCTV in China.*

Source: First photo courtesy of Sony.

FIGURE 3.15

This computer software "control room" allows a director to direct four cameras and add graphics and limited special effects, as well as monitor audio. This is a very low-cost option for streaming productions on a very limited budget.

Source: Photo courtesy of Volar Video.

FIGURE 3.16 *Control rooms and studios take all different forms. In this situation, Yahoo located their control room area in the same room as their studio.*

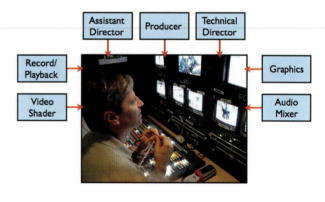

FIGURE 3.17 *The director has a large group of people to deal with during a production.*

and microphone). In smaller productions, the entire crew is on one intercom channel, with everyone hearing everyone else's instructions from the director. Larger productions utilize multiple channels of intercom, allowing fewer voices to be heard, which can reduce confusion.

Although some directors may prefer to switch for themselves, most directors utilize a technical director (TD). TDs are responsible for switching between the various video and graphic inputs on the switcher (Figure 3.18). The TD enables the director to concentrate on controlling the many other aspects of the show. The TD may oversee the

engineering aspects of the production such as aligning effects, checking shots, ensuring source availability, and monitoring quality.

Depending on the size of the production, there are a variety of other personnel who may be involved in the production, such as the lighting director, producer, and so on (see the list of personnel in Chapter 2).

Special effects may be added live or, if the show is not live and is being recorded, it may be more convenient to leave all video effects and image manipulation until a postproduction session rather than attempt them during production.

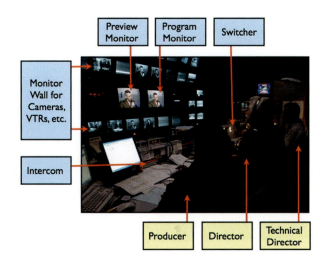

FIGURE 3.18 *Areas of the control room (blue) and the control room personnel (yellow).*

Source: Photo courtesy of PBS/Department of Defense.

As far as cameras are concerned, each studio camera's cable is routed via a wall outlet to its separate CCU (camera control unit), where a video operator (shader) monitors the picture quality, checking and adjusting the video equipment as necessary (see Chapter 2 for job descriptions). The video operator is also responsible for color correcting the recorders and other image sources to match (color balance, exposure, and contrast) with the cameras. Video operators generally monitor multiple sources at the same time. The video operator's position may be located in a nearby master control room or within the production control room itself. Other equipment, such as recording equipment, may be located in or outside the actual control room. These details differ from company to company.

Complex dramas and musical productions require an audio mixer that not only has a good ear, but also a great deal of dexterity and split-second operations. Incoming sources will include not only multiple studio microphones, but disks, audio and video recordings, and remote feeds. At the same time, the audio mixer guides the sound crew on the studio floor by using the intercom. The audio assistants may need guidance to avoid a mic appearing in a shot and to avoid boom shadows, and may need action reminders (e.g., when talent is going to move to a new position).

Services and Support Areas

Most studios have a variety of storage and service facilities nearby that help in the smooth running of day-to-day production. Their size and scale vary, but studios typically have the following:

- Makeup rooms (for individual makeup and in-program repairs) (Figure 3.19).
- A green room with restrooms where talent or guests can wait during production breaks.
- Dressing rooms (where performers can dress, rest, and await their calls).
- Prop and set storage space (Figure 3.20).
- Technical storage: all the portable technical equipment is housed here ready for immediate use, such as camera mounts, lighting gear, audio equipment, monitors, cables, and so on. This not only helps to protect the equipment, but keeps the studio floor clear.

FIGURE 3.19 *Makeup rooms or areas are essential in maintaining the talent's look.*

FIGURE 3.20 *Prop rooms store a collection of set decorations, furniture, and items handled by the talent.*

FIGURE 3.21 *Set shops are used by the larger studios to construct sets on-site.*

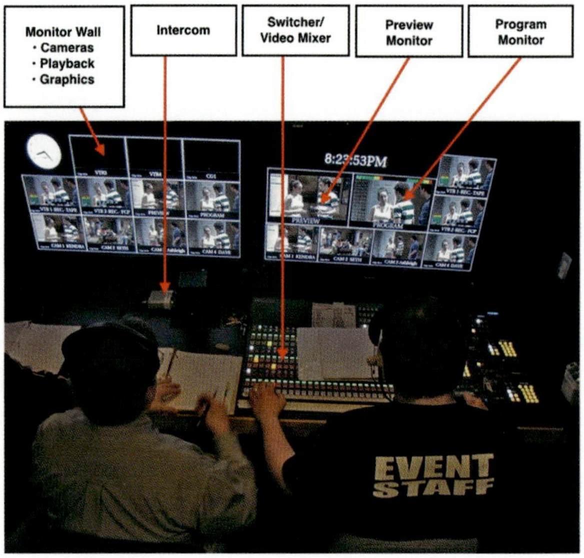

Various technical areas for electronic and mechanical maintenance are also usually located near the studios. The larger studios will even include a set shop, in which sets can be constructed (Figure 3.21).

FIGURE 3.22 *The production area of a remote truck includes the same areas as the studio control room. It is the area where the director, producer, and technical director (vision mixer) are located to create the production.*

Remote Production Facilities

Compared to studios, remote productions face a myriad of difficulties, such as dealing with changes in weather, parking the remote unit, venue non-production personnel who may not understand the operations of a remote crew, electrical power, and unwanted audio.

However, as every event cannot happen in a television studio, mobile units have been designed to go on the road and capture the event where it is happening. They are really just mobile, self-contained television production control rooms. Basically, they are the same as the control rooms found within studios.

The Remote Production Truck

Remote trucks, sometimes called OB (outside broadcast) or production trucks, are the ultimate in mobile control rooms. These units not only contain the control room, but also storage for cameras, microphones, and other related production equipment (Figures 3.22 and 3.23).

Although the remote truck may include the same areas as the studio control room, they are generally much more

compact. These units can take many different shapes and sizes, from 12 to 53 feet long—all depending on the need. Sports production units have a tendency to be very large, requiring large crews (Figure 3.24). Remote news production units are generally small and can be run by a few people.

Portable Flypack Control Systems

Remote units do not always have to be contained in a truck; sometimes they come in shipping cases. Flypacks, or portable control rooms that can be shipped or flown into a location, can take a number of different shapes. Although a small system can be a simple one-piece unit (Figures 3.14 and 3.15), these are generally a custom-ordered set of equipment that can allow you to do anything a remote truck can provide. These systems are built into portable shipping cases that can be shipped to any location, avoiding the issues of driving a large mobile unit to the location. Like building blocks, the units can be wired together with a pre-designed wire harness. Because these units can be shipped by a standard shipping company, they can be a very cost-effective alternative to a full remote truck. However, they do take more time to build once they arrive on-site (Figure 3.25).

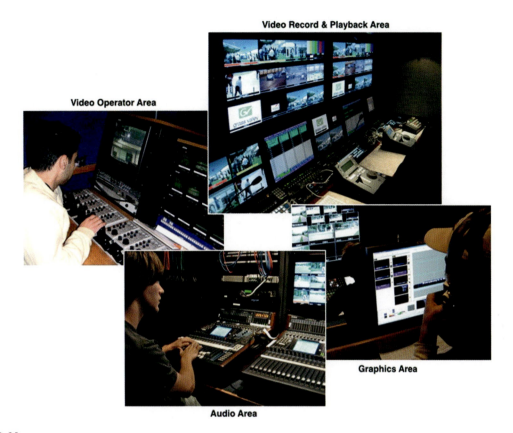

FIGURE 3.23

Besides the production area shown in Figure 3.22, remote trucks also have a graphics area, a video record and playback area, a video operator area, and an audio area.

FIGURE 3.24 *Remote trucks come in all different sizes and shapes. The blue truck is one of the largest mobile units available, a triple-wide trailer. The small car is considered to be the smallest mobile unit in the world.*

Source: Second photo courtesy of Newtek.

FIGURE 3.25 *Flypacks vary in size. The flypack shown is a made-to-order large system that includes many video and audio sources, large speakers, large monitor wall, test equipment, and so on.*

THE PRODUCTION SWITCHER

The production switcher allows the director to edit live between the various program sources (cameras, graphics, satellite feeds, and playback units). Today's switchers are quite varied, from high-end broadcast switchers, computer-based one-piece units, to limited-ability video switchers available as an app for tablets and some mobile phones (Figures 3.27–3.29). The primary means of transition are:

- *Cut* or *take*. This instant change from one image to another is the most used transition during productions. It has been said that 99 percent of all transitions are cuts.
- *Dissolve*. Dissolves are a gradual transition from one image to the next and are usually used to show a change of time or location.
- *Wipe*. The wipe is a novel transition that can take many different shapes (Figure 3.29). It shows a change of time, subject, or location. Although the wipe can inject interest or fun in the sequence of shots, it can be easily overused.
- *Fade*. A fade signifies a dissolve transition to or from black (see Chapter 17 for more information about switching).

The Wipe

Wipes, used as a unique transition between incoming images on the switcher, can be quite flexible. Most medium and higher-end switchers allow wipes to be customized.

PRODUCTION: THE 2D-3D APPROACH

After the initial surge in 3D production, many production companies decided that 3D production was not sustainable in the long term. The audience was not excited about having to wear 3D glasses and the cost of production was high. Originally, two sets of cameras were used for each position. 3D cameras had to be placed near the 2D cameras. A number of networks experimented with using one single mobile unit to obtain both 2D and 3D production programming. Using one director, they use the left-eye feed from the 3D production for the 2D telecast while using the left- and right-eye feed for the 3D telecast. This has been called shooting in 5D. "The concept of using the left-eye feed from 3D to make 2D is hardly an extraordinary technical challenge in itself," says Phil Orlins, ESPN coordinating producer. "The real challenge was accommodating the shots and the graphics in a way that works for both."

FIGURE 3.26 *3D production trucks are usually equipped for both 2D and 3D.*

Although the results are always geometrical, you can change them in several ways:

- Pattern size can be adjusted—expanded or contracted.
- Pattern shapes can be adjusted. For example, a square can be made rectangular; a circle can become an ellipse.
- The pattern can be moved around the frame: up/down, left/right, diagonally.
- You can control the speed at which the pattern changes or moves, using a fader lever or an auto-wipe button.

FIGURE 3.27 *A very complex production switcher.*

FIGURE 3.29 *Some all-in-one switchers use a computer screen interface, requiring that you use a cursor to make your selections. Most of these companies also have devices available that allow you to switch on an interface switcher as well.*

Source: Image courtesy of NewTek.

FIGURE 3.28 *An app called CollabraCam allows one iPad or iPhone (director's device) to accept wireless video streams from up to six other iPhones or iPads. The director's device can also send out camera instructions and sends a "tally" signal to each camera when it is selected. Although the final program is a bit rough, it is easy to see where this technology can go in the future.*

Source: Image courtesy of CollabraCam.

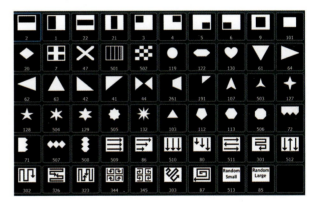

FIGURE 3.30 *This illustration shows some of the wipe patterns that are available on production switchers. Video editors also have a wide variety of wipes available.*

Source: Image courtesy of Thomson/Grass Valley.

- The symmetry of the pattern can be adjusted.
- The pattern edges can be made hard (sharp) or soft (diffused). If sources are inter-switched instantaneously, there will be a sharply defined division.
- A border can be placed around a pattern insert, in black, white, or color.

There are literally dozens of wipe patterns, but some typical examples are shown in Figure 3.30.

Chromakey

A significant option on many switchers is chromakey. Chromakey allows the director to insert one image onto another (Figure 3.31). It has endless applications, especially when combined with other video effects. Chromakey can simulate total reality or create magical, stylized, decorative displays (additional information concerning chromakey virtual sets is provided in Chapter 13).

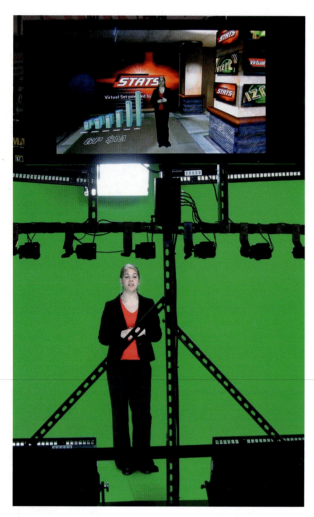

FIGURE 3.31 *With a subject positioned in front of a green background, a secondary image (computer-generated news set) is inserted into whatever is green in the camera shot. The combined result makes it look as though the subject is actually in front of the background scene. The secondary background image can be anything: colors, still images, graphics, and/or recorded or live video images. Note the talent in front of the green screen and then the completed chromakey image in the monitor above her.*

For decades, chromakey has been used to insert studio action "into" photographs, graphics, and video images. It has been used in all types of productions, from interviews to opera, from "soap operas" to musical extravaganzas, from kids' shows to serious discussions, and from newscasts to weather forecasting.

Synchronized Movements with the Virtual Set

In order to use a moving or zooming camera during a production that uses a virtual (chromakey) set, the inserted imagery must move in synchronization with the camera. Otherwise, the set (background) would stay the same size, no matter which camera was used, which creates a very unrealistic-looking set situation. Some form of a servo link is required to ensure that both operate in exact synchronism. Then, as the camera moves, proportional changes will be seen in both subject and background pictures. The combined results in the composite are completely realistic. Precise information is required about the camera's floor position, height, shooting angle, and lens angle, and must be fed into a computer, which correspondingly adjusts the background source. Even slight changes in the subject camera require the entire background picture to be instantly recalculated and redrawn to correspond if the composite is to be compatible. Entire scenic environments can be created artificially to enhance these shots. The background can be derived from a camera shooting, a photographic display, or a three-dimensional scale model, or computer-generated backgrounds can be used.

Digital Video Effects

Digital video effects (DVEs) can be designed into the switcher or may be a separate DVE piece of equipment. Like wipe patterns, some digital effects are rarely used, and others have become a regular part of production.

As you can see from Figure 3.32, an increasing range of visual effects is available. Some have a direct production value, some add an interesting new (for now) dimension to presentation, and others are for novelty.

It is not always obvious that an effect is being used. For instance, a digital effect may be used to fill the screen with a cropped section of a video segment, or to trim the edges of its image. A "graphic" showing a series of portraits could be displaying images from a DVE, combined with digital effects.

FIGURE 3.32 *Here is a small sampling of the digital video effects that are available.*
Source: Images courtesy of Grass Valley.

INTERVIEW WITH A PROFESSIONAL: BRYAN JENKINS

Briefly define your job. I am a consultant to educational institutions, government, and corporations regarding their video production efforts in their facility, with their personnel, and among their practice.

What do you like about your job? I enjoy having a direct impact on the bottom line and assisting a facility to function better. I like helping clients make good equipment decisions and assisting them to create better procedures for effective operation.

FIGURE 3.33
Bryan Jenkins, video production consultant.

What are the types of challenges that you face in your position? The challenges that I face are for my clients to give me an honest perspective of their intent. What is it that must happen? What is the time frame for the action? Sometimes clients are not open to the evaluation and do not want to make changes based on the recommendations. Finally, there is always the budget. Oftentimes the budget is just too low to accomplish the objectives. Many times the objectives can still be accomplished if we can be flexible.

How do you prepare for a production? My production preparedness begins with the vision of the executive production, producer, line producer, and director. I must have a thorough understanding of what the objectives are and then what the budget is; then I can determine how to move forward (crew, equipment, etc.).

What suggestions or advice do you have for someone interested in a position such as yours? My suggestions for an interested person are to be honest, resourceful, and respectful of the needs of your client, the crew, and the facility. All require different needs and attention in order to facilitate the needs of your production and bring success to the entire team.

Bryan Jenkins has over 20 years of experience in education, cable, and broadcast television. He has worked on multiple Olympic broadcasts, as well as projects for BET, Turner Broadcasting, and WSB-TV. He is the owner of Jenkins Video Associates, Inc.

REVIEW QUESTIONS

1. Compare a studio and a remote production, and explain the advantages and disadvantages of each one.
2. Explain the basic multi-camera audio and video path from the camera and microphone in the studio to the final recorder.
3. What are the primary components of a control room?
4. What is the role of the director during a production?
5. List the four main switcher transitions and explain how each is used.
6. How can chromakey be used effectively in television production? Give examples.

CHAPTER 4

THE PRODUCTION PROCESS

"The reality in the production world in which budget, bottom lines, timelines and Murphy's Law bump into creativity, character and story is what the production process is all about. Television projects are made in pre-production. They may be executed in production, but they are made in the pre-production phase. There is no substitution for pre-production before you get into production. That is one of the toughest things for a director or producer to learn when creating television and film projects."
—adapted from Myrl Schreibman, Producer

"If it doesn't look as if it's going to work on paper, then it probably won't work. Put enough effort into planning and then communicate those plans to everyone."
—Claire Popplewell, Senior Producer and Director (BBC)

TERMS

Dolly: The action of moving the whole camera and mount slowly toward or away from the subject.

Goals: Broad concepts of what you want to accomplish in a production.

I-mag: Image magnification; refers to video on large television screens next to a stage in order to help the viewers see the stage action.

Objectives: Measurable goals.

Postproduction: Editing, additional treatment, and duplication of the project.

Storyboard: A series of rough sketches that help someone visualize and organize the desired camera treatment.

As you would expect, in the real world there is a broad spectrum of approaches to television production. One end of the spectrum is a "get me a good shot" approach, largely relying on the initiative of the camera team to find the best shots. At the other extreme are directors who know precisely what they want, and arrange the talent and cameras to get exactly that.

THE PRODUCTION PROCESS

"In TV we've used something that I love ... it's called 'process.' I love process."

—Jerry Bruckheimer, Producer

Most television productions go through three main stages:

1. *Planning and preparation*. The preparation and organization of the production and the rehearsal before the production begins. Ninety percent of the work on a production usually goes into the planning and preparation stage (Figure 4.1).

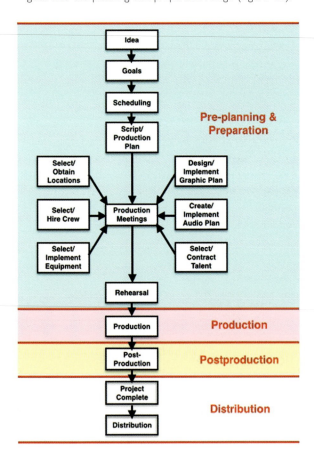

FIGURE 4.1 *During the flow of the video production process, roughly 90 percent of the work goes into the planning and preparation stage.*

2. *Production*. Actually shooting the production.
3. *Postproduction*. Editing, additional treatment, and distribution.

The amount of work at each stage is influenced by the nature of the subject. One that involves a series of straightforward "personality" interviews is generally a lot easier to organize than one on an Arctic exploration or a historical drama. But in the end, a great deal depends on how the director decides to approach the subject.

Working at the highest quality level, directors can create incredible programming by using simple methods. Treatment does not have to be elaborate to make its point. If a woman in the desert picks up her water bottle, finds it empty, and the camera shows a patch of damp sand where it rested, the shot has told us a great deal without any need for elaboration. A single look or a gesture can often have a far stronger impact than lengthy dialogue attempting to show how two people feel about each other.

It is important to understand the complexity of the production. Some ideas seem simple enough, but can be difficult or seem impossible to carry out. Others look very difficult or impracticable, but are easily achieved on the screen.

STAGE 1: PLANNING AND PREPARATION

"A production's quality will be in direct proportion to the quality and quantity of pre-production. Every single element in the video or audio channel of a production must be controlled, because every single element will affect the audience's reaction."

—Ronald Hickman, Director

Why Plan?

Some people find the idea of planning very restrictive. They want to get on with the shooting. For them, planning somehow turns the thrill of the unexpected into an organized commitment.

But many situations must be planned and worked out in advance. Directors need to get permission to shoot on private property, to make appointments to interview people, to arrange admissions, and so on. They might occasionally have success if they arrive unannounced, but do not assume this. However, directors do need to be prepared to take advantage of unexpected opportunities. It is worth taking advantage of the unexpected, even if you decide not to use it later.

The television production process is sometimes linear and sometimes nonlinear. Sometimes the concept starts with a piece of music and sometimes a script. Other times the concept may originate with a writer, the producer, the director, or even a production assistant. So, even though the steps in the process may vary at times, the most common process is described in the following sections—and it always starts with an idea.

The Idea: Starting with a Concept

Something triggers the idea. Usually, it comes from an interesting personal experience, a story you heard, something you read in a book or newspaper—some interesting incident that gave you the idea for your production. After the idea, you have to begin to formulate your goals and objectives.

Setting the Goals and Objectives

What do you really want your audience to know after they have viewed your production? It could be that you want the program to be educational, to entertain, or to inspire. The answer to this question is essential, as it guides the entire production process. The goals and objectives will determine what is used as a measuring stick throughout the rest of the production process. *Goals* are broad concepts of what you want to accomplish. Here's a sample goal: *I want to explain how to field a Formula One racing team.*

Objectives are measurable goals. That means something that can be tested to see whether the audience reacted the way you wanted them to react to the program. Take the time to think through what the audience should know after seeing your program. Following are some sample objectives.

When the viewers finish watching the program, 50 percent of them should be able to:

- Identify three types of sponsorship.
- Identify four crew positions.

Both of these are objectives because they are measurable. The number of objectives will be determined by your goals. This means that sometimes only one objective is needed; other times, five may be required.

"Forty-six percent of our television audience is not watching our network's shows online. We are finding that online viewing is complementary to broadcast viewing, so making our programming more accessible to people drives awareness, interest, and ratings both online and on-air."
—David Botkin, Senior Vice President, Research & Audience Analytics, CBS Interactive

The Target Audience

Whether your program is a drama, sitcom, news program, commercial, or sports production, it is essential to determine *whom* the program is for, and its chief *purpose*:

- Who is the viewing audience? Are they senior citizens, teens, or children?
- Is it for the general public, for specific groups, or for a local group?
- What level of content is required: basic, intermediate, or advanced? If the audience consists of children who cannot read, relevant images may be required.
- Is any specific background, qualification, language, or group experience necessary for the audience? If the audience is new to the language, carefully chosen words will need to be used.
- Are there specific production styles that this audience favors? Teens lean toward an MTV style, yet older viewers may appreciate a CBS Sunday Morning News production style. The target audience should determine your program's coverage and style. It is self-evident that the sort of program you would make for a group of content experts would be very different than one made for young children (Figure 4.2).

The conditions under which the audience is going to watch the program are important too. Today, most video programs are not made for broadcast television. They are sometimes seen on DVDs, but are most often streamed into homes, classrooms, corporate offices, and many other locations. So the director has to anticipate these conditions

FIGURE 4.2 *The intended audience should determine how the director covers the subject. In this situation, generally a younger crowd would favor a different style of coverage than elderly people.*
Source: Photo by Paul Dupree.

Cell Phone

Television

Internet

Large Screen

FIGURE 4.3
How will your audience view your production? Historically, television productions were created for a moderately sized television set. Today, you need to know what medium your audience will be watching on, as it can significantly impact how the production is made.

Source: Photos by Panasonic, Jon Cypher, and YouTube.

because they can considerably affect the way the program is produced. How and where is your audience going to see the program (Figure 4.3)? Consider:

- Will they be watching a traditional television screen in their home?
- Will they be a seated group of students watching a large screen in a darkened classroom?
- Will the viewer be watching a streamed video over the Internet while at the office?
- Will the program be viewed on an iPod while riding in a car?
- Will the video be projected on large screens next to a stage in order to help the viewers see the stage action? This is known as image magnification (i-mag).

Try to anticipate the problems for an audience watching a distant i-mag screen at a concert. Long shots have correspondingly little impact since the viewer is already far from the stage. Closer shots are essential, as they add emotion and drama. Small lettering means nothing on a small distant screen. To improve the visibility of graphics, keep details basic and limit the information.

FIGURE 4.4 *Close-up shots add emotion and drama to a production and add needed detail to small-screen productions, such as those seen on an iPod-type device.*

If your target audience is watching on an iPod or other small-screen device, directors should lean toward more close-ups than usual, as the long shots may not be as discernable on the small screen (Figure 4.4).

Here are a number of reminder questions that can help you anticipate your audience's potential problems:

- Does the program rely on their previously established knowledge?
- How much does the audience already know about the subject?
- Will the program be watched straight through, or will it be stopped after sections for discussion?
- Will there be other competing, noisy attractions as they watch, such as at an exhibition?
- Will the program soon be out of date?
- What is the time limit for the program?

The Budget

It is understandable that most directors are more creative-minded than business-minded. However, you have to be financially savvy in order to stay within the budget constraints—and every production has budget constraints. Figure 4.5 shows a sample basic television budgeting sheet.

It is important to understand what you have available financially at the beginning of the project. Once the total budget has been established, it needs to be broken down into categories. The categories may include, but are not confined to, the following:

- Transportation.
- Staff/crew.
- Talent/actors.
- Script.
- Equipment costs (rental or purchase).
- Postproduction.
- Props.
- Permits.
- Food.
- Lodging.
- Supplies.

An estimate needs to be made for each category. Once the estimates are completed, you can see if your project is going to fit the assigned overall budget. Most of the time, you will need to trim in order to fit the budget. However, occasionally you will see that you have some extra money in the budget, allowing you to increase a category or two (Figure 4.6).

Once the budget is final, it is important to begin tracking each expenditure. This enables you to keep an eye on the categories, as well as the overall budget. If you go over in one

category, it means that you have to take money from a different category—or you will go over budget.

Building a track record of being able to stay within budgets will increase the trust that clients have in you, knowing that you can responsibly create productions.

Limitations/Restrictions

There are a number of obvious limiting factors that determine what you can do, and how you go about the production, including budget restrictions, the amount of time available, money and legal issues, the time of day or season, weather, limitations or shortages of equipment or personnel, the experience and adaptability of the talent, and any local intrusions, such as location noises.

The Production Plan

Every program needs a production plan. As mentioned earlier, sometimes the program may be an unscripted event unfolding live, right in front of you. Although some live events cannot be scripted, a production plan is still a necessity. Going through almost the same process as creating a script, the director must know the event and create the best coverage plan. After reviewing the various outcomes of the event, it is important to come up with enough contingency plans that will allow you to continue covering the event, no matter what happens (Figure 4.7).

Production Methods

Great ideas are not enough. Ideas have to be worked out in realistic, practical terms. They have to be expressed as images and sounds. In the end, the director has to decide what the camera is going to shoot, and what the audience is going to hear. Where do you start?

There are two very different approaches to video production:

- The *unplanned* method, in which instinct and opportunity are the guides.
- The *planned* method, which organizes and builds a program in carefully arranged steps.

The Unplanned Approach
Directors following the unplanned approach get an idea, then look around for subjects and situations that relate to it. After shooting possible material, they later create a program from

TV PRODUCTION BUDGET SUMMARY SHEET		
Name of Program:		
Number of TV Episodes & duration:		
Previous Funding		
Development	$	$
Production	$	$
TV DEVELOPMENT/SCRIPT		
Concept & Rights:	$	
Research:	$	
Story/Script/Writers Fees:	$	
Other (specify):	$	
Development Subtotal	$	
TV PRODUCTION		
Producer Fees (total incl. EP)	$	
Director Fees (total):	$	
Presenters / Actors / Talent:	$	
Production Staff & Crew:	$	
Studio / Locations:	$	
Equipment Hire:	$	
Wardrobe/Make-Up/Art Department:	$	
Travel/Accommodations/Living:	$	
Production Office/Admin:	$	
Other (specify):	$	
Production Subtotal	$	
TV POST PRODUCTION		
Music & Copyright:	$	
Library Footage & Copyright:	$	
Film/ Tape Stock:	$	
Picture Post Production:	$	
Audio Post Production:	$	
Titles/Graphics:	$	
Post Production Labor:	$	
Other (specify):	$	
Post Production Subtotal	$	
TV MARKETING & ADMINISTRATION		
Marketing/Delivery:	$	
Administration/Overheads:	$	
Legal:	$	
Insurance:	$	
Sundry (eg. finance, ACC etc)	$	
Other (specify):	$	
Marketing/Admin Subtotal	$	
Total Above The Line:	$	
Total Below The Line:	$	
Contingency:	$	
Production Company Overhead:	$	
TOTAL TELEVISION PRODUCTION BUDGET	$	
Total cost per episode	$	

KristofCreative.com

FIGURE 4.5 *An example of a simple television budget sheet.*

Source: Courtesy of Kristof-Creative, Inc./KristofCreative.com.

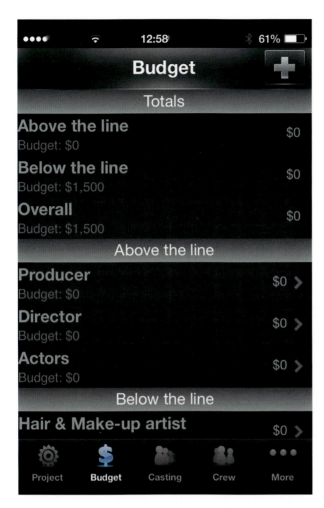

FIGURE 4.7 *A sports event is an example of an unscripted event that still requires a production plan.*
Source: Photo by Josh Taber.

material, probably writing a commentary as a "voice-over" to match the edited pictures.

At best, this approach is fresh, uninhibited, improvised, makes use of the unexpected, avoids rigid discipline, and is very adaptable. Shots are interestingly varied. The audience is kept alert, watching and interpreting the changing scene.

At worst, the result of such shot hunting is a haphazard disaster, with little cohesion or sense of purpose. Because the approach is unsystematic, gaps and overlaps abound. Good coherent editing may be difficult. Opportunities may have been missed. The director usually relies heavily on the voice-over to try to provide any sort of relationship and continuity between the images (Figure 4.8).

FIGURE 4.6 *Budgeting software keeps track of production expenses, comparing the estimated costs to the actual costs. Computer and mobile software is available to keep a very detailed budget. Mobile software that works on a portable device is helpful when in the field. Smartphone screenshot shown is Producer.*

whatever they have found. Their inspiration springs from the opportunities that have arisen.

An example would be when a director decides to make a program about "safety at sea." Using the unplanned approach, the director might go to a marina and develop a production based on the stories heard there. Or the idea could be discussed with the lifeguards, and the director might then decide to follow an entirely different plan. A commercial dock could also be visited, and the director might discover material there of an entirely different kind.

After accumulating a collection of interesting sequences (atmospheric shots, natural sound, interviews, etc.), the content is reviewed and then put into a meaningful order. A program could then be created that fits the accumulated

FIGURE 4.8 *Some documentaries are shot using the unplanned approach, in which instinct and opportunity are the guides.*
Source: Photo by Bill Miller.

The Planned Approach

The planned method of production approaches the problem quite differently, although the results on the screen could be similar. In this situation, the director works out, in advance, the exact form he or she wants the program to take and then creates it accordingly.

Fundamentally, you can either:

- Begin with the environment or setting and decide how the cameras can be positioned to get the most effective shots (Figure 4.9).
- Envision certain shots or effects that you want to see, and create a setting that will provide those results (Figure 4.10).

FIGURE 4.9 *Production and engineering staff review a venue for a television production using the planned approach. They begin with the setting and then decide where to place the cameras.*

FIGURE 4.10 *Narrative directors use the planned approach to production by utilizing a storyboard. However, they create the setting to fit the story.*

A lot will depend on which of the following scenarios applies:

- *Interpreting an existing script (as in drama).* This will involve analyzing the script: examining the storyline and the main action in each scene and visualizing individual camera shots.
- *Building a treatment framework (as in a sports production or breaking news event).* That is, considering how you are going to present a specific program subject and working out the kinds of shots you want.

At best, the planned approach is a method in which a crew can be coordinated to give their best. There is a sense of systematic purpose throughout the project. Problems are largely ironed out before they develop. Production is based on what is feasible. The program can have a smooth-flowing, carefully thought out, persuasive style.

At worst, the production becomes bogged down in organization. The program can be stodgy and routine, and lack originality. Opportunities are ignored, because they were not part of the original scheme, and would modify it. The result could be a disaster.

In reality, an experienced director uses a combination of the planned and unplanned approaches, starting off with a plan and then taking advantage of any opportunities that become available.

Schedule

It is imperative to sit down and establish a schedule that includes the essential deadlines. These deadlines usually revolve around the following issues: When does the script need to be completed? When do locations need to be chosen? When does the talent need to be selected and contracted? Deadlines are also selected for the crew, equipment, rehearsals, graphics, props, rehearsals, and rough production and postproduction schedule. Keep in mind that, at this stage, you are not actually doing any of these things; you are determining the schedule and setting the deadlines as to when they must be completed.

Any time the schedule is not met, there will be a ripple effect on the other areas of the production, because one of two things is going to happen: the production will either go over budget or some elements will need to be cut in order to make up for lost time and/or money.

Building a Program Outline

The program outline begins with a series of *headings* showing the main themes that need to be discussed.

Let's use the example of an instructional video about "building a wall." In this case, the topics that might be covered could include: tools needed, materials, foundation, making mortar, and methods of laying bricks and pointing. We can now determine how much program time can be devoted to each topic. Some will be brief and others relatively lengthy. Some of the topics will need to be emphasized; others will be skipped over to suit the purpose of the program.

The next stage is to take each of the topic headings and note the various aspects that need to be covered as a series of *subheadings*. Under "tools," for instance, each tool that must be demonstrated should be listed. Now there is a structure for the program and the director can begin to see the form it is likely to take.

Research

Some programs, such as dramas (especially period dramas), documentaries, news, interviews, and others, need to complete research in order to create the program's content or make sure that the existing content is accurate. This research may involve going to the library, doing online research, or contacting recognized experts in the content area of the show. Travel may even be required.

The important thing to remember here is that research is time-consuming and may affect the production budget—especially if a content expert wants an appearance fee or flights and lodging are included for the crew or guest.

Coverage

What do you want to cover in the available time? How much is *reasonable* to cover in that time? If there are too many topics, it will not be possible to do justice to any of them. If there are too few, the program might seem slow and labored. There is nothing to be gained by packing the program full of facts, for although they may sound impressive, audiences rarely remember more than a fraction of them. Unlike with the printed page, the viewer cannot check back to confirm something, unless the program is designed to be stopped and rewound or is frequently repeated.

The kind of subject that is being covered, as well as who makes up the audience and the content that needs to be featured, will influence how the camera is utilized in the

FIGURE 4.11 *This producer is determining the best camera angles for coverage in a stadium.*

program, where it concentrates, how close the shots are, and how varied they are. As a director, here are some of the areas that need to be considered in advance:

- What are the content areas that need to be covered?
- Is the subject (person, event, or object) best seen from specific angles? Does a specific angle help communicate the message more effectively (Figure 4.11)?
- Would the addition of graphics help the audience understand the content of the production?
- If possible, do a rehearsal so that the best viewpoints and shots can be determined.
- Give the talent and crew the vision and goals for the production and then help them know what they can do to assist in attaining the goals.

Thinking through the Shots

There are a number of different shots that are used in productions: interviews, panel discussions, piano and instrumental performances, singers, and newscasts. These are usually familiar to a studio production crew, and meaningful shooting variations can sometimes be limited. What the talent is saying (or playing) is more important than straining to achieve new and original shots. Consequently, these productions follow normal patterns so that the director often starts off with planned angles and then may introduce a specific treatment as it becomes desirable (Figures 4.12 and 4.13).

Planning for regular productions is primarily a matter of coordinating staff and facilities, ensuring that the video

FIGURE 4.12
In a formal interview, there are relatively few shots that you can get without the shots becoming a distraction from the content.

Cam 1

Cam 2

Cam 3

packages or clips are available (with known timing and cue points), graphics, titles, and similar features are prepared, and any additional material organized. The production itself may be based on a series of key shots and some spontaneous on-the-fly decisions.

Treatment Breakdown

For many shows, the action is predetermined by the program format—the talent is going to enter by walking out onto the stage and then sitting in a specific chair to be interviewed, then move over to a table to talk about the product on display. The shots are straightforward. The director checks over the plans, and, with an eye to the program script, places cameras in regular positions that fit the treatment (long shots, group shots, singles, reaction shots). Any variations can be worked out during rehearsals.

However, if you are directing a more complicated production, planning will need to be more detailed. You begin

by reading through the script, with the sets and layouts of the location. As you read, you visualize each scene and consider issues such as whether talent is going to make an entrance or exit, what they are going to do, where they will be doing it, and whether they will need props such as a map. Are they going to speak to the camera? Are other people in the scene, and what are they doing? Are they going to move around (look out a window on the set)? Although these questions sound very obvious, they don't just happen. Everything has to be planned, arranged, and learned.

The next step is to systematically block out the action. For example, you might ask yourself: Would it be better if the host is seen entering the room and moving over to the desk, or should we discover her already seated? Because the show is about dress design, would it be preferable if the audience saw her clothing as she enters rather than opening with a head and shoulders shot at the desk? Is the idea of a desk too formal anyway? Would an easy chair be preferable? And so on.

FIGURE 4.13

There is often a broad range of shot options.

Source: Photos by Austin Brooks.

Top Shot

Long Shot

Reaction Shot

Medium Long Shot

Hands

Over the Shoulder

Through the Lid

After determining the action, consider what type of shots would be appropriate and where the cameras should be to get them. If one camera moves into the set to get a close shot of the item they are discussing, will the camera now be in the second camera's shot? Would it be better to record some shots of the product after the formal interview and then edit in the insert shots during postproduction? This is how a director builds a continuous succession of shots, noting them in the script margin, either as abbreviated reminders or as tiny sketches. The same situation can often be tackled in various different but equally successful ways. Let us look at an example for a dramatic production:

> *Returning husband enters door, wife has an unexpected guest.*

This could be broken down as:

Shot 1: Husband enters door, hangs up hat . . . medium shot shows who it is and orientates the audience.

Shot 2: Wife looks up, greeting him . . . medium two-shot showing her with guest.

Shot 3: Husband turns and sees the guest . . . close-up shot of husband's reaction.

Shot 4: The guest rises, walks to greet the husband . . . from medium two-shot; pans with guest, dollying in for tight two-shot of guest and husband.

Developing the Camera Plan

Depending on the type of production, the director may need to determine the type of lenses required for a specific type of shot. In order to plan the specifics, it is very helpful to have a scale-accurate floor plan or ground plan. The other option is to visit the location and measure it. These issues arise especially often when working at remote sites. One of the ways to calculate the lenses needed is to use a lens calculator (a printed chart or tablet/smartphone software) (Figure 4.14).

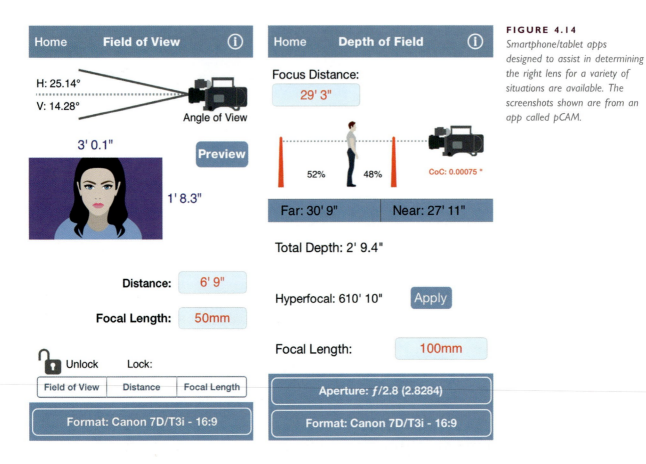

FIGURE 4.14
Smartphone/tablet apps designed to assist in determining the right lens for a variety of situations are available. The screenshots shown are from an app called pCAM.

You can also judge the types of shot (close-up, full-length, etc.) that a certain lens angle will give.

The resultant production plan (camera plan), together with margin action notes or sketches, will form the basis for all technical planning and subsequent rehearsals. Even the biggest productions can be analyzed into shots or sequences in this way.

Storyboards

Directors need to think through each scene in their minds, capturing the images and turning them into a storyboard. A storyboard is a series of rough sketches that help the director visualize and organize his or her camera treatment. It is a visual map of how the director hopes to arrange the shots for each scene or action sequence (Figure 4.15). Storyboards will be covered in more detail in Chapter 6.

Production Aspects

Once the script and/or production plan, and possibly even storyboards, are finished, there is still quite a bit of planning

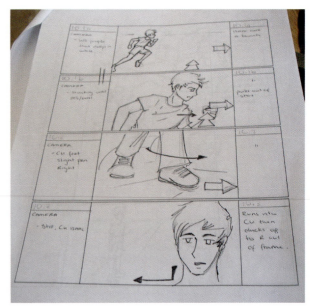

FIGURE 4.15 *A storyboard on the set of a production. The storyboard roughly visualizes what the program will look like.*
Source: Photo by Taylor Vinson.

and preparation that needs to be completed. These tasks are not linear; that is, the production staff can be working on all of them at the same time. The producer and director often act as coaches, reviewing the work by each segment of their staff, encouraging, giving feedback, providing quality control, and challenging them to move forward at a brisk pace. The producer or the director has to juggle all of these aspects at the same time.

Select and Obtain Locations

Shoot locations must be chosen. A location may be a studio, a sports venue, a house, or some other location. Permits or contracts may need to be obtained. Later in this chapter, we will discuss the location site survey.

Determine Camera Locations

By determining the specific camera locations, you can calculate cable lengths, contract equipment if needed, and plan other steps.

Select and Contract the Talent

Anchors, narrators, actors, or anyone else who will be appearing in front of the camera must be chosen and then contracted. Once the talent has been chosen, then you begin working on contracts, costuming, makeup, and similar related details.

Select and Hire the Crew

The appropriate production and engineering crew must be hired. This may or may not involve contracts. Each person needs to know their start and end dates, as well as their pay, instructions, and a multitude of other details.

Create and Implement the Audio Plan

Audio specifications must be determined. For example, what needs to be heard? Once you know what must be heard, an audio plan is created. This plan would include the placements for each microphone, needed sound effects, and any other required audio inputs.

Design and Implement the Graphic Plan

Working with designers and decorators, graphics will need to be designed to create a mood or look for the production. These graphics may include text, animations, and even set graphics. It is imperative that the graphics are consistent.

Equipment Must Be Chosen and Contracted

The equipment needs must be determined. This may include cameras, audio gear, lights, and so on. This equipment may already be owned, could be purchased, or may need to be rented, which would then require contracts.

The Production Meeting

Production meetings are an essential part of the planning process. These meetings should include representatives from each of the production areas listed previously, as well as engineering, the production manager, and location personnel. Each person participating in the meeting will then have the opportunity to stay informed and ask questions, with all of the other area representatives present. This approach allows them to work out complicated situations that arise when changes are made (Figure 4.16).

FIGURE 4.16 *Production meetings allow the staff from the various areas to get together to be updated on a regular basis. These meetings can be regularly scheduled in a meeting room or on location as needed.*

Source: First photo by Andrew Wingert.

Once the meeting is complete, each individual follows up on the issues raised in the meeting. This may include documentation, budgeting, staffing, scheduling, and so on. This production planning meeting forms the basis for efficient teamwork. Problems anticipated and overcome at this stage prevent last-minute compromises. Camera cable routing, for example, is a typical potential hazard. Cables get tangled and can impede other cameras, or drag around noisily.

Remote Location Surveys (Recce)

Fundamentally, there are two types of shooting conditions: your *base location* and *remote location*. Your base is wherever you normally shoot. It may be a studio, theater, room, or even stadium. The base is where you know exactly what facilities are available (equipment, supplies, and scenery), where things are, the amount of room available, and so on. If you need to supplement what is there, it usually can be easily done (Figure 4.17).

A remote location is anywhere away from your normal shooting location. It may be just outside the building or way out in the country. It could be in a vehicle, down in a mine, or in someone's home. The main thing about shooting away from your base is to find out in advance what you are going to deal with. It is important to be prepared. The preliminary visit to a location is generally called a *remote survey*, *site survey*, or *location survey*. It can become anything from a quick look around to a detailed survey of the site. What is found on the survey may influence the planned production treatment. The following remote survey checklist gives more specifics.

Setup

All of the production equipment—which may include the cameras, mounting gear, lights, microphones, graphic generators, and related cabling—must be set up early enough to leave time for troubleshooting. This means that the setup usually begins hours before it is needed. If everything works, then the crew gets a break before the production begins. If there are problems, the crew may still be troubleshooting when the production begins. Rehearsals may begin before everything is operational if needed.

The Rehearsal

Depending on the type of production, one or many rehearsals may occur. These rehearsals give the director, crew, and talent the chance to see how everything flows, whether the equipment works the way it was planned, and whether changes need to be made. When the rehearsals are complete, it is generally time to begin the production (Figure 4.18).

FIGURE 4.17 *Site surveys allow you to check out the actual location to make sure that it will meet the production needs.*

FIGURE 4.18 *Rehearsals allow the crew to understand how the event will unfold and how the director wants to cover it.*

CHECKLIST: THE REMOTE SURVEY

The amount of detail needed regarding a specific location varies with the type and style of the production. Information that may seem trivial at the time can prove valuable later in the production process. Location sites can be interiors, covered exteriors, or open-air sites. Each has its own problems.

Sketches	☐	Prepare rough maps of route to site that can ultimately be distributed to the crew and talent (include distance, travel time).
	☐	Rough layout of site (room plan, etc.).
	☐	Anticipated camera location(s).
	☐	Designate parking locations for truck (if needed) and staff vehicles.
Contact and Schedule Information	☐	Get location contact information from primary and secondary location contacts, site custodian, electrician, engineer, and security. This includes office and mobile phones, as well as email.
	☐	If access credentials are required for the site, obtain the procedure and contact information.
	☐	Obtain the event schedule (if one exists) and find out whether there are rehearsals that you can attend.
Camera Locations	☐	Check around the location for the best camera angles.
	☐	What type of camera mount will be required (tripod, Steadicam, etc.)?
	☐	If it's a multi-camera production, cable runs must be measured to ensure that there is enough camera cable available.
	☐	What lens will be required on the camera at each location to obtain the needed shot?
	☐	Any obstructions? Any obvious distractions (e.g., large signs, reflections)?
	☐	Any obvious problems in shooting? Anything dangerous?
Lighting	☐	Will the production be shot in daylight? How will the light change throughout the day? Does the daylight need to be augmented with reflectors or lights?
	☐	Will the production be shot in artificial light (theirs, yours, or mixed)? Will they be on at the time you are shooting?
	☐	Estimate the number of lamps, positions, power needed, supplies, and cabling required.
Audio	☐	What type of microphones will be needed?
	☐	Any potential problems with acoustics (such as a strong wind rumble)?
	☐	Any extraneous sounds (elevators, phones, heating/air conditioning, machinery, children, aircraft, birds, etc.)?
	☐	Required microphone cable lengths must be determined.
Safety	☐	Are there any safety issues that you need to be aware of?
Power	☐	What level of power is available and what type of power is needed? The answer to this question will differ greatly between single-camera and multi-camera productions.
	☐	What type of power connectors are required?
Communications	☐	Are walkie-talkies needed? If so, how many?
	☐	How many mobile phones are needed?
	☐	If a multi-camera production, what type of intercom and how many headsets are required?
Logistics	☐	Is there easy access to the location? At any time, or at certain times only? Are there any traffic problems?
	☐	What kind of transportation is needed for talent and crew?
	☐	What kind of catering is needed? How many meals? How many people?
	☐	Are accommodations needed (where, when, how many)?
	☐	If the weather turns bad, are there alternative positions/locations available?
	☐	Have a phone number list available for police, fire, doctor, hotels, and local (delivery) restaurants.
	☐	What kind of first-aid services need to be available (is a first-aid kit enough, or does an ambulance need to be on-site)?
	☐	Is location access restricted? Do you need to get permission (or keys) to enter the site? From whom?
	☐	What insurance is needed (against damage or injury)?
Security	☐	Are local police required to handle crowds or just the public in general?
	☐	What arrangements need to be made for security of personal items, equipment, props, etc.?
	☐	Do streets need to be blocked?

STAGE 2: THE PRODUCTION

We have finally reached the production! Hard work in the planning and preparation stage should diminish the number of problems that occur during the production. It does not mean that you won't have problems, but it should at least reduce the number of problems.

The Director during the Production

During the production is when the director finally gets to direct the crew to capture the audio and video needed to communicate the message. The director visually interprets the script or event, motivates the crew to do their best work, and guides the talent to get the best performances (Figures 4.19 and 4.20).

FIGURE 4.19 *The director directing a sports production.*
Source: Photo by Morgan Irish.

FIGURE 4.20 *The director of this sitcom is directing from the floor of the studio, reviewing the camera shots on a quad-split monitor.*

The Producer during the Production

The producer's responsibilities vary based on the type of production. However, most of the time he or she has to keep an eye on the budget and the production schedule and ensure that the production is meeting the originally stated goals.

Creating the Show

You can use images and sounds to report events simply, undramatically, and unobtrusively. For some types of production, the best kind of staging and camera treatment provides a quiet, sympathetic background to the performance—such as an opera or an interview. However, there are program subjects that need as much "hype" as possible—flashing, swirling light effects, arresting color, unusual sound quality, bizarre camera angles, and unpredictable cutting—to create an exciting production.

Although some productions have a relatively loose format, others require split-second timing, with accurately cued images coming from live remote locations. Some programs effectively use a compilation of prerecorded material woven together by commentary and music. Other types of production concentrate on action; others on reaction. Dialogue may be all-important—or quite incidental.

Away from the studio, productions can be anything from a one-camera shoot trailing wildlife to a large-scale remote (outside broadcast) covering a vast area with many cameras. Unforeseen problems are inevitable, and the restrictions of the environment and the weather affect the director's opportunities.

The Tools

The camera and microphone do not behave like our eyes and ears, but substitute for them. Our eyes flick around with knowledge of our surroundings, providing us with an impression of unrestricted stereoscopic vision; in fact, we can detect detail and color over only a tiny angle, and our peripheral vision is monochromatic and quite blurred. In daily life, we build up an impression of our environment by personally controlled sampling: concentrating on certain details while ignoring others. The camera and microphone, on the other hand, provide us with only restricted segments. And the information provided in these segments is modified in various ways, as we have seen, by the characteristics of the medium (distorting space, proportions, scale, etc.).

Selective Techniques

If you simply set up your camera and microphone overlooking the action and zoom in close for detail, the viewer progressively loses the overall view and struggles to understand the big picture. If you go with a wide shot, showing more of the scene, some detail will become indiscernible. The choice of suitable viewpoint and shot size provides the concept of guided selection—the beginnings of techniques.

Good production techniques provide variety of scale and proportion; of composition pattern; of centers of attention and changing subject influence. You achieve these things by variation in shot size and camera viewpoint, by moving the subject and/or the camera, or by altering the subject that is seen.

Although you may sometimes encourage the viewer to browse around a shot, you will generally want him or her to look at a particular feature, and to follow a certain thought process.

The Screen Transforms Reality

The camera and microphone can only convey an impression of the subject and scene. Whatever the limitations or inaccuracies of these images, they are the only direct information our viewer has available to them. And, of course, interpretation must vary with one's own experience and previous knowledge. Whether you are aiming to convey an accurate account (newscast) or to conjure an illusion (drama), the screen will transform reality.

You could fill the screen with a shot of a huge aircraft, or with a diminutive model. The pictures would look very similar. Yet neither conveys the subjective essentials—that is, how you feel standing beside the giant plane or handling the tiny model. Introducing a person into the shot would establish scale, but it would still not include our characteristic responses to such a situation: the way we would be awed by the huge size, or intrigued by the minute detail.

The camera can be used to select detail from a painting or a photograph and the television screen puts a frame around it, transforming this isolated area into a new complete picture, an arrangement that did not originally exist, or an arrangement that, if sustained in a close-up, can easily become detached and dissociated in the audience's mind from the complete subject.

When a solid sculpture is shot in its three-dimensional form, it becomes reproduced as a flat pattern on the television screen. Planes merge and interact as they cannot do when we examine the real sculpture with our own eyes. Only on the flat screen can a billiard ball become transformed into a flat disk under diffused lighting. In practice, you can actually make use of this falsification of reality. The very principles of scenic design heavily rely on it. Keep in mind that the camera and microphone do inevitably modify the images they convey, and that these images are easily mistaken for truth by the viewer.

Interpretative Production Techniques

It is one of those production paradoxes that although your camera can show what is happening, it will often fail to convey the atmosphere or spirit of the occasion. You can often achieve more convincing representative results by deliberately using selective techniques than by directly shooting the event.

Straightforward shots of a mountain climber may not communicate the thrills and hazards of the situation. But use low camera angles to emphasize the treacherous slope—show threatening overhangs, straining fingers, slipping feet, dislodged stones, laboring breath, slow ascending music—and the illusion grows. Even climbing a gentle slope can appear hazardous if strong interpretative techniques are used.

Sometimes the audience can be so strongly moved by this subjective treatment that sympathetic bodily reactions set in when watching such scenes—even dizziness or nausea. Even situations outside the viewer's personal experience (e.g., the elation of free fall or the horror of quicksand) can be conveyed to some degree by carefully chosen images and audio.

Techniques can also be introduced obtrusively for dramatic effect, or so unobtrusively that the effect appears natural, and the viewer is quite unaware that the situation is contrived:

- *Obtrusive*: The camera suddenly drops from an eye-level shot to a low-angle shot.
- *Unobtrusive*: The camera shoots a seated actor at eye level. He stands, and the camera tilts up with him. We now have a low-angle shot.

When situations seem to occur accidentally or unobtrusively, they are invariably more effective. For example, as an intruder moves toward the camera, he becomes menacingly underlit by a nearby table lamp.

Many techniques have become so familiar that we now regard them as the norm—a natural way of doing things. But they are really illusions that help us convey specific concepts:

- "Chipmunk" voices (high-pitched audio) for small creatures.
- Echo behind ghostly encounters.
- Rim light in "totally dark" scenes.
- Background music.

If the program is live, the production process generally ends at the completion of the production segment of the process. However, if it is recorded for later distribution, the next stage is postproduction.

STAGE 3: POSTPRODUCTION

Everything that was shot earlier is now assembled together in a sequential fashion. Mistakes can be corrected and visual effects, sound effects, and music added. Postproduction will be covered in detail in Chapter 17.

The goal is a final show that is polished, without any noticeable production issues. If the production meets the originally stated goals, then it is a success!

INTERVIEW WITH A PROFESSIONAL: JAMES STUART

Define your job. Coordinate overall logistics, facilities, and vendors for network programs and studio shows. Financial control of show cost budgets. Maintain crew logs and daily schedules. Liaison with various sports leagues, teams, and venues.

What do you like about your job? Even though I may be working on a specific sport for the entire season, the work is never the same. Each city, venue, or game presents its own unique challenges, whether it be logistically, operationally, or production/technical wise. I may be able to have a baseline on how I approach each event, but I need to adjust for the wrinkles each event presents. That keeps it fresh and prevents getting in a cookie-cutter rut.

FIGURE 4.21
James Stuart, production manager.

What are the types of challenges that you face in your position? Challenges come both from work and personal life. You need to be able to balance both. There are times when you will be traveling four to five days a week with little time for home life. If you are single, this may not be much of an issue, but if married with kids you need to be sure everyone is on board with your chosen profession. Work challenges will be all the unforeseen last-minute changes that come up. This could be anything from a flight being delayed, bad weather, delay or postponement of a game, power outage, adding a camera to the broadcast, or breaking news, whether it's event-related or other national news.

How do you prepare for a production? The start to each event begins with a production meeting so you know what the goal of the broadcast is and what is expected. From there, I am able to begin the logistical planning and work with the teams, venues, etc. on preparing for our arrival at the venue. If it's a studio show I am working on, it's finding out the info on what games we will focus on and need additional production support at in order to make our show stand out.

What suggestions or advice do you have for someone interested in a position such as yours? Organization and multitasking are key traits to possess if wanting to get into broadcast production management. You will not just be working on the game being broadcast that week. You will be working on games three to four weeks out and also other events as well. You could be working on NFL, NHL, and figure skating all at the same time. A person needs to be able to budget his or her time and keep everything in sync in order to be successful.

James Stuart is a production manager for NBC Sports Group. He has worked on multiple Olympics, Dateline, and specials, as well as other sports events.

REVIEW QUESTIONS

1. Explain the three stages of a television production.
2. Where do concepts come from for a production?
3. How do setting production goals and objectives impact the way you shoot a program?
4. Describe the differences between the planned and unplanned approaches to production.
5. What is the value of a production meeting? Who attends this meeting?

CHAPTER 5

THE SCRIPT AND PRODUCTION PLAN

"Stories can take us places of which we can only dream. An audience will suspend their disbelief in order to be swept into a story. We have to put aside the idea that 'men can't fly' to enjoy *Superman*."

—Barry Cook, Director (*Mulan* and *Arthur Christmas*)

"The stage loves words; television and the cinema love movement. The goal is to say what you want to say in the briefest way possible. If that means taking out an entire speech and replacing it with an arched eyebrow, do so. If you have to choose between the two, the arched eyebrow probably packs the greatest punch. A nod speaks volumes; a facial tick can bring down an empire. This approach can guide writers to write more effective dialogue in all genres. The dialogue can be qualitative, not quantitative."

—adapted from Sebastian Corbascio,
Writer and Director

TERMS

Breakdown sheet: An analysis of a script, listing all of the production elements listed in order of the schedule.

Camera script: A revised script for camera rehearsals, including the details of the production treatment: cameras and audio, cues, transitions, stage instructions, and set changes.

Fact sheet/rundown sheet: Summarizes information about a product or item for a demonstration program, or details of a guest for an interviewer.

Outline script: Usually includes any prepared dialogue, such as the show opening and closing.

Preliminary script/writer's script: Initial submitted full-page script (dialogue and action) before script editing.

Rehearsal script: Script prepared for television and used for pre-studio rehearsal. Script details the settings, characters, action, talent directives, and dialogue.

Running order: In a live production, the program is shot in the scripted order.

Shooting order: When taping a production, the director can shoot in whatever order is most convenient for the crew, actors, and/or director.

Show format: Lists the items or program segments in a show, in the order in which they are to be shot. It may show durations, who is participating, shot numbers, and so on.

Synopsis: An outline of the characters, action, and plot. This synopsis helps everyone involved in the production understand what is going on.

Treatment: A film treatment, or script treatment; it is more than an outline of the production and less than a script. It is usually a detailed description of the story that includes other information such as how it will be directed.

Once an idea is conceived, it must be transformed into a message, a script, and/or a production plan. Generally, the script must be created before anything else is done, because it will be the source that every other area draws from.

THE SCRIPT'S PURPOSE

Planning is an essential part of a serious production, and the script forms the basis for that plan. Scripts do the following:

- Help the director clarify ideas and develop a project that successfully communicates to the viewers.
- Help the director coordinate the entire production team.
- Help the director determine what resources will be needed for the television production (Figure 5.1).

Although some professional crews on location (at a news event, for instance) may appear to be shooting entirely spontaneously, they are usually working through a tried-and-true process or pattern that has been proven successful in past situations.

For certain types of production, such as narratives, the script usually begins the production process. The director then reads the draft script, which usually contains general information on characters, locations, stage directions, and dialogue. He or she then visualizes the scenes and assesses the possible treatment. The director must also anticipate the script's potential and possible difficulties. At this stage, changes may be made to improve the script or make it more workable. Next, the director goes on to prepare a camera treatment.

Another method of scripting begins with an outline. In this method, you decide on the various topics you want to cover and the amount of time that you can allot to each topic. A script is then developed based on this outline and decisions are made concerning the camera treatment for each segment.

When preparing a documentary, an extended outline becomes a shooting script, showing perhaps the types of shots that the director would like. It usually also includes rough questions for on-location interviews with participants. All other commentary is usually written later, together with effects and music, to finalize the edited production.

In order to have a concise, easy-to-read script, abbreviations are usually used. These abbreviations significantly reduce the amount of wordage on the script, which also reduces the number of required pages (Table 5.1).

THE PRODUCTION PLAN

The Unscripted Production Plan

There are some types of program that cannot be scripted. For example, sports events cannot be controlled; you never know where they are going to go. However, the director still needs to think through a quasi-script, or what is often known as a *production plan*. These production plans are designed to map out the general flow of the production, with contingency plans taking into consideration that the event could take many unexpected turns along the way.

The Outline Script: Semi-Scripted Production

The type of script used will largely depend on the kind of program being made. There will be some production situations—particularly where talent improvise as they speak or perform—when the "script" simply lists details of the production group, facilities needed, and scheduling, and shows basic camera positions, and so on.

An *outline script* usually includes any prepared dialogue, such as the show's opening and closing. When people are going to improvise, the script may just list the order of topics to be covered. During the show, the list may be included on a card held by the talent, a cue card positioned near the camera, or a teleprompter in order to remind the host. If the show is complicated, with multiple guests or events occurring, a show format is usually created (Table 5.2). This lists the program segments (scenes) and shows the following:

- The topic (such as a guitar solo).
- The amount of time allocated for this specific segment.
- The names of all talent involved (hosts and guests).

FIGURE 5.1 *Director on the set of a sitcom discussing the script with writers. Scripts help coordinate the production team.*

TABLE 5.1 *Television Script Abbreviations*

Many organizations duplicate the rehearsal script on white paper. Operational information is added to the original to provide the camera script on yellow paper. Often, script revisions change color with every update.

All camera shots are numbered in a fully scripted show. Inserts are not numbered (stills, prerecorded video, graphics, etc.), but are identified in the audio column. Where possible, cutting points are marked in the dialogue with a slash mark: _____ /

Equipment	CAM	Camera	CG	Character generator
	MIC	Microphone	BOOM	Mic boom
	G	Graphic	VT, VTR	Video recorder
	ESS	Electronic still store	CP, CAP	Caption*
	DVE	Digital video effect		
Position	L/H, R/H	Picture left hand, right hand	B/G, F/G	Background, foreground
	C	Center	X	Move across
	U/S, D/S	Upstage, downstage	POV	From the point of view of person named
	O/C	On camera		
Cueing	Q, I/C, O/C	Cue, in cue, out cue	RUN VT	Play recording
	S/B	Stand by	F/X	Cue effects
Cameras	ECU, CU, MS, LS, XLS (see Chapter 8 for more shot details)		DI	Dolly in
	2-S, 3-S	Two-shot, three-shot	P/B, D/B	Pull back, dolly back
	O/S	Over-the-shoulder shot	FG/BG	One person foreground; another background
Switching (vision mixing)	CUT	Not marked, but implied unless other transition marked	LOSE GRAPHICS	Cut out graphics
	MIX (DIS)	Mix (dissolve)	FI, FO	Fade-in, fade-out
	WIPE	Wipe	T	Take
	S/S	Split screen	KEY (INSERT)	Electronic insertion (chromakey)
	CK	Chromakey		
Audio	F/UP	Fade-up audio	Spot FX	Sound effect made in studio
	FU	Fade audio under	STING	Cue strong musical chord emphasizing action
	P/B	Playback	OS, OOV	Over scene, out of vision (audio heard where source not shown)
	REVERB	Reverberation added	ANNCR	Announcer
	ATMOS	Atmosphere (background sounds)	SOT	Sound on (tape) recorder
General	TXN	Transmission	ID	Station identification
	ADD	An addition (to a story)	MOS, VOX, POP	Man-on-street interview
	EXT	Exterior	INT	Interior
	PROP	Property	LOC	Location
	RT	Roll titles		

* U.K. term.

- Facilities (cameras, audio, and any other equipment and space needed).
- External video content sources that will be required (such as tape, digital, satellite, and so on).

When segments (or edited packages) have been previously recorded to be inserted into the program, the script may show the opening and closing words of each, as well as the package's duration. This step assists accurate cueing.

TABLE 5.2 *Sample Show Format*

The show format lists the items or program segments in a show in the order in which they are to be shot. It may show durations, who is participating, shot numbers, and the like.

Example		
	CARING FOR THE ELDERLY	**Total duration: 15 mins**
1.	OPENING TITLES AND MUSIC	00:10
2.	PROGRAM INTRO	00:30
3.	PROBLEMS OF MOBILITY	2:20
4.	INJURIES	02:15
5.	DIET	02:45
6.	DAILY ACTIVITIES	03:40
7.	EXERCISES	01:20
8.	AIDS THAT CAN HELP	01:15
9.	CLOSING	00:25
10.	END TITLES	00:10
		15:00

Fully Scripted Shows

When a program is fully scripted, it includes detailed information on all aspects of the production, as described in the following subsections.

Scenes

Most productions are divided into a series of *scenes*. Each scene covers a complete continuous action sequence and is identified with a number and location (Scene 3—office set). A scene can involve anything from an interview to a dance routine, a song, or a demonstration sequence.

Shots

When the director has decided how he or she is going to interpret the script, each scene will be subdivided into a series of *shots*; each shot shows the action from a specific viewpoint. The shots are then numbered consecutively for easy reference on the script, in the order in which they will be screened.

In a live production, the program is shot in the scripted order (*running order*). When taping a production, the director can shoot in whatever order is most convenient (*shooting order*) for the crew, actors, and/or director. The director may decide to omit shots ("drop shot 25") or to add extra shots (shots 24A, 24B, etc.). He or she may decide to record shot 50 before shot 1 and then edit them into the correct running order at a later time.

Dialogue

The entire prepared dialogue, spoken to the camera or between people. The talent may memorize the script or read it off teleprompters or cue cards.

Equipment

The script usually indicates which camera/microphone is being used for each shot (Cam. 2 Fishpole).

Basic Camera Instructions

Details of each shot and camera move (Cam. 1 CU on Joe's hand; dolly out to long shot).

Switcher (Vision Mixer) Instructions

For example: cut, fade.

Contributory Sources

Details of where video recordings, graphics, remote feeds, and so on appear in the program.

When Is It Necessary to Fully Script a Production?

- When the dialogue is to follow a prescribed text that is to be learned or read from a prompter or script.
- Where action is detailed, so that people move to certain places at particular times and do specific things there (this can affect cameras, sound, and lighting treatment).
- When there are carefully timed inserts (prerecorded materials) that have to be cued accurately into the program.
- When the duration of a section must be kept within an allotted time slot, yet cover certain agreed upon subject

points (the speaker might otherwise spend too much time on one point, and miss another altogether).

- Where there are spot cues—such as a lightning flash and an effects disk sound of thunder—at a point in the dialogue.

Script Stages

The fully scripted show is developed in several stages, as described in the following subsections.

Draft/Preliminary Script/Outline Script/ Writer's Script

The initial submitted full-page script (dialogue and action) before script editing.

Rehearsal Script

A script prepared for television and used for pre-studio rehearsal. The script details the locales (settings), characters, action, talent directives, and dialogue (Table 5.3).

Camera Script

A revised script for camera rehearsals, augmented with details of production treatment: cameras and audio, cues, transitions, stage instructions, and set changes (Table 5.4).

Full Script

The *full script* is not, as some people believe, an artistically inhibiting document that commits everyone concerned to a rigid plan of procedure. It can be modified as the need arises. It simply informs everyone about what is expected at each moment of the production. Rehearsal time is too precious to use up explaining what is expected of everyone as you go. It is far better to have a detailed script that shows the exact moment for the lighting change, to cue the graphics, or to introduce a special effect. The full script is a changeable plan of how the production will proceed that has details added to it as the production develops.

Fully scripted approaches can be found in newscasts, drama productions, operas, situation comedy shows, documentaries, and commercials. When dialogue and/or action are spontaneous, there can be no script—only an outline of where the show is headed (e.g., discussions). A formal talk is usually scripted.

The more fragmentary or disjointed the actual production process is, the more essential the script becomes. It helps everyone involved to *anticipate*. And in certain forms of production (such as chromakey staging), anticipation is essential for tight scheduling.

In a complex production that is recorded out of sequence, the production crew may be unable to function meaningfully

TABLE 5.3 *The Rehearsal Script*

SCENE	INT/EXT	LOCATION	TIME OF DAY
3	INT	LOUNGE	NIGHT
		(GEORGE ENTERS, WALKS TO TABLE, SWITCHES ON LAMP.)	
GEORGE: (CALLS) The lights are OK in here. It must be your lamp.			
		(GEORGE TAKES GUN FROM DRAWER.) SLIPS IT INTO HIS POCKET: PULLING OUT TELEGRAM. HOLDS IT UP.	
"SORRY CANNOT COME WEEKEND, BRIAN SICK, WRITING . . . JUDY."			
		(HAND WADS IT UP, THROWS IT INTO FIRE.) (DOOR OPENS: EILEEN ENTERS.)	
EILEEN: Really, these people are not good. They promised to be here tonight. Look how late . . . GEORGE: It's probably the storm that has delayed them. They'll be here all right.			
		(EILEEN SITS ON COUCH: GEORGE JOINS HER.)	
EILEEN: If Judy knows you're here, it'll take wild horses to keep her away. GEORGE: How many more times do I have to tell you . . . EILEEN: Why do you keep pretending?			
		LIGHTNING FLASH	
GEORGE: I've warned you. You'll go too far.			

TABLE 5.4 *The Camera Script*

SHOT	CAM.	(POSITION)	SCENE	INT/EXT	LOCATION	TIME OF DAY	F/X
			CAMS. 2A, 3A, 4B. SOUND ROOM BI				
			3	INT	LOUNGE	NIGHT	RECORDER:
10.	F/U 2.	A					WIND RAIN
		LS PAN/ZOOM on GEORGE to MS	(GEORGE ENTERS, WALKS TO TABLE, SWITCHES ON LAMP.)				
11.	3.	A.	GEORGE: (CALLS) The lights are OK in here. It must be your lamp.				
12.	2.	A.	(GEORGE TAKES GUN FROM DRAWER.)				
13.	4.	MS B	SLIPS IT INTO HIS POCKET: PULLING OUT TELEGRAM, HOLDS IT UP				
		CU of telegram	"SORRY CANNOT COME WEEKEND. BRIAN SICK. WRITING . . . JUDY."				
14.	3.	PULL FOCUS on fire as he throws A	(HAND WADS IT UP, THROWS IT INTO THE FIRE.)				
		LS	(DOOR OPENS: EILEEN ENTERS.)				
			EILEEN: Really, these people are not good. They promised to be here tonight. Look how late . . . GEORGE: It's probably the storm that has delayed them. They'll be here all right.				
		DOLLY IN to MS as EILEEN X's to couch.	(EILEEN SITS ON COUCH: GEORGE JOINS HER.)				
			EILEEN: If Judy knows you're here, it'll take wild horses to keep her away. GEORGE: How many more times do I have to tell you . . . EILEEN: Why do you keep pretending?				
			LIGHTNING FLASH				
15.	2.	A					
		CU	GEORGE: I've warned you. You'll go too far.				DISK: THUNDER-CLAP

without a full script, which makes it clear how shots/sequences are interrelated and reveals continuity. The lighting director may, for example, need to adjust the lighting balance for a scene so that it will cut smoothly with the different shots that were previously captured.

The full script can be a valuable coordinating document, enabling you to see at a glance the relationships between dialogue, action, treatment, and mechanics. During planning, of course, it helps the team estimate how much time there is for a camera move, how long there is for a costume change, whether rearranging shooting order will give the necessary time for a makeup change, the scenes during which the "rain" should be seen outside the windows of the library set (i.e., the water spray turned on and the audio effects introduced), and the thousand and one details that interface in a smooth-running show.

The full script is used differently by various members of the production team. For the director, the script has two purposes: as a reference point when developing treatment, estimating the duration of sequences, planning camera moves, and so on, and to demonstrate to members of the team what he or she requires. The director's assistant(s) follows the script carefully during rehearsals and taping, checking dialogue accuracy, noting where retakes are needed, timing sections (their durations, where a particular event occurred), and perhaps readying and cueing contributory sources, as well as "calling shots" on the intercom—for example, "Shot 24 on 2. Coming to 3." The person operating the production switcher follows the script in detail, preparing for upcoming transitions, superimpositions, effects, and so on, while checking the various monitor pictures.

TABLE 5.5 *Dramatic Show Format/Breakdown Sheet/Running Order*

Color Code	Script Breakdown Sheet	Date: _____
Day Ext.—Yellow	Production Company:	Breakdown Page # _____
Night Ext.—Green	_____	Page Count _____
Day Int.—White	Production Title:	
Night Int.—Blue	_____	

Scene # _____	Scene Name	Int. or Ext.
	_____	

Description:		Day or Night

CAST Red	STUNTS Orange	EXTRAS/ATMOSPHERE Green
	EXTRAS/SILENT BITS Yellow	
SPECIAL EFFECTS Blue	PROPS Purple	VEHICLE/ANIMALS Pink
WARDROBE Circle	MAKEUP/HAIR Asterisk	SOUND Brown
SPECIAL EQUIPMENT Box	PRODUCTION NOTES Underline	

Others in the team for whom a script would be too distracting (e.g., camera and boom operators) use it as a detailed reference point when necessary, but are guided by simplified outlines such as *breakdown sheets* and *camera cards* as they memorize their operations (Table 5.5).

Basic Script Layout Formats

There is not one standard script layout. Script layout styles can vary widely. Some prefer a single-column cinematic format, with transitions in a left margin, and all video and audio information in a single main column. Other versions use two vertical columns, with picture treatment (cameras, switching) on the left, and action and dialogue on the right, together with studio instructions and lighting/effects cues. Directors often mark up their script by hand with their own instructional symbols to indicate transitions and shots. Software is available that provides multiple script formats.

Single-Column Format

Although there are different variations of the single-column format, all video and audio information is usually contained in

TABLE 5.6 *Sample Single-Column Shooting Script/Single-Camera Format*

FADE-IN:		
1. EXT: FRONT OF FARMHOUSE—DAY		
Front door opens. FARMER comes out, walks up to gate. Looks left and right along road.		
2. EXT: LONG SHOT OF ROAD OUTSIDE FARM (Looking east)—DAY		
POV shot of FARMER looking along road, waiting for car.		
3. EXT: FARM GATE—DAY		
Medium shot of FARMER leaning looking over gate, looking anxiously. He turns to house, calling.		
FARMER:		
I can't see him. If he doesn't come soon, I'll be late.		
4. INT: FARMHOUSE KITCHEN—DAY		
Wife is collecting breakfast things. Sound of radio.		
WIFE:		
You're too impatient. There's plenty of time.		
5. EXT: FARM GATE—DAY		
Medium shot of FARMER, same position. He looks in other direction. Sound of distant car approaching. Sudden bang, then silence.		

a single main column. Before each scene, an explanatory introduction describes the location and the action.

Reminder notes can be made in a wide left-hand margin, including transition symbols (for example, X = cut; FU = fade-up), cues, camera instructions, thumbnail sketches of shots or action, and so on.

This type of script is widely used for narrative film-style production and single-camera video, in which the director works alongside the camera operator. It is perhaps less useful in a multi-camera setup, in which the production team is more dispersed, with everyone needing to know the director's production intentions (Table 5.6).

Two-Column Format

Like the one-column format, there are many variations of the two-column format. This traditional television format is extremely flexible and informative. It gives all members of the production crew shot-by-shot details of what is going on. They can also add their own specific information (e.g., details of lighting changes) as needed (Table 5.7).

Two versions of the script are sometimes prepared. In the first (*rehearsal script*), the right column only is printed.

TABLE 5.7 *Sample Two-Column Shooting Script/Multi-Camera Shooting Script*

SHOT	CAM (Position)	SCENE/ACTION/AUDIO
CAMS:	1B, 2D, 3A	SOUND: BOOM POLE Scene 4. INT. BARN—NIGHT
15.	FU 2D LS DOORWAY Zoom in to MS	(FARMER ENTERS, HANGS TAPE 7: WIND LAMP ON WALL-HOOK DISK 5: RAIN BESIDE DOOR) FARMER: It's getting late. How is the poor beast doing?/
16.	1B O/S SHOT	SON: I don't think she'll last the night. SON'S POV She has a high fever./
17.	3A LS FARMER He comes	(FARMER WALKS FORWARD TO THE STALL) FARMER: I called Willie. He's on his way. (FARMER KNEELS BESIDE COW)/
18.	CU SON	SON: D'you think he'll be able to get here?
19.	1C CU FARMER	FARMER: If the bridge holds. But the river's still rising./

Abbreviations used:
CU: Close-up
MS: Medium shot
LS: Long shot
FU: Fade-up
O/S: Over the shoulder
POV: Point of view
___ /: Indicates point to "cut to next shot"

Subsequently, after detailed planning and preproduction rehearsals, the production details are added to the left column to form the *camera script*.

The Dramatic Script

The dramatic full script may be prepared in two stages: the *rehearsal script* and the *camera script*.

The *rehearsal script* usually begins with general information sheets, including a cast/character list, production team details, rehearsal arrangements, and similar details. There may be a synopsis of the plot or storyline, particularly when scenes are to be shot/recorded out of order. The rehearsal script generally includes the following types of details:

- Location: the setting where the scene will be shot.
- Time of day and weather conditions.
- Stage or location instructions: (The room is candlelit and a log fire burns brightly.)
- Action: basic information on what is going to happen in the scene, such as actors' moves (Joe gets in the car).
- Dialogue: speaker's name (character) followed by his or her dialogue. All delivered speech, voice-over, voice inserts (e.g., phone conversation), commentary, announcements, and so on (perhaps with directional comments such as "sadly" or "sarcastically") (Figure 5.2).
- Effects cues: indicating the moment for a change to take place (lightning flash, explosion, Joe switches light out).
- Audio instructions: music and sound effects.

FIGURE 5.2 *Actors review the script as they rehearse for a production.*
Source: Photo by Luke Wertz.

The *camera script* adds full details of the production treatment to the left side of the rehearsal script, and usually includes:

- The shot number.
- The camera used for the shot and possibly the position of the camera.
- Basic shot details and camera moves (CU on Joe. Dolly back to LS as he rises.)
- Switcher instructions (cut, dissolve, etc.).

SUGGESTIONS FOR SCRIPTWRITING

"The biggest misconception about writing for television is that it's just like *The Dick Van Dyke Show*. Writers are almost never home by 5:30. Only the actors have those hours."

—Carmen Finestra, Co-Creator and Executive Producer, *Home Improvement*

There are no shortcuts to good scriptwriting, any more than there are to writing short stories, composing music, or painting a picture. Scriptwriting techniques are learned through observation, experience, and reading. But there are some general guidelines that are worth keeping in mind as you prepare your script, as explained in the following sections.

Be Visual

Although audio and video images are both very important in a television production, viewers perceive television as primarily a visual medium. Material should be presented in visual terms as much as possible. If planned and shot well, the images can powerfully move the audience—sometimes with very few words. Other times programs rely almost entirely on the audio, using the video images to strengthen, support, and emphasize what is heard. Visual storytelling is difficult but powerful when done well.

When directors want their audience to concentrate on what they are hearing, they try to make the picture less demanding. If, for instance, the audience is listening to a detailed argument and trying to read a screen full of statistics at the same time, they will probably not do either successfully.

Pacing the Program

Keep in mind that writing for print is totally different than writing for television. In print, the reader can read at his or her own pace, stop, and reread whenever he or she wants to. In a television program, the viewing audience generally has to watch the program at the director's chosen pace. An essential point to remember when scripting television is the difference between the rates at which the viewer can take in information. A lot depends, of course, on how familiar the audience already is with the subject and the terms used. Where details are new to the audience, and the information is complicated, more time will be required to communicate the information in a meaningful way. Ironically, something that can seem difficult and involved at the first viewing may appear slow and obvious a few viewings later. That is why it is so hard for directors to estimate the effect of material on those who are going to be seeing it for the first time. Directors become so familiar with it that they know it by heart and lose their objectivity.

The key to communicating complex subjects is to simplify. If the density of information or the rate at which it is delivered is too high, it will confuse, bewilder, or just encourage the audience to switch it off—mentally, if not physically.

As sequences are edited together, editors find that video images and the soundtrack sometimes have their own natural pace. That pace may be slow and leisurely, medium, fast, or brief. If editors are fortunate, the pace of the picture and the sound will be roughly the same. However, there will be occasions when they find that they do not have enough images to fit the sound sequence, or they do not have enough soundtrack to put behind the amount of action in the picture.

Often when the talent has explained a point (perhaps taking five seconds), the picture is still showing the action (perhaps 20 seconds). The picture, or action, needs to be allowed to finish before taking the commentary onto the next point. In the script, a little dialogue may go a long way, as a series of short pieces cut into the program, rather than a continual flow of verbiage.

The reverse can happen, too, when the action in the picture is brief. For example, a locomotive passes through the shot quickly in a few seconds, taking less time than it takes the talent to talk about it. So, more pictures of the subject are needed, perhaps from another viewpoint, to support the dialogue.

Even when picture and sound are more or less keeping the same pace, do not habitually cut to a new shot as soon as the action in the picture is finished. Sometimes it is better to continue the picture briefly, in order to allow time for the audience to process the information that they have just seen and heard, rather than move on with fast cutting and a rapid commentary.

It is all too easy to overload the soundtrack. Without pauses in a commentary, it can become an endless barrage

of information. Moreover, if the editor has a detailed script that fits in with every moment of the image, and the talent happens to slow down at all, the words can get out of step with the key shots they are related to. Then the editor has the choice of cutting parts of the commentary, or building out the picture (with appropriate shots) to enable picture and sound to be brought back into sync.

Style

The worst type of script for television is the type that has been written in a formal literary style, as if for a newspaper article or an essay, where the words, phrases, and sentence construction are those of the printed page. When this type of script is read aloud, it tends to sound like an official statement or a pronouncement, rather than the fluent everyday speech that usually communicates best with a television audience—not that we want a script that is so colloquial that it includes all the hesitations and slangy half-thoughts one tends to use, but certainly one that avoids complex sentence construction.

It takes some experience to be able to read any script fluently, with the required natural expression that brings it alive. But if the script itself is written in a stilted style, it is unlikely to improve with hearing. The material should be presented as if the talent were talking to an individual in the audience, rather than proclaiming on a stage, or addressing a public meeting.

The way the information is delivered can influence how interesting the subject seems to be. The mind boggles at: "The retainer lever actuates the integrated contour follower." But we immediately understand if it is written this way: "Here you can see, as we pull this lever, the lock opens."

If the audience has to pause to figure out what is meant, they will not be listening closely to what is being said immediately afterwards. Directors can often assist the audience by anticipating the problems with a passing explanation, or a subtitle (especially useful for names), or a simple diagram.

Hints on Developing the Script

How scripts are developed will vary with the type of program and the way individual directors work. The techniques and processes of good scriptwriting are a study in themselves, but we can take a look at some of the guiding principles and typical points that need to be considered.

Writers Often Collaborate

Many television shows are written by a team of writers who collaborate on scripts. For example, on a sitcom, writers cannot be funny every day. So, a team of comedy writers work together to get the best script possible.

Writer's Block

There will be times when you "hit a wall" when writing. When you have been working for awhile and realize that you are not getting any traction on the script, take a break and do something else for a while to help clear your mind. Then go back with a fresh perspective.

The Nature of the Script

The Script May Form the Basis of the Entire Production Treatment

Here the production is staged, performed, and shot as indicated in the script. As far as possible, dialogue and action follow the scripted version.

The Scriptwriter May Prepare a Draft Script (i.e., a Suggested Treatment)

This is studied and developed by the director to form a shooting script.

The Script May Be Written after Material Has Been Shot

Certain programs, such as documentaries, may be shot to a preconceived outline plan, but the final material will largely depend on the opportunities of the moment. The script is written to blend this material together in a coherent storyline, adding explanatory commentary/dialogue. Subsequent editing and postproduction work is based on this scripted version.

The Script May Be Written after Material Has Been Edited

Here the video editor assembles the shot material, creating continuity and a basis for a storyline. The script is then developed to suit the edited program. Occasionally, a new script replaces the program's original script with new or different text. For example, when the original program was created in a different language from that of the intended audience, it may be marketed as an *M&E version*, in which the soundtrack includes only "music and effects." All dialogue or voice-over commentary is added (dubbed in) later by the recipient in another language.

Scriptwriting Basics

A successful script satisfies two important requirements:

- *The program's main purpose*: to amuse, inform, intrigue, persuade, and so on.
- *It must be practical.* The script must be a workable vehicle for the production crew.
  Fundamentally, we need to ensure that:

- The script meets its deadline. When is the script required? Is it for a specific occasion?
- The treatment is feasible for the budget, facilities, and time available. An overambitious script will necessarily have to be rearranged, edited, and have its scenes rewritten to provide a workable basis for the production.
- The treatment usually must fit the anticipated program length. Otherwise, it will become necessary to cut sequences or pad the production with added scenes afterwards to fit the show to the allotted time slot.
- The style and the form of presentation are appropriate for the subject. An unsuitable style, such as a lighthearted approach to a very serious subject, may trivialize the subject.
- The subject treatment is suitable for the intended audience. The style, complexity, concentration of information, and other details are relative to their probable interest and attention span.

Ask Yourself

Who Is the Program For? What Does Your Audience Already Know?

Analyzing your audience is covered in Chapter 4.

What Is the Purpose of This Program?

Examples: entertainment, information, instruction, or persuasion (as in advertising, program trailers, propaganda). Is there a follow-up to the program (such as publicity offers or tests)?

Is the Program One of a Series?

Does it relate to or follow other programs? Do viewers need to be reminded of past information? Does the script style need to be similar to previous programs? Were there any omissions, weaknesses, or errors in previous programs that can be corrected in this program?

What Is the Length of the Program?

Is it brief? Must it make an immediate impact? Is it long enough to develop arguments or explanations for a range of topics?

How Much Detail Is Required in the Script?

Is the script intended to be complete with dialogue and action (actual visual treatment depends on the director)? Is the script a basis for improvisation (e.g., by a guide or lecturer)? Is it an ideas sheet, giving an outline for treatment?

Are You Writing Dialogue?

Is it for actors to read, or inexperienced performers? For the latter, keep it brief, in short "bites" to be read from a prompter or spoken in their own words. Is the dialogue to be naturalistic or "character dialogue"?

Is the Subject a Visual One?

If the subjects are abstract, or no longer exist, how will you illustrate them?

Have You Considered the Script's Requirements?

It takes only a few words on the page to suggest a situation, but to reproduce it in pictures and sound may require considerable time, expense, and effort (e.g., a battle scene). You may have to rely on available stock library video. Does the script pose obvious problems for the director (e.g., a script involving special effects, stunts, and the like)?

Does the Script Involve Costly Concepts That Can Be Simplified?

An intercontinental conversation could be covered by an expensive two-way video satellite transmission, or can it be accomplished by utilizing an online telephone call accompanied by previously acquired footage or still images?

Does the Subject Involve Research?

The script may depend on what researchers discover while investigating the subject. Do you already have information that can aid the director (have contacts, know of suitable locations, availability of insert material, etc.)?

Where Will the Images Come From?

Will the subjects be brought to the studio? This allows maximum control over the program treatment and presentation. Or will cameras be going on location to the subjects? This may include situations such as shooting in museums. Script opportunities may depend on what is available when the production is being shot.

Remember

Start Scripting with a Simple Outline

Before embarking on the main script treatment, it can be particularly helpful to rough out a skeleton version, which usually includes a general outline treatment, covering the various points that need to be included, in the order in which the director proposes dealing with them.

Visualize

- Sometimes pictures alone can convey the information more powerfully than the spoken word.
- The way a commentary is written (and spoken) can influence how the audience interprets a picture (and vice versa).
- Pictures can distract. People may concentrate on looking instead of listening!
- Avoid "talking heads" wherever possible. You may want to show the subject being talked about, rather than the person who is speaking.
- The script can only *indicate* visual treatment. It will seldom be specific about shot details, unless that is essential to the plot or situation. Directors have their own ideas!

Avoid Overloading

- Keep it simple. Don't be long-winded or use complicated sentences. Keep to the point. When a subject is difficult, an accompanying diagram, chart, or graph may help make the information easier to understand.
- Do not give too much information at a time. Do not attempt to pack too much information into the program. It is better to do justice to a few topics than to cover many inadequately.

Develop a Flow of Ideas

- Deal with one subject at a time. Generally, avoid cutting between different topics.
- Do not have different information on the screen than in the commentary. This can be very distracting and confusing to the viewer.
- Aim to have one subject or sequence lead naturally into the next.
- When there are a number of separate different topics, think through how they are related and the transitions necessary to keep the audience's interest.

Consider the Pacing

- Vary the pace of the program. Avoid a fast pace when imparting facts. It conveys an overall impression, but facts do not sink in. A slow pace can be boring or restful, depending on the content.

- Remember that the audience cannot refer to the program later (unless it is interactive or they have a video player). If they miss a point, they may fail to understand the next—and will probably lose interest.

Watch Your Style

- Use an appropriate writing style for the intended viewer. Generally, aim for an informal, relaxed style.
- There is a world of difference between the style of the printed page and the way people normally speak. Reading from a prompter produces an unnatural, stilted effect.
- Be very careful about introducing humor in the script!

Scripting Tools

Most writers use one of the numerous scriptwriting software programs to create their scripts. These programs usually have a variety of formats, allowing the writer to select the most appropriate script format that fits the production he or she is working on. Some of this software will allow the writer to create the full script designed in a way that other scripts, such as a camera script, can be easily printed with little additional effort.

Scriptwriting apps have also been created that allow mobile phones and tablets to be used for scriptwriting. These tools are especially helpful to update a script while in the field or write down ideas when you are not close to your computer (Figure 5.3).

Storyboards

Many directors need to think through each production scene in their minds, capturing the images and turning them into a storyboard. The storyboard is primarily used for dramatic productions, but is used for other events as well. An example would be the opening ceremonies at the Olympics (Figure 5.4).

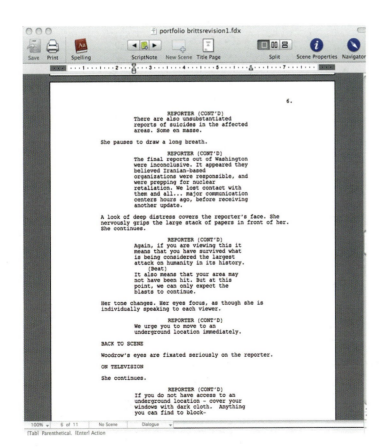

FIGURE 5.3 *There are many different computer-based scriptwriting programs. The first photo shows a screen shot of Final Draft software and a script. The second photo shows a smartphone scriptwriting app.*

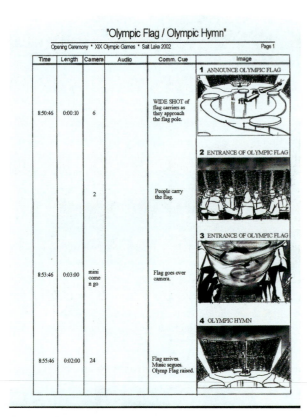

FIGURE 5.4 *This storyboard was a page from the opening ceremony at the Olympics. Hand-drawn, it served as a guide to the staff involved in the broadcast of the production.*

Source: Image courtesy of International Sports Broadcasting.

FIGURE 5.5 *This storyboard was created for a television commercial.*

Source: Image courtesy of Jim Mickle.

The storyboard is a series of rough sketches that help the director visualize and organize his or her camera treatment. It is a visual map of how the director hopes to arrange the key shots for each scene or action sequence (Figures 5.5 and 5.6).

There are many advantages to using a storyboard:

- Assists the director in thinking through the shot sequences.
- Helps the director think about transitions between scenes.
- Visualizes the sequence so that others can give helpful feedback.
- Assists the director in making budget decisions.
- Helps the director be more efficient while working on the set.
- Allows the crew to know the number of camera angles and camera placements.
- Helps maintain scene continuity.

Directors find that the storyboard can be a valuable aid, whether they are going to shoot action:

- Continuously, from start to finish.
- In sections or scenes (one complete action sequence at a time).
- As a series of separate shots or "action segments," each showing a part of the sequence.

Storyboards can be designed a number of different ways. There are software programs that assist the director in visualizing ideas, someone can roughly sketch them out, or a storyboard artist can create detailed drawings that can even be animated to be shown during the fundraising period (Figures 5.7–5.9).

Hand-drawn storyboards usually begin with a grid of frames. Start to imagine your way through the first scene, roughly sketching the composition for each shot. You don't have to be able to draw well to produce a successful storyboard. Even the crudest scribbles can help you organize your thoughts and show other people what you are trying to

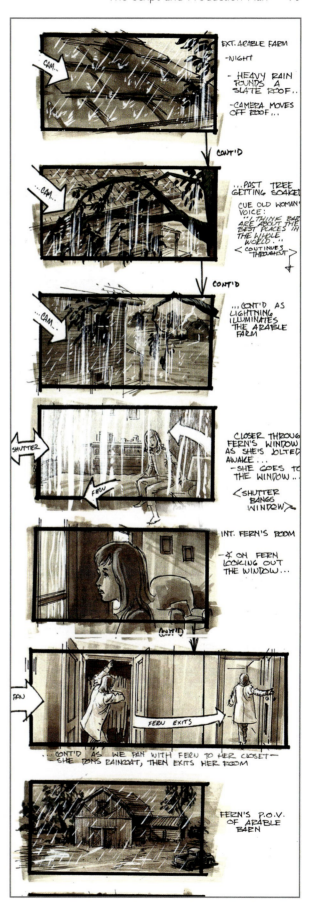

FIGURE 5.6 *These storyboards were created by artist Josh Sheppard for dramatic productions.*

FIGURE 5.7 *(Left) A frame from a storyboard. (Right) How the camera shot that specific scene.*

Source: Photo by Taylor Vincent.

FIGURE 5.8 *Storyboard programs are available on all types of devices. Both of these screenshots were taken from a storyboard program on a smartphone.*

Source: Software by Cinemek.

Figure 5.9
Computer-based storyboard programs allow the non-artist to create professional-looking storyboards.

Source: Software by Storyboard Quick.

do. If the action is complicated, you might need a couple of frames to show how a shot develops. Let's look at a very simple storyline, to see how the storyboard provides you with imaginative opportunities.

The young person has been sent to buy her first postage stamp.

There are dozens of ways to shoot this brief sequence. You could simply follow her all the way from her home, watching as she crosses the road, enters the post office, goes up to the counter . . . the result would be totally boring.

Analyzing Action

Let's think again. We know from the previous scene where she is going and why. All we really want to register are her reactions as she buys the stamp. So let's cut out all the superfluous footage, and concentrate on that moment (Figure 5.10).

- The child arrives at the counter, and looks up at the clerk.
- Hesitatingly, she asks for the stamp.
- She opens her fingers, to hand the money to the clerk.
- The clerk smiles and takes the money and pulls out the stamp book.
- A close shot of the clerk tearing the stamp from a sheet.

You now have a sequence of shots, far more interesting than a continuous "follow-shot." It stimulates the imagination. It guides the audience's thought processes. It has a greater overall impact. However, if this type of treatment is carried out poorly, the effect can look very disjointed, contrived, and posed. It is essential that the treatment matches the style and theme of the subject.

The whole sequence could have been built with dramatic camera angles, strong music, and effects. But would it have been appropriate? If the audience knows that a bomb is ticking away in a parcel beneath the counter, it might have been. It is all too easy to overdramatize or "pretty-up" a situation (such as star filters producing multi-ray patterns around highlights, and diffusion filters for misty effects). Resist the temptation to add too much to the scene; keep it appropriate.

This breakdown has not only helped you to visualize the picture treatment, but allowed you to begin to think about how one shot is going to lead into the next. You can start to deal with practicalities. You see, for example, that shots 1 and 3 are taken from the front of the counter, and 2, 4, and 5 need to be taken from behind it. Obviously, the most logical approach is to shoot the sequence out of order. The storyboard becomes a shooting plan.

1

2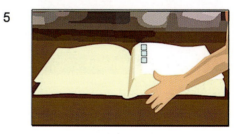

3

4

5

FIGURE 5.10 *When edited together correctly, a sequence looks natural. But even a simple scene showing a person buying a stamp needs to be thought through. Note how shots 1 and 3 are taken on one side of the counter, and shots 2, 4, and 5 on the other.*

Source: Illustration created by StoryBoard Quick software.

To practice storyboarding, review a motion picture carefully, making a sketch of each key shot. This way, you will soon get into the habit of thinking in picture sequences, rather than in isolated shots.

Additional Production Plan Information

In addition to the script and possibly a storyboard, there are a number of other production/scripting tools that may be helpful:

- *Synopsis*: An outline of characters, action, and plot. Appended to a dramatic script, particularly when shot out of sequence, or to coordinate a series. This synopsis is helpful to assist everyone involved in the production to understand what is going on.
- *Fact sheet/rundown sheet*: Summarizes information about a product or item for a demonstration program, or details of a guest for an interviewer. Provided by a researcher, editor, or agency for guidance.
- *Show format (figure/breakdown sheet/running order)*: Lists the events or program segments in order, allowed durations, participants' names, cameras, and audio pickup allocated, setting used, video and audio inserts, and so on. This is sometimes also called a rundown sheet, but is not to be confused with the fact sheet. Invaluable for unscripted, semi-scripted, and scripted shows that contain a series of self-contained segments (sequences, scenes). Also as a summary of a complex dramatic script to show at a glance: shot numbers for each scene, operational details, inserts, break for major resetting (clearing props, redressing set, moving scenery), costume, makeup changes, and so on (Figure 5.11 and Table 5.5).

Individual camera information:

- *Shot sheet*: The shot sheet is a list of all of the individual cameras' required shots. Shot sheets reduce the amount of communication between the director and the camera operators during the actual production, as the shots are already on the sheet (Figure 5.12).
- *Team roster*: In sports production, it is often helpful for camera operators to have a list of team members with their jersey numbers. That way, if the director tells them to shoot a specific person (by name), they can see what his or her number is (Figure 5.13).

Camera 1 Shot Sheet

1. CU of host

2. 2-S of host and guest

3. CU of host

4. CU of host

5. O-S of host (move cam)

6. CU of host

7. XLS of set (include cams)

FIGURE 5.12 *Shot sheets list all of the individual cameras' required shots. Shot sheets reduce the amount of communication between the director and the camera operators during the actual production.*

Item	Source	Audio		Time	Total
	10:14		**Segment 1**		10:00:00a
1.	VT 32	SOT	Show Open & Tease	4:28	10:04:28a
2.		Bill v/o music	Opening graphic	0:15	10:04:43a
3.		Bill v/o	Shots of players lining up	0:20	10:05:03a
4.		Bill v/o	Wide push to team shooting	0:10	10:05:13a
5.		Sam v/o	Host welcome	0:12	10:05:25a
6.		Ray v/o	Interview with guest coach Throw to booth	0:14	10:05:39a
7.		Sam v/o	Home team intro	0:13	10:05:52a
8.		Paul	Paul o/c	0:52	10:06:44a

FIGURE 5.11 *This network remote sports show format lists the program segments in order, durations, names, audio and running time.*

FIGURE 5.13 *Camera operators often use a shot sheet or a team roster, to assist them in knowing the players on the field of play.*

INTERVIEW WITH A PROFESSIONAL: ANDREA NASFELL

"Being a good television screenwriter requires an understanding of the way film accelerates the communication of words."

—Steven Bochco, Producer

FIGURE 5.14
Andrea Gyertson Nasfell, screenwriter.

Briefly define your job. I create the story for movies and TV shows and turn it into the script that communicates the artistic vision, as well as the practical instructions to the production crew. Usually I am hired by a producer who has a particular story in mind (based on an existing property such as a book, or a concept that he or she likes), and my job is to understand that vision and create a story and script that meets the need. Sometimes I write my own projects ("spec scripts") and take those finished projects to producers who might buy them.

What do you like about your job? I like the flexibility. I can write when and how I want to, as long as I am meeting deadlines. This helps me juggle my family and other responsibilities with my work. I like the fact that if I'm not on a work-for-hire project, I can still write my own projects to sell. I like the variety in projects—for months, I may be writing an intense drama, but when it's done I can write comedy for a while. There is always a new challenge. I love the creative aspect of it, combined with the problem-solving aspect of it. While you can create totally new and original people and worlds, you are also working for the needs of the producer and the market, so you have to know how to make the art and commerce come together, which is its own kind of creativity. I love to see a finished project, remembering that I was sitting in my kitchen wondering how to make that character or plot point work, and then months or years later a team of amazingly creative people brought it all to life, and it went out into the world and touched an audience. That's an exhilarating feeling.

What are the types of challenges that you face in your position? Writers are considered easily expendable, so there is a lot of pressure to deliver a perfect project in order to stay hired for multiple drafts and keep writing credit (which can determine residual payments). If you're not a person that can listen to criticism and execute notes, then you can face a lot of challenges finding work or staying employed on a project. Producers want writers to listen, understand, and be flexible. But there are numerous ways to get fired, from just not hitting the mark, to the producers wanting a different "feel," to the producer wanting to hire a buddy as a favor. Another major challenge is that screenwriting can pay big money, but the big jobs only come along once in a while. In features, you can make seven figures one year, and then not work for the next three or four years. Television is a little more stable, if you are "staffed" on a show, but if the show gets cancelled it's a similar problem. There are a lot of writers vying for the same jobs, and only "A-list" writers get hired on the top projects. You have to live cautiously and know how to make money last over long periods of time.

How do you prepare for a production? All of my work comes before production. I'm usually off writing something else by the time my script goes into production, but if it's a smaller project I'm sometimes brought back for production rewrites for specific locations they've found or actor challenges they might face. To prepare for writing in general, I studied a lot of story structure and character theory, and watched a lot of movies and TV shows. For specific projects, I'll do heavy research into the locations, the characters' careers, or any other unique element of the story. I'll also watch a lot of other films in the genre—especially any that the producer wants to use as a template for tone, structure, etc.

What suggestions or advice do you have for someone interested in a position such as yours? Read everything you can about writing for film and television, and then read the scripts of movies or shows you love. You can find a lot of scripts online now, if you search, and it's very educational to see how it went from the page to the screen. After that, write, write, write. And show people what you are writing. All of the jobs I've landed came from connections I made and from showing my scripts around. It takes several scripts before you really get the hang of formatting, screenplay style, structure, and character. I know professionals in the business who won't read a new writer's work unless they've written three (and even up to 10!) screenplays already. Write, write, write.

Andrea Gyertson Nasfell is a partner at Sodium Entertainment in Los Angeles.

INTERVIEW WITH A PROFESSIONAL: HAKEM DERMISH

FIGURE 5.15
Hakem Dermish, sports reporter/anchor.

Briefly define your job. I work as an anchor on ESPN's SportsCenter.

What do you like about your job? I love having the opportunity to entertain people with great video and stories. My greatest passion is telling stories of people who have overcome great adversity. It's a privilege to have the opportunity to share stories of inspiration.

What are the types of challenges that you face in your position? Challenging myself to always write creative fresh stuff that engages and informs our viewers. Managing time is crucial.

How do you prepare for a production? Including watching the games that night, I research, research, and research. I always look for that interesting nugget or fact that makes the story stronger. The exciting part is using that bit of information and crafting a lead to set up the highlights. Another great thing is the opportunity to learn something new every day, whether it be about a specific team or player. Being prepared is critical because it proves your credibility as a journalist. I like to think that because I'm prepared, our viewers trust me and the information I provide.

What suggestions or advice do you have for someone interested in a position such as yours? Work hard. It's said all the time, but you really have to work hard to obtain a position in television. There's so much competition, so while you're in school, go the extra mile, do the extra work, challenge yourself. If your professors don't have a particular curriculum that allows you to do something you'd like, ask them if you can go "outside the box" and you might surprise yourself with the end result. Be focused when you're in school; it's a time to have fun, but really turn your attention on your future. Be aggressive. Go after things; don't hesitate because you think it's too difficult. I enjoy my job because I get paid to talk about sports every day, which is very satisfying for a sports junkie such as myself; find something you're really passionate about, sports, weather, or news. As a senior in college, get a portfolio together and begin sending it out in the spring before graduation. Waiting until after you graduate just makes it more difficult. It should have a montage of you in different settings (i.e., in studio, in the field, etc.). Follow that with a sportscast/weather cast/newscast, then a story or two that shows your writing. Also, get an internship; that's really where you'll learn the most about the business. Observing what goes on and having the chance to get your hands on a camera or editing equipment will help develop your skills. Having the ability to do several "things" will help you go further.

Hakem Dermish is an anchor on ESPN's SportsCenter.

REVIEW QUESTIONS

1. What does "running order" refer to when shooting a script?
2. What are the different types of scripts and how are they used in a production?
3. Why is a show format used during a production?
4. When is it necessary to fully script a production?
5. How is a storyboard used in a production?

CHAPTER 6

WHAT THE CAMERA CAN DO

"While cameras with lots of features are nice, it is not the camera that creates incredible images—it is what you do with the camera that is important."

—Jeff Hutchens, Photographer

TERMS

3D: Images that appear to have three dimensions, height, width, and depth.

Convertible camera: Can be used in a variety of configurations. This type of camera generally starts out as a camera "head." A variety of attachments can then be added onto the camera head, including different kinds of lenses, viewfinders, recorders, and so on, to suit a specific production requirement.

EFP (electronic field production) camcorders: Used for non-news productions such as program inserts, documentaries, magazine features, and commercials. Can also be used for a multi-camera production.

ENG (electronic newsgathering) camcorders: Generally used for newsgathering. Often these cameras are equipped with a microphone and camera light, and are used to shoot interviews and breaking news.

Depth of field: The distance between the nearest and farthest objects in focus.

Dolly shot: The action of moving the whole camera and mount slowly toward or away from the subject.

DSLR: Digital single lens reflex still camera with video capabilities.

Focal length: An optical measurement—the distance between the optical center of the lens and the image sensor, when you are focused at a great distance such as infinity. Measured in millimeters (mm) or inches.

Focus puller: The person responsible for keeping the camera in focus when using a follow-focus device.

Follow-focus device: A device that is attached to a film-style camera's lens, allowing a focus puller to adjust the focus on the camera.

POV (point-of-view) camera: These small, sometimes robotic cameras can be placed in positions that give the audience a unique viewpoint.

Prime lens (primary lens): A fixed coverage, field of view, or focal length.

Telephoto lens: A narrow-angle lens that is used to give a magnified view of the scene, making it appear closer. The lens magnifies the scene.

Wide-angle lens: A lens that shows us a greater area of the scene than normally seen. The subject looks unusually distant.

Video/television cameras have become increasingly user-friendly over the years. Various controls that previously needed watchful readjustments can now be left to circuitry that tweaks them automatically. On the face of it, there may seem to be little point in getting involved with the technicalities of its controls. It looks as if you only need to point the camera and zoom in and out to adjust the size of the shot—leaving circuitry to take care of all other issues such as exposure, light quality, and so on. Why do we need to learn about lens apertures, depth of field, lens angles, and other details?

Frankly, it really depends on how critical you are of the results. When you want successive pictures to match in brightness, contrast, and color quality, with consistent perspective and controlled focusing, you need to understand the effects of the various camera controls. Although automatic compensatory circuits can certainly ease pressures when working under difficult conditions and are valuable fallback devices that can be helpful, they do have their limitations. Good camerawork involves making subtle artistic judgments, and new technology simply can't do that for you.

Knowing about the various camera adjustments and the effects that they can have on your images will not only enable you to make the best judgments, but will also prepare you for problems that arise under everyday conditions.

TELEVISION CAMERAS

A wide range of television/video cameras are available today, from modestly priced designs for consumers (even smartphone cameras) to very sophisticated state-of-the-art professional cameras. The market spread of models suits a

FIGURE 6.1 *Cameras have been designed for a wide variety of users and situations.*
Source: Photos courtesy of Sony, Panasonic, and Thomson.

THE VIDEO SIGNAL PATH THROUGH THE CAMERA

The Video Signal

The image from the lens is focused onto an image sensor within the video camera. A pattern of electrical charges forms on the light-sensitive area of the sensor, which corresponds in strength at each point to light and shade in the lens image. As special scanning circuits systematically read across this charge pattern in a series of parallel lines, a varying signal voltage or video is produced relating to the original picture tones. After amplification and electronic corrections, the video signal from the camera (with added synchronizing pulses) can be distributed.

Looking at the apparently endless range of colors and shades in the world around us, it seems incredible that they can be reproduced on a television screen simply by mixing appropriate proportions of red, green, and blue light. Yet, that is the underlying principle of color television.

The sensors in the television camera can only respond to variations in brightness. They cannot directly detect differences in color. However, if a color filter is placed in front of the light sensor in the camera, the video signal that it produces will then correspond to the proportions of that color in the scene.

Some camera systems use a multicolored striped filter fitted over a single light sensor. These single-chip cameras have grown in quality and popularity. However, most professional cameras, which are required to provide more accurate color and detail, use three separate sensors. In these, the image from the lens passes through a special prism block with dichroic filters (Figure 6.2). These filters produce three color-filtered images corresponding to the red, green, and blue proportions in the scene. The video signals from the respective sensors correspond to the three primaries needed to reproduce a picture in full color.

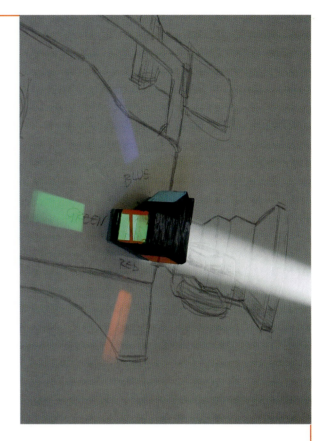

FIGURE 6.2 *A prism block with dichroic filters produce the three colors (red, green, and blue) that are needed to reproduce an image in full color.*

Light and Shade

Although the sensors and filters provide the full range of colors for the television image, what about its light and shade and its brightness variations? If the red, green, and blue are added together, they equal white. If they are all equal but very weak, they will result in a very weak white, or what we would call a dark gray. If the signals are stronger, the area on the screen will be brighter. If most parts of the screen are energized by strong video signals, we interpret this as a bright picture, and vice versa.

To summarize: the effective color (hue) results from the actual proportions of red, green, and blue, and the brightness of each part of the screen depends on the overall strength of the mixture.

variety of applications; as you would expect, both design and performance vary with cost. Although cameras at the lower end of the range can provide very satisfactory image quality under optimum conditions, the more advanced equipment designs produce consistently excellent pictures for long periods, even in difficult circumstances (Figure 6.1).

A number of factors can help you determine which camera is best for your project:

- Cost: initial and running costs.
- Physical aspects: weight, portability, method of mounting, and reliability.
- Operational features: available options, controls, handling, and flexibility (e.g., zoom range).
- Image performance: resolution, color quality, any picture impairment (artifacts), performance stability, and sensitivity.

Determining the features that are most important to you depends on how you are going to use the camera. Are you shooting live or recording? Are you working with a single-camera setup or as part of a multi-camera production? Are you editing what you shoot as you go or will the material be edited later? Some camera systems are more appropriate than others for specific situations.

One factor can strongly outweigh others. If, for instance, a news unit is working under hazardous conditions where the likelihood of equipment loss or damage is high, it may be wiser to use a small, low-cost consumer/small-format camcorder rather than a larger, expensive camera!

TYPES OF CAMERAS

Historically, there have been many different types of cameras, generally sorted into one of three categories: consumer, prosumer/industrial, and professional. Generally, the categories were assigned according to the level of quality and the features each camera provided. Of course, there are still some high-end professional cameras and some low-resolution/low-feature consumer cameras that are aimed at specific markets. However, the quality of today's cameras has blurred the lines between these three categories, sometimes allowing lower-end/lower-cost cameras to be used in high-end professional productions. Some of the different styles of cameras are discussed in the following sections.

Camcorders

Camcorders are the main type of camera used today in television productions. As high-quality camcorders have become available at lower prices, these have been adopted by many organizations throughout the world, for both studio and location production (Figures 6.3–6.7). Although camcorders can record their own signal, depending on the model, they can also be used in multi-camera productions. These cameras have proven to be not only convenient, but adaptable, with lower operating costs than many other cameras.

Studio Cameras

Studio productions make use of a wide range of camera designs, from handheld cameras in a studio configuration to the large traditional studio cameras (see top camera in Figure 6.1, and Figures 6.10 and 6.11). However, this type of camera is not limited to the studio; you will also see it at sports events and other multi-camera productions. The large "studio" viewfinder is usually intended to make it easier for the camera operator to accurately focus and compose the images.

Studio cameras are usually mounted on heavy-duty wheeled dollies—pedestals or rolling tripods—on a pan-and-tilt head, which enables it to turn (pan) and tilt (Figure 6.12). Its focus and zoom controls are usually fixed to the panning handles (pan handles), which position the head.

Miniature or Point-of-View Cameras

Miniature or POV (point-of-view) camera use has grown considerably over the last few years. Originally, they were used only by extreme sports types of programming. However, they are now regularly being utilized by a wide variety of productions. Part of the reason that they have been so widely accepted for use is their relatively low cost. This allows consumers to use them to shoot family videos and professionals to use them in extreme situations, knowing that they can be easily replaced. Even studio productions are utilizing them at times.

POV cameras are used by directors who need to get a camera into a location that would be extremely difficult for a normal camera. These difficulties may be size (the camera may not fit into a tight location), safety (it may not be safe to have a camera and operator in a dangerous location) or visibility (the director may not want the camera to be seen). For multi-camera professional productions, these cameras are generally remotely controlled by a camera operator. POV

FIGURE 6.3 *With their high-resolution sensors and HD capabilities, cell phones are becoming increasingly popular as video cameras. The iPhone 6 shown above is attached to a grip that includes tripod/light mounts, a professional mic and an add-on wide-angle lens. One of the main advantages of these phones is that the video can be edited on the phone and then transmitted back to a news station, client, or directly onto a website, right from the phone.*

Source: Photo courtesy of ALM Media.

FIGURE 6.5 *High-end professional cameras are used in situations where absolute control is needed over every aspect of the image.*

Source: Photo courtesy of Panasonic.

FIGURE 6.4 *High-quality wide-angle and telephoto lenses can be easily attached to smartphones, making it easier to capture professional videos.*

Source: Photo courtesy of Olloclip.

FIGURE 6.6 *Professional 3D cameras require two lenses.*
Source: Photos courtesy of Panasonic.

FIGURE 6.7 *Digital cinema 4K video cameras have grown in popularity.*
Source: Photo courtesy of Canon.

CASE STUDY: GUIDING LIGHT

Guiding Light, the longest-running scripted television series in broadcast history, strategically enhanced its look in 2008 by becoming the first series of its kind to be filmed exclusively with small, handheld, highly portable digital camcorders, which enabled the show to be shot inside actual homes and offices, or on location practically anywhere. The camcorder chosen for this new production model was a Canon prosumer camera.

The key to *Guiding Light*'s new production model of exclusively using compact, handheld Canon camcorders is what Janet Morrison, the producer of the show's digital department, described as "Four walls and a ceiling. Its purpose is to make the show more intimate for the viewer and to really bring them into Springfield, so they can be a part of these characters' lives in a way they haven't been before." Springfield,

FIGURE 6.8 *The camera used to shoot Guiding Light.*
Source: Photo courtesy of Canon.

Guiding Light's fictional locale, is portrayed by a suburban New Jersey town several miles west of the series' Manhattan studios. "In our old production model [using pedestal-style cameras to shoot actors performing in traditional three-sided sets], our two studios in Manhattan had room for seven sets at a time," Morrison added. Now, however, using these compact camcorders, producers of *Guiding Light* can shoot in as many locations as they wish. "This has opened up our 'canvas' in ways we weren't able to imagine before. Our writers have so many more places where scenes can happen. They can write people in the park, or at the municipal building, or using cars that actually drive, as opposed to cars that just sit on a studio floor. This new production model has completely changed the way the show looks and the way stories can be told" (Figure 6.8).

—Jennie Taylor, *Highdef* magazine

cameras have high-quality images (up to 4K), are inexpensive, have extreme wide-angle perspectives, are rugged, and are so small that they can fit almost anywhere (Figures 6.10–6.13).

One of the unique things about the POV camera is that it can often provide the audience with an angle of view that is not seen with the human eyes. It provides the director with the opportunity to show the audience something they cannot see without you, to experience something that only you can provide. This means that you place the camera in very unique positions, creating a very unique shot. GoPro, one of the leading POV camera manufacturers, has stated that its mission is "to help people capture meaningful experiences in an engaging immersive manner, allowing them to relive that experience."

One of the latest improvements to some of the POV cameras is the ability to transmit images using Wi-Fi. This allows the user to adjust the camera and to see what is being shot by remote control.

CAMERA BASICS

Let's now take a closer look at the various features found on cameras (Figures 6.18–6.21) and how they can affect the way you use them:

- The camera's viewfinder.
- The camera's main controls.
- The camera lens and how it behaves.
- The techniques of adjusting exposure for the best picture quality.
- Methods of supporting the camera.

The Viewfinder

The viewfinder enables you to select, frame, and adjust the shot, to compose the picture, and to assess focus adjustment. When you are working alone, with a portable camera, the viewfinder will usually be your principal guide to picture quality and exposure, as well as providing continual reminders

THE CONVERTIBLE CAMERA

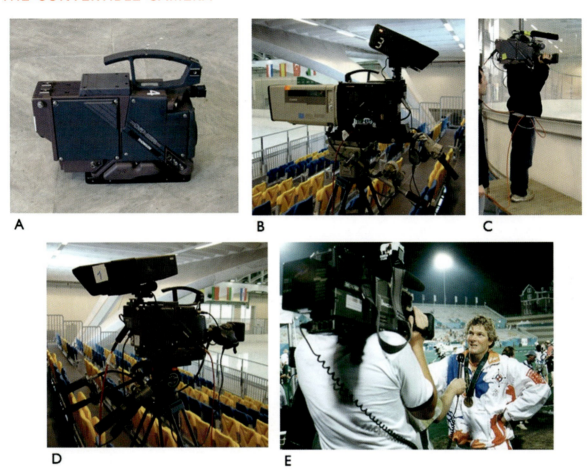

FIGURE 6.9 *Convertible cameras provide maximum flexibility for the camera crew, allowing them to configure the camera for the specific needs of the production.*

Convertible cameras can be used in a variety of configurations. This type of camera generally starts out as a camera "head." The camera head is the part of the camera that creates the image—primarily the image sensors. Figure 6.9A shows a camera head with a studio back attached, which allows it to be used in a multi-camera situation. A variety of attachments can then be added onto the camera head, including different kinds of lenses, viewfinders, and recorders, to suit a specific production requirement. Here are a few examples:

- *Studio, stationary, or hard camera*: This configuration utilizes a long lens, a studio back, a large studio/EFP viewfinder, and remote controls for the zoom and focus. Due to the weight of the large lens, this camera is fairly stationary unless also attached to a dolly (Figure 6.9B).
- *Handheld camera*: When maximum mobility is required, a shoulder-supported handheld camera can be used. This camera utilizes a small eyepiece viewfinder, the EFP/ENG lens, and no remote controls for the zoom and focus (Figure 6.9C). A convertible handheld camcorder can also be created by replacing the studio back with a recorder (Figure 6.9D).
- *EFP camera*: The electronic field production camera can be set up in a number of different ways. In Figure 6.9E, the camera is configured as part of a multi-camera production with an EFP lens, zoom controls, and the large viewfinder. This type of camera is lighter to transport than the studio/stationary camera. However, the EFP's lens is not as long.

FIGURE 6.10 *This studio configuration utilizes a small camcorder with studio back, large studio monitor, remote controls, and an ENG lens.*

Source: Photo courtesy of JVC.

FIGURE 6.11 *These lower-cost lightweight HD-4K studio cameras have no recording ability. They are niche cameras designed specifically for multi-camera production.*

Source: Photo courtesy of Blackmagic Design.

FIGURE 6.13 *This POV camera was mounted onto a referee's helmet in order to give viewers a chance to see what the referee was seeing.*

FIGURE 6.12 *Wheeled dollies are often used with studio cameras.*

FIGURE 6.14 *Camera operators can use a POV camera to get into small areas. The photo on the left shows an operator using a POV camera attached to a pole to get a shot in a bull pen at a rodeo. The photo on the right shows a small POV camera on a stand to capture dramatic shots of a drummer.*

Source: Photos by Ben Miller and Will Adams.

FIGURE 6.15 *The Discovery Channel's Mike Rowe uses a small POV camera taped to a light to capture tight areas in a cave. In this situation, Mike is both the host and one of the camera operators for the show Dirty Jobs.*

Source: Photo by Douglas Glover.

FIGURE 6.16 *Traditionally thought of as consumer cameras, small POV cameras, such as the GoPro shown, are now being used by television networks, as well as documentaries, for specialty shots. These cameras have gained popularity as a camera that can be connected to anything. For a relatively low cost, this high-quality little camera comes with a wide-angle lens and a waterproof case. The photo on the right was shot with this camera.*

THE DSLR (PROS AND CONS)

Digital single lens reflex cameras (DSLRs) have become a formidable force in the television world today (Figure 6.17). While originally they were designed as a still camera that had some video capability, they have also become the low-budget filmmaker's camera of choice due to their high quality and lower cost. Directors are using them to shoot corporate videos, commercials, television network shows, and even a few feature films. Some camera models' high megapixel sensors and 1080p quality, combined with their small size and low cost, make them a cost-effective option in video production. However, there are positives and negatives about the current class of DSLR cameras. While they are continually improving, the current models are good for some projects and not good for other projects. Here are some of the pros and cons:

FIGURE 6.17 *The DSLR has become a popular choice as a video camera.*

Source: Photo courtesy of Canon.

Advantages

- *Low cost*: When compared to video cameras with similar image quality, the DSLR is very cost-effective.
- *Depth of field*: The large sensor size can provide a very shallow depth of field.
- *Weight*: Its light weight makes it easier to move around.
- *Low profile*: When shooting documentaries, the camera is less obvious.
- *Low light*: DSLRs can shoot in very low light situations and still maintain their quality.

Disadvantages

- *Recording time*: Some of the current DSLR models have around a 12-minute maximum continuous recording time. While the camera can be immediately restarted, this does create some recording limitations.
- *Audio*: DSLRs usually have automatic gain control, unbalanced inputs, and no phantom power. All of these mean that if you want high-quality audio, you have to record your audio on a separate audio device and then sync it later. Another audio disadvantage is that most DSLRs do not include a headphone input to allow you to monitor the audio quality, and some of the ones that do include a headphone jack don't allow monitoring in real time. There are a number of software options available that assist in syncing the audio.

- *Stability*: Due to their small size, DSLRs are handheld, not shoulder-mounted. This is not a great design for stability while recording motion. Most camera operators use a shoulder-mounted support when handholding them.
- *Time code*: The lack of time code can create some problems when attempting to sync audio in the postproduction process.
- *Quality*: In order to obtain the 1080 image from their sensor, some models use a type of line skipping when capturing the image. While this type of down-resing reduces the amount of processing for the camera, it can cause some serious aliasing issues, which causes problems when shooting highly detailed patterns.

Other DSLR issues

- The terms used on a still camera are not the same as the terms on a video or digital cinema camera. For example, still cameras use "ISO" while video cameras use "gain."
- There are few video controls on a DSLR compared to a prosumer or professional video camera. This limits the camera operator's ability to make fine adjustments to the image.
- The DSLR does not always look professional if you are working for a client. Compared to a typical-size professional camera, it may look amateurish.

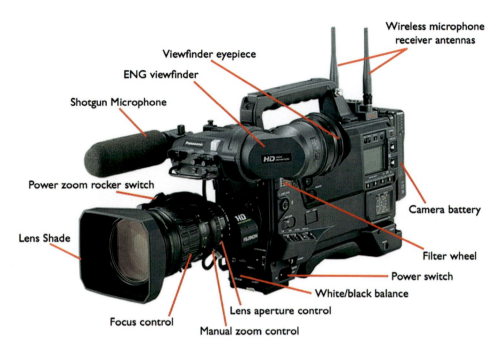

Wireless microphone receiver antennas

Viewfinder eyepiece

ENG viewfinder

Shotgun Microphone

Power zoom rocker switch

Lens Shade

Camera battery

Filter wheel

Power switch

White/black balance

Focus control

Lens aperture control

Manual zoom control

FIGURE 6.18 *Parts of an ENG/EFP camera.*

Source: Photo courtesy of Panasonic.

Tally light

Zoom lens

Studio viewfinder

Focus control

Camera cable

Zoom control

Camera release plate

Tripod or pan head

Pan handle

FIGURE 6.19 *Parts of a studio-configured camera.*

Source: Photo courtesy of JVC.

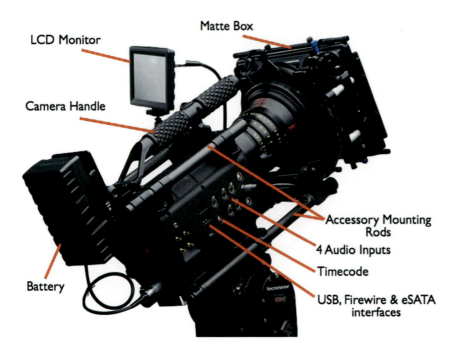

FIGURE 6.20
Parts of a high-quality (4K) HD camera.
Source: Photo courtesy of Red.

LCD Monitor

Matte Box

Camera Handle

Battery

Accessory Mounting Rods

4 Audio Inputs

Timecode

USB, Firewire & eSATA interfaces

Production monitor/viewfinder

Film-style viewfinder

Focus puller

Matte box

FIGURE 6.21 *A film-style video camera configuration may include two viewfinders. The film-style viewfinder is used by the camera operator. The production monitor or viewfinder can be used by the operator or by the focus puller. The focus puller is a person who uses a follow-focus device to keep the image sharply focused. These camera configurations generally also include a matte box used to hold filters and shield the lens from light glares.*

Source: Photo courtesy of Panasonic.

about recording, the battery's condition, and other relevant factors.

Some camera systems show just a little more than the actual shot being transmitted so that the camera operaor can see whether there is anything just outside the frame that might inadvertently come into the picture. Cameras equipped with this type of viewfinder allow the camera operator to see if an unwanted subject (e.g., a microphone, a bystander) is about to intrude into the shot.

Because the viewfinder is a monitoring device, any adjustments made to its brightness, contrast, sharpness, or switching will not affect the camera's video output. The viewfinder on a camcorder not only shows the video image being shot, but it can also be switched so that the camera operator sees a replay of the newly shot images. Camera operators can check for any faults in camerawork, performance, or continuity and can then reshoot the sequence if necessary.

Viewfinders often have a number of indicators that aid in framing the image, such as a safe area line around the viewfinder's edges to remind the camera operator how important information and titles near picture edges can be lost inadvertently through edge cutoff on overscanned TV receiver screens. Other viewfinders may keep the camera operator informed about the camera or recorder's settings and status. These may include light or audio meters, shutter speeds, a tally light showing when the camera is recording or "on-air," zoom lens settings, "zebra" (refers to a camera feature that displays all overexposed areas of the image with diagonal stripes warning the operator to reduce the exposure), battery status, and other displays. Various indicators keep you informed about the camera and recorder's settings and status. Some monitors even place a red line around all sharply focused subjects.

The Camera's Controls

Television cameras have three different categories of controls:

- Those that need to be continually readjusted while shooting, such as focus.
- Occasional adjustments such as compensating for changing light.
- Those involved in aligning the camera's electronics in order to obtain optimum consistent performance.

Once these are set up, they should not be casually tweaked! There are really two quite distinct aspects to picture making: camerawork and image quality control. These can be

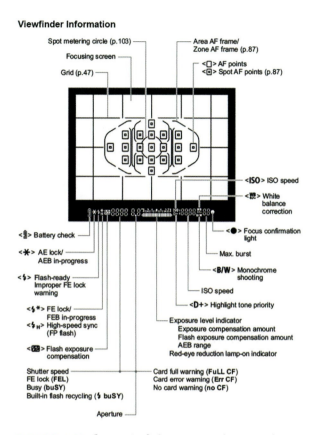

FIGURE 6.22 *Camera viewfinders can assist the operator/ photographer by providing helpful information. This diagram shows the information available in a Canon DSLR's viewfinder. Each manufacturer's viewfinder is different.*
Source: Photo courtesy of Canon.

controlled manually, semi-automatically, or even completely automatically. Let's look at some practical cases.

Multi-Cameras

Camera operators who are part of a multi-camera production concentrate on the subtleties of camerawork. They spend their time selecting and composing the shots, selectively focusing, zooming, and controlling camera movements. The quality of the video image is remotely controlled by a video operator utilizing a camera control unit (CCU) or remote control unit (RCU), as shown in Figure 6.24. There are basically two types of camera adjustments made by the video operator:

- *Preset adjustments*: During the setup period, camera circuitry is adjusted to ensure optimum image quality. This may include color and tonal fidelity, definition, and shading (adjusting the aperture). These adjustments can be done

VIDEO CAMERA TERMINOLOGY

White balance: The most common technique used to color balance a camera. It is the process of calibrating a camera so that the light source will be reproduced accurately as white. In order to balance the color, the camera is aimed at a white subject, usually a white card lit by the scene's light source, and the white balance button is pushed, adjusting the color (Figure 6.23). By capping the camera or closing the iris, *black balance* can also be used to provide a reference for black after white balancing.

Black level: The intensity of black in the video image. If set incorrectly, picture detail quality will be poor in dimly lit areas. Lowering the picture's black level moves all image tones toward black and crushes the lowest tones. Raising the black level lightens picture tones but does not reveal detail in the blackest tones.

Lens aperture: The opening in the lens (iris) that lets light into the camera. It allows the operator to darken or brighten the video image, depending on the light in the scene.

Gamma: Adjusts the tone and contrast of a video image. High gamma settings produce a coarser, exaggerated

FIGURE 6.23 *Cameras are zoomed in on the white card, allowing accurate white balance.*
Source: Photo by Luke Wertz.

contrast; lower settings result in thin and reduced tonal contrast.

Video gain: The amplification of the camera's video signal, usually resulting in some video noise. Boosting the gain allows the camera to shoot in lower light situations than normal. However, it also provides a lower-quality image.

manually using test signals displayed on test equipment, or, in digital systems, the process can be semi-automated by inserting programmable memory cards.

- *Dynamic adjustments*: During the actual television production, the video operators, often known as "shaders," continually adjust the camera's images (in terms of exposure, black level, gamma, color balance) to optimize the subjective quality of shots and match the various cameras' images. This approach not only leaves the camera operator free to concentrate on effective camerawork, but results in the highest and most consistent image quality.

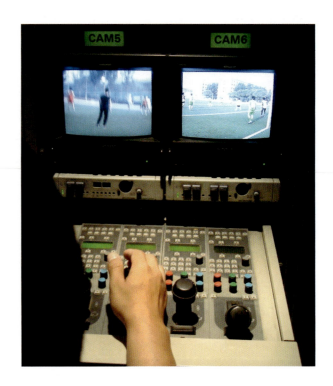

FIGURE 6.24 *A video operator uses a CCU to remotely adjust a camera in a multi-camera production.*

Single Cameras

Single-camera operators are not only required to ensure that they obtain well-composed images, but must also adjust the camera to obtain the highest-quality image possible. There are a number of options for adjusting the camera:

- The camera's built-in automatic circuitry can be used to readjust the camera's performance as conditions change. Although this is a good option when in a hurry to capture the images, it gives the least amount of quality control to the camera operator—and may not provide the optimum quality in every situation.
- Preview the scene before shooting and adjust the controls until pictures appear optimum in the viewfinder or a nearby monitor.
- Readjust the camera controls while shooting.

THE CAMERA LENS

"Never make the mistake of thinking that a lens is a mere accessory to the camera—especially in HD imaging. The HD lens has the primary role in shaping the image and giving it critical performance attributes that, in turn, will determine the ultimate digital performance of your camera."

—Larry Thorpe, Canon USA

Lens Angle

Lens angles have a distinct impact on the final image. Each lens focal length has its own type of distortion. You need to understand the types of lenses to know what lens is best for your specific situation.

The term *focal length* is simply an optical measurement—the distance between the optical center of the lens and the image sensor when you are focused at a great distance, such as infinity. It is generally measured in millimeters (mm).

A lens designed to have a *long focal length* is called a *telephoto lens* or a *narrow-angle lens*. The subject appears much closer than normal. When the lens has a short focal length, usually called a *wide-angle lens*, it takes in much more of the scene. However, subjects will look much farther away.

How much of the scene and subject the lens shows will depend on several areas:

- The size of the subject itself.
- How far the camera/lens is from the subject.
- The focal length of the lens being used (wide angle, normal, or telephoto lens).
- The size of the camera's image sensor.

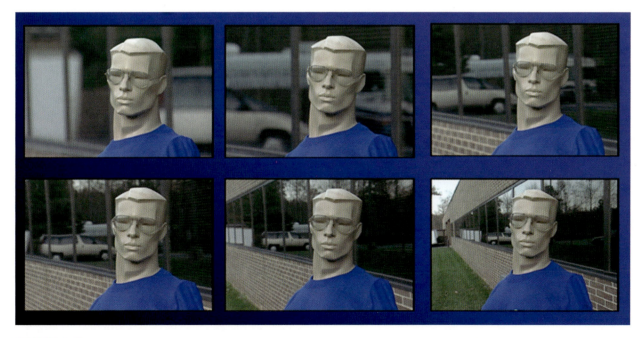

FIGURE 6.25 *The background changes in size are due to the lens used. Note that the first photo was shot with a telephoto lens and the final shot was taken with a wide-angle lens. The camera had to be moved closer to the subject for each shot, as a wider lens was attached so that the subject would stay the same approximate size.*

Source: Photos by K. Brown.

The focal length of your lens is not a mere technicality. A lens's focal lengths will affect:

- How much of the scene your shot shows.
- The apparent proportions, sizes, and distances of everything in the shot (perspective).
- How much of the scene is in focus and how hard it is to focus accurately.
- Camera handling, such as how difficult it is to hold the shot steady.

There are three broad types of camera lens angles: normal, telephoto (narrow angle), and wide angle (Figure 6.25).

Normal Lens

A normal lens gives a viewpoint that is very close to what is seen by the human eye. This focal length lens is used more than any other lens since it has the least amount of distortion.

Telephoto Lens

A telephoto lens is designed to have a long focal length. The subject appears much closer than normal, but you can see only a smaller part of the scene. Depth and distance can look unnaturally compressed in the shot (Figures 6.26 and 6.27).

Using a telephoto lens angle has some major advantages by allowing you to get closer to the subject:

- When you cannot get the camera near the subject because of obstructions.
- When there is insufficient time to move the camera closer to the subject.
- When the camera is fixed in position.
- When the camera operator is unable to move around unobtrusively.

There are often occasions, both in the studio and in the field, when the only way that you can hope to get an effective shot of the subject is by using a telephoto lens.

However, the telephoto lens also has some drawbacks (Figure 6.28):

- Depth can appear unnaturally compressed—foreground-to-background distance seems much shorter than it really is. Solid objects look shallower and "depth-squashed."
- Distant subjects look much closer and larger than normal. Things do not seem to diminish with distance as we would expect.
- Anyone moving toward or away from us seems to take an interminable time to cover the ground, even when running

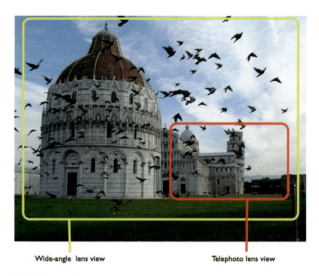

Wide-angle lens view   Telephoto lens view

FIGURE 6.26 *The angle of the lens determines what is captured in the camera.*

FIGURE 6.27 *The red lines show a wide-angle lens while the blue lines show a narrow-angle/telephoto lens.*

Source: Camera photo courtesy of JVC.

Total Scene

FIGURE 6.28
Lens angles distort the perspective of the scene. Wide-angle lenses expand the subjects and telephoto lenses compress the subjects.

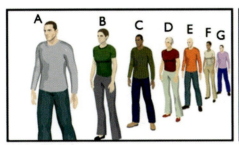

Wide-Angle Lens

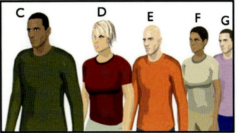

Normal Lens

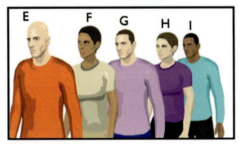

Telephoto Lens

fast. At the racetrack, horses gallop toward us, but appear to cover little ground despite their efforts.

- When the camera uses a telephoto lens to shoot a subject head-on lengthwise, the subject will look much shorter than it really is. For example, a large oil tanker may look a few yards long.

If a long telephoto lens is utilized to capture close-up shots of people, you will notice that their features are unpleasantly flattened and facial modeling is generally reduced. Shot from the front (full face) or when slightly angled (three-quarters face), this effect can be very obvious.

Camera handling becomes increasingly more difficult as the telephoto lens focal length is increased. Even slight camera movement, caused by a shaky handheld or dollying over an uneven floor, causes image jerkiness. It is no easy feat to follow a fast-moving subject, such as a flying bird, in a close shot with a telephoto lens and keep it accurately framed throughout its maneuvers! Apart from handling difficulties, there is the additional focusing hazard of working with a shallow depth of field!

When using a telephoto lens, it is difficult to hold the camera still for any length of time without shake from one's breathing, heartbeat, or tiring muscles. It is very difficult to handhold a camera with a long telephoto lens and get steady shots. Telephoto lenses usually require the camera operator to use a tripod or other camera support to keep the image steady.

Wide-Angle Lens

A wide-angle lens has a short focal length that takes in correspondingly more of the scene. However, subjects will look much further away and the depth and distance appear exaggerated. The exact coverage of any lens will depend on its focal length relative to the size of a camera's image sensors. As you might expect, wide-angle lenses have the opposite characteristics of telephoto lenses.

Wide-Angle Lens Distortion

The wide-angle lens exaggerates perspective. Depth and space are overemphasized. Subjects in the image seem further away than they really are (Figures 6.29 and 6.30). Anyone approaching or walking away from a camera with a wide-angle lens seems to be moving much more quickly than normal.

This distortion can be used to your advantage. A wide-angle lens can give a wide overall view of the scene, even when it is relatively close. This can be a great help when

FIGURE 6.29 *This image was shot with a very wide-angle lens. Note the distortion that takes place when shooting with a wide angle-lens toward a large vertical subject.*

FIGURE 6.30 *Although super-wide-angle shots can give a good view of a very confined area, faces can become incredibly distorted.*
Source: Photo by Guillaume Dargaud.

TABLE 6.1 *Lens Characteristics*

	Advantages	Disadvantages
Normal lens	• Perspective appears natural. Space/distance appears consistent with what the eye sees. • Camera handling feels natural and is relatively stable. • Focused depth is generally satisfactory, even though it can be fairly shallow for close-up shots.	• May not provide the image needed when shooting in confined spaces. • Camera must move closer to the subject in order to capture detail; may be seen by other cameras taking longer shots of the same subject.
Telephoto lens	• The field of view is restricted. • Brings distant subjects closer. • Permits "closer" shots of inaccessible subjects. • Compresses objects into a more cohesive group, such as a line of cars. • Defocuses distracting background. • Camera handling and focusing become more difficult.	• The compressed image may be distracting. • Can flatten the subject's modeling, which can appear unnatural. • Camera handling is much more difficult (pan, tilt, smooth dollying) due to image magnification. • The restricted depth of field makes focusing highly critical.
Wide-angle lens	• Can obtain wide views at closer distances, especially in confined space. • Lens reduces the "jerkiness" of a handheld shot. • Provides considerable depth of field. • The extended depth of field makes focusing less critical. • Space, distance, and depth of field are exaggerated.	• Depth appears exaggerated. • Produces geometric distortion (such as buildings) at the widest lens angles. • Susceptible to lens flares.

shooting in restricted spaces. Even a small room can look quite spacious when shot on a wide-angle lens—a very useful deception when you want to emphasize space, or make small settings appear more impressive.

There are times when the wide-angle lens distortion can be an embarrassment. If, for instance, you are shooting within a very confined area such as a cell, only a wide-angle lens will give you a general view of the location (any other lens angle would provide only close shots). But at the same time, the wide-angle lens exaggerates perspective and makes the surroundings seem very spacious. In these circumstances, the only alternative is to use a normal lens angle, and move outside the room, shooting in through a window or door.

If you take close shots of people through wide-angle lenses, you will get extremely unreal distortions such as an odd-shaped head (Figure 6.30).

Never use wide-angle lenses for close shots of geometrical subjects such as pages of print, graph paper, or sheet music. The geometrical distortion can become very noticeable. When panning a shot containing straight vertical or horizontal lines, wide-angle lenses bend the lines near the picture edges. The wider the lens, the more obvious these effects become.

Wide-Angle Camera Handling

On the plus side, camerawork is much easier when using a wide-angle lens. Camera movements appear smoother, and any slight bumps as the camera moves are less likely to cause jerky shots. As an extra bonus, focusing is less critical, as the available depth of field is much greater.

The advantages and disadvantages of the three types of lenses are outlined in Tables 6.1 and 6.2.

Supplementary Lenses

By adding an extra, "supplementary" lens onto the front of a camera lens, you can alter its focal length. These supplementary lenses can be used with many types of lenses and can change the lens so that it becomes more of a telephoto, wide-angle, or close-up lens (Figure 6.31).

Lens Controls

Most lens systems have three separate adjustments that can be made manually or semi-automatically:

- Focus—adjusting the distance at which the image is sharpest.
- F-stop—adjusting a variable iris diaphragm inside the lens system.

TABLE 6.2 *Why Change between Telephoto and Wide-Angle Lenses?*

Why use lenses other than a normal lens?	• When you want to exclude (or include) certain foreground objects, and repositioning the camera or subject would spoil the proportions. • Where a normal lens would not provide the required size or framing without repositioning the camera or subject.
A telephoto lens can be used:	• When shooting subjects that are a long distance away. • When the camera is isolated on a camera platform (tower) or shooting through some foreground subject. • When the camera cannot be moved—stationary camera—or the camera cannot be moved closer.
A wide-angle lens can be used:	• When the normal lens does not provide a wide enough shot of a scene. • To maintain a reasonably close camera position (so talent can read the prompter) yet still provide wider shots.
The perspective of an image can be adjusted as follows:	• By changing the lens from telephoto to wide angle, and changing the camera distance to maintain the same subject size, which alters relative subject/background proportions and effective distances. • Wide-angle lenses increase the depth of field and enhance the spatial impression. • The telephoto lens reduces the spatial impression, compressing the depth in the image. • To increase production flexibility. • When dollying the camera would distract the talent or obscure the action from the audience. • When using only one camera.

FIGURE 6.31 *This supplementary wide-angle lens is screwed onto the front of the camera lens.*

Source: Photo courtesy of VF Gadgets.

- Zoom (if utilizing a zoom lens)—altering the lens focal length to adjust how much of the scene the shot covers.

Generally, how the lens controls are adjusted will affect the following:

- How sharp the detail in the image is (focusing).
- Exactly how much of the image appears sharply defined in the shot (depth of field).
- How much of the scene appears in the shot (focal length/angle of view).
- The impression of distance, space, and size that the picture conveys.
- The overall brightness of the picture, and the clarity of lighter tones and shadows (exposure).

FIGURE 6.32 *A director's viewfinder can be used to determine the camera position and the appropriate lens.*
Source: Photo courtesy of Alan Gordon Enterprises.

LENS CARE

Camera lenses are very sensitive tools. Although we may begin by handling new camera equipment with tender loving care, familiarity can result in casual habits. Experienced camera operators develop a respect for their lens systems. They appreciate that a moment's distraction can cause a great deal of expensive damage to these precision optical devices, especially when working under adverse conditions. Liquid, dust, or grit can in a moment wreck the lens or degrade its performance.

Working Practices

There are several ways of working out the type of shot you will get at various distances:

- *Trial and error*: A "try it and see" approach, in which the camera is moved around to potential positions, changing the lens focal length setting until you get the result you want. This is a very laborious process.
- *Director's viewfinder*: In this approach, the director stands in the planned camera position and checks out the scene through a handheld portable viewfinder. After its zoom lens has been adjusted to provide the required shot size, the corresponding focal length is then read off the viewfinder's scale and the camera's lens is set to this figure (Figures 6.32 and 6.33).
- *Computer calculators*: There are a number of tools available that will calculate lens angles, depth of field, and other measurements. Some, such as the one shown in

Figure 6.34, work on a PDA device, making it very easy to take into the field. These computer-based calculators are especially good for preplanning a shoot.

- *Experience*: When working regularly in certain surroundings, such as a news studio, you soon come to associate various shot sizes with specific camera positions and focal length settings. However, these proportions will be altered if you happen to change to a camera with a different sensor size.

TYPES OF LENSES

Zoom Lens

Most television/video cameras come with a zoom lens, which is a remarkably flexible production tool. A zoom lens system has the great advantage that its focal length is adjustable. You can alter its coverage of the scene simply by turning a ring on the barrel of the lens. This change gives the impression that the camera is nearer or farther from the subject, at the same time modifying the apparent perspective. At any given setting within its range, the zoom lens behaves like a fixed focal length lens, called a prime lens.

The zoom lens's main advantage is that you can select or change between any focal lengths within its range without interrupting the shot. You can zoom out to get an overall view of the action and zoom in to examine details (Figures 6.34 and 6.35). The fact that, in the process, you are playing around with the spatial impressions your picture conveys is usually incidental to the convenience of getting shot variations from a camera viewpoint.

FIGURE 6.33 *This mobile phone app, Artemis, uses the internal video camera to show the various lenses that can be used to shoot a scene. The image can then be saved for future use.*

FIGURE 6.34 (Above right) *This screenshot is from the software program pCAM, which runs on a smartphone. The software helps the director or cameraperson (video, film, and photography) calculate the correct lens, depth of field, and settings.*

Source: Photo by David Eubank.

FIGURE 6.35 *Zooming in progressively fills the screen with a smaller segment of the scene.*

FIGURE 6.36 *Parts of a zoom lens.*
Source: Photo courtesy of Canon.

FIGURE 6.37 *The fingers on the left are positioned on a rocker switch, allowing the camera operator to zoom in or out.*
Source: Photo by Josh Taber.

Zoom Lens Controls

There are basically two methods of adjusting the focal length of a zoom lens:

- *Direct control:* By turning a ring or a lever on the lens barrel. The major advantage of this type of control is that it does not make any noise, unlike motorized systems. On most cameras, you can override the motorized lens servo-switch control and operate the zoom action manually.
- *Servo control:* The lens zoom action is controlled by a small forward/reverse variable-speed electric motor.

On most cameras, a rocker switch forms part of the lens housing. When the camera is handheld or shoulder-mounted, fingers on the right hand can easily operate the two-way rocker switch to adjust the zoom. Pressing W (wide angle) causes the lens to zoom out toward its widest angle; when the T (telephoto) is pressed, the lens zooms in to the longest focal length. By releasing the rocker switch, the zoom can be stopped at any point within the zoom range. Most zoom lenses also have an adjustable zoom speed (Figure 6.36).

On large cameras, such as the larger studio and field cameras, a servo remote control is usually located on one of the tripod panhandles. These controllers provide a smooth, even change, particularly during a very slow "creep-in." The spring-loaded thumb-lever control is turned to the left to zoom out, and to the right to zoom in. Light thumb pressure results in slow zooming, and heavier pressure gives a fast zoom. You soon learn to adjust thumb pressure to control the zoom rate (Figure 6.37).

Zoom Lens Advantages

Zoom lenses have unquestionably added an important dimension to camerawork:

- Zooms provide a wide selection of lens's focal length.
- Camera operators can rapidly select any focal length within the lens's range.
- Zooming can be controlled manually or by a motor driven at various preset speeds.
- They allow the camera operator to slightly adjust the size of the shot to improve composition/framing without having to move the camera nearer or farther away from the subject.
- Zooming is much easier and more precise than dollying (although the final effect may not be as good as a dolly shot).

Zoom Lens Disadvantages

Although a zoom has many advantages, it also has some disadvantages:

- Depth of field, perspective, and camera handling vary as the lens is zoomed in and out.
- If you are continually varying the zoom settings during a rehearsal, it is important to repeat the same angles and camera positions when recording. Suppose, for instance, that you are now further away than previously and you zoom in to compensate. This "corrected" shot will not have the same relative subject proportions. You might now be in the way of another camera or a boom mic in the

studio. To avoid this type of discrepancy, note the details of the lens angle on a card, shot sheet, or script, and mark or tape the camera's floor positions during rehearsal.

- Zooming can easily be misused as a lazy substitute for more effective camera movement. Sometimes a dolly shot would be much more effective than a zoom, but the zoom is easier.

Zooming the Lens versus the Dolly Shot

When needing to move closer or farther away from your subject during a shot, you have two options:

- Use a zoom lens and zoom in or out.
- Dolly, or actually move the physical camera toward or away from your subject.

You may wonder, what's the big deal? Don't they look the same? If you are shooting a flat surface, such as a painting, the effect of zooming is identical with that of dollying—an overall magnification or reduction of the image. However, there is a significant visual difference between a zoom shot and a dolly shot. Although the zoom shot can show the magnification or reduction of the image, a dolly shot shows a different perspective; it emphasizes the dimensional aspects of the scene. In Figure 6.38, you can actually see that the camera

Zoom Lens Shot

1a

1b

1c

1d

Dolly Shot

2a

2b

2c

2d

FIGURE 6.38 *Zooming simply magnifies and reduces the picture. It does not produce the changes in the scene that arise as the dolly is moved through a scene. Increasing the focal length narrows the lens angle, filling the screen with a smaller and smaller section of the shot. The photographs illustrate the difference between the zoom and dolly. Notice that there is not much of a difference between images 1b and 2b. However, as the dolly continues in 2c, it begins to show a different perspective, showing that there is a table between the two chairs on the left. Photo 1d zooms past the table, while the dolly perspective in 2d takes the viewer up to the table and allows him or her to see the top of it.*

moved down the aisle, showing more information to the viewer than is possible with a zoom shot. Zooming is seldom a completely convincing substitute for moving the camera.

Prime Lens

A very popular lens with narrative or film people is a prime (primary) lens. Prime lenses have a fixed focal length, which means that the scene coverage cannot be varied without physically moving the camera and lens. The prime lens is particularly useful when:

- The highest optical quality is necessary.
- Creating a special optical effect such as an extremely wide-angle fisheye lens.
- Shooting in low-light situations. Primes have lower light losses, and thus are able to obtain a quality image in low light.

Because its focal length is fixed, the prime lens will cover only a specific angle of view, according to whether it is designed as a narrow-angle (or telephoto or long) lens, a wide-angle (or short) lens, or a normal lens system. You can only change a prime lens's coverage by adding a supplementary lens. Prime lenses are available with fixed focal lengths ranging from just a few millimeters (for wide-angle work) to telephoto lenses with focal lengths of more than 1,000 mm (Figure 6.39).

FIGURE 6.39 *Prime lenses are used when the highest optical quality is required.*

Source: Photo courtesy of Ziess/BandPro.

Extender Lens

Some zoom lenses include a built-in supplementary lens that is called a 2x extender. The extender can be flipped in or out of use and is available in sizes other than 2x, but that is the most common configuration. The 2x extender doubles the focal length of the lens, allowing the camera operator to zoom in much closer to the subject (Figure 6.40).

There are a few drawbacks to using an extender lens:

- The extender lens cannot be switched in or out when shooting because it disrupts the visual image. It must be changed between shots.
- There can be a substantial loss of light when utilizing an extender, which causes camera operators to adjust their aperture.
- The minimum focusing distance on the zoom lens can substantially change when switching in the extender. A zoom lens usually focuses on closer subjects when not using the extender.

FIGURE 6.40 *A 2x extender doubles the focal length of the zoom.*

Source: Photo courtesy of Canon.

FOCUSING

Why Focus?

When you focus the camera, you are adjusting the lens to produce the sharpest possible image of the subject. Simple photographic cameras do not have a focus control, yet everything in the shot looks reasonably sharp. Why do television cameras need to be continually focused? There are several reasons why fixed-focus systems do not work with television cameras:

AUTOFOCUS PROBLEMS

There are a number of inherent problems that happen with autofocus. Here are some of the most common issues:

- If the subject is not central, autofocus may adjust on whatever happens to be there, leaving your subject defocused.
- Although you may want two subjects at different distances from the camera to appear equally sharp, the autofocus may indiscriminately sharpen on one or neither.
- If you are shooting through a foreground framework of, for example, branches, a grille, netting, or railings, the system will focus on this, rather than on your subject beyond (Figure 6.43). If the subject is behind glass or beneath the surface of a pool, the autofocus system can also be fooled by this.
- If someone or something moves in front of the camera, the system may automatically refocus on it, defocusing your subject. If, for instance, you pan over a landscape, and a foreground bush comes into the shot, the lens may defocus the distant scene and show a well-focused bush.
- When shooting in rain, snow, mist, or fog, autofocusing may not be accurate.
- Subjects that are angled away from the camera may produce false results.

Although these problems may not occur, it is good to be aware of their possibility. When in doubt, switch to manual focusing.

FIGURE 6.42 *Autofocus struggles to focus if shooting through other objects such as this cage or branches, railings, and in similar situations.*

Remote Focus Control   Remote Zoom Control

FIGURE 6.43 *Larger cameras incorporate zoom and focus controls that are located on the tripod's panhandles.*

FIGURE 6.41 *Focusing with a shallow depth of field makes the rose stand out against a defocused background.*

- They do not provide maximum sharpness of the subject.
- Because the lens aperture is reduced (small stop) in order to provide a large depth of field, high light levels are needed for good video images.
- Everything in the shot is equally sharp. That means there is no differential focusing in which subjects can be made to stand out against a defocused background (Figure 6.41).
- The camera cannot focus on close subjects (although a supplementary lens may help).

If needed, a camera can be put into fixed focus by using a lens with as wide an angle as possible and using the smallest possible aperture. This combination will give you the best images your camera can provide in that situation.

Focusing Methods

Two different methods are generally used to focus the camera's lens system:

- *Focus ring*: Lens systems on handheld and lightweight cameras can be focused by turning a ring on the lens barrel, until subject details are as sharp as possible (Figure 6.44).

FIGURE 6.44 *This cameraman is focusing an ENG/EFP camera by adjusting the lens's focus ring with his left hand.*

FIGURE 6.45 *There are a variety of remote focus controls that attach to the panhandle. The type in this photo requires that the operator turn the spoked wheel, often called a capstan control, in order to focus the lens.*

FIGURE 6.46 *In situations where the camera operator needs to concentrate on framing, a focus puller (in the yellow shirt) may be used. This person attaches a follow-focus mechanism to the lens in order to adjust the focus from the side of the camera.*

- *Remote focus control*: When a camera is attached to a tripod (or other mount), panhandles are usually attached to assist the camera operator in panning and tilting the camera. A variety of different focus controls are available to be clamped onto one of the panhandles to adjust the lens focus by remote control. The remote focus control is linked to the lens focusing mechanism through a flexible cable (Figure 6.45).

Dramatic projects often shoot with a focus puller (Figure 6.46). The focus puller is in charge of the sharpness of the image, letting the camera operator concentrate on framing the subject.

Consumer cameras often have automatic focusing features. Ideally, you just point the camera, and it focuses on the subject. Autofocus is particularly useful when moving around with a handheld camera, as it maintains focus wherever you move, allowing you to concentrate on framing the shot and checking out your route. However, it is best to think of the autofocus feature as an aid instead of thinking that it relieves you of all focusing worries.

Focusing the Zoom Lens

If a shot is correctly focused, zooming in or out should have no effect. Focusing during a wide-angle shot is easy, as there is plenty of depth. However, when you zoom in, the focused depth becomes progressively shallower as the lens angle narrows; then, the focus becomes much more critical. It is

important to zoom in on the subject as close as you can, focus, and then zoom back out. The image should stay in focus now when zooming in and out on that subject.

Adjusting the Zoom's Back Focus

If the image does not stay focused during the zoom, once the lens has been focused during the close-up as mentioned earlier, it usually means that the lens needs the back focus adjusted. To ensure that the zoom lens maintains focus as it is zoomed (referred to as "correct focus tracking"), the lens system has to be carefully adjusted:

- With the lens set at the maximum aperture, zoom in close and adjust the lens for maximum sharpness using the focus ring on the lens.
- Zoom out to the widest angle, and adjust the internal back focus (flange) control for the sharpest image (Figure 6.47).

Now the zoom lens should be working effectively. If not, repeat the adjustment for optimum results.

Back Focus Flange

FIGURE 6.47 *The back focus flange needs to be adjusted whenever the zoom does not stay in focus when zooming in or out on the same subject. We are assuming that the lens was focused on the close-up of the subject.*
Source: Photo courtesy of Panasonic.

LENS APERTURE (*F*-STOP)

If you look directly into a lens, you will see an adjustable circular diaphragm or iris made up of a number of thin overlapping metal blades. The size of the hole formed by these plates is carefully calibrated in graduated stops (Figure 6.48). These *f*-stops are usually marked around a ring on the lens barrel. Turning the ring alters the effective diameter of the lens opening over a wide range (Figure 6.49).

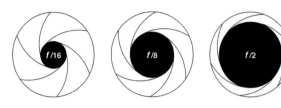

FIGURE 6.48 *The aperture is the opening in the lens that allows light to fall on the sensors. Each opening is carefully calibrated into a series of f-stops. These f-stops significantly affect the scene being shot by varying the amount of light and the depth of field. The larger the number (smaller the opening in the lens), the larger the depth of field.*

Focus ring   Aperture ring

FIGURE 6.49 *The aperture ring is used to adjust the f-stops.*
Source: Photo courtesy of Zeiss/BandPro.

When the lens aperture is adjusted, two quite separate things happen simultaneously:

- It changes the brightness of the lens image falling on the image sensor (exposure of the image).
- It alters the depth of field in the shot—which is also affected by the lens focal length.

F-Stops (F-Numbers)

The larger the opening in the lens, the more light the lens will let through to the sensor. As the lens aperture is reduced (referred to as stopping down), the corresponding f-stop number increases. The smallest lens aperture may be around f/22, and the largest aperture opening may be around f/1.4. For each complete stop that you open the lens, its image brightness light will double (halved when stopping down). The standard series of lens markings is:
f/1.4 2 2.8 4 5.6 8 11 16 22 32

In reality, many intermediate points such as f/3.5, f/4.5, and f/6.3 are also used.

EXPOSURE

There is a lot of misunderstanding about "correct exposure." A picture is considered to be correctly exposed when the subject tones you are interested in are reproduced the way you want them. Obviously, this may vary in different situations and is ultimately up to the director.

Because the camera's image sensor can respond accurately over only a relatively restricted range of tones, brighter areas will crush out (become white) above its upper limits; darker tones are lost (turn black) below the sensor's lower limits. The goal is to adjust the lens aperture to fit the director's vision (Figure 6.50).

If more detail is needed in the shadows, the lens aperture will need to be opened in order to increase the exposure. But, of course, this will cause all the other tones in the picture to appear brighter, and the lightest may even crush out to a blank white. Occasionally, the director may choose to deliberately overexpose a shot to create a special high-key effect with bleached out lighter tones.

Conversely, if you want to see better tonal gradation in the subject's white dress, you may need to reduce the exposure by closing down to a slightly smaller f-stop. This will, however, cause all the other tones in the image to appear darker, and the darkest might now merge or even be lost as black.

As you can see, exposure is an artistic compromise. But it is also a way in which the image can be manipulated. The director may choose to deliberately overexpose or underexpose the image to achieve a certain effect such as stormy skies, mystery, or night effects. It is quite possible to set the exposure to suit the prevailing light levels, and shoot everything in the scene with that stop. But for more sensitive adjustment of picture tones, one really must be alert and continually readjust the exposure for the best results.

Automatic Iris

Cameras are sometimes fitted with an auto-iris. When turned on, it automatically adjusts the lens aperture to suit the factory-set average light levels. The auto-iris can actually be quite helpful at times. The auto-iris can be used as a "light meter" to get an accurate reading, as long as you know about some lighting situations that can fool it. Without it, you would

Underexposed   Correctly Exposed   Overerexposed

FIGURE 6.50 *Exposing the image appropriately is incredibly important.*

have to continually check the viewfinder and then open the iris until the scene was correctly exposed. The auto-iris can also be especially helpful if the camera operator is very mobile, moving around quickly and concentrating on framing the shot.

Unfortunately, like any automatic function, the auto-iris cannot make artistic judgments, and can be easily fooled. Ideally, you want the exposure of your main subject to remain constant. But an auto-iris is activated by all other tones in the

picture. Here are some examples of problem issues that will require the camera operator to manually adjust the aperture:

- Move in front of a light background, bring a newspaper into shot, or take off a jacket to reveal a white shirt, and the iris will close and reduce the exposure even though it is not important that there is detail in those white areas. The problem is that all of the other picture tones will now appear darker (Figure 6.51).

FIGURE 6.51 *The auto-iris was used for both of these images. (Left) The bright white sky threw off the iris, making the subjects too dark. (Right) However, by adjusting the shooting angle and getting rid of the bright sky, the auto-iris was able to capture a well-exposed image.*

- If a large dark area comes into shot, the iris may open up "to compensate," and all picture tones appear lighter.
- Walk through trees, where areas of sky and foliage come and go, and exposure can open and close, becoming very distracting to the audience.
- If shooting a concert where the lead singer is in a very bright spotlight, the auto-iris will usually adjust the camera so that the stage is black and the lead singer is pure white, with no detail.

Auto-iris systems work by measuring the brightness of the lens image falling on the image sensor. Most auto-iris systems concentrate on the center area of the frame. Some are designed to avoid being over-influenced by the top of the frame, where bright skies, for example, could falsely reduce the exposure. The best systems judge exposure by sampling all parts of the image.

Manual Iris Adjustment

Turning the iris/diaphragm/aperture ring on the lens barrel (Figure 6.48) allows the camera operator to decide exactly how the image should be exposed. The overall picture brightness can be adjusted and compensation can be made for features in the scene that would fool the auto-iris system.

Remote Iris Control

When a camera is connected to a CCU (Figure 6.24), usually in a multi-camera situation, its lens aperture is usually set to suit prevailing light conditions by the video operator, who is also known as a shader. During a production, the lens aperture of each camera can be remotely controlled to vary exposure subtly for the best image quality.

SHUTTER SPEEDS

Selecting the correct shutter speed is very important in capturing the best video image. In some instances, a fast shutter speed should be used in order to get the clearest images of a speeding car. However, a slower shutter speed may be very appropriate for the same situation—although the final project will have some blurriness to it. In other situations, a specific shutter speed is needed in order to increase or reduce the depth of field in an image. Adjustment of shutter speeds is one of the creative tools available for the camera operator (Figure 6.52).

VIDEO GAIN ADJUSTMENT

Video gain is the amplification of the video signal in order to shoot in extremely low light. Although there are times when

FIGURE 6.52 (Left and above) *Adjustable shutter speeds can be used for creativity in images. In the first photo, a very slow shutter speed was used to cause a sense of action about the child. The second photo shows blurred snow, caused by a slow shutter speed.*

FIGURE 6.53 *Filters come in a variety of shapes, sizes, colors, and effects.*
Source: Photo courtesy of Tiffen.

FIGURE 6.54 *There are a number of companies that have created software that can apply filters to your video image during postproduction.*
Source: Photo courtesy of Tiffen.

this may be necessary in order to capture an image, it can substantially deteriorate the image. Some cameras allow the camera operator to set different levels of gain, such as 13, 19, and 118. However, it is important to know that the higher the gain level, the poorer the quality of the final video image. In summary, the gain should be adjusted only in situations where you must capture an image—even if you have to compromise some of the quality.

CAMERA LENS FILTERS

Filters of various kinds can be attached to the front of the lens, located in the camera's internal filter wheel, or inserted into a matte box on the front of the camera (Figure 6.53). Another option for filters is applying them to recorded video during the editing process. The following subsections describe some of the most common lens filters (Figure 6.54).

Neutral-Density Filters

When a scene is too bright for the aperture you want to work at, a neutral-density (ND) filter is used to cut down the overall intensity. These transparent gray-tinted (neutral) filters do not affect the colors—just the overall brightness. Most cameras come with one or more of these filters built into the camera. If the camera does not have them, they must be attached to the front of the lens.

ND filters may be used when shooting in very strong sunlight to prevent overexposure. Should you want to open

up the aperture of the lens to restrict the focused depth for artistic reasons, an ND filter can be used to bring down the light level, allowing you to open up the aperture, which will give a smaller depth of field.

Corrective Filters

When there are changes in the color temperature of light sources, corrective filters can be used to compensate. For example, when moving from a daylight scene to a tungsten-lit area, a correction filter should be used (Figure 6.55). Some cameras do include built-in corrective filters. This will be discussed in the "Camera Filter Wheel" section ahead.

Star Filters

Star filters are clear disks with closely scribed grid patterns. These diffraction-effects filters can produce four- to six-point stars from bright points of light, including flames, reflections, and lamps. The star's directions change as the filter is turned (Figure 6.56).

Diffusion Disks

Available in various densities, these filters provide general image softening through fine surface scratches or dimpling on a clear disk. Sharp detail is reduced and highlights develop glowing halos (Figure 6.57).

FIGURE 6.55 *The blue color-correction filter shown can be used to get rid of the warm tungsten light in the scene.*

Source: Photos courtesy of Tiffen.

FIGURE 6.57 *A diffusion filter softens the image.*

Source: Photos courtesy of Tiffen.

FIGURE 6.56 *The scribed grid pattern on the filter creates a star pattern at each point of bright light.*

Source: Photos courtesy of Tiffen.

FIGURE 6.58 *Polarizing filters are used to reduce or remove distracting reflections, such as this window.*

Source: Photos courtesy of Tiffen.

UV (Haze) Filters

One of the most common filters on lenses, the UV filter reduces haze blur (due to ultraviolet light) when shooting daylight exteriors and is often used to protect the actual lens surface.

Polarizing Filters

Polarizing filters are occasionally used to reduce strong reflections or flares from smooth or shiny surfaces such as glass or water (they have very little effect on rough materials). Polarizing filters can be used to darken an overly bright sky without affecting overall color quality, although there will be some light loss. By rotating the filter, you can selectively reduce or suppress specific reflections (Figure 6.58).

Graduated Filters

When shooting in the field, there are times when the main subject is properly exposed, but distant skies are far too bright and distracting. This can be a particular problem when the foreground subjects are dark-toned. A graduated filter can often overcome this dilemma. Its upper section has a neutral gray tint that reduces the brightness of the image in that part

of the picture. So it will "hold back" the overly bright skies while leaving anything in the clear lower section unaffected.

Graduated filters have a gradual tonal transition, giving a soft blend between the treated and untreated areas. There are also graduated filters than can be used to create a deliberate effect. One half of the filter may be orange and the other half clear or blue-tinted. Some color filters have a central horizontal orange or yellow band, which, with care, can simulate the effect of a sunset (Figure 6.59).

Camera Filter Wheel

Professional cameras generally include an internal filter wheel. The filter wheel can be rotated, placing the desired filter in front of the image sensors. The filters in the wheel usually include a 5,600° daylight color-correction filter (used when shooting outdoors), a 3,200° tungsten color-correction filter (for shooting indoors under tungsten light), and a couple of different ND filters. There are also usually blanks in the wheel so that additional filters can be added.

FIGURE 6.59 *(Left) This is a graduated filter. (Center) This image is a scene without a filter. (Right) In this shot, the camera operator used a graduated filter to enhance the sky and mountains.*
Source: Photos courtesy of Tiffen.

SUPPORTING THE CAMERA

When shooting while moving over uneven ground, climbing stairs, or in the middle of a crowd, the audience expects pictures to bounce around a bit. At times, this "point-of-view" style of shooting can even add to the mood of the program. However, images that weave around, bounce up and down, or lean to one side soon become tiring for the audience to watch. Smooth subtle movement and rocksteady shots are usually essential for effective camerawork, and there are various forms of support to help achieve this.

What Type of Support?

Before beginning any project, consider whether you have an appropriate camera mount (or a suitable substitute). Otherwise, you may not be able to get the kind of shot that the director would like. The kind of camera support you need will depend on a number of very practical factors:

*   The size and weight of your camera. Do you intend to handhold or shoulder-mount the camera? Will the shots be brief or are you shooting sustained action?
*   Are you shooting from a fixed position, moving only when not recording, or are you moving while shooting? Will there be any quick moves to other camera positions?

*   Will you want high/very high or low/very low shots? Will you be raising/lowering the camera to these positions or even swooping or gliding within the action?
*   Your surroundings can influence the type of mounting you use—the floor surface, operational space, height. Does the mounting have room to move around within the scene, around furniture, through doorways, and between trees?
*   Is the camera likely to be unsteady? Are you shooting while walking/running or from a moving vehicle?
*   In some situations, the solution is a remotely controlled camera mounted on a rail system, a robotic pedestal, or even a self-propelled camera car.

The Handheld Camera

If you are handholding an ENG/EFP camera, the goal is to support it firmly, but not so tightly that your slightest movements are transmitted to the shot. Your right hand is usually inserted through an adjustable support strap near the zoom control. Your left hand usually holds the lens barrel. Holding the eyepiece viewfinder against an eye can even help to steady the camera (Figure 6.60). Whether you make hand adjustments or rely on auto-controls (autofocus and auto-iris) will largely depend on the camera design and shooting conditions.

FIGURE 6.60 *The shoulder-mounted handheld camera is steadied by the right hand, positioned through the strap on the zoom lens. That same right hand also operates the record button and the zoom rocker (servo zoom) switch. Note that her camera cable is tied to her belt in case the cable gets snagged; it will pull her, not jerk the camera. She also keeps both eyes open so that she can see what is happening around her.*

The lighter the camera, the more difficult it is to hold it steady. Make use of a nearby stable structure such as a wall, if you can (you'll see various regular ways of steadying a handheld camera in Figure 6.61). Some camera lens systems include an image stabilizer that is built in or attached to the lens system to steady slight picture movements.

Camera Stabilizers

The widespread technique of supporting the camera on one shoulder has its limitations. Its success largely depends on the camera operator's stamina! Some people can continue

shooting over long periods, without their images tilting or becoming unsteady. But arms tire and back muscles ache after a while, and it is not easy to sustain high-quality camerawork, particularly when shooting with a telephoto lens.

Camera Tripods

Tripods are the most common camera support. Although the tripod can't always be repositioned quickly, it does have advantages. It is simple, collapsible, and easily transportable, and provides solid camera support. It can be used in a wide variety of situations: on rough, uneven, or overgrown surfaces, on stairs, and so on.

Basically, a tripod has three legs that each have independently adjustable length. The legs are spread apart to form a stable base for the camera. When moving to a new location, the tripod legs can simply fold together. The camera can be screwed directly to the tripod pan and tilt head or a quick-release mounting plate can be used, which allows the camera to be detached in a moment (Figure 6.62).

Tripods are a great help when used properly, but they are not foolproof. Here are some useful warnings:

- Don't be tempted to use the tripod legs partly open, as it can easily fall over. Instead, always adjust the camera height by altering the leg length, not by changing the spread.
- Fully extended tripods can be a bit unstable at times, especially in windy conditions. It may be necessary to use a sandbag to provide stability or fasten the tripod to the ground or a platform.
- Don't leave the camera standing unattended on a high tripod. People (or animals) may knock it over or a pulled cable may overbalance it. It is much safer to drop the tripod to its lowest level (Figure 6.63).

Tripods usually have two types of feet: retractable spikes for use on rough surfaces and rubber pads for floors. Spikes can easily damage carpets and wooden floors.

The feet of a tripod can be fitted onto a "spreader" or "spider" in order to prevent its feet from slipping (Figure 6.63).

A tripod dolly, which can be folded for transportation, is added to the bottom of a tripod in order to allow it to roll across a floor (Figure 6.64). Although the dolly moves around quite easily on a flat, level floor, uneven surfaces will cause a jerky image, especially when a telephoto lens is used.

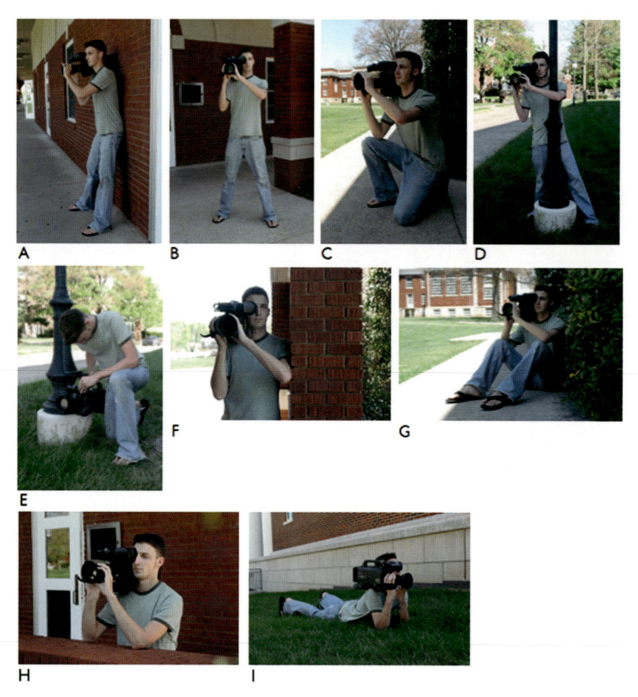

FIGURE 6.61 *Keeping the handheld camera steady takes practice. Here are some techniques to handhold a camera: (A) rest your back against a wall; (B) bracing the legs apart provides a better foundation for the camera; (C) kneel, with an elbow resting on one leg; (D) rest your body against a post; (E) lean the camera against something solid; (F) lean your side against a wall; (G) sit down, with elbows on knees; (H) rest your elbows on a low wall, fence, railings, car, etc.; and (I) elbows resting on the ground.*

Source: Photos by Josh Taber.

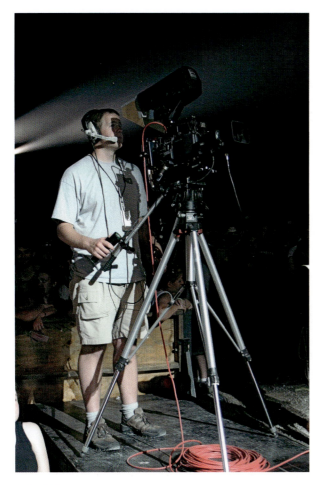

FIGURE 6.62 *The collapsible tripod, a three-legged stand with independently extendable legs.*

Source: Photo by Paul Dupree.

FIGURE 6.63 *When not in use, the camera operator drops the tripod to its lowest position in order to provide maximum stability for the camera.*

FIGURE 6.64 *A tripod dolly is attached to the legs of the tripod to enable it to roll over smooth surfaces.*

Camera Pedestals

A pedestal (ped) is the most widely used studio camera mount. Fundamentally, it consists of a central column of adjustable height, fixed to a three-wheeled base that is generally guided by a steering wheel.

The rubber-tired wheels can be switched into either:

- A "crab" mode, in which all three wheels are interlinked to move together.
- A "steer" or "dolly" mode, in which a single wheel steers while the other two remain passive.

Pedestal designs range from lightweight hydraulic columns on casters to heavyweight designs for large cameras. The ideal pedestal is stable, easy to move, and quickly controlled by one person (Figure 6.12).

FIGURE 6.65 *A Gorillapod's legs can be twisted around to hold a camera securely in place.*

Source: Photo courtesy of Joby.

Gorillapod

The Gorillapod is actually a type of tripod. However, the difference is that its flexible, jointed arms allow you to wrap it around a railing, twist it around a tree branch, or loop it on the handle of a shopping cart. This type of camera support is extremely light and easy to travel with, as well as providing great camera support (Figure 6.65).

Beanbag

Beanbag camera supports allow the camera to be positioned in all types of angles, while holding the camera steady. This type of support is very light and flexible. There are many different brands available. These bags are especially helpful when you need to use your camera on a rough or uneven surface. Most bags are filled with foam balls while small bags may contain sand (Figure 6.66).

FIGURE 6.66 *Beanbags can securely hold a camera in place without the need for a tripod.*

Source: Photos courtesy of The Pod and Cinesaddle.

Monopod

The monopod can be easily carried and is a very lightweight mounting. It consists of a collapsible metal tube of adjustable length that screws to the camera base. This extendable tube can be set to any convenient length. Braced against a knee, foot, or leg, the monopod can provide a firm support for the camera, yet allow it to be moved around rapidly for a new viewpoint. Its main disadvantage is that it is easy to accidentally lean the camera sideways and get sloping horizons. And, of course, the monopod is not self-supporting (Figure 6.67).

Body Brace or Shoulder Mount

A body brace or shoulder mount can be attached under the camera to make it more comfortable to support and keep the camera steady (Figure 6.68).

FIGURE 6.67 *The monopod shown has additional support with the base.*
Source: Photo courtesy of Manfroto.

FIGURE 6.68 *A body brace helps to firmly support the camera.*
Source: Photo by Phil Putnam.

Steadicam/Glidecam Support

One of the most advanced forms of camera stabilizer, the Steadicam- or Glidecam-type system, uses a body harness with ingenious counterbalance springs. Stabilizers of this kind will not only absorb any camera shake, but actually allow you to run, climb stairs, jump, and shoot from moving vehicles—while still providing smooth, controlled shots! The operator uses a small electronic viewfinder attached to the stabilizer. Near-magical results are possible that are unattainable with other camera mountings. But underneath it all, there is still a vulnerable human operator, and extended work under these conditions can be very tiring. Smooth camera work with a Steadicam-type system is only attainable with lots of practice (Figures 6.69–6.71).

FIGURE 6.69 *Small cameras can be supported by a handheld steady device such as this Steadicam.*
Source: Photo courtesy of Steadicam/Tiffen.

Jib Arms

As filmmakers have demonstrated so successfully over the years, a large camera crane offers the director an impressive range of shot opportunities. It can hover, then swoop in to join the action. Or it can draw back, rising dramatically to reveal the broader scene. It allows the camera to travel rapidly above the heads of a crowd, or to sweep around near floor level as it follows dancers' movements. But such visual magic is achieved at a price! Larger camera cranes are cumbersome, need a lot of room to maneuver, and require skilled and closely coordinated crews. Today, relatively few TV studios make use of such camera cranes. Instead, the modern jib arm can satisfy most directors' aims (Figures 6.72–6.75).

Smaller lightweight jibs are easily disassembled and transported, and have proved to be extremely adaptable both in the studio and in the field. All the camera controls,

FIGURE 6.70 *Larger cameras need to be attached to a brace on the body in order to spread the weight over the body and not just the camera operator's arms.*

Source: Photo by Katie Oostman.

FIGURE 6.71
Steadicam-type devices can be used in conjunction with other equipment such as dollies, snowmobiles, vehicles, and Segways.

including focus, iris, zoom, tilt, and pan, are adjusted by hand or by a remote control.

A jib is more compact than the traditional camera crane, much more portable, and a lot less costly to buy or rent. The camera on a jib arm may be handled by a single operator. It can stretch out over the action (like a crane), reaching over any foreground objects. It can support the camera at any height within its range, moving smoothly and rapidly from just above floor level up to its maximum height, and swing around over a 360-degree arc.

FIGURE 6.72 *Camera jibs provide directors with an impressive range of shot opportunities, including movement up, down, left, and right.*

FIGURE 6.73
Jibs can be used at remote events and in studios. Large jibs require a specially trained operator.

FIGURE 6.74 *The smaller jib shown in the foreground of this photo is much easier to use than the large jib shown in Figure 6.73.*

Source: Photo by Paul Dupree.

FIGURE 6.75 *Cranes, different from the jib, usually provide a seat for the camera operator.*

Source: Photo courtesy of Sodium 11.

Camera Dollies

Camera mountings that have been widely used in filmmaking have been equally successful in television production. Camera dollies work incredibly well at capturing "dolly in," "dolly out," and "tracking" shots. Dollies can use wheels that work on a smooth floor or can be designed to run on a track.

There are a variety of dollies available, from very small dollies to larger systems designed to be ridden on (Figures 6.76–6.79).

FIGURE 6.78 *This dolly utilizes a pivoting seat and camera support (or turret), allowing the camera operator to control camera pans by moving his or her feet.*

Source: Photo by Will Adams.

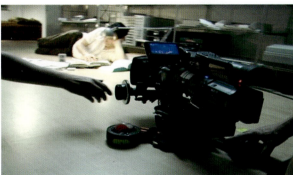

FIGURE 6.76 *This "skater" dolly provides an extremely low-angle dolly shot (above) The top of the skater shows that it uses three skateboard-type wheels in order to glide silently and smoothly over the floor.*

Source: Photos courtesy of PIS Technik.

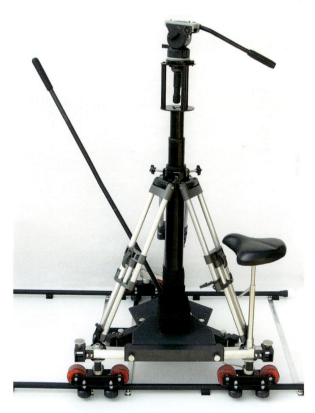

FIGURE 6.79 *Dollies come in all shapes and sizes, based on the camera weight and movement requirements.*

Source: Photo courtesy of Indie-Dolly Systems.

FIGURE 6.77 *The camera operator is standing on this dolly, which is rolling on a dolly track.*

FIGURE 6.80 *The director of photography for this music video attached a DSLR to an off-the-shelf remote-controlled truck.*

FIGURE 6.81 *The pan head allows the camera to pan left/right and tilt up/down.*

Create a Camera Mount

Use your imagination; don't be limited by the mounts available commercially. Today's lightweight cameras can be mounted on almost anything. Look for creative ways that will allow you to move your story forward (Figure 6.80).

The Pan-and-Tilt Head

As you might expect, there are several types of camera head designed to suit different types and weights of camera.

If you simply bolted the camera onto a mounting, it could be held firmly, but would be unable to move around to smoothly follow the action. Instead, a pan-and-tilt head/camera head is needed, which goes between the camera and the top of the camera mount. It firmly anchors the camera, yet allows it to be turned (panned) and tilted, or fixed at any required angle (Figure 6.81).

Panning handles are attached to either side of the pan-and-tilt head for the camera operator to support and guide the camera. Zoom and focus controls can be clipped to the panning handles (Figure 6.62).

If the camera head moves around too easily, it can be difficult to make smooth pans and tilts. So drag controls allow the camera operator to introduce a controlled amount of friction to steady movements. They should never be over-tightened to "lock off" the camera. The separate headlocks should be used to prevent panning or tilting (such as when leaving the camera).

Balance adjustments ensure that the camera remains level. Careful balancing is absolutely essential when you have large zoom lenses, prompting devices, and camera lights attached to the camera; otherwise, the camera will be front-heavy and very difficult to operate. In some situations, it might even overbalance the mount, causing the camera to fall.

INTERVIEW WITH A PROFESSIONAL: NATHAN WHITE

What do you like about being a news videographer? Few positions within television production can put a person in the middle of the action quite like being a news photographer. Whatever is unfolding in the production is literally taking place directly right in front of me. An example is that of the fall of the Berlin Wall. When looking back in history, the image that demonstrates what happened is not an anchor speaking; few remember what they said. It is the video of liberated West Germans tearing down the wall and standing above the crowd with arms raised in triumph. A news videographer's work lives forever in archives and the minds of viewers. There is nothing like having people all over the world coming to a stop to look at my video; it is an experience that few others will ever have.

FIGURE 6.82
Nathan White, news videographer.

How do you decide what shot to use? Shot selection for a videographer is an art. If the right angles are not shot, then the whole piece may not edit together very well at all. When I was first starting out, editors complained that my video was a series of shots that were well composed, but did not fit together in a sequence. Every shot conveys something different to the audience. When a videographer understands what each shot can do, he or she can better know what shot to get. Long shots establish context, close-ups convey intimacy and emotion, and medium shots propel the piece along. Long shots should only be used long enough to lead the viewer into a scene or away from it. If they are used too often, the audience will become bored. Dramatic vistas look better as a painting than as video. Close-ups bring viewers in very close to the action. A CU of a pair of hands putting something together, or of gears in a machine, can show how something is done and they are easy to use as cutaway shots. A CU of a person's face is an excellent way to show emotion. These make excellent cutaways. Shoot as many close-ups as possible, because they are easily edited and add intimacy. Medium shots are the most used shot in videography. They are easily edited, close enough to show action and intimacy, and wide enough for the viewer to still see some context. If there is a single event that can only be captured once, the MS is the best shot to use.

Do you ever use auto settings on your camera? It is best to learn how to use a camera without using any of the auto settings. It is rare that auto settings can completely ruin a shot, but manual settings put the control in the hands of the videographer. Auto settings can be very useful in situations where there is little time to make adjustments (on breaking news, for example). Or, they can be used as a reference point when setting up the shot.

Nathan White is a videographer for a local television station.

REVIEW QUESTIONS

1. What is the difference between an EFP camera and an ENG camera?
2. Why use a POV camera?
3. How does a wide-angle lens optically adjust the scene?
4. What are some of the challenges experienced when using a telephoto lens?
5. What are some of the advantages of using a zoom lens?
6. What are the advantages and disadvantages of automatic focus?
7. What are two of the types of camera supports and what are their advantages?

CHAPTER 7

USING THE CAMERA

"In the hands of even a moderately skilled photographer who knows the capabilities of the camera, even a cell phone can produce reasonable pictures. Unfortunately, owning a camera and knowing which button to push doesn't make you a photographer any more than owning an automobile makes you a Formula One race-car driver."

—Andy Ciddor, *TV Technology*

TERMS

CU: Close-up shot.

ECU/XCU: Extreme close-up shot.

ELS/XLS: Extreme long shot.

***f*-stop**: The *f*-stop regulates how much light is allowed to pass through the camera lens by varying the size of the hole the light comes through.

I-mag: "Image magnification" is when you are shooting for large video screens positioned near a stage to allow the viewers to get a better view of the stage.

LS: Long shot.

MS: Medium shot.

Tally light: Cameras, or their viewfinders, usually have a tally light on the front of the camera and in the viewfinder. The front light is to let the talent know that the camera is recording. The back tally lets the camera operator know when his or her camera has been chosen to be recorded by the director.

Now that we have examined the camera's features, let's take a look at the techniques that are the foundation of good camerawork.

STANDARD SHOTS

As filmmaking developed, a fairly universal system for classifying shots evolved. These provide convenient quick reference points for all members of the production team, especially for the director and the camera operator.

This series of shot terms are relative to the size of the subject. However, the overall concept works for any subject. Figure 7.1 demonstrates the various terms with each shot. Terminology does vary from place to place, but the most widely recognized ones are included here. When framing, it is important to avoid shots that cut through the body at natural joints.

Exactly how you get a specific shot does not affect the term used. For instance, you can take a close-up shot with a camera that is close to the subject with a wide-angle lens or it can be shot with a camera that is quite a distance from the subject, utilizing a telephoto lens. Of course, there will be differences in perspective distortion and camera handling, depending on which method you use.

Selecting the Shot

Part of the issue when selecting a shot is knowing where that shot is primarily going to end up. For example, if your final

FIGURE 7.2 *Image magnification (i-mag) requires a different type of shot sequence than a standard television. Note the screen on the right side of the stage that is used to help the audience get a better view of the guitar.*

project will primarily be seen on a home television, a variety of shots from extreme long shots to extreme close-ups are appropriate. If you are shooting for large video screens positioned near a stage to allow the viewers to get a better view of the stage action (also known as image magnification or i-mag) (Figure 7.2), medium shots and close-ups are used since the viewer can already look at the stage and get his or her long shot. If the main use is for something such as a video iPod or a small video area for the Internet, the small viewing screen lends itself more to close-ups, so the viewer can really understand the nuances of what is going on.

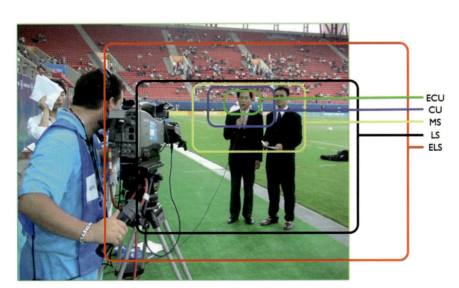

FIGURE 7.1 *When shooting any subject, shots are classified by the amount of the subject taken in. Below are the shots of a person:*

- Extreme close-up (ECU or XCU) or big close-up (BCU) is a detail shot.
- Close-up (CU) is generally framed just above the head to the upper chest.
- Medium shot or mid shot (MS) cuts the body just below or above the waist.
- Long shot (LS) or wide shot (WS) generally features the entire person in the frame, just above and below the body. The European term for a long shot of a person is a full shot (FS).
- Extreme (or extra) long shot (ELS or XLS) or very long shot (VLS) shows significant space above and/or below the subject.

Checking the Shot

Once a shot has been established, a review of the overall scene can be very helpful:

- Check for potential problems such as a light stand that will come into the shot if you pan right. It can let you know if someone is going to move into the shot, and that you may need to recompose the picture to include that person.
- Check your viewfinder image to see if something is about to move out of shot, or is going to be partly cut off at the edge of the shot.
- Check the composition of the shot (framing, headroom, etc.), subtly correcting for changes that develop such as people moving to different positions in the shot. Composition will be discussed in a later chapter.
- Watch for the unexpected, such as distracting objects in the background (Figure 7.3). Are microphones, cameras, lamps, or their shadows appearing in the shot? You can often reframe the shot slightly to avoid them (Figure 7.4).

CAMERA OPERATION

Focusing

Focusing is not always as straightforward as it looks. When the subject has well-defined patterns, it is fairly easy to detect maximum sharpness. However, with less-defined subjects you may rock focus either side of the optimum, and somehow they may still look soft-focused.

The exact point at which you focus can matter. There is usually more focused depth beyond the actual focused plane than there is in front of it. So in closer shots, there can be advantages in focusing a little forward of the true focusing point (nearer the camera) to allow for subject movement (Figure 7.5). If you are focused too far back (away from the camera), the problem worsens. When shooting people, the eyes are a favorite focusing point.

Depth of field continually changes as you focus at different distances, select different lenses, or zoom in or out. This is something you quickly become accustomed to, but it can't be ignored. Focusing is much easier in longer shots and more complicated with close-up shots. You shoot two people speaking, yet can only get a sharp image of one of them at a time. In very large close-ups, focusing can be so critical that only part of a subject is sharp, while the rest is completely defocused. Figure 7.6 illustrates some of the solutions to this dilemma.

FIGURE 7.3 *The incredibly busy background makes it difficult to distinguish the athlete. The image can be improved by moving the camera to a different location that provides a better background.*

FIGURE 7.4 *Camera operators need to be careful not to include a boom mic in the frame. Boom operators have to be careful not to allow the mic to drift downward into the frame.*

Camera Moves

Whenever you are going to move the camera, there are a number of things camera operators need to think about:

- Always check around you to make sure that you don't run over cables, bump into the set or props, move in front of other cameras, or run into people.
- When part of a multi-camera production, make sure that you have enough cable by ensuring that you have sufficient slack before the move begins. Never pull a cable that has a tight loop in it. Cables can be easily damaged.

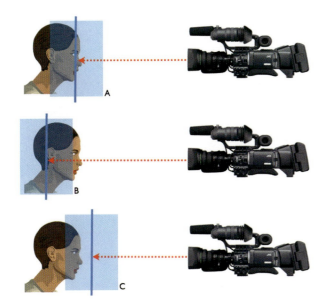

FIGURE 7.7 *ENG/EFP productions require the camera operator to improvise in many situations.*

Source: Photo courtesy of Sony.

FIGURE 7.5 *When accurately focused (A), the most important part of the subject should be the sharpest. The specific focus point (as represented by the darker blue line in the illustrations), should have a large enough depth of field (represented by the light blue field) to keep the subject in focus as she moves a little forward or backward. Note that the incorrect focus points, as demonstrated in B and C, would make it easier for the subject to move out of focus.*

Source: Camera photo courtesy of JVC.

PRODUCTION FORMAT STYLES

Camera operators working on a single-camera production have many different responsibilities than those working on a multi-camera shoot. Multi-camera camera operators usually have someone in the control room adjusting their camera's aperture (so that the image quality is the same as the other cameras involved in the production) and they have a director telling them exactly what he or she wants, in real time. Single-camera production camera operators have to adjust their own cameras in many different ways since they are usually not being monitored in a control room or production truck. The single-camera operator has more individual control over the image, making shot decisions on a continual basis.

The Single-Camera Shoot

If you are on a shoot for a documentary or a news story, there will always be quite a bit of improvisation (Figure 7.7). You need to make the most of any opportunity that presents itself. Clearly, a lot depends on whether you are working indoors or in the open, whether you are on your own, or

backed by a team. The weather, audio, and light conditions will affect the shoot. It is very important to ensure that your camera is in good working order. Single-camera productions are usually referred to as ENG (electronic newsgathering) or EFP (electronic field production) equipment. See Table 7.1 on suggestions of things to look at when checking out ENG/EFP equipment.

Some people on a single-camera shoot prefer to "travel light," setting off with just a camera, spare batteries, and recording media. However, for peace of mind, there's a lot to be said for using a systematic checklist routine. This list can be as brief or comprehensive as you like, including whatever gear you personally find invaluable on that kind of project (the item you'll need in an emergency is sure to be the one you've left behind)! Lists also help to avoid losses when hurriedly repacking equipment after a shoot (Table 7.2).

The Multi-Camera Shoot

Whether you are working out on a remote location or in a television studio, you will find that all multi-camera setups have a common theme. Unlike a single-camera unit, which often relies on a spontaneous and improvised approach, a multi-camera team is essentially a closely coordinated group, working to a planned pattern. We are going to be looking here at the ways these teams are organized and managed (Figure 7.8).

DEALING WITH A LIMITED DEPTH OF FIELD

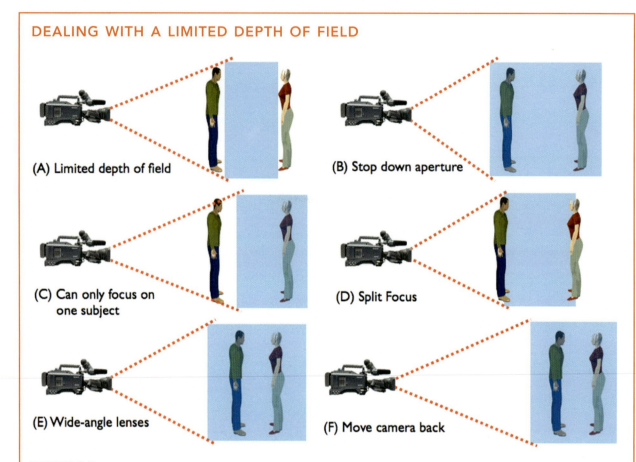

(A) Limited depth of field

(B) Stop down aperture

(C) Can only focus on one subject

(D) Split Focus

(E) Wide-angle lenses

(F) Move camera back

FIGURE 7.6

Source: Camera photo courtesy of Panasonic.

When a limited depth of field creates problems (A), here are a few solutions:

- If the aperture is stopped down (a higher f-stop number), the depth of field will increase (B).
- If there is not enough depth to cover two people (C), it is possible to split focus (D). However, both subjects may be slightly soft-focused.
- Wide-angle lenses have a much wider depth of field than a telephoto lens (E).
- The closer the camera is to the subject, the shallower the depth of field. By moving the camera farther back (F), a wider depth of field will be created. However, the subject will be smaller.

The Camera Operator in a Multi-Camera Production

Productions vary considerably in their complexity. Some follow a familiar routine, with cameras taking standard shots from limited angles. In other shows, cameras are mobile and follow very complicated action, such as sports productions (Figure 7.9). These action productions can require the skilled, highly controlled camerawork that only comes with experience. Each camera operator is on his or her own, required to get the best shots possible, despite having to find ways around specific problems. At the same time, the operator is part of a crew, working together to achieve a coordinated, excellent program for the viewing audience. In some productions, the camera operator can help the director by offering potential shots, while in others this would be a distraction from a planned treatment since you don't really know what shots other cameras have.

TABLE 7.1 *Before You Take out the Camera*

Although the exact procedures will depend on the design of your specific camera, here are the most common areas to review before taking a camera into the field.

Camera support: If using a quick-release plate, make sure it is present. Otherwise, make sure that all other supports are operational.

Lens: Examine the lens surface for dust, dirt, and fingerprints. Clean and dust off surfaces with a lens brush or an air can. Then breathe gently on surfaces and clean them with clean lens tissue. If the lens is regularly removed from the camera, the rear element of the lens will need to be inspected as well.

Power: Batteries and AC power adapters are the most common power supplies. Check to ensure that they are fully charged and operational.

Viewfinder: Most viewfinders that come with an ENG/EFP camera are called ENG viewfinders, although studio viewfinders are also occasionally used. Adjust the viewfinder position to your comfort level. The eyepiece on the ENG viewfinder may need to be adjusted to suit your eyesight. Check the viewfinder's performance to make sure the image is good (sharp detail, brightness, and contrast).

Preset Controls

- Check the macro position to make sure it is operational.
- If the lens is equipped with a 2x extender, make sure that it is working properly.
- Check aperture settings in manual and automatic settings. Manually underexpose picture, and check if auto-iris mode compensates.
- Watch the exposure indicator (zebra) for overexposure or underexposure.
- Check the gain (leave at 0 or OFF). When shooting in low-light areas, increase the gain, and check changes in iris settings.
- Check shutter speeds to make sure they are operational. The default shutter is usually set at 1/60th.
- Test the white balance setting to make sure it responds to the lighting situation. A white card or surface is used for this test.
- Adjust the black balance to see if it is working. The iris is completely closed or an opaque lens cap is used.
- Check the audio inputs (mics fit firmly and the monitors respond to the audio).
- Do a test video recording, making sure that an image and audio are recorded.

TABLE 7.2 *Have You Forgotten Anything?*

Camera	Standard zoom lens Supplementary lens Filters (UV, sky, ND, fluorescent, etc.) Matte/filter box Lens cleaning kit Viewfinder (ENG or studio) White card
Camera protection	Carrying case Weather and/or water protection (rain, sand, dust, etc.) Underwater camera housing
Batteries (fully charged)	Camera battery (on-board) Battery charger AC power (mains) adapter
Recording media	(ready to use)
Camera mount equipment	Tripod Pan-and-tilt head and quick release Panning handle(s) Remote zoom/focus controls Shoulder support Body brace Monopod
Audio equipment	Camera microphone Handheld (interview) mic Shotgun mic Wireless microphone and receiver Headphone Portable audio mixer Audio cables
Lighting equipment	Camera light Lamp power supply (batteries or AC power supply) Portable light kit Additional lights? Reflectors (white, silver, gold) Sandbags Lighting cables with extensions Spare lamps Diffusers (spun glass, scrim) Color filters (daylight, tungsten) ND filters
Supple-mentary	Portable video monitor with light hood Teleprompter Gaffer tape Small tool kit

FIGURE 7.8 *Multi-camera productions may utilize a camera operator for each camera or, as shown in this image, they may be robotically controlled by one remote operator.*

Source: Photo by Jon Greenhoe.

FIGURE 7.10 *Shot sheets help the camera operators know how the director needs them to compose specific shots. Some sheets, such as the photo sheets shown above, help the camera operator identify specific people.*

FIGURE 7.9 *Sports multi-camera productions often require the skilled highly controlled camerawork that only comes with years of experience.*

FIGURE 7.11 *It is important that camera operators check out their cameras before the production begins.*

If the production has been planned in meticulous detail, the camera operator's role may be to reproduce exactly what the director drew in storyboard sketches weeks before. On the other hand, the entire show may be "off the cuff," with the director relying on the camera crew to find the best shots of the action.

Usually directors will have a camera meeting before the production to explain the types of shows they want from each cameraperson. During the breaks, the director may discuss specific shot problems or new shots. Many times, directors will use a shot sheet showing the shots that will be needed throughout the production (Figure 7.10). From then on, the camera crew relies on intercom instructions to guide their camerawork.

Preparing for Rehearsals

It is possible to simply move your camera into the opening position, focus, and wait for the director's instructions. However, if you did that you would probably get burned. The camera must be checked before the production, or even the rehearsal, begins. Table 7.3 lists the majority of the issues that need to be thought about and tested. Once this checkout becomes routine, it can be quickly accomplished (Figure 7.11).

TABLE 7.3 *Multi-Camera Pre-Rehearsal Checklist*

Camera	☐ Is the camera switched on?
	☐ Are the cables tightly attached on both ends?
	☐ Is there sufficient cable for the camera moves?
	☐ Check to see that the lens is clean.
Viewfinder	☐ Check to see if the viewfinder's focus, brightness, contrast, and picture shape (aspect ratio) is correct.
	☐ See if the tally lights are working.
Lens	☐ Check the smoothness of focusing action from close-up to infinity.
	☐ Check focus when fully zoomed in, and when zoomed out.
	☐ Prefocus on a distant subject and then slowly zoom out, checking that focus does not wander.
Zoom	☐ Is the zoom action smooth throughout the range?
	☐ Check the function of the 2x extender (if your lens has one).
Filters	☐ Check the filter wheel, making sure that the correct filter is used.
Camera mount	☐ Is the camera firmly attached to the pan-and-tilt head?
	☐ Are the pan-and-tilt head handles firmly attached and at a comfortable angle?
	☐ Is the pan-and-tilt head accurately balanced?
	☐ Check and adjust drag/tilt friction.
	☐ If using a tripod, check the legs to make sure that they are firmly in place.
	☐ If using a pedestal, raise/lower it slowly. Is it smooth and easy to move? Check to make sure that it does not drift up or down.
	☐ If the camera support has wheels, make sure that it can be easily steered.
Intercom	☐ Test your mic and make sure that you can hear the director and others on the intercom.
Tele-prompter	☐ If a teleprompter is fitted, check that it is secure and working.
Shot sheets	☐ Read through the shot sheet and make sure that you understand what the director wants.

THE CAMERA OPERATOR DURING THE PRODUCTION

Although there are no specific rules for being an effective camera operator, there are some basic guidelines that can make life a lot easier, and produce more consistent results.

You can tell an inexperienced cameraperson at a glance; standing poised rigid and tense, eyes glued to the viewfinder, gripping controls tightly—relax (Figure 7.12)! The best posture is an alert, watchful readiness, very aware of what is going on around. Not so relaxed as to be casual, but continually waiting to react.

In a multi-camera show, you need to keep a watchful eye on the tally light on the camera. It shows that your camera has been selected on the production switcher. It means that your camera is "on-air." When the light is out, you can move to new positions, adjust zooming or focus, check composition, and so on (Figure 7.13). It is important not to move the camera until the tally light goes off. It is also useful to check other cameras' lights when moving around, to ensure that you are not going to get into their shots.

"Studio" viewfinder tally light

FIGURE 7.13 *Viewfinders usually have a tally light on the front of them and on the back. The front is to let the talent know that the camera is on-air. The back tally lets the camera operator know when his or her camera is on-air.*

The camera viewfinder must be set at a comfortable angle and easy to see. If it is set too low, it requires the operator to bend over a little, which is very uncomfortable after a while. The viewfinder also must be adjusted to give optimum detail in both the shadows and the lightest tones.

Always prefocus the lens whenever you move to a new position. Zoom in, focus sharply, and then zoom out to the shot that was requested by the director. You will then be ready to zoom in from a wide shot to a close-up if needed, staying in focus the whole time.

For most shots utilizing talent, the lens should be around the talent's eye level—unless the director wants higher or lower angles for some reason. The eye-level shot is a very neutral position (Figure 7.14). If the camera is slightly lower, at chest height, this tends to give the talent a more authoritative look. A higher camera can make the talent look a bit inferior.

If you have received a shot sheet, be ready to move to the next position once each shot is completed. With action happening all around you, this often requires a very focused attention to details (Figure 7.10).

Listen to all of the intercom instructions, including those for other cameras. This will help you understand how the production is progressing and prepare you for what your role requires.

Problems often occur during rehearsals that only the director can solve by making changes or reorganizing the camera shots. These problems may include where one camera

FIGURE 7.12 *Although handheld cameras can get very heavy, this new camera operator's back will be hurting due to bad posture.*

FIGURE 7.14 *The camera operator shooting this NBC cooking segment has his camera at the talent's eye level.*

FIGURE 7.15 *This BBC camera rehearsal allows the director to test his or her preplanned camera shots to make sure that they work with the stage action.*

blocks another's shot, insufficient time for camera moves, compositional problems, focusing difficulties, etc. Let the director know the issues; it may even mean stopping rehearsal to work things out. Don't assume that it will be all right the next time. It may be worse (Figure 7.15).

Occasionally an experienced camera operator can help the director by suggesting slight changes in the camera angle that would improve a sequence, or simplify a complicated situation. What is obvious on the floor may not be evident on monitors in the control room. However, you must avoid appearing to "sub-direct" the show. If the director does not choose to go with your suggestion, don't take it personally; the director may be dealing with other issues you don't know about.

Things go wrong during a busy production. When that happens, you have to quickly determine whether someone will handle it quickly or whether you need to inform others (such as about a lamp knocked out of position by moving scenery). In most cases, the floor manager is your on-the-spot contact. Teamwork is the essence of good production.

PRODUCTION TECHNIQUES THAT IMPACT THE CAMERA OPERATOR

Directors can tackle productions in a variety of ways. The techniques vary with the personality and experience of the director, and with the content and who the intended audience will be. These techniques may include the following:

- An improvised spontaneous approach.
- An outlined shooting plan that cannot be rehearsed before taping or transmission, such as a dance group who are arriving during airtime.
- A rehearsal with stand-ins in place of the actual performers. This allows you to check the shots in advance.
- A closely planned show in which the director explains the action/shots to the camera and sound crew before beginning rehearsals.
- A "stop-start" approach in which the action is rehearsed on camera while the director guides shots, stopping and correcting errors/problems as they arise. After a complete run-through rehearsal, the show is usually recorded or transmitted from beginning to end.
- A "rehearse and record" approach, in which each shot, segment, or scene is rehearsed and recorded before going onto the next. Shots will not necessarily be in the final program order since they will be edited together later.

As you will see, the problems and opportunities for the production crew can be very different with each method, ranging from a "one-chance-only" situation to a "retake-until-we-get-it-right" approach.

AFTER THE SHOW

At the end of a production, when any retakes have been completed and checks made, the director and the technical director will announce on intercom that the crew is cleared. The floor manager will repeat this for everyone, including performers. No one should leave their position until they have been released.

At the end of a studio production, as the camera operator, you should lock off the camera's pan-and-tilt head. Make sure that the cable has been arranged in a neat pattern so that it will not tangle the next time and place the lens cap on the camera. Some multi-camera production camera operators, especially on remote productions, may need to tear down and put away their camera. Their job also may include rolling up long cables and then loading all of the gear into a truck. Once you have been released, you are done.

INTERVIEW WITH A PROFESSIONAL: THOMAS CRESCENZO

Briefly define your job. I operate a Steadicam and cameras on a freelance basis.

What do you like about your job? As a freelancer, I have the privilege of working on a variety of projects with all different types of gear. I work in a lot of different locations so the scenery is always changing.

What are the challenges you face in your position? Being a freelancer in this economy is challenging; work can be uncertain and hard to find. With limited projects taking place, many pros are cutting their rates. Employers, knowing that work is limited, often push for lower rates because someone will still fill the positions.

How do you prepare for a production? I prepare for a production by checking over all of the equipment I will be using to make sure it is in top operating condition and charged up before packing it. Beyond that, I just get a good night's rest and set redundant alarms to make sure that I'm on time.

FIGURE 7.16
Thomas Crescenzo, Steadicam operator.

What suggestions or advice do you have for someone interested in a position such as yours? The best advice I can give to someone interested in a position such as mine is to build a solid reel. If you want to work on a camera crew, your reel will define you as much or more than your resume. If you are in school studying media, now is the best time to tailor your reel to the type of projects you want.

Thomas Crescenzo is a freelance Steadicam owner/operator and camera operator.

INTERVIEW WITH A PROFESSIONAL: JON LORD

Briefly define your job. My role is to deliver all aspects of camerawork for a program, from single operator ENG to large multi-camera productions, live or recorded. The job description is probably best summed up as planning, support, and delivery to any customer's camera needs. Interpersonal skills, thinking quickly, and reacting to issues are a cameraperson's qualities.

What do you like about your job? I love achieving a shot that others struggle with. I enjoy building a team and having a crew who I know will deliver terrific results and have fun doing it. Even when things go completely off script, a great team still delivers, remembering the viewer's passion.

FIGURE 7.17
Jon Lord, senior cameraman.

continued

Interview with a Professional: Jon Lord—*continued*

What are the types of challenges that you face in your position? I need to know enough about the equipment and techniques so that I can answer production questions and save the company time and money. Safety concerns are always very high on my list. There is always a safe way to achieve any shot. At the end of the shoot, I am ultimately responsible for equipment breakages and losses.

How do you prepare for a production? Preparation is the key to a good production. Find out as much about the event, the production, your crew, and the budget as possible. Then find out about what the audience needs to see, what are the program's key points, and what is the "must-see" action. You may miss shooting the ballerina's entry if you don't know which side of the stage she enters.

What suggestions or advice do you have for someone interested in a position such as yours? For me, it's about passion and respect. You must want to put in the effort; the financial reward isn't great, but the life experiences are. Unless you are willing enough to listen and respect the guidance of others, you probably will not be invited to join a team.

Jon Lord is a senior cameraman and lighting camera supervisor at Satellite Information Services in London. He previously worked in the same position with the BBC.

REVIEW QUESTIONS

1. How do you select the right shots?
2. Why does the medium (iPod vs. i-mag screen) make a difference in the shots that are used?
3. What significance does the tally light have?
4. Why is it important to look around the scene (away from the viewfinder) before recording?
5. What are some ways of dealing with a limited depth of field?
6. How does a shot sheet help the camera operator?

CHAPTER 8

THE PERSUASIVE CAMERA

"A film is never really good unless the camera is an eye in the head of a poet."

—Orson Welles

TERMS

Arc shot: A camera shot that moves around the subject in a circle or arc.

Deep focus: A very wide depth of field.

Depth of field: The distance between the nearest and farthest objects in focus.

Dolly shot: Moving the whole camera and mount toward or away from your subject. This shot does not require the use of an actual dolly.

Dutch: Tilting the camera is called a "Dutch" or a "canted" shot. This movement increases the dynamics of the shot.

Eye-level shot: Provide an image that is roughly at the eye level of the talent (in a studio show) or the average viewing audience.

High-angle shot: Provide a view from above the subject.

Low-angle shot: Provide a view from below the subject.

Pan shot: The pivoting of a camera to the left or right.

Tilt: Moving the camera up or down.

Truck (track) shot: The truck, trucking, or tracking shot is when the camera and mount move sideways (left or right).

Communicating your ideas visually can be challenging. This chapter will cover some of the most common techniques.

SHOOTING STYLE

The simplest way to cover a subject is to aim the camera at the subject and then zoom between long shots showing the general action and close-up shots showing detail. However, this mechanical, less-than-stimulating routine soon becomes very boring to watch. Creative techniques add to the subject's appeal and help hold the viewer's interest.

When pointing a camera at a scene, you are doing much more than simply showing your audience what is going on there. You are selecting specific areas of the scene (Figure 8.1). You are drawing their attention to certain aspects of the action. The way you use your camera will influence the impact of the subject on the audience.

In an interview, for example, the guests can be shot from a low angle, which will give them a look of importance or self-confidence. From a high angle, they might look diminished and unimportant. Concentrating on detailed shots of their nervous finger movements helps build a sense of insecurity.

The camera interprets the scene for the audience. How the camera is used affects the audience's responses. If the scene is just shot with no understanding of the impact of techniques, the result will be a haphazard production.

Whenever a camera is pointed at action, you have to make a series of fundamental decisions, such as:

- Which is the best angle? Can the action be seen clearly from there?
- Which features of the scene need to be emphasized at this moment (Figure 8.2)?
- Do you want the audience to concentrate on a specific aspect of the action?
- Do you want to convey a certain impression?

SCREEN SIZE

As discussed in previous chapters, the size of the screen on which the audience watches the production can influence how they respond to what they see there. It is more difficult to distinguish detail on a small screen (or a larger screen at a distance). The picture is confined and restricted, and we tend to feel detached as we closely inspect the overall effect. On the other hand, when watching on a large screen we become more aware of detail. Our eyes have greater freedom to roam around the shot. We feel more closely involved with the action. We are onlookers at the scene.

At typical viewing distances, most television receivers allow us to effectively present a wide range of shot sizes, from vistas to microscopic close-ups. Although wide shots of large-scale events and panoramic views are not particularly impressive on television, this limitation is not too restrictive in practice.

FIGURE 8.1

The camera isolates. The camera shows only what is going on in its frame of the scene. The audience does not really know what is outside the field of view.

Source: Photo by Josh Taber.

SHOOTING FOR THE INTERNET

Compressing video to be used on the Internet can deteriorate the overall quality of the video. Any time footage is compressed, a bit of the quality has to be sacrificed. Here are a few things to keep in mind when shooting something that will be streamed on the Internet:

- Do not use more camera motion than needed. Whenever there is camera motion, the result is more compression.
- Use a tripod to give the most stable shot possible (this usually should be done anyway). Camera pans and tilts should be limited and, when used, slow and smooth.
- Light the subject well.
- Keep the background simple. The more detailed it is, the more compression needs to happen.

SELECTING THE SHOT

Each type of shot has its specific advantages and disadvantages. Some are best for setting the scene; others allow the audience to see intense details and emotions. For example, long shots can be used to:

- Show where all of the action is taking place by establishing the scene.
- Allow the audience to follow broad movements.
- Show the relative positions of subjects.
- Establish mood.

However, long shots do not allow your audience to see details, and they may be frustrated at what they are missing. For example, in a wide shot of an art gallery, audience members may feel that they are being prevented from seeing individual paintings clearly (Figure 8.3).

Closer shots are usually used to:

- Show detail.
- Emphasize certain areas.
- Reveal people's reactions/emotions.
- Dramatize the event.

If too many close shots are used, the effect can be very restrictive. The audience can be left feeling that they were prevented from looking around the scene, from seeing the responses of other performers, from looking at other aspects of the subject, and from following the general action.

It is usually important to shoot both wide-angle shots and close-ups. This enables the audience to see the whole situation and establish the scene, as well as the close-up detail, showing drama and emotion. The key is to focus on the part of the scene that best moves your content forward.

A shot that is appropriate at one moment could be very unsuitable the next. In fact, there are times when an inappropriate or a badly timed shot can totally destroy an entire sequence.

The Extreme Long Shot (ELS or XLS)

The extreme long shot enables you to establish the location and to create an overall atmospheric impression. It can be used to cover very widespread action, or to show various activities going on at the same time. It could be a high shot from a hilltop or an aerial view, such as from a blimp at a sports venue (Figure 8.4). An extreme long shot of a person would show a lot of the background, which is often done to show context or solitude. With extreme long shots, the audience takes a rather detached, impersonal attitude, surveying the scene without any sense of involvement. The XLS is generally wider than the long shot. It usually shows much more than just the "field of play."

The Long Shot (LS)

Often used at the start of a production, the long shot immediately shows where the action is happening. This establishing shot sets the location and atmosphere. It allows the audience to follow the purpose or pattern of action (Figure 8.5).

As the shot is tightened, and shows less of the scene, the audience is influenced less by the setting and the lighting. The people within the scene have a greater audience impact; their gestures and facial expressions become stronger and more important.

Medium Shots (MS)

Medium shots are generally mid shots, although they may be framed a little larger or smaller. Their value lies in the idea that you are close enough to a person to see his or her expressions and emotions but far enough away to understand some of the context. Gestures can usually be captured in this type of shot. The MS is thought to be the one shot that "tells the story" (Figure 8.6).

FIGURE 8.2 *"Pulling focus" can be used to guide the viewer through a scene. Notice the change of focus in the two photos. The audience can also be directed where to look, because the eye is drawn to what is in focus.*

Source: Photos by Josh Taber.

FIGURE 8.3
Long shots let the viewer see where everything is taking place.

FIGURE 8.4 *Although the extreme long shot (ELS or XLS) does not provide detail, it definitely establishes the scene for the viewer. Most of the time, in sports, the ELS shows the viewer the entire field of play, plus the audience.*

The Close-Up (CU)

An extremely powerful shot, the close-up concentrates interest. With people, it draws attention to their reactions, responses, and emotions. Close-ups can reveal or point out information that might otherwise be overlooked, or only discerned with difficulty. They focus attention or provide emphasis (Figures 8.7 and 8.8).

The Extreme Close-Up (XCU or ECU)

The extreme close-up adds drama to the situation or clarifies a situation. By filling the screen with the face, it easily communicates the emotion of the situation. A close-up of an object allows the viewer to understand the detail a bit more (Figure 8.9).

FIGURE 8.5 *Both of these shots are long shots (LS). They set the scene. (Left) A long shot of the venue. (Right) A long shot of a person. In contrast, an extreme long shot of the first photo would have probably included the entire venue.*

Source: Derby photo by Josh Taber.

FIGURE 8.6 *These medium shots (MS), one from a sitcom and the other from a concert, demonstrate that the shot roughly cuts the subject in half. Generally, it is framed above or below the waistline of a person.*

Source: Concert photo by Paul Dupree.

FIGURE 8.7 *Close-ups (CU) of people are generally framed from mid-chest up. Note that you can see the framing of the interview on the stage's i-mag screen to the far right. It is a close-up.*

FIGURE 8.8 *Close-up shots of a scene allow the viewer to see the intricate details as compared to the long shot of this same scene.*

FIGURE 8.9

The extreme close-up (ECU or XCU) generally cuts into the face and is a great tool for adding drama by communicating the emotion of the moment.

Source: Photo by Paul Dupree.

When using a close-up, you have to ensure that the audience wants to look that close, and do not feel that:

- They have been cheated of the wider view, where something more interesting may be happening.
- They have been thrust disconcertingly close to the subject—the audience may become overly aware of facial blemishes in enlarged faces.
- Detail that is already familiar is being overemphasized.
- Through continually watching close-up fragments, they have forgotten how these relate to the main subject, or have become disorientated.

Depth of Field

The distance between the nearest and farthest areas of focus is called the *depth of field*. In the right situation (such as lots of bright light or a wide-angle lens), the depth can be huge. In other situations, the depth can be incredibly shallow (low light or shooting, a close-up shot, or a telephoto lens), requiring the camera operator to readjust the focus every time the subject moves forward or backward (Figure 8.10).

The depth of field varies with the following factors:

- The *distance* at which the lens is focused.
- The size of the *image sensors*.
- The *focal length* of the lens.
- The lens *f*-stop (*aperture*) (Figure 8.11).

Alter any of these elements, and the depth of field changes.

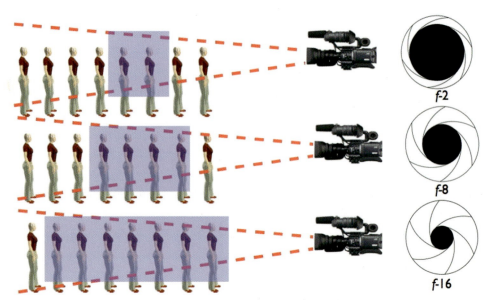

FIGURE 8.10
By adjusting the aperture (f-stops), the depth of field can be increased or decreased. The larger the f-stop number, the larger the depth of field.

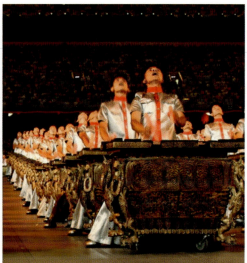

FIGURE 8.11
The lens aperture is adjustable in this illustration from a maximum of f/2 to a minimum of f/16. The larger the aperture (opening), the smaller the f-number (such as f/2), the shallower the depth of field. Less light is needed in this situation. The smaller the lens aperture, the larger the f-number (such as f/16), the larger the depth of field. More light is needed to shoot at higher number f-stops.

Large Depth of Field (Deep-Focus) Techniques

As you saw earlier, the depth of field in a scene varies with the lens f-stop, type of lens (wide-angle/telephoto), and focused distance. You can change it by altering any of these parameters.

Stopping down the lens (such as f/11 or f/16), everything from foreground to far distance appears sharply focused.

The camera has no problems in following focus and there is little danger of subjects becoming soft-focused. There is an illusion of spaciousness and depth, enabling shots to be composed with subjects at various distances from the camera. However, higher light levels are necessary (Figure 8.12).

One weakness of this technique is that when there is little camera movement or few progressively distant planes in the

FIGURE 8.12 *A large depth of field was achieved by using a wide-angle lens or a large-numbered f-stop, such as f/22. The bright sunlight in this photo allowed the large-numbered aperture.*

picture, it can appear unattractively flat. Surfaces or subjects at very different distances can merge or become confusing to the audience.

Shallow Depth of Field (Shallow-Focus) Techniques

Using a wider lens aperture (e.g., f/2) restricts focused depth. It enables you to isolate a subject spatially, keeping it sharp within blurred surroundings, and avoids the distraction of irrelevant subjects. You can display a single sharply focused flower against a detail-free background, concentrating attention on the bloom and suppressing the confusion of foliage. Sharply defined detail attracts the eye more readily than defocused areas.

By deliberately restricting depth of field, you can soften obtrusive backgrounds, even if these are strongly patterned, so that people or other subjects stand out from their surroundings. On the other hand, restricted depth can prove embarrassing when essential details of a close object are out of focus. The camera must continually refocus in order to keep a close-up of a moving subject sharp. For instance, close-up shots along a piano keyboard can demand considerable dexterity when trying to follow and focus on quickly moving fingers.

Occasionally, by changing focus between subjects at different distances (pulling focus, throwing focus), you can move the viewer's attention from one to another; however, this trick easily becomes disturbing unless coordinated with action. Blurred color pictures can be frustrating (Figures 8.13 and 8.14).

FIGURE 8.13 *The shallow depth of field was achieved by using a telephoto (narrow-angle) lens, along with a fairly small-numbered f-stop such as f/4.*

Source: Photo by Josh Taber.

Figure 8.14 The shallow depth of field shown here was achieved with the use of a telephoto (narrow angle) lens along with a fairly small numbered f-stop such as f/4.

Source: Photo by Chad Crouch.

MOVING THE CAMERA HEAD

In everyday life, we respond to situations by making specific gestures or movements. These reactions and actions often become very closely associated. We look around with curiosity, move in to inspect an object, and withdraw or avert our eyes from a situation that we find embarrassing, distasteful, or boring.

It is not surprising to find that certain camera movements can evoke associated responses in the audience, causing them to have specific feelings toward what they see on the screen. These effects underlie the impact of persuasive camera techniques.

SHOOTING IN 3D

"When shooting in 3D, everything is multidimensional. That means all the objects seen on the screen are typically in constant focus as they are with human vision. This is one of the reasons a rack focus effect in 3D looks uncomfortable to the eye and should be avoided."

—David Kenneth, President, I.E. Effects

Panning the Camera

The pan shot is the smooth pivoting of the camera from left to right on the camera support, which might be a tripod or even a person (Figures 8.15 and 8.16). Panning shows the audience the spatial relationship between two subjects or areas. Cutting between two shots does not provide the same

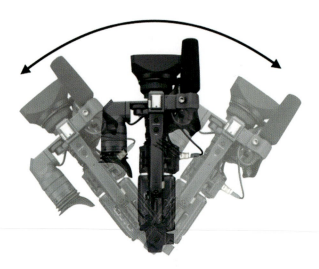

FIGURE 8.15 *The pan shot is the pivoting of a camera to the left or right.*

FIGURE 8.16 *The pan shot, as shown, smoothly moves from the red frame to the black frame, providing one continual shot when the action moves beyond the original red frame.*

Source: Photo courtesy of Dartfish.

sense of continuity. When panning over a wide area, the intermediate parts of the scene help us to orient ourselves. We develop an impression of space. However, it is important to avoid panning across irrelevant areas, such as the "dead" space between two widely separated people or subjects.

Unless you are creating a special effect, panning should be smooth—neither jerking into action nor abruptly halting. Erratic or hesitant panning irritates the audience. If the pan-and-tilt head is correctly adjusted, problematic pan shots usually only occur when using long telephoto lenses or when the subject makes an unpredicted move.

Follow Pan

The follow pan is the most common type of camera move. The camera pans as it follows a moving subject. In longer shots, the viewer becomes aware of the interrelationship between the subject and its surroundings. Visual interaction can develop between the subject and its apparently moving background pattern, creating a dynamic composition. In closer shots, the background becomes incidental, or even often indecipherably blurred (Figure 8.17).

Survey Pan

In the survey pan, the camera slowly searches the scene (a crowd, a landscape), allowing the audience to look around at choice. It can be a restful anticipatory action—providing that

FIGURE 8.17 *When following the subject in a wide arc with a handheld camera, the feet should face the midpoint of the arc. This will allow the camera operator to smoothly pan with the action.*

there is something worth seeing. It is not enough to pan hopefully.

The move can also be dramatic, building anticipation: the shipwrecked survivor scans the horizon, sees a ship . . . but will it notice him? But the surveying pan can build to an anticlimax, too; the fugitive searches to see whether she is being followed.

Interrupted Pan

The interrupted pan is a long, smooth movement that is suddenly stopped (sometimes reversed) to provide visual contrast. It is normally used to link a series of isolated subjects. In a dance performance, the camera might follow a solo dancer from one group to the next, pausing for a short while as each becomes the new center of interest.

In a dramatic application, you might see escaping prisoners slowly trek through treacherous marshland. One man falls exhausted, but the camera stays with the rest. A moment later it stops and pans back to see what occurred to the person.

Whip Pan

The whip pan (also known as the swish, zip, or blur pan) moves so rapidly from one subject to the next that the intermediate scene becomes a brief, streaking blur. Whether the effect generates excitement or annoyance is largely determined by how the preceding and following shots are developed. As our attention is dragged rapidly to the next shot, this pan gives each subject transitory importance. The whip pan usually produces a dynamic change that continues the pace between two rapidly moving scenes. A whip pan has to be accurate and appropriate to be successful: no fumbling, reframing, or refocusing at the end of the pan.

Tilting the Camera Head

Tilting refers to moving the camera up or down (Figure 8.18). Tilting, like panning, allows you to visually connect subjects or areas that are spaced apart. Otherwise, you would need to intercut different shots, or use a longer shot to include both subjects.

Tilting can be used:

- To emphasize height or depth—tilting up from a climber to show the steep cliff face to be climbed.
- To show relationships—as the camera tilts from the rooftop spy down to the person in the street below, or from the person in the street up to the rooftop, revealing that he is not alone.

FIGURE 8.18 *A camera tilt is when the camera is pointed up or down.*

Camera Height

Camera height can have a significant influence on how the audience perceives your subject. How you get the angle is immaterial. You can use a jib, suspend a camera on wires, or use any other mechanism. The key is whether the image does what you want it to do. There are three general categories that deal with camera height:

- *Eye-level* shots provide an image that is roughly at the eye level of the talent (in a studio show) or of the average viewing audience. This is the most common shot used in television and provides a sense of normalcy (Figure 8.19).

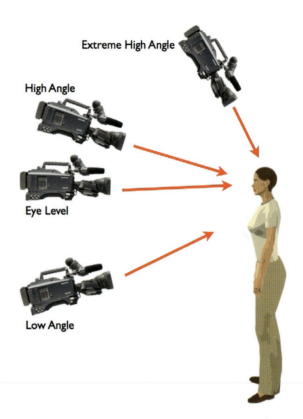

FIGURE 8.19 *The eye-level angle is used more than any other angle. The other angles are used when appropriate.*

FIGURE 8.20 *Although high-angle shots can help the audience see the big picture, they can also make the subjects look inferior or unimportant.*

FIGURE 8.21 *All high-angle shots do not make the subject look inferior. This shot allows the audience to see the drummer's movements.*

Source: Photo courtesy of Sennheiser.

- *High-angle* shots provide a view from above the subject. This high vantage point can provide the viewer with additional information, such as showing the actions and context of the subject. However, these shots can also give the viewer an impression that the subject is not important, or even inferior (Figure 8.20–8.22).

- *Low-angle* shots make the subject appear more important and very strong. These shots make the viewer feel inferior (Figures 8.23 and 8.24).

FIGURE 8.22 *High-angle shots can be obtained in a number of ways, such as holding the camera high, using a jib, climbing onto a higher area, or shooting from a helicopter.*

Source: Photos courtesy of Sony and Jeff Hutchens.

Figure 8.24 Low-angle shots are obtained by placing the camera lower than the subject. Even jibs, primarily used for high-angle shots, can be used to obtain low-angle shots.

Source: Photo by Paul Dupree.

Figure 8.23 Low-angle shots generally can make the subject look important, strong, or powerful.

Source: Photo by Josh Taber.

Extreme Camera Angles

Extreme angles can be creative and attention-grabbing. If appropriate, they can add a lot to the production. However, they sometimes draw attention to the abnormality or ingenuity of the camera's position. If the audience is wondering how we got that shot, techniques have obscured artistic purpose (Figure 8.25).

Where extreme angles appear naturally, viewers accept them readily: looking down from an upper-story window; looking up from a seated position; even an eavesdropper peering through plank flooring to the room below. But an unexplained extreme shot usually becomes a visual stunt.

FIGURE 8.26 *Dollying in causes increased interest and a buildup of tension. However, the close-up may result in diminished interest. Dollying back usually results in lowered interest or relaxed tension— unless unseen subjects are revealed, or when curiosity, expectation, or hope has been aroused. Attention tends to be directed toward the edges of the picture. Dollying appears faster than it really is on a wide-angle lens, and slower on a narrower lens angle.*

FIGURE 8.25 *Extreme angles sometimes draw attention to the novelty of the shot, distracting the audience from the main subject.*
Source: Photo by Tyler Hoff.

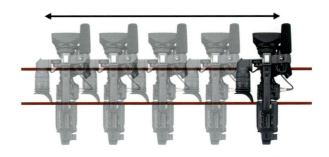

FIGURE 8.27 *A truck, track, or tracking shot moves the camera and mount sideways (left or right).*

MOVING THE CAMERA

How freely the camera can be moved around is determined by the type of camera mount used. Although a jib offers incredible flexibility, it may not be able to relocate as rapidly as a handheld camera. Well-chosen camera moves add visual interest, plus influence certain audience reactions. However, camera movement needs to be motivated, appropriate, smoothly controlled, and done at a suitable speed, or it can become distracting and/or disturbing. The most common camera moves can be seen in Figures 8.26–8.29.

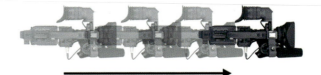

FIGURE 8.28 *A dolly shot is the action of moving the whole camera and mount slowly toward or away from the subject (whether an actual dolly is used or not).*

FIGURE 8.29 *The arc shot moves the camera around the subject in a circle or arc path.*

Subjective Camera Treatment

When you use the camera subjectively, you are portraying an individual's perception of the event. The subjective camera makes it look as if the viewer is actually walking around the scene, following the action. The camera can walk through a crowd, move up to inspect something, and then glance up at nearby details. We encounter this approach regularly when shoulder-mounted or Steadicam-type cameras are used instead of shooting from a stationary position on a tripod.

Subjective camera movement creates a participatory effect for the audience. But if the director moves the camera when the audience is not ready, or fails to show them something they wish to see, the audience will probably feel resentful. The skilled director persuades the audience to want a change of view or a move. The unskilled director thrusts it upon the audience.

THE MOVING CLOSE-UP

Close-ups of people allow the subject to dominate the image; its strength is determined by the position of the camera. Although the influence of the environment can be limited with the close-up, the pace varies with dynamic composition. When the subject is slightly off-center in the direction of movement, a sense of anticipation and expectancy is created. The following are some thoughts about close-ups of people:

A. Profiles of people are weak against a plain background. However, there is an impression of speed and urgency if shot against a detailed background.

B. A three-quarter frontal shot can be dramatically strong. By preventing the viewer from seeing the subject's route or destination, a sense of anticipation can be built.

C. A high-angle elevated frontal shot weakens the subject but still allows him or her to dominate the environment.

D. A low-angle shot makes the subject look powerful and dominating.

E. A three-quarter shot from behind the subject can be subjective, because the viewer moves with the subject, expectancy developing during the movement.

F. A high-angle shot from behind is not only highly subjective, but also produces an increased anticipation—almost a searching impression.

G/H. In level and low shots, there is a striking sense of depth. The subject is strongly linked to the setting and other people, yet remains separate from them.

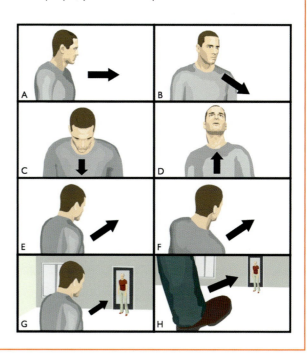

FIGURE 8.30 *The moving close-up shot.*

Imitative Camera Movement

Cameras can be moved to suggest jogging or the rolling movement of a ship. The camera can significantly affect the dynamics of the subject. If used effectively, camera movements can even provide subjective influence on the action itself (Figure 8.31).

Movement Using the Zoom Lens

As discussed earlier, the zoom lens brings both advantages and pitfalls for the unwary. It is too easy to change the lens angle just to change subject image size. There is always the temptation to stand and zoom, rather than move around with a normal lens. Zooming demands little of the camera operator or director. There is just the need for a pre-zoom focus check before zooming in for the shot.

Zooming is extremely convenient. However, the zoom only simulates camera movement. The zoom optically isolates a section of the scene. There are no natural parallactic changes as you zoom in; scale, distance, and shape become distorted through zooming. A slow zoom made during panning, tilting, or subject movement may disguise these discrepancies. A rapid zoom during an exciting fast-moving ball game can make the image more dynamic. Much depends on the occasion.

Zooming can provide a visual bridge from the wide view to the close-up, without the time and effort involved in dollying or the interruption (and possible disorientation) of cutting (Figure 8.31). A rapid zoom-in produces a highly dramatic swoop onto the subject. An instant (snap) zoom-in

flings subject detail at the audience (Figure 8.32). Such effects can be incredibly dynamic, or just plain annoying—it has to be done right.

Zoom should be smooth and decisive. Use zoom shots to direct attention, to increase tension, to give powerful emphasis, or to restrict the coverage. But the zooming action itself should be used discriminately for specific occasions.

Directors use different terms when calling for a zoom in or out. Some of the most commons terms for zooming out are "zoom-out," "pull-out," or "widen." When needing a zoom-in, the director may ask for a "zoom-in," "push-in," or "tighten."

FIGURE 8.32 *Zooming can provide a bridge between close-ups and long shots. In these shots, zooming out from the CU to an LS provides the context for the subjects.*

FIGURE 8.31 *In order to increase the dynamics of this shot, the camera was tilted, which is usually called a "Dutch" or "canted" shot.*

FIGURE 8.33 *A snap zoom quickly moves the audience from a CU to an LS, or vice versa.*
Source: Photo by Will Adams.

INTERVIEW WITH A PROFESSIONAL: MATT GRIMM

FIGURE 8.34
Matt Grimm, producer.

Briefly define your job. I work closely with the executive producer and/or series producer to deliver a finished program that will meet goals/expectations. I coordinate production staff (camera, audio, graphics, editors) to develop and ultimately deliver finished programs. In recent years, I have worked mostly on long-form documentaries.

What do you like about your job? I enjoy the variety in my job. A career in television has opened doors of opportunity and access that I may not have been given otherwise. Projects have allowed me to travel overseas, have access to restricted areas, and interview many interesting people. I also enjoy the sense of accomplishment that comes after spending a year or two on a production, then being able to see a project complete. You learn all you can about a subject, pour your time and energy into examining it, and then share it with your audience. It is wonderful to see it all come to fruition.

What are the types of challenges that you face in your position? Every project brings its own unique challenges to overcome. Budget constraints, production scheduling, and shooting logistics can all complicate a production. Universally, you are challenged to connect with your audience. If your audience isn't enlightened, moved emotionally, or entertained, then you have likely missed the mark. As a producer, I am charged with reaching our audience. However, I'd like to not just share a story with them, but share it in an impactful way.

How do you prepare for a production? Preproduction is huge. When planning a program, I first consider the goals/expectations for it. How is it going to be different than others already created? Who is the audience to be reached? Before ever picking up a camera, I like to meet with potential interview subjects. Get to know them a bit. This allows them to become comfortable with me and gives me a better feel for what they know. The shape of a production will present itself during this planning stage. Preproduction meetings with the camera operator and audio tech ensure that everyone is on board with the look and feel for the production. Creative decisions and potential challenges are discussed and agreed upon before ever heading into the field for shooting.

What suggestions or advice do you have for someone interested in a position such as yours? It may sound cliché, but work your way up. Learn all you can during school and in your entry-level positions. Hands-on experience is so valuable and cannot be replicated. For someone to become a great producer or director, I think he or she needs to be a great videographer or editor first. Television production is a collaboration, bringing together so many different specialties. Having an understanding of each of them will help you eventually master one.

Matt Grimm is a producer for Kentucky's PBS station KET. Some of his productions have also aired nationally on PBS.

REVIEW QUESTIONS

1. Each type of shot (CU, MS, and LS) conveys different information. Explain those differences.
2. How does the camera interpret the scene for the audience?
3. What are some of the challenges of shooting for the Internet?
4. How is deep focus obtained in an image?
5. How do you change the perspective of an image without moving the camera?
6. What are two of the types of pan shots and how can they be used?
7. What are the advantages and disadvantages of camera movement?

CHAPTER 9

CREATING AN
EFFECTIVE IMAGE

"There are three main responsibilities for the camera operator.
First is the ability to work well with others. Second, you must
know your equipment well. Third, the application of the
operator's own talents, physical ability, intuition, and patience."
—Martin Goldstein, Camera Operator

TERMS

Animation (image composition): The video images should give the audience the same emotional response that you had while shooting.

Axis of action line: Also known as an "eyeline," the "180 line," or the "proscenium line," this is the line along the direction of the action in a scene. It is the line that separates the "stage" from the audience. Cameras should only shoot from one side of this line.

Composition: Creating an image that is attractive or that captures and keeps the audience's attention and effectively communicates the production's message.

Context: Making sure that the content of the image allows the viewer to understand the subject better; composing the shot in such a way that it includes a background or foreground that adds additional information or context to the image.

Continuity: Making sure that the shots will edit together in the final production to avoid ending up with a series of shots that do not fit together smoothly. This happens especially often when repositioning the camera to shoot a repeated scene from a different location.

Cutaway shot: Used to cover edits when any video sequence is shortened or lengthened. Generally, it is a shot of something outside of the current frame.

Headroom: The amount of space above the head. This changes proportionally with the length of the shot, lessening as the shot tightens.

Symbolism (image): Creating an image that is meaningful to the viewer.

At the end of the day, it's how you see the world—the camera, no matter how simple or complex, is just the means to record your vision, not the producer of it.

Why do some pictures look very attractive, while the eye passes over others disinterestedly? Why do some draw the eye to a specific subject, while in others we look around? Effective images are much more than just a pretty picture; they are images that effectively communicate the mood, emotion, meaning, symbolism, and/or context needed to communicate the intended message. This chapter will explore what makes an effective image.

BEHIND THE PICTURE

Most people think that the process of creating an image is the responsibility of the camera operator. For the single-camera operator, that is largely true. But in the studio, where the entire environment is contrived, the final image is usually the outcome of a combination of a number of different people's talents. The design of the setting, the way it has been lit, and the angle selected by the director can all be controlled to provide the appropriate conditions for maximum impact on the audience.

The camera operator's opportunities to create a meaningful composition depend directly on the way the scene has been developed and how action has been arranged. Stand someone in front of a flatly lit plain background, and the prospects for interesting shots are very limited. However, sit the same person in a well-designed, attractively lit setting, and the camera can explore the situation, producing effective shots.

Persuasive images are the result of successful planning. If the lighting is inappropriate, or the director selects an ineffective camera position, the result will be inferior images—and lost opportunities.

COMPOSING THE PICTURE

The goal of composition is to create an image that is attractive or that captures and keeps the audience's attention and effectively communicates the production's message. It is a way of arranging pictures so that the viewer is directly attracted to certain features (Figure 9.1). You can influence how viewers respond to what they are seeing. The image can be composed to create anticipation, unease, apprehension, excitement, and restful calm. The mood can be anywhere from depressing to exciting—and anywhere in between.

It is always tempting to devise shots that are different— shots that make the eye stop and wonder. Wildly distorted perspectives from a close wide-angle lens, very low-angle

FIGURE 9.1 *Why be concerned with composing the image? The unguided eye will wander around the scene, finding its own areas of interest.*

shots, or pictures using weird reflections are fine when you need them for a dramatic or comic effect. But extremely unusual viewpoints don't just make a picture look different; they also draw attention to themselves. They may distract the audience from the real subject.

Composition principles or "rules" are really guides. Composition is up to the person creating the image. It is his or her responsibility to create an image that meets the needs of the intended audience. That means that effective composition can be translated in different ways based on the interests, styles, and age of the viewers.

Practical Composition

Camera operators or directors can adjust an image's composition in a number of ways:

* *Adjust framing*: Positioning the shot to deliberately include/exclude parts of the scene, or to alter the subjects' position in the frame.

* *Increase or decrease the lens angle of view*: The lens angle of view (wide angle or telephoto) will determine how much of the scene appears in the picture from that viewpoint.
* *Adjust the camera position*: As the camera moves up/down or sideways, foreground objects change position in the frame more noticeably than distant ones. So even slight readjustments can considerably alter the compositional relationships.
* *Change the shot proportions*: By altering the lens angle, and changing the camera distance to compensate, you can keep the same size shot but adjust proportions within it (Figure 6.25).

The Director and Composition

Directors vary on how they deal with composition. For many productions, the director is so preoccupied with what is being said, and with performance, continuity, and techniques, that he or she does not arrange the specific composition of shots. Instead, the director indicates the shot size required (CU,

---

COMPOSING THE SHOT

Good composition does not have to be difficult. However, it does take careful planning to get the best image. Here are some key factors used to shoot images that effectively communicate the message of the production:

* **Symbolism**: Does the image have meaning to the viewer? When the viewer sees the image, what does he or she immediately think of? Is that what you are trying to communicate?

* **Context**: The content of the image should allow the viewer to understand the subject better. Compose the shot in such a way that it includes a background or foreground that adds additional information or context to the image (Figure 9.2).

* **Animation**: The video images should give the audience the same emotional response that you had while shooting. Does the image portray emotion or motion in some way?

FIGURE 9.2 *Since this image was shot within a meaningful context, the viewer knows that it was shot in Italy.*

Source: Photo by Sarah Owens.

two-shot, group shot) and leaves the details of lens, exact framing, and so on to the camera operator.

In other types of production, the director deliberately groups actors to provide specific compositional arrangements for the camera—for dramatic effect, or to direct the audience's attention. In some cases, the director may have prepared a storyboard sketch (see Chapter 5) showing the detailed composition of certain key shots.

Composition Principles

Composition principles are not laws. They are indications of how people respond to a specific design of the image. These are important, in that if you do not organize images appropriately, your audience may react by looking at the wrong things, interpreting the picture inappropriately, or becoming bored by unattractive shots. Composing shots is not just a matter of stunning images, but a method of controlling the continuity of thought.

THE EFFECT OF THE PICTURE FRAME

The camera does much more than just "put a frame" around a segment of the scene. It inherently modifies whatever it shows. Because the screen totally isolates its subjects so that the viewer cannot see whatever else is happening (Figure 9.3), and because the resulting image is flat, unique relationships develop within it that are not present in the actual scene.

Few shots directly portray reality. Former U.S. President Richard Nixon once said that "while a picture does not lie, it does not necessarily tell the truth." In many cases, our own experience enables us to rationalize and interpret, so that we make a pretty accurate assessment of what we are seeing.

Unfortunately, there is no formula for the perfect image. However, there are some basic composition guidelines, described in the following sections.

Framing

You choose exactly what the viewer is going to see: what is to be included within the picture, and what is to be excluded from it. You may be selecting to concentrate attention, to avoid distractions, or to show more subject detail. You might even omit information deliberately—and then reveal it in a later shot.

As all parts of the frame do not have equal pictorial value, the effect of the image changes depending on where you

FIGURE 9.3 *Because you can isolate the viewer from the rest of the scene, the first photo looks like a normal house. However, the audience has no idea that the scene is actually part of a backlot of a film studio and is part of a soundstage, as shown in the second photo.*

place the main subject. How the shot is framed will not only alter compositional balance, but can also influence the audience's interpretation of events. Framed in a certain way, a two-shot might lead the viewer to expect that someone is about to enter the room, or that an eavesdropper is right outside the door.

Headroom (the amount of space above the head) changes proportionally with the length of the shot, lessening as the shot tightens. In a multi-camera production, the headroom can vary considerably between different cameras. So it is important for the director to check that comparable shots match (Figure 9.4).

One of the problems with monitors across the world is that they are not all the same size. Different manufacturers' technologies differ and the images on the monitors sometimes drift with age. With that in mind, it is important that the

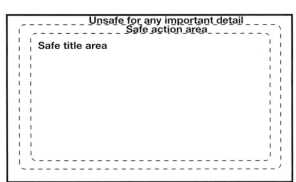

FIGURE 9.4 *The first image has too little headroom, the second image has too much headroom, and the third one is correct.*
Source: Photo by Josh Taber.

audience does not miss any important details, such as titles. Always compose the shot so that everything essential fits into the "safe area" shown in Figure 9.5. The safe area should include all graphics and video images that are essential for the viewers.

Framing the image within the video screen is a framing guideline. Framing occurs when the camera operator selects something that will appear in the foreground, creating a frame made of a fence, a building, a window or doorway, a tree, and so on. The frame object must not distract the viewing audience from the primary subject on the screen (Figure 9.6).

Pictorial Balance

Good camera operators and directors strive for balanced composition; not the equal balance of formal symmetry, as that can be boring, but an image with equilibrium.

Balance in an image is affected by:

- The size of a subject within the frame.
- Its tone.
- Its position within the frame.
- The relationships of the subjects in the shot.

A balanced picture unifies the subjects within a shot. Although images can be deliberately arranged so that they are unbalanced to create a dynamic tension, make sure this is done sparingly. Balanced arrangements do not have to be static. Shots can be continually readjusted to balance the image by moving a person, altering the framing, and so on. This readjustment allows you to redirect attention to a different subject, or to alter the picture's impact. Balance is a very subjective effect. You cannot measure it. But there are a number of useful guiding principles:

- While centering the object in the center of the frame is OK and safe, it can be very dull to view (Figure 9.9).
- A subject or object on one side of the frame usually requires some type of counterbalancing in the remainder of the shot. This can be an equal opposite mass, giving

FIGURE 9.5
The picture edges are sometimes lost on the screen due to different sizes of monitors and the aging of monitors. To ensure that no important action or titling is lost, keep it within the borders shown.

FIGURE 9.6 *Framing the image can take place many different ways: windows, trees, foreground objects, and so on.*

Source: First photo by Zachary Brewer and third photo by Sarah Owens.

symmetrical balance, or a series of smaller areas that together counterbalance the area (Figure 9.10).

- Tone significantly influences visual weight.
- The darker-toned subjects look heavier and smaller than light-toned subjects.
- A small darker area, slightly offset, can balance a larger light-toned one further from the picture's center.
- Darker tones toward the top of the frame produce a strong downward thrust—top-heaviness, or a depressed closed-in effect. At the bottom of the frame, they introduce stability and solidity (Figure 9.11).
- People's eyes are always drawn to the brightest area of the image. That means that the audience can be told where to look by making that area a little brighter (Figure 9.12).
- Regularly shaped subjects have greater visual weight than irregular ones.
- Warmer colors (red, orange) appear heavier than cooler ones (blue, green); bright (saturated) hues look heavier than desaturated or darker ones.

FIGURE 9.9 *(Top) The subject is centered, causing the subject to be unbalanced and thus a bit boring for the viewer. (Bottom) By moving the subject slightly to the left, or panning the camera to the right, the subject has more room to talk and the image has more scene depth and image balance.*

There are many ways to change the balance and emphasis in an image:

- Change the lens (zoom in or out).
- Adjust the aperture to adjust the focus (depth of field).
- Alter the camera position, which also may require changing the lens (zoom).
- Move or adjust the subjects.

- Change the camera height.
- Adjust the lighting.

Unity (Order)

When the composition of a shot is unified, all its components appear to fit together and form part of a complete pattern (Figure 9.13). Without unity, one has the impression of randomness, of items being scattered around the frame.

FRAMING PEOPLE

- Don't let the frame cut people at natural joints; intermediate points are much more attractive (Figure 9.7).
- Avoid having people seem to lean or sit on the edge of the picture.
- If the shot is framed too close to contain the subject's motion, the subject will keep moving in and out of the picture. The result is very disconcerting and distracting (Figure 9.8).

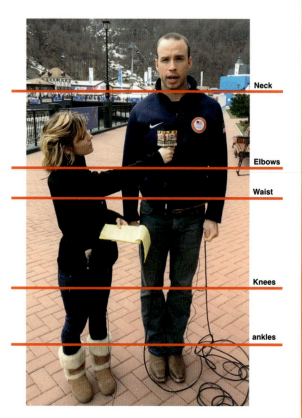

FIGURE 9.7 *Avoid framing people at natural joints. It looks more natural to go above or below bends of the body.*

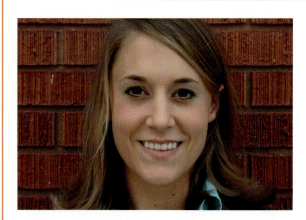

FIGURE 9.8 *If shooting too close, the camera operator may not be able to keep up with the movement of the subject.*
Source: Photo by Josh Taber.

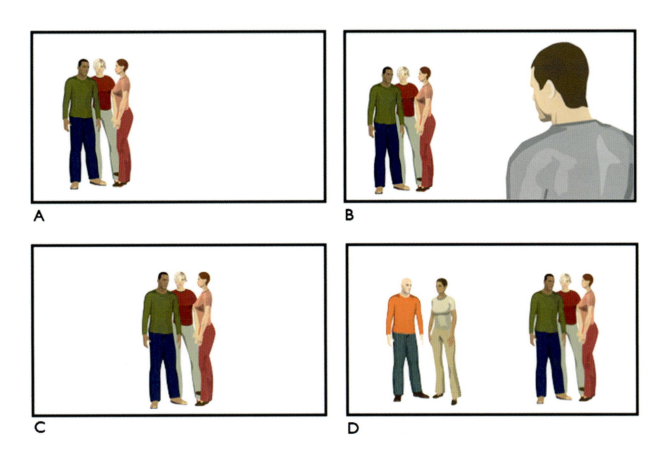

FIGURE 9.10 *A group that would look lopsided and unbalance the picture (A) can be counterbalanced by another mass in another part of the screen (B). If centered (C), the picture is balanced, even without other subjects, but continual centering gets monotonous. Different-sized masses can balance each other, but take care not to split the audience's attention (D).*

FIGURE 9.11 *Darker areas at the bottom of a picture project stability and solidity.*

FIGURE 9.12 *Note that your eye is always drawn to the brightest spot in the photo.*

FIGURE 9.13 *By slightly moving the talent and/or changing the camera's position, the separated people in the image on the left became more of a unified subject, as shown on the right.*

Visual Patterns

The eye is attracted to a variety of patterns. A solemn, quiet mood is suited by slow, smooth-flowing lines; a rapid, spiky staccato would match an exciting, dramatic situation (Figures 9.14–9.16).

Lines can create specific feelings:

- Horizontal lines can portray calm and tranquility.
- Vertical lines show strength and dignity.
- Diagonal lines can show movement and speed.
- Curved lines can portray serenity.
- Converging lines show depth.

Leading Lines

Leading lines occur when the lines within the image lead the viewer's eyes to whatever the director wants them to look at. The eyes naturally follow the lines to the subject at the convergence point (Figure 9.17).

Rule of Thirds

If the screen is divided into even proportions (halves, quarters), either vertically or horizontally, the result is generally pretty boring. For example, a horizon located exactly halfway up the frame should be avoided.

FIGURE 9.14 *Horizontal and curved lines provide a feeling of rest, leisure, beauty, and/or quiet.*

FIGURE 9.15 *Vertical lines give a feeling of strength and dignity.*

FIGURE 9.17 *Converging leading lines portray depth in this image.*

Source: Photo by Chris Jensen.

FIGURE 9.16 *Dynamic (diagonal) lines provide a feeling of speed, vitality, excitement, or drama.*

FIGURE 9.18

The rule of thirds suggests that the main subject should not be in the exact center of the image.

Source: Photo by Sarah Owens.

FIGURE 9.19 *While placing the main subject in the exact center of the image allows formal balance, it can be boring. By placing the subject slightly to the left or right of the frame's centerline, the image becomes more dynamic.*

Source: First photo by Chad Crouch.

The rule of thirds is a useful aid to composing the picture. Divide the screen into thirds both horizontally or vertically (Figure 9.18). The main subject should be on one of those lines, or ideally on the intersection of two of the lines. The rule of thirds suggests that the main subject should not be in the middle of the image (Figures 9.19–9.21). Instead, it should be placed before or after the center of the image, depending on the effect the director would like. Keep in mind that the rule is merely a guideline—sometimes it may be closer to a fifth or somewhere in between. Good camera operators instinctively compose shots with these features at the back of their mind.

FIGURE 9.20 *When shooting a moving subject, it is important to put space before or after it, providing subtle meaning and balance. Image A shows the subject in the center, which lacks meaning. By moving, putting space in front of the subject (B), it looks as though the subject is going somewhere. Placing space behind the subject (C) makes the audience believe he or she is coming from somewhere, or being followed.*

Context

By providing contextual information within the image, the audience will better understand the subject. For example, context can include a background that shows where the event is taking place, or some tools that help the audience realize what the person's job is or what he or she is doing. Context provides visual cues to the audience that do not need to be scripted (Figure 9.22).

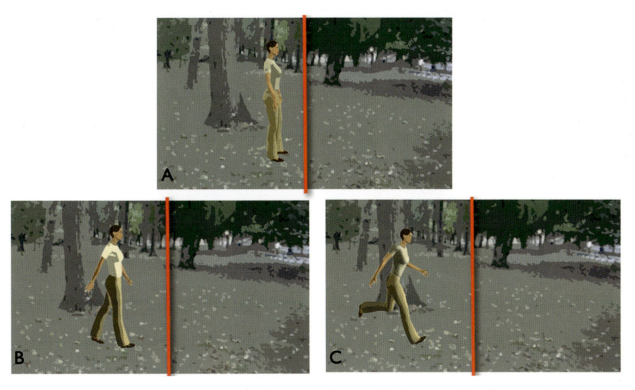

FIGURE 9.21 *When a subject moves, it should be kept slightly behind the center of the frame (the red line). The amount of offset should increase as the speed increases.*

FIGURE 9.22 *The buildings in the background give context to the image. They provide additional information to help the audience understand more about the individual.*

Source: Photo by Lynn Owens.

Scale

We judge how large or small a thing is by comparing its size and proportions with familiar items that we recognize in the picture. Without a comparative scale, we can only guess. A finger moves into shot, and we realize that a chair we are looking at is a skillfully made miniature. The eye is easily fooled, as you can see in Figure 9.23.

Perspective lines can influence our impression of scale and relative size. Again, although there is no relationship between shot size and the size of the subject, we often assume when an unfamiliar subject fills most of the frame that it is larger than it really is.

Subject Prominence

A subject's surroundings have a considerable influence on our attitude toward it. Consider the difference between a coin imposingly displayed on a velvet cushion and one heaped with others in a rusty junk box. Depending on how you present a subject, it can appear important or trivial. It can look powerful or weak; interesting or incidental. It can even be overlooked altogether (Figure 9.24)!

Isolation gives a subject emphasis. You can create this emphasis in many ways:

- By contrasting tones.
- By the camera height (Figure9.25).
- By the composition of the picture (Figure 9.26).
- By the subject's position and size relative to its surroundings.
- By using background pattern or form to make the subject look more prominent (Figure 9.27).

Impressions of the Subject

How the subject is shown by the camera can modify its on-screen strength as understood by the viewer—whether they look forceful, scared, or submissive. A person's general posture is significant, too. Weak attitudes include side or rear views, lying down, looking down, bowed, stooping, clasped hands, and slow movements. Strong attitudes include frontal view, up-tilted head, hands clenched, stamping feet, and fast movement.

FIGURE 9.23 *If shooting something unusually small or large, show it next to something that the audience is familiar with so that the audience can understand that size.*

FIGURE 9.24 *Simplify the background. The first photo shows an airplane in front of a ship—the airplane is difficult to see. By shooting the plane with a simple background, the audience is drawn to the airplane.*

Source: Photos by Sarah Owens.

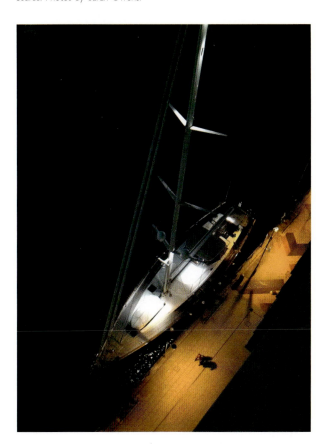

FIGURE 9.25 *The height of the camera emphasizes the boat by isolating it.*

FIGURE 9.26 *While the tree is beautiful, the portable restrooms in the background distract the audience. By simply moving the camera to another angle, the attention is drawn back to the tree.*

FIGURE 9.27 *(Left) By isolating the pin, all attention is drawn to it. (Right) Even though the pin is still prominent, the coins are distracting to the audience.*

FIGURE 9.28 *The low-angle shot of this horse makes it look incredibly strong and powerful.*

Source: Photo by Sarah Owens.

Camera treatment can have a considerable influence on our reactions to a subject. Shooting a ranting dictator in a high-angle long shot would make his or her gestures appear futile, weak, and ineffective. However, a low-angle mid shot would give him or her a powerfully dramatic force (Figure 9.28).

When you use a strong camera treatment on a weak character, you actually strengthen him or her. For example, an elderly woman making weak submissive gestures could seem to have a dignity and an inner strength against adversity when shot from a low viewpoint.

Composition and Color

The shades of color in an image can significantly impact the viewer: color and emotion are inextricably interlinked. It is important to think about color as images are composed, graphics created, and sets designed. Here are a few examples (Figures 9.29 and 9.30):

- Red—warmth, anger, excitement, power, strength.
- Green—spring, macabre, freshness.
- Yellow—sunlight, treachery, brilliance.
- White—snow, delicacy, purity, cold.

Composition and Motion

Composition that appears forceful in a "frozen moment" still photograph often goes unnoticed during motion. For example, a photograph of a horse leaping a fence can look dynamic, strong, and graceful. However, a moving image of the same action may seem quite commonplace and unexciting.

FIGURE 9.29 *White and blue are generally associated with coldness. Viewers take that stereotype with them when viewing other images that are white and blue, allowing the director to manipulate a shot so that the audience reacts the way he or she wants them to react.*

FIGURE 9.30 *The color yellow is often associated with sunshine, brilliance, and sometimes treachery.*

A still photograph seldom holds the attention for long; the moving picture offers continued interest. Movement attracts. Through change, the director can introduce variety into the image, holding the audience's attention on the chosen subject. Changes enable a director to:

- Alter a subject's prominence.
- Redirect the viewer's attention.
- Add or remove information as the audience watches.
- Transform the mood of the scene.
- Show movement, growth, and development.

This brief list reminds us of the potential persuasive power of the moving image.

The still photograph allows viewers freedom to scrutinize and assess; the moving image often gives them time to grasp only the essentials, before it is replaced by the next—particularly if action is fast or brief.

If the information in a succession of images is obvious, simple, or familiar to the audience, changes can be rapid. An informed audience may actually find the fast pace exciting as they rapidly assess each new shot. However, if the pace is too fast for the audience, they are likely to get frustrated as each shot disappears before they have finished looking at it. In the long run, their attention and interest will deteriorate.

Movement can be dynamic and stimulating—or confusing and tiring. It depends on the subject, and on your audience.

A Theory of Dynamic Composition

In daily life, we continually make very subjective judgments as we assess the movements and speed of things around us. As we drive through flat, featureless country, we feel that the car is traveling at a much slower speed than the speedometer shows. Driving along a tree-lined road, the reverse happens, and we tend to overestimate speed (Figure 9.31).

Watching moving pictures, we carry over these arbitrary interpretations. We cannot see speed and movement in the picture; we can only make interpretations from the clues that are present in the scene. Against a plain white sky, a fast-moving aircraft appears stationary. When we see the landscape beneath streaking past, we have an impression of speed (Figure 9.32).

Here are a few typical ways in which we react to movement in an image:

- *Effort*: Slower speeds can suggest effort, or that motion is difficult, especially if it is accompanied by sounds that are similarly associated (low-pitched, forceful, percussive). By replaying a normal action in slow motion, the impression of effort involved can be increased. We also interpret the amount of effort, from signs of strain, tension, slipping, and so on, even when these have been faked.
- *Relative speeds*: We assess the speed at which someone is running toward us by the rate at which his or her image size grows (shot with a long telephoto lens, this increase is slow, so we lose our sense of speed).

- *Gravity*: Although gravity is irrelevant, we subconsciously associate movement and position within the frame with gravitational forces. Something moving from top to bottom appears to be moving downward, sinking, falling, or collapsing. Moving from bottom to top of the shot, it is rising against gravity, floating, or climbing.
- *Fixation point*: The visual impact of movement can depend on where we happen to fix our attention. Looking

skyward, we see moving clouds and static buildings—or static clouds and toppling buildings.

- *Strength*: Something that is large in the frame and is moving toward the camera appears to grow stronger and more threatening. Seen from a side viewpoint, the same action (e.g., a truck backing up) can seem quite incidental.

DIRECTION AND SPEED

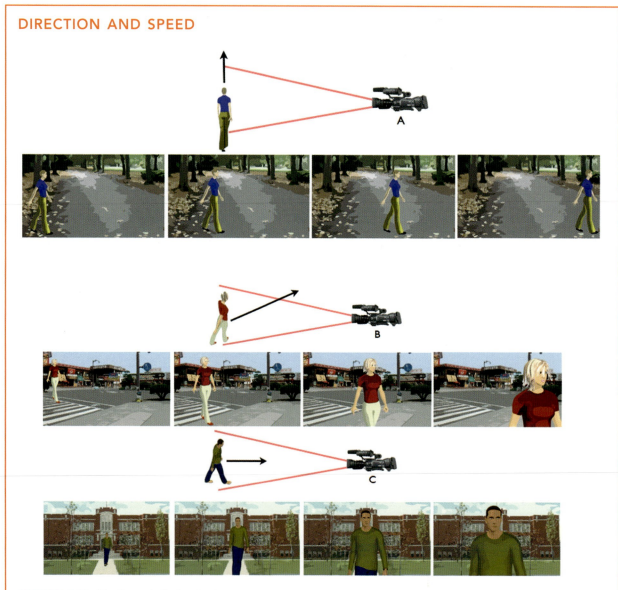

FIGURE 9.31 *Direction and effective speed:*

A. Talent moving across the screen quickly passes out of the shot.

B. Diagonal moves are more interesting (and take longer).

C. Moves toward (or away from) the camera are sustained the longest. However, they may take too long.

FIGURE 9.32 *Seeing that the background is blurring by, the viewer understands that there is fast action.*

Using Dynamic Composition

Dynamic composition is used by the media daily. Some widely accepted working principles have emerged:

- *Direction of movement*: Like vertical lines, vertical movement is stronger than horizontal.
- A left-to-right move is stronger than a right-to-left action (Figure 9.33).

- A rising action is stronger than a downward one. For example, a rise from a seated position has greater attraction than a downward sitting movement.
- An upward move generally looks faster than a horizontal one.
- *Diagonal movement*: Like diagonal lines, this is the most dynamic movement direction (Figure 9.34).

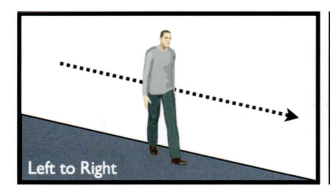
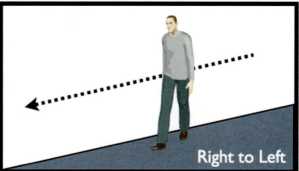

FIGURE 9.33 *The direction of the slope can alter an image's attractiveness to the viewer. The left-to-right version has a tendency to be more interesting than a right-to-left version.*

Figure 9.34 Diagonal movement is the most dynamic movement direction.

Figure 9.35 Movements toward the camera are usually more powerful than movements away from the camera.

- *Movement toward the camera*: All forward gestures or movements are more powerful than recessive action away from the camera: a glance, a turned head, a pointing hand. Similarly, a shot moving toward a subject (dolly/zoom-in) arouses greater interest than one withdrawing from it (dolly back/zoom-out) (Figure 9.35).
- *Continuity of movement*: A moving subject attracts attention more readily than a static one, but continuous movement at constant speed does not maintain maximum interest. When action is momentarily interrupted or changes direction, the impact is greater than one carried straight through. Converging movements are usually more forceful than expanding ones.

Crossing the Line

Camera angles can easily confuse the audience's sense of direction and their impression of spatial relationships if care is not taken when selecting camera positions. For example, during a basketball game, if cameras are placed on both sides of the court, it is confusing to see a player running toward the left side of the screen and then, when the director cuts to the camera on the other side of the court, to see the player running toward the right side of the screen.

To avoid this happening, draw an imaginary line along the direction of the action (called the action line, axis of action, or eyeline), as shown in Figure 9.36. Then be careful that the cameras shoot from only one side of this line—generally it is not crossed. It is possible to dolly across the line, shoot along it, or change its direction by regrouping people, but cutting between cameras on both sides of this imaginary line produces a reverse cut or jump cut (Figure 9.37).

ANTICIPATING EDITING

It is very important to think about the editing process when shooting. If proper preproduction planning has taken place, the director and camera operator should have a good idea of how the shots should be created.

Continuity

Every time the camera is set up to take the next action shot, think about future editing. Otherwise, it is possible to end up with a series of shots that do not fit together smoothly. This happens often when repositioning the camera to shoot a repeated scene from a different location.

The most frequent problems are:

- Part of the action is missing.
- The action shot from another angle does not match that from a previous shot of the subject.
- The direction of the action has changed between successive shots.
- The shot sizes are too similar or too extreme.
- Action leaves the frame, and re-enters it on the same side.
- Successive shots show continuity differences, such as with and without eyeglasses, different attitudes/expressions, or different clothing.

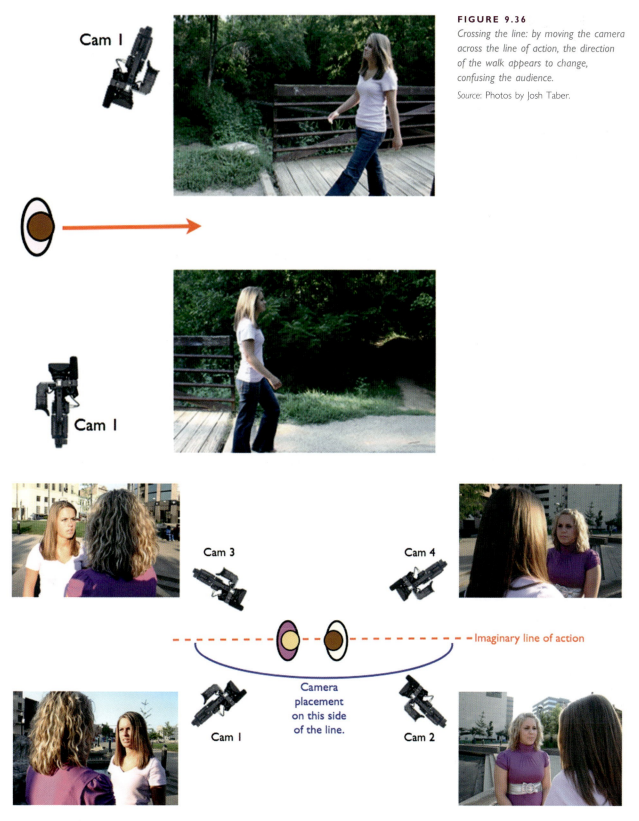

FIGURE 9.36
Crossing the line: by moving the camera across the line of action, the direction of the walk appears to change, confusing the audience.
Source: Photos by Josh Taber.

FIGURE 9.37 *The line of action: shots can be cut between cameras located on the same side of the imaginary line of action—between 1 and 2, or 3 and 4. Cutting between cameras on opposite sides of this line could cause a jump cut (1 and 3, 1 and 4, 2 and 3, 2 and 4).*

Source: Photos by Josh Taber.

Improving Editing Flexibility

When shooting, editing flexibility can help in various ways.

Avoid Brief Shots

- Always record the beginning and end of the action, to allow editing flexibility. Do not wait to record until the instant the action starts. Extra footage on each side of the action, also known as heads or tails, will give the editor the opportunity to utilize dissolves and fades, and allow him or her to trim the footage in an optimal way.
- Where possible, start and finish a long panning shot with a still shot.

Extra Editing Material (Usually Called "B" Roll)

- Always shoot some cutaways showing the surroundings, general scene, bystanders' reactions, and so on.
- Determine whether specific reaction shots would be appropriate.
- It is sometimes helpful to shoot the same scene at a leisurely pace and at a faster pace (slow pan and faster pan) to allow editing flexibility.

Faulty Takes

- Do not record over an unsuccessful shot. Parts of it may be usable during the editing process.
- If an action sequence goes wrong, it is sometimes better to retake it entirely. At other times, just change the camera angle (or shot size), and retake the action from just before the error was made (called a pickup shot).

Establishing Shots

- Always begin shooting with a wide shot or long shot of the scene (called an establishing shot), even if it is not used eventually. This shot allows the editor to understand the placement of the other shots.
- Consider taking a long shot (cover shot or master shot) of all of the action in a scene, then repeat the action while taking closer shots.

INTERVIEW WITH A PROFESSIONAL: KEITH BROWN

FIGURE 9.38
Keith Brown, director of photography.

What do you like about shooting? The camera is a tool that allows me to be a storyteller. Whether a concept has a storyboard or simple notes on a napkin, it's the art of turning that concept into something that conveys a visual message. It comes through pre-visualization, collaboration with other crew members, and deciding how to manipulate the camera to tell that story. What focal length will best illustrate this point, how does depth of field contribute or distract, what do I include or exclude from the frame? It's the art of allowing viewers to experience the feelings I had at that particular moment, showing them more than just what the experience was factually. It's the challenge of eliciting the same emotional response to an event or story that I had as the camera operator. There's also the joy of the unexpected moment when light, subject, and movement, combined with framing, lens selection, creative use of white balance, and being in the right place at the right time, all come together to create something magical.

How do you decide what camera to purchase? It's not the camera that creates a quality image; it's the person behind the camera, whether it's a video or DSLR rig. With that said, there are certain things we look for in a camera. We do a lot of overseas travel with limited crew so a small camera package is essential. I use a small camera with matte box and a basic filter package of UV, polarizer, graduated NDs, and a few graduated colors.

What are the features that you believe are essential for professional production?

- Quality—quality camerawork, lighting, audio, and scripting when applicable.
- A complete understanding of the production.

continued

Interview with a Professional: Keith Brown—*continued*

- What's the age group? Helps determine the style and approach of your presentation.
- How familiar are they with the topic? This might determine how much background information you have to give.
- Do they all speak the same language—both literally and figuratively?
- How long can you hold their attention?
- What's the walk away? What's the main thing you want the audience to remember?
- What's the best way to deliver the message—live, on a DVD, over the Web, or some other method?

What kind of support do you use for your camera or do you primarily handhold it? Why? Due to our style of production and small crew, we don't have the luxury of using our dolly or jib very often. Therefore, most of our work is on a tripod or handheld. The style of the show obviously dictates whether it's shot handheld or on a tripod. Shooting pace, setting (is there even room for a tripod or is one allowed), and the feeling one's trying to convey all come into play. There are dozens of reasons or situations that help determine the choice; these are a few:

- **Tripod**: A tripod is employed if it's a slower-paced show, shots are held longer, scenics, long focal length lenses are used, or it is the traditional interview situation.
- **Handheld**: Handheld is employed for a faster pace, mobility, frenetic, walking and talking (without a Steadicam, unfortunately), intensity, and run-and-gun filming.

Keith Brown is a director of photography for an international nonprofit organization. He travels the world shooting documentaries that highlight world issues.

REVIEW QUESTIONS

1. When dealing with composition, what are the three key factors, and how do they impact the final image?
2. Why is composition important?
3. What are two common faults while shooting?
4. How can shooting improve the editing process?
5. What are some of the practical ways to adjust composition?
6. How is balance in an image affected by the subject?
7. Visual patterns can significantly affect the image. Give three examples and explain their impact.

TABLE 9.1 *Common Faults While Shooting*

Wrong color temperature (the image is bluish or yellowish)	Make sure that the camera was white-balanced properly and that the appropriate color-correction filter was used.
Soft focus	The camera operator must take the time to make sure that the lens is properly focused.
Camera shake, unsteady camera	The camera may need to be secured with some type of a camera mount. It is a good habit to use a tripod, or some other mount, as much as possible. Most people cannot handhold a camera very steadily, especially for an extended period of time.
Sloping horizons	Leveling the camera may be time-consuming but is essential.
Headroom wrong (too little or excessive)	Each time the camera is repositioned, the headroom must be reviewed.
Cut off top of head or feet	Make sure that the subject is properly framed. Do not cut on the natural bend of a body.
Subject size (too distant or close to see properly)	The size of the subject may be problematic when successive shots are too similar, inappropriate camera height for the situation occurs, disproportionate number of long shots or close shots, or focus is on the wrong subject. Advanced planning, including storyboarding, is essential.
All subjects composed center-frame	Keep the rule of thirds in mind when shooting.
Shots do not interrelate (too many isolated subjects)	Think "transitions" as you shoot the project.
Subject obscured (foreground intrusion), background objects sprouting from heads, background distractions (posters, traffic)	Carefully review the foregrounds and backgrounds of a scene for distraction objects.
Shots too brief (or too lengthy) or start of action missed	It is better to have a clip that is too long than too short. Make sure that you begin shooting before the action begins and continue shooting briefly after the action is complete.
Zooming excessive or distracting	Zoom shots must be motivated; that is, they need to be there for a reason.
Panning not smooth, too fast, or overshoots	The tripod may require some fine adjustment.
Person handling objects badly (obscuring, moving around)	Practicing with the talent is important so that they understand what you are trying to do.
Poor lighting (such as black eyes, half-lit face)	Lighting may need to be adjusted for maximum impact.

CHAPTER 10

TELEVISION GRAPHICS

"Trendy title styles are often hot today but stone cold tomorrow."

—Morgan Paar, Producer

TERMS

Bug: A small logo or score/clock/time period graphic.

CG (character generator): A generic name for any type of television graphic creation equipment. CGs can change fonts, shape, size, color, and design of the lettering.

Crawl: The movement of the graphics horizontally across the screen.

Credits: Recognize those appearing in and contributing to the program.

End titles: Draw the program to its conclusion.

Font: A specific size, weight, and style of typeface.

Lower third (L/3rd): A graphic that appears in the lower third of the screen.

Opening titles: Introduce the show to the audience.

Roll: The movement of the graphics up or down the television screen.

Subtitles: Identify people and places.

First impressions are important. And the first impressions the audience gets from your production may come from the opening graphics. They help set the style and ambiance of the program; they inform; they guide. Well-designed graphics make a direct contribution to the success of any production. Poorly designed graphics immediately discredit the entire production. Graphics don't have to be elaborate—they just need to clearly communicate, and help grab the audience's attention. However, they do need to be brief, clear, and appropriate in style.

Effective television graphics require the graphic operator or designer to think through a number of stages in the production process:

- How does this graphic help the audience understand the subject or story better?
- What is the purpose or goal of the graphic?
- Would words, illustrations, photographs, or video imagery work best to communicate to this audience?

While we will primarily be discussing on-screen graphics, graphics do extend into becoming part of the set. Background designs, monitors shown with graphics, logos; all of these are part of television graphics (Figure 10.1).

TELEVISION GRAPHIC GOALS

Television graphics should:

- *Convey information clearly and directly.* They should be prepared for maximum communication impact. This means that television graphics should be simply created, not elaborate. Because television graphics move quickly and cannot be studied for a long period of time, the font should be bold and straightforward.
- *Establish the show's overall mood and tone through the graphic style.* The font and presentation style can do much to advance the "story" being told. These can set the scene for the rest of the program (Figures 10.2 and 10.3).
- *Present fact, concepts, or processes visually so that the viewer will understand the program content.* Keep the graphics organized and presented in a way that holds the audience's attention and makes it simple for them to follow the process or understand the concept being presented (Figure 10.4).

Aim to keep graphic information to a minimum, particularly if it is combined with a detailed background. A screen full of written information can be daunting to most viewers and

Figure 10.1 Graphic designers created a projected graphics moving background for the studio set.

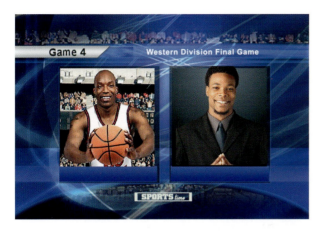

FIGURE 10.2 *Television graphics can help establish the mood and tone for the whole program.*

Source: Photo courtesy of Compix.

FIGURE 10.3 *Graphics for television often create the look and style of the program. Note that the logo in this photo was used as a set graphic; it is showing up (probably animated) on the set's video screen in the back and a graphic generator is being used so that it appears as part of the credit design.*

Source: Photo courtesy of KOMU-TV.

FIGURE 10.4 *Graphics need to be organized in a way that can be easily understood by the viewer.*

Source: Photo courtesy of KOMU-TV.

tiring to read. People are easily discouraged from reading rapid graphics. Leave information on the screen long enough to allow it to be read aloud twice, so that even the slowest reader can assimilate it.

Types of Graphics

Graphics add clarity to a show's presentation. They are used to announce the place or time, to identify a plant, to display data, to clarify how food should be cooked, and so on. There are a number of different types of graphics (Figure 10.5):

- *Opening titles* announce the show, generally by showing the title of the show.
- *Subtitles* identify people and places.
- *Opening credits* identify some of the personnel from the production. These may include producers, director, actors, writer, and composer.

TIPS ON MAKING GREAT TV GRAPHICS

(Adapted from Al Tompkins, Poynter Institute)

1. What is the context of this graphic? How does the information compare with previous studies or surrounding areas? Is the information really important for the viewer to know?

2. Think shapes, not just numbers. It is sometimes difficult to read with the relative nature of numbers when they are presented quickly on the TV screen. But it is easy to understand the growth of a budget when a bag of money is shown growing on the screen. Imagine that you have no words—that the graphic is all the viewer can see. How clearly would he or she understand what you are trying to show?

3. Think clearly about the purpose of the graphic. Ask yourself: What exactly do I want the viewer to learn from this graphic?

4. Write after you make the graphic, not the other way around. This will ensure that the copy and graphic match exactly.

5. Get other people to look at it. Let them tell you what the graphic conveys to them. It is no different than copyediting.

FIGURE 10.5 *Some of the most common graphic terms are crawl, roll, bug, opening title, lower thirds, and credits.*

- *Lower third* (L/3rd) is a graphic that appears in the lower third of the screen.
- A *bug* is a small logo or score/clock/time period graphic.
- A *roll* is the movement of the graphics up or down the television screen.
- A *crawl* is the movement of the graphics horizontally across the screen.
- Credits recognize those appearing in and contributing to the program.

FORMS OF GRAPHICS

Graphics can make a valuable contribution to all types of television programs:

- Statistical graphics in the form of bar graphs and charts can show, in a moment, information that would be hidden in columns of figures. They enable you to simplify complex data, to compare, to show developments, to demonstrate relationships, and so on (Figure 10.6).
- Pictorial graphics can be used to illustrate a children's story, to set the scene in a drama, to explain scientific principles, to provide an atmospheric background to titling, and so on (Figure 10.7).

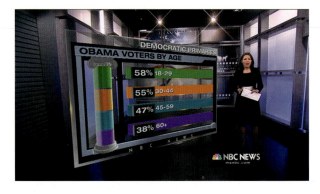

FIGURE 10.6 *Charts can simplify complex statistical information to compare or demonstrate relationships.*
Source: Photo courtesy of NBC News.

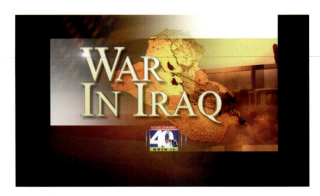

FIGURE 10.7 *Backgrounds can provide a context for the titles.*
Source: Photo by Tristan Bresnen.

Animated Graphics

Animation can bring a graphic to life. Even the simplest movement, such as panning over it from one detail to another, zooming in/out on details, or cutting between sections of it, can sustain interest in what would otherwise be a static display (Figure 10.8). Such techniques are an effective way of illustrating a documentary, or any program that relies heavily on graphics or photographs (or maps or paintings in historical sequences).

Animation can take place in a number of different ways. For instance, you can build a graphic on the air by progressively adding details or sections. Character generators can usually save the animation, which can be replayed at a later time.

Interactive Graphics

There are a number of interactive graphics that are used to help the audience understand situations and also hold their attention. Newscasters have used interactive monitors to provide data to the audience. Sports productions have created 3D characters, which represent actual players, to illustrate the various plays in sports. The goal of these graphics is to illustrate the nuances and variations in a play that could have occurred or has occurred (Figure 10.9).

DESIGNING GRAPHICS

"The screen can contain some graphic elements but the focus should be on communicating the information, not the pretty background or the design elements."
—Gerald Millerson, Former BBC Director

Video and television productions today may use either of two screen formats. Standard-definition television has an aspect ratio of 4:3 (4 units across and 3 units high). Most countries have adopted a high-definition television (HD) system that has an aspect ratio of 16:9 (16 units across and 9 units high). If viewers have both types of formats, all graphics need to be designed so that they fall into the 4:3 area. Otherwise, 16:9 viewers may not be able to see important graphics (Figures 10.10 and 10.11).

FIGURE 10.8 *Animated graphics hold the viewer's attention better than static graphics.*

Source: Photo courtesy of WLEX-TV.

FIGURE 10.9 *Interactive graphics that allow touchscreen graphics or graphic representations of players/products to be placed within any scene are becoming common.*

Source: Photos courtesy of Orad Hi-Tec Systems.

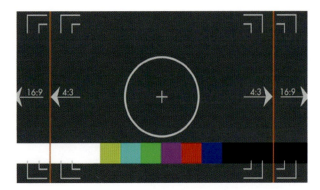

FIGURE 10.10 *Today, most viewing audiences use HD 16:9 televisions. However, some still use 4:3 standard definition. Graphics often need to be created so that they work on both formats.*

FIGURE 10.11 *Essential graphics should stay within the middle of the television's scanning area. Note that this graphic can be used on 4:3 or 16:9 televisions.*

Source: Photo courtesy of KOMU-TV.

Television graphics are usually designed in layers in order to provide the production personnel with the most flexible use of the graphics. One layer may include the station's logo, another layer may include scores, another weather, and so on. The idea is that any one, any combination, or all of the layers can be used as needed (Figure 10.12).

Some things to consider when designing graphics are as follows:

- Keep titling well away from the edge of the frame to avoid edge cutoff. Graphics should be designed so that they fall within the middle 80 percent of the television's scanning area. This center area of the screen is referred to as the safe title area (Figure 10.13).
- Simple and bold typefaces are usually best. Generally avoid thin-lined, elaborate lettering. Although HD's resolution can handle the thin lines, some of the viewing audience's televisions still struggle with thin lines (Figure 10.13).
- Limit the number of different fonts within a program (Figure 10.13).
- Lettering smaller than about 1/10 screen height is difficult to read. It is important for directors to determine what media the audience will use to see the final production, or at least what the dominant media will be.
- Outlining and drop shadows often makes lettering easier to read by preventing bleeding and providing contrast (Figures 10.13 and 10.14). However, avoid placing a black-edged outline around smaller letters, because it can become hard to read. The holes in the letters B, O, A, and R tend to fill in.
- Punctuation is not normally used, except in the following instances: quotations, hyphens, apostrophes, possessives, and names.

FIGURE 10.12 *Television graphics are usually designed in layers to provide maximum flexibility.*

Source: Photo courtesy of Compix.

- Abbreviations are rarely punctuated on television graphics. However, if abbreviations make the title ambiguous, use three lines if necessary and spell out the words.
- Leave a space between title lines of around 1/2 to 2/3 the height of capital letters.
- Lettering should generally contrast strongly with its background. Television lettering is usually much lighter than the background (Figure 10.15).
- Don't fill the screen with too much information at a time. It is often better to use a series of brief frames, or to use a crawl (continuous information moving vertically into the frame and passing out at the top).
- Warm bright colors will attract the most attention (Figure 10.15).

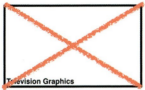

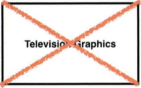

FIGURE 10.13 *Basic television graphic design principles.*

FIGURE 10.14 *Outlining the letters often makes them easier to read.*

Source: Photo courtesy of Compix.

FIGURE 10.15 *Lettering will be more readable if it contrasts with the background.*

Source: Photo courtesy of Compix.

FIGURE 10.16 *The backgrounds behind the graphics should not compete with the graphics. Note how the graphics operator here blurred and darkened the area behind the text in order to keep the attention on the graphics.*

Source: Photo courtesy of Compix.

Backgrounds for Graphics

Choosing a suitable background for a graphic can be as important as the foreground graphic. When creating full-screen graphics, graphic operators need to be careful when choosing graphic backgrounds. If the wrong background is used, it may compete for attention with the graphic. For example, don't use a sharply focused shot of a group of people in the background. Viewers will look through the words and look at the people (Figure 10.16). There are a number of different strategies that can be successfully used for backgrounds:

- Create a simple color background.
- Freeze the video background in order to not have a moving background.
- Defocus the video image so that it is blurry.
- Select a single-color background (grass, water, sky, etc.).

When using a scenic background, such as the closing shots of a drama, the background content or meaning may actually help determine the style and weight of the lettering that can be used.

Plain backgrounds can prove very effective, as they are unobtrusive and emphasize the lettering. However, they can also be dull and uninteresting. Ornamental backgrounds, which include patterning, texture, and abstract designs, may increase the graphic's visual appeal. However, they can look confusing. Clearly, background selection requires careful choice.

Lettering against a multi-hued or multi-toned background is invariably harder to read. If graphics are inserted over location shots, such as a street scene, the eye may have some difficulty in discerning information, and may also be tempted to wander around the background instead.

In most cases, by using larger type in light tones (white or yellow) with strong borders or shadows, legibility is considerably improved. As a general rule, avoid introducing lettering over backgrounds of similar tones or hues, or over printed matter (e.g., titles over a newspaper page). Light lettering is usually more easily read than dark, and pastel or neutral backgrounds are preferable to saturated hues.

GRAPHIC EQUIPMENT

Character generator (CG) is a generic name for any type of television graphic creation equipment. CGs can change the fonts, shape, size, weight, color, and design of lettering. They can make graphics flash, flip, crawl, roll, and animate. Once the graphic is created, it can be rearranged, stored, and kept ready to appear on the screen at the press of a button (Figure 10.17).

Stand-alone graphic generator systems used to hold 99 percent of the market in professional television. They are still widely popular in larger markets and sports production. However, computers with graphic generation software have cornered a significant portion of the market. Today, computers are used in all markets and provide sophisticated on-screen graphics. Some mobile production crews have moved to laptop systems (Figures 10.18 and 10.19).

FIGURE 10.18 *A stand-alone character generator.*

FIGURE 10.19 *Laptop CGs have become popular for graphic operators who are on the road a lot.*
Source: Photo courtesy of Compix.

FIGURE 10.17 *This is the composition screen of a high-end graphics system. The system allows almost unlimited manipulation of the graphics.*
Source: Photo courtesy of Chyron.

THE GRAPHIC OPERATOR

The graphic operator is often located behind the director. This provides graphics with the opportunity to see what is currently on-screen and what is coming next. In many ways, they need to anticipate where the production is going so that they can already be building (creating) a graphic before the director calls for it (Figure 10.20).

FIGURE 10.20 *Graphic operators often sit directly behind the director and technical director in the control room when working on a multi-camera production.*

INTERVIEW WITH A PROFESSIONAL: SCOTT ROGERS

FIGURE 10.21 *Scott Rogers, sports graphic operator.*

What do you like about being a graphics operator? I like being part of a team that is providing supporting information to a story. I also enjoy the challenge of trying to do a mistake-free show. When you go into a break, the producer will often say, "What should we do coming out of the break?" A graphics team has about 10 seconds to fill or make the graphic content decision for that void—that is my specialty.

What challenges do you have to deal with as a graphics operator? Working graphics has the opportunity to easily make mistakes. You have to be very detail-oriented in order to eliminate mistakes when dealing with very limited time. A lot of people watch sports events without being able to hear the announcers, so graphics fill in the crucial gaps. You have to know your graphic equipment. You also have to deal with a lot of "looks" that you didn't build and therefore have to reverse-engineer sometimes.

What is your philosophy about graphics? When do you think they should be used, or not? I think graphics should be used to highlight storylines. I don't like graphics over replays. If you are going to put in a graphic, make sure you leave it in long enough to read it! Less is more with graphics. You don't write TV graphics like you would write for a book or newspaper story. Most of the time, conjunctions are not needed.

What suggestions do you have when designing graphics? Focus much more on readability and much less on animation. *Way too much* emphasis is put on animating graphics. Special graphics should animate (open, sponsors, and so on), but animating the basic stuff just takes up time you need to read the graphics (which is usually in too short of a time anyway).

Scott Rogers has been involved in productions for the NBA and multiple Olympics, as well as local television. He currently works as a producer and as a graphic operator.

REVIEW QUESTIONS

1. How do graphics help producers meet their intended goals for a production?
2. What are two of the issues that need to be considered when designing a graphic for television?
3. How do screen ratios affect television graphics?
4. What can be done to letters to make them "pop" on the screen?
5. What are two ways that backgrounds can make graphics/letters more readable?

CHAPTER 11

LIGHTING FOR TELEVISION

"Pore over artwork and other films. Study the lighting and
consider how it serves the story being presented."
—Thomas McKenney, Director of Photography

"When it comes to image acquisition, lighting is still the art that
separates the amateurs from the pros."
—Jay Holben, Producer

TERMS

Barn doors: Metal flaps attached to the front of a lighting instrument; used to control where the light falls on a scene.

Base light: The minimal amount of light that allows the camera to see the subject.

Diffuser: Some type of a translucent material (wire mesh, frosted plastic, or spun-glass sheet) that diffuses and reduces the light's intensity.

Floodlight: Also known as soft light; the scattered, diffused, shadowless illumination that in nature comes from a cloudy overcast sky, and is reflected from rough surfaces of all kinds.

Gaffer: The head of the electrical department, many times in charge of lighting on a television set.

Gels: Color gels (filters) can be placed over lights to enhance the color of the light.

Incident light: Also known as direct lighting; what's measured with a handheld light meter—that is, the amount of light falling on a subject. When measuring incident lighting, the light meter must be positioned next to the subject, pointed at the light sources.

Key light: The main light, usually a spotlight, that reveals the shape and surface features of the subject.

Lighting plot: A plan that shows where each light will be placed on the set.

Reflected light: Measured with a handheld light meter or a camera's built-in meter; the light bouncing off of a subject. In this situation, the meter is aimed directly at a subject.

Spotlights: Spotlights are a directional or hard illumination that produces sharp shadows.

Three-point lighting: Also known as photographic lighting and triangle lighting; use of two spotlights (key light and backlight) and a floodlight (fill) to illuminate the subject.

White balance: The process of calibrating a camera so that the light source will be reproduced accurately as white; the most common technique of color-balancing a camera.

Zebra: Some video cameras include a zebra indicator in the viewfinder. The zebra allows camera operators to evaluate the exposure of the image in the viewfinder by showing all overexposed segments of a scene with stripes.

Effective lighting makes a vital contribution to a television production. We quickly recognize bad lighting when we see it, but good lighting is so unobtrusive and "natural" that we usually take it for granted. In this chapter, you will learn some successful lighting techniques.

THE GOALS OF LIGHTING

Lighting in television and film is about much more than just making things visible. It has to also satisfy a number of often-conflicting objectives:

- The lighting must be bright enough to enable all the television cameras to produce pictures of the highest quality. This is usually referred to as base lighting.
- Lighting should convey to the viewer the time, mood, and atmosphere of the story (Figure 11.1).
- The lighting must provide a consistent look, as chosen by the director. That means it must look consistent from each camera angle.
- The lighting must fit in with the other components of the studio or location: scenery, camera placement, mic placement, and so on. Badly positioned lights can prevent the talent from reading the teleprompter, cause flares in a camera lens, result in shadows, spoil a skillfully designed setting, and degrade makeup.
- Good lighting creates a three-dimensional illusion in a flat image. It provides an impression of solidity and depth in subjects and surroundings.
- Lighting should lower the contrast ratio between the light and dark areas. By adding more light to the dark areas or dimming the extremely light areas, it is possible to get more detail in both areas.

FIGURE 11.1 *Lighting helps create the mood of the event.*

FIGURE 11.2 *Your eyes are drawn to the brightest area of the screen. The director can guide the audience's attention by lighting specific areas brighter than others.*

- Successful lighting guides the audience's interest. It directs their attention toward important features, because the viewer's eyes are generally attracted to the brightest area of the screen. Lighting can create compositional opportunities for the camera (Figure 11.2).
- Lighting is used to increase or reduce the picture's depth of field.

WHY IS LIGHTING NECESSARY?

Technical Reasons

As you know, the camera needs a certain amount of light reflected from the scene to be able to produce a good tonal range in the image. If there is too little light, the lens aperture has to be opened up to compensate, and the available depth of field is considerably reduced. When there is too much light, the images become overexposed, unless the lens is stopped down (the depth may now be too great) or a neutral-density filter is used to reduce the light. In the studio, excess light wastes power, can cause ventilation problems, and is unpleasant to work in. Light levels need to be related, wherever possible, to the preferred working *f*-stop.

Unlike film, the television camera can handle and reproduce only relatively limited tonal contrasts. If the lighting is contrasty, details in the lightest and darkest tones will be lost.

FIGURE 11.3 *Light was added to these daylight scenes so that the talent had even lighting while shooting a segment for Access Hollywood and on the set of ESPN.*

Source: Second photo by Josh Taber.

On location, the existing lighting may not be suitable. From the camera position, it may prove to be:

- Too bright: strong sunlight causing performers to squint their eyes.
- Too dim: insufficient for well-exposed shots.
- Too flat: diffused light in which subjects lack form or definition.
- Too contrasty: lighter tones burned out (pure white) and shadows too black.

Under these conditions, we need to augment or replace the natural lighting—or tolerate the results. Sometimes the only solution is to alter the camera position (Figures 11.3 and 11.4).

Artistic Reasons

Lighting plays a major part in how we interpret what we see. Even when structure and outline give us leading clues, the play of light and shade strongly influences our judgments of size, shape, distance, surface texture, and contours.

Lighting is strongly associated with mood. Through carefully chosen light direction and contrast, you can change a scene's entire atmosphere. It can portray fun, fantasy, mystery, or dramatic tension. Lighting can enhance a setting and create

FIGURE 11.4 *Two LED lights were used to balance lighting in the room with the arena lighting in the background for this interview with William Shatner.*

FIGURE 11.5 *Light and shadows strongly influence our judgments of size, shape, distance, surface texture, and contours.*

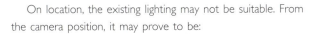

pictorial beauty, or it can deliberately create a harsh, unattractive setting.

You can use light selectively to emphasize certain aspects of the scene while subduing others, or avoiding or reducing distracting features. You can exaggerate form, and draw attention to texture or suppress it. Through shadow formations, lighting can suggest structures that do not exist, or hide what is there (Figure 11.5).

THE NATURE OF LIGHT

Light can be applied with large "brush strokes," or with fine delicate attention to detail. It can be washed across the scene, or used to pick out and emphasize certain features. But to exercise this control, you must be able to appreciate the subtleties of light itself. Let's take a look at the practical basics of illumination:

- *Light intensity*: The strength of light that we require on the subject and the surroundings will determine how powerful the lamps need to be, relative to the area they have to cover.
- *Color temperature*: The color quality of the light. Light and camera performance should be matched to avoid poor color quality.
- *Light dispersion*: Some light sources produce hard light, which casts strong shadows; others create soft light, which is diffused and has few shadows. This range of tools offers us the choice of bold brush strokes or subtly graded halftones.
- *Light direction*: The direction of the light affects the way light and shade fall on a subject. It determines which features are highlighted, and which fall into shadow.

Light Intensity

The amount of light needed to illuminate a set or location and the action within it is partly a technical decision and partly an artistic one. You might, for example, use one strong key light to cover both the action and the background, or a series of restricted lamps, each lighting a carefully chosen area.

Camera systems are usually quoted as having a specific sensitivity, requiring a certain light level (intensity) for a given *f*-stop. But this is only a general guide. A lot depends on the nature of the surroundings and the mood the director wants to create. To reveal detail in the walls of a dark-paneled room will require much more light than would be needed with light-toned walls.

Lighting intensities can be influenced by the surface finish: whether walls are smooth or rough textured, and whether they are plain or strongly contoured. The contrast range of the set dressings used can also affect the amount of light needed to illuminate a situation effectively.

Interestingly, a spacious light-toned setting may require less-intense lighting than much smaller dark-toned surroundings. Lighting quality cannot be judged by the watts of the lights. The most helpful way to review the lighting is on an actual monitor, on which we can assess its artistic effect and judge its technical qualities.

There are two primary ways of measuring the lighting in a scene (Figure 11.6):

- *Incident* light measurement helps to assess the relative intensities of lighting from various directions. When measuring incident lighting, the light meter must be positioned next to the subject, pointed at the light sources.

FIGURE 11.6 *Stand-alone light meters are used to measure incident or direct lighting. Cameras have built-in reflected light meters.*

Source: Photos courtesy of Sony and Mole-Richardson

ZEBRA EXPOSURE INDICATOR

Higher-level video cameras generally include a zebra indicator in the viewfinder (Figure 11.7). The zebra allows camera operators to evaluate the exposure of the camera in the viewfinder by showing all overexposed segments of the scene. It offers a very simple way to ensure that the lens's iris is set correctly. Most zebras are set at between 102 and 105 IRE (a unit used in the measurement of video signals) in order to show which areas are overexposed.

FIGURE 11.7 *When looking through the camera viewfinder, some cameras have zebra stripes that show the overexposed areas of the image.*

Source: Photo by Austin Brooks.

It is measuring the amount of light that is falling on the subject, from the subject's perspective.

- *Reflected* light measurement provides a general indication of the amount of reflected light reaching the camera. Reflected light measurements average the amount of light reflected from the scene, arriving at the camera's lens. In this situation, the meter is aimed directly at the subject. Television cameras use reflected light metering. Today's more advanced cameras can adjust the meters to average the light using multiple sensors and center-weight the sensor (use the center of the image to meter the lighting), in addition to other options.

The Color Quality of Light

The eye and brain are astonishingly adaptable. We appear to see effects that are not really there (as in optical illusions), and we overlook effects that can be clearly measured by an instrument. An everyday example of the latter is the way we accept a very wide range of quite different light qualities as representing "white" light. If we analyze "white" light, we find that it is really a mixture of a range of colors. The spectrum is red, orange, yellow, green, blue, indigo, and violet light in similar proportions. In many forms of illumination, some parts of the spectrum are much more prominent than others, and the result is far from true white. For example, light from candles or dimmed tungsten lamps actually has a warm yellowish-red color quality, or, as we say, a low color

temperature. Daylight can vary considerably, from cold, bluish north-sky light to the warm quality of light around sunset. Yet our brains compensate and accept all these sources of illumination as white light. Television cameras are not fooled in this way. They do not self-compensate. If the lighting has bluish or yellowish characteristics, the images will show this as a pronounced color cast. There are basically two ways to correct the color when shooting:

- Adjust the camera to compensate for the color variations of the prevailing light, by using a suitable color-correction filter (in the filter wheel) and/or by adjusting the white balance. The white balance is usually adjusted by aiming the camera at a white surface while pressing the white balance button.
- Adjust the color temperature of the light to suit the camera's color balance. If you are using tungsten lights when shooting in daylight to illuminate shadows, blue filter material can be used to raise their color temperature from 3,200 K (tungsten) to around 5,600 K (daylight). Conversely, if daylight is illuminating a room in which you are using quartz or tungsten lamps, a large sheet of amber-orange filter can be stretched over the window to reduce the effective color temperature of the daylight to match the interior lighting (Figure 11.8).

Cameras are usually balanced for the dominant light source in the scene. However, different light sources have different qualities. Tungsten-halogen lamps (quartz lights) usually are at

3,200 K, and tungsten lamps can have a noticeably lower color temperature of around 3,000 K. High-powered HMI (hydrargyrum medium-arc iodide) lamps produce light of 5,600 K, which blends well with daylight but does not mix with that of tungsten sources.

Another issue is that as tungsten and quartz lamps are dimmed, their color temperature can change—the light lacks blue, and develops a "warmer" yellowish-red quality.

Light Dispersion

Floodlight

Floodlight, also known as *soft light*, is the scattered, diffused, shadowless illumination that in nature comes from a cloudy overcast sky, and is reflected from rough surfaces of all kinds (walls, sand, snow). You can create soft light artificially by using the following:

- Large-area light fittings (e.g., multi-lamp banks) (Figure 11.9).
- Diffusion material in front of light sources (Figure 11.10).
- Internal reflection light (Figures 11.11 and 11.12).
- Light bounced from large white surfaces (e.g., matte reflector boards, or a white ceiling).

Floodlights include scoops, broads, floodlight, banks, internally reflected units, strip lights, and cyclorama lights. They are used mainly for fill light and for broad lighting of backgrounds. They can be hung, supported on light stands, or rested on the ground.

Adjustment of soft light sources is limited. Egg-crate shields and barn doors are usually used to restrict the coverage to some extent. Otherwise, diffusers or dimmers are used to control their light intensity (Figure 11.13).

Soft lights are usually used to:

- Illuminate shadows without creating additional shadows.
- Avoid emphasizing modeling and texture.

However, soft light does have practical disadvantages:

- Really diffused light can be difficult to control, because it spreads around and is not easy to restrict.
- Badly used soft light can produce flat unmodeled illumination. Texture and form can be difficult to see in the picture. If used as an overall base light, it can reduce the impression of depth, over-light the walls of settings, destroy atmosphere, and produce flat, uninteresting pictures.

FIGURE 11.8 *Gels (filters) can be placed over lights in order to color-correct a scene.*
Source: Photo courtesy of Litepanels.

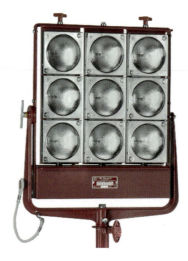

FIGURE 11.9 *Multi-lamp bank.*
Source: Photo courtesy of Mole-Richardson.

Spotlights

A *spotlight* is a directional or hard illumination that produces sharp shadows. This type of light comes from any concentrated light source, such as the sun or a spotlight (Figure 11.14). Spotlights produce well-defined shadows and are used to:

- Create well-defined modeling.
- Cast pronounced shadows (e.g., tree-branch shadows).
- Localize light to specific areas.
- Produce coarse shading or an abrupt brightness fall-off.

FIGURE 11.10 *Diffusion material can be placed in front of a light in order to create a softer light. Lights can also be aimed at a white ceiling in order to bounce light onto a scene.*

FIGURE 11.11 *Weighing three pounds, the LED (light-emitting diode) light panel projects a bright soft light. It is extremely lightweight, offers low power consumption, provides accurate color reproduction, is heat-free, and can be dimmed.*

Source: Photo courtesy of Litepanels.

FIGURE 11.12 *The LitePad is nicknamed the "Everywhere Light." At 1/3 inch (2.5 cm) thick, it can be placed anywhere to add a little soft light.*

Source: Photos courtesy of Rosco.

FIGURE 11.13 *Egg-crate shields can be used to restrict the light coverage—to a certain extent.*

Source: Photo courtesy of Westcott.

FIGURE 11.14 *Spotlights provide distinct shadows. This fresnel is a non-defined spotlight. It is lightweight, less expensive than an ellipsoidal, and has an adjustable beam.*

Source: Photo courtesy of Mole-Richardson.

- Project light over some distance at a reasonably constant intensity. The light from focused spotlights does not "fall off" quickly with distance.

Spotlights are generally used as key lights, as backlights, as background/set lighting, and for effects (sunlight, broad decorative patterns). The light spread of the spotlight's beam is adjustable and can be focused. Ideally, the beam intensity should remain even overall as the beam width is altered. However, many designs do develop hot spots or "dark centers." A couple of specialty types of spotlights include:

- *Effects/pattern projectors:* These spotlights, which include the ellipsoidal spotlight, can project patterns on the set or scene to simulate windows, branches, and so on.
- *Follow spots:* Used for isolating static subjects or following moving performers (singers, skaters, dancers) in a confined pool of light. These large spotlights are carefully balanced for continuous accurate handling and are usually mounted on a stand.

However, spotlights do have some practical disadvantages:

- The shadows from a spotlight may prove to be unattractive, inappropriate, or even distracting.
- Spotlights can overemphasize texture and surface modeling.

- High-contrast lighting can produce harsh, unsubtle tones.
- Multiple shadows arise when the subject is lit by more than one hard light source.

Clearly, for effective lighting treatment, you need a suitable blend of spotlights and soft diffused light. Usually the spotlight reveals the subject's contours and textures, and the soft light reduces undue contrast or harshness and makes shadow detail visible. You may deliberately emphasize contour and texture, or minimize them, depending on the blend and direction of the illumination.

The Direction of the Light

The effects of lighting will vary as you alter the angle at which it strikes the subject. Raising or lowering a lamp, or moving it round the subject, will change which parts are lit and which are thrown into shadow. It will affect how contours and texture are reproduced. Surface markings and decoration become more or less obvious. If you reposition the light, or alter the camera's viewpoint, the appearance of various features of the subject can change.

Three-Point Lighting

Three-point lighting, also known as *triangle* or *photographic lighting*, utilizes both directional and diffused lighting to obtain the best results (Figure 11.15).

The main light, or key light, is positioned slightly above and to one side of the camera. This is normally a spotlight, and it reveals the shape and surface features of the subject. The key light produces distinct, harsh shadows.

The *fill light* is a floodlight that is placed on the opposite side of the camera from the key light. It reduces the shadows (made by the key light) but should not eliminate them. The fill light also reduces the lighting contrast. The more the key light is offset, the more important this soft fill light (also known as *filler* or *fill-in*) becomes. If the key is nearly frontal, you may not need fill light at all. Note that in the subject's image in Figure 11.15, the shadows on her face have not been eliminated. This helps give the face texture and shape.

Finally, a backlight is angled down onto the subject from behind to give some separation between the subject and the background. The backlight emphasizes the shape of the subject. The key light and backlight are generally the same intensity. However, the backlight may need to be reduced due to light hair color. The fill light is usually one-half or three-quarters the intensity of the key light and backlight.

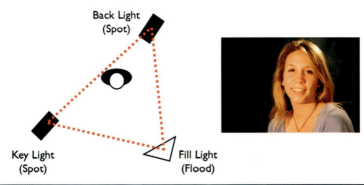

FIGURE 11.15
Three-point lighting is also known as triangle lighting or photographic lighting. Three lights are used to create the lighting treatment: the key, fill, and backlights.

Source: Photo by Josh Taber.

LIGHTING TERMS

Background light or set light: Light illuminating the background alone.

Based light or foundation light: Diffused light uniformly flooding the entire setting. Used to prevent underexposure of shadows or excessive contrast.

Bounce light: Diffused illumination obtained by reflection from a strongly lit surface such as a ceiling or a reflector.

Camera light: Small light source mounted on a camera to reduce contrast for close-up shots.

Contrast control light: Soft fill light from camera position illuminating shadows and reducing lighting contrast.

Edge light: Light skimming along a surface, revealing its texture and contours.

Eye light or catch light: Eye reflection (preferably one only) of a light source, giving lively expression. Sometimes from a camera light.

Hair light: Lamp localized to reveal hair detail.

Kicker, cross-back, or 3/4 backlight: A backlight that is roughly 30 degrees off the lens axis.

Modeling light or accent light: A loose term for any hard light revealing texture and form.

Rim light: Illumination of the subject's edges by backlight.

Side light: Light located at right angles to the lens axis. Reveals subject's contours.

Top light: Vertical overhead lighting (edge lighting from above). Undesirable for portraiture.

Wherever possible, additional lights can be used to illuminate the background behind the subject. But where space or facilities are limited, spill illumination from the key lights and fill lights may be used to cover the background areas.

While three-point lighting is a great place to start when lighting your subject, it is not the answer to every situation. You'll need to adapt it, using the basic concepts, to each new lighting challenge.

Basic Guidelines for Lighting People

There are a number of generally accepted guidelines when lighting people:

- Place the key light within about 10–30 degrees of a person's nose direction.
- Avoid steep lighting (above 40–45 degrees).
- Avoid a very wide horizontal or vertical angle between the fill light and the key light.
- Do not have more than one key light for each viewpoint.
- Use properly placed soft light to fill shadows.

Lighting Groups of People

Check the direction in which each person is looking, and arrange their key/back/fill lights to fit. One appropriate lamp may suit a whole group, but problems can arise if there are big differences in the relative tones of their clothing, skin, and hair. Light intensity may be too strong for one, yet too weak for another. In such situations, localize the light reductions with diffusers or shading light off with a barn door or flag.

Where subjects are close, or you have few lamps, one light may have to serve two different purposes, such as a key light for one and a backlight for someone else (Figure 11.16).

Where there are groups of people (audience, orchestra), you can either light them as a whole or in subdivided sections, still using three-point lighting principles and keeping overlaps to a minimum.

Lighting Areas of the Scene

It is best to avoid flooding areas with light. The most attractive picture quality usually comes from analyzing the performance area into a series of locating points (by the table, at the door, looking out the window) and tailoring the three-point lighting at each to suit the action. One lighting arrangement will often suit other shots or action in that area (perhaps with slight light adjustments). Given sufficient lighting

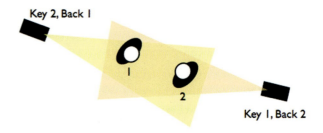

FIGURE 11.16 *Lights can be shared when practical. In the illustration, one spotlight is used as a key light for subject 1 and a backlight for subject 2. The second spotlight is used as the key light for subject 2 and the backlight for subject 1.*

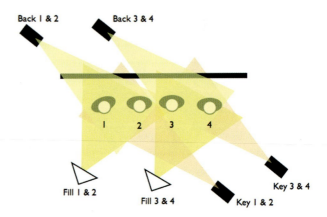

FIGURE 11.17 *Sets can be lit with a series of three-point lighting setups.*

facilities (enough lights, dimmers, etc.) and adequate time to readjust lights, you could light to suit each individual shot. But such elaboration is not normally feasible. Where action is more general or widespread, you must cover the area in a systematic pattern of lamps (Figure 11.17).

LIGHT SOURCES

There are many different types of light sources. Each one can be used in different situations. Here are some of the most common:

- **Tungsten (or incandescent)**: The regular tungsten filament lamp is relatively cheap, has a reasonably long life, exists in a wide range of intensities (power ratings), is generally reliable, and can be mounted in many types of fittings. However, tungsten lamps waste much electrical energy as heat. Their typical color temperature is around 3,000–3,200 K.

FIGURE 11.18 *Quartz lights put out a high level of light but are extremely hot.*

FIGURE 11.19 *HMI lights have an extremely high light output. This HMI is being used to simulate daylight coming through a window. Note that it must be attached to a ballast unit (right), which makes it quite bulky.*

• **Quartz (tungsten-halogen)**: In these lamps, the tungsten filament is enclosed within a quartz or silica envelope filled with a halogen gas. This restricts the normal filament evaporation, providing a longer lamp life and/or a higher, more constant light output, of increased color temperature—around 3,200 K. The bulb must not be handled, as body acid attacks the surface, and the lamp becomes brittle and extremely hot in use (Figure 11.18).

• **HMI**: The HMI is also known as a type of gas discharge lamp. These lamps are extremely efficient light sources, using a mercury arc ignited within argon gas. The HMI lights' abilities result in near-daylight illumination (5,600–6,000 K). The highly efficient HMI lamp is particularly convenient for use on location to fill shadows in exteriors and to light within large day-lit interiors, because its color temperature blends well with daylight. It provides about three to five times as much light as a quartz light of equivalent power, while producing less heat. A single 2.5 kilowatt (kW) HMI lamp can give as much light as two color-corrected 10 kW tungsten lamps. The lights require auxiliary circuitry (igniters and ballast units), and there is usually a 1.5–3 minute buildup time from switch-on to full light. Dimming methods are restricted, and a lamp cannot usually be turned on again quickly after switch-off, which is a disadvantage for "shoot-and-run" productions (documentaries, ENG) (Figure 11.19).

• **Fluorescent tubes**: The traditional tubular fluorescent lamp consists of a sealed gas-filled glass tube with a phosphor-coated inner surface. When switched on, the mercury vapor within the tube ionizes, causing the phosphor coating to glow brightly (fluoresce); the color depends on the specific materials used. The fluorescent lamp is three to four times more efficient than a tungsten source (more light per watt), so power consumption is correspondingly lower for the same light intensity. There is little radiant heat in the light beam. (Half of the power used by tungsten light sources may be wasted as heat!) Because fluorescent lights provide a broad light source, the illumination is relatively soft and easier on the eyes than intense spotlights. Fluorescent tubes have a relatively long life.

However, the lights can be somewhat bulky and rather fragile. The main shortcoming of this type of source lies in the light spread, for even with shutters and grilles (egg-crates, louvers), it is not easy to direct or confine light and avoid spill or over-lit backgrounds (Figure 11.20).

• *LED*: LED lighting instruments are the newest addition to television lighting. They are made up of a series of LED lights, available as a flood or spotlight. LED lights have become much brighter, they are extremely energy efficient (cool to the touch), lightweight, and durable. Some LED

FIGURE 11.20 *Fluorescent lights provide a broad light source.*
Source: Photo courtesy of Mole-Richardson.

FIGURE 11.21 *Some of the newest LED lights are now available up to 2K.*
Source: Photo courtesy of Litepanels.

lights can be powered by a camera battery when needed. These lights generally have a knob that allows the light intensity to be easily adjusted (Figures 11.8, 11.11, and 11.21). LED lights are among some of the most expensive television lights.

LIGHTING INSTRUMENTS

"You light with the lights you have, not the lights you wish you had. The difference between a good gaffer and a great gaffer is knowing how to best use the lights you have."

—Jefrey M. Hamel, Lighting Director

Camera Light

A small portable light can be attached to the top of a video camera. These camera lights are generally powered by an AC adapter, an exterior battery pack, or the camera's battery. Its main advantage is that it always illuminates whatever the camera is shooting, and does not require another pair of hands. Portable lighting of this sort can provide a very convenient key light when shooting under difficult conditions, especially when following someone around. They are probably most popular with television news crews. The light can also provide modeling light for close exterior shots on a dull day, or fill light for hard shadows when shooting someone in sunlight.

The disadvantages of these lights is that they add to the camera's overall weight, and their light is extremely frontal and thus tends to flatten out the subject. This light will reflect in glasses and shiny surfaces near the subject as an intense white blob. People facing the camera may also find the light dazzling.

Some camera lights have fixed coverage; others are adjustable. The illumination is invariably localized, and when using a wide-angle camera lens subjects may move out of its light beam. Another problem is that the illumination may not really be appropriate for the scene. Though anything near the camera is easily over-lit, anything farther away remains virtually unlit, which can be very obvious in longer shots. Some camera light systems even have an auto-sensor intensity control, which nominally adjusts their intensity to match exposure to the prevailing lighting conditions. Like all automatic systems, its performance is variable (Figures 11.26 and 11.27).

Fresnel Spotlights

In television studios, where the lights have to be positioned a fair distance from the subjects, the large heavy-duty *fresnel* spotlight is universal, suspended from ceiling bars or battens. It is lighter than the other studio spotlight (the ellipsoidal), sometimes has an adjustable beam, and provides an unfocused spotlight beam. The fresnel is probably used more in studios than any other light (Figure 11.14).

LOW-BUDGET LIGHTING

If you are just starting out and need some basic lights to get started, you may not want to invest in expensive lights in order to save your budget. Visit your local hardware store, home build store, or other home retail store. They'll have lots of options, from quartz and LED work lights to very inexpensive clamp lights with light bulbs. Some of these lights even come with light stands/tripods (although they may be under five feet in height).

Clamp lights are the lightest weight, are the least expensive, and provide you with multiple bulb options. Since they clamp onto about anything, you may not need light stands. Some of the bulbs can be very fragile and can get hot. One of the advantages of these clamp lights is that they use a standard home light bulb receptacle. This allows you to use any type of bulb, including screw-in incandescent, quartz, fluorescents, and LED bulbs. There are also spotlight and floodlight bulbs available. These fixtures can be purchased from US$5–10 plus the cost of the type of bulbs that you choose (Figure 11.22). Clamp lights can be supplemented with a fixture such as the one pictured in Figure 11.23. These fixtures can be screwed to a board and placed on the floor to provide you with low light when needed. Some of these types of fixture,

FIGURE 11.22 *Clamp lights are the least inexpensive low-budget lighting instrument.*

FIGURE 11.23 *These light fixtures can be attached to a board and used for low-budget lighting along with the clamp lights.*

FIGURE 11.24 *(Left) These quartz work lights are available at your local home repair store.*

FIGURE 11.25 *(Above) LED work lights can often be run off of rechargeable batteries.*

especially those made for Christmas lights, come with a stake support that can be pounded into the ground. You just need to make sure that you match the light bulbs between the instruments.

Quartz work lights are generally very hot, which can raise the temperature, making the crew and talent uncomfortable if you are working in a small room. It can also be a burn hazard if someone accidentally touches the light. The advantage is that they can be very bright, depending on which lights you purchase. These can be purchased from US$20–85, including bulbs (Figure 11.24).

The LED work lights are cooler and lightweight, but are the most expensive. They often can operate on a rechargeable battery, which helps in situations where you need a light but don't want the cord showing. LED lights are usually the least bright, making them impractical for lighting large areas. They generally cost in the range of US$50–100. Of course, you don't have to purchase bulbs for these lights (Figure 11.25).

Make sure that you purchase matching light sources if you plan to use two or more at the same time. This will give you a better chance at white balancing.

Even with low-budget lighting, you can still hone your lighting skills. While these are good options when working with a very limited budget, professionally designed television lights will always give you the most control over your lights and consistency in the light output.

FIGURE 11.26 *On-camera lights can draw their power from an external battery pack, the camera's battery, or an AC power supply.*

Source: Photo courtesy of Grass Valley/Thomson.

FIGURE 11.27 *This LED 6,000 K energy-efficient on-camera ring light is designed to provide an overall shadowless shot or can utilize individual sections of the ring.*

Source: Photo courtesy of VF Gadgets.

Ellipsoidal Spotlights

The *ellipsoidal* light is a sharply focused spotlight. The focusing ability allows it to project patterns on the set. It is heaver, includes internal shutters (like barn doors), is more expensive than other spotlights, and has an adjustable beam (Figure 11.28).

Scoop

The *scoop* is an inexpensive and simple light instrument, requiring little maintenance, and works well when a floodlight (fill) is required. However, it can be inefficient and bulky. Unfortunately, the light from most scoops spreads uncontrollably, spilling around over nearby scenery (Figure 11.29).

Broad Light

The lightweight *broad light* (also known as a *V-light* or *broadside*) has a short trough containing a reflector and a tubular quartz light of usually 500–1,000 W. The bulb may have a frontal shield to internally reflect the light. Although the broad is widely referred to as a soft light source, due to its small area, it produces quite discernible shadows. Nevertheless, it is an extremely useful wide-angle broad light source that can be hung conveniently in various ways, supported on stands, or laid on the floor.

Two-leaf or four-leaf barn door shutters that can be closed to reduce the spread of the light are often fitted to broads (Figure 11.30).

Soft Light

Soft lights are available in two primary types: the *studio soft light* and the *portable soft light*. Soft lights generally utilize a central lamp that is reflected off of the back of the lighting instrument. Although this device spreads its illumination uncontrollably, it is a very handy lighting tool (Figure 11.31).

The portable soft light is designed to be carried easily into the field, but is still often used in the studio. It provides a large amount of soft light. The portable soft light is available in different models. Some of these diffusion attachments fit on standard lights; others can be purchased with special lighting instruments (Figure 11.32).

FIGURE 11.28 *The ellipsoidal light is a sharply focused spotlight. Its focusing ability allows it to project patterns on the set.*

Source: Photo courtesy of Mole-Richardson.

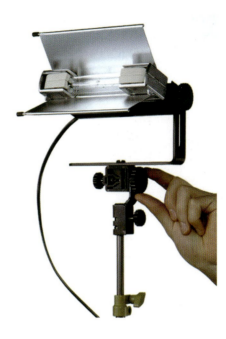

FIGURE 11.30 *The V-light, or broad light, is a very compact light source. This powerful light source can be used as a key or fill, and folds small enough to fit into a camera case.*

Source: Photo courtesy of Lowell.

FIGURE 11.29 *The scoop is a simple floodlight. It is inexpensive, usually not adjustable, lightweight, and does not have a sharp outline. The scoop works well as a fill light and is great for lighting large areas on a set.*

Source: Photo courtesy of Mole-Richardson.

FIGURE 11.31 *Studio soft light.*

Source: Photo courtesy of Mole-Richardson.

FIGURE 11.32 *The portable soft light is lightweight, may work with existing lighting instruments, and provides a large level of soft light.*
Source: Photos by Mole-Richardson and Taylor Vincent.

Cyclorama Light

Cyclorama lights, or "cyc" lights, are used to illuminate backgrounds with broad lighting. There are basically two types of cyc lights:

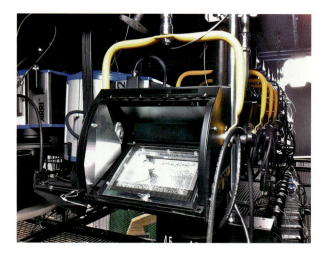

- The *floor light*, also known as a *cyclorama light* or *strip light*, is used to light backgrounds from the floor up (Figure 11.33).
- A hanging cyc light that projects a consistent broad light down on the background (Figure 11.34).

Multi-Lamp Sources

Several soft light sources use groups of lamps, which combine so that the shadows cast by each are "lit out" by its neighboring lights.

A *strip light* or cyc light consists of a row of light units joined in a long trough. Each unit has a bulb with a curved

FIGURE 11.34 *The hanging cyclorama light illuminates the background from the light grid.*

FIGURE 11.33 *Strip lights or cyc lights are used primarily to light sets from the floor up.*
Source: Photo courtesy of Mole-Richardson.

metal reflector. As mentioned earlier, the strip light can be used to illuminate backgrounds or translucent screens from the floor (Figure 11.33).

Multi-lamp banks are excellent soft light sources. A typical design has multiple panels of grouped internal reflector lamps. Each panel can usually be independently switched and turned to adjust the brightness and spread of the unit. The floodlight bank is mainly used as a booster light for exteriors and for illumination of large areas (Figure 11.9). Large side-flaps may be fitted to restrict the light spread. Soft light sources that rely on internal reflection to produce light scatter produce quite diffused light, but are relatively inefficient.

Large units fitted with a bank of fluorescent tubes are favored by some people as a soft light source. Although these lights can be a bit fragile in use, one of their main advantages is that they produce little to no heat, use much less energy than a normal television light, and put out a large amount of light (Figure 11.20).

Open-Face Adjustable Light

The *open-face adjustable light* is widely used in the field. It has a variety of names, including *lensless spotlight*, *open-bulb spot*, *external reflector spot*, and *reflector spotlight*. This light has many advantages. It is extremely portable, compact, and efficient. Diffuser and/or corrective color gels are easily clipped to its barn doors (Figure 11.35).

FIGURE 11.35 *This open-face adjustable light unit can be used as a spot or floodlight.*
Source: Photo courtesy of Mole-Richardson.

FIGURE 11.37 *An umbrella reflector can be attached to a light source in order to create a soft lighting instrument.*
Source: Photo by Josh Taber.

Reflectors

The least expensive way to improve a subject's lighting when shooting in sunlight is to use a reflector. This is simply a surface such as a board, screen, cloth, or even a wall that reflects existing light onto the subject from another angle. The quality of the reflected light depends on the surface you use.

There are many commercial reflectors available, such as the ones shown in Figures 11.36 and 11.37. These lightweight

FIGURE 11.36 *Commercially produced cloth reflectors can be purchased that have a variety of colors. This specific reflector is designed with six different colors, for six different lighting effects.*
Source: Photo courtesy of Wescott.

cloth reflectors, sewn onto a spring-metal frame, can be easily folded and transported. Available surfaces include silver, gold, white, and combination reflectors. A mirrored surface, such as metal foil, will reflect a distinct beam of light from a hard light source, creating sharp, well-defined shadows. The reflected light travels well, even when the subject is some distance away. (A mirrored surface can even reflect soft light to some extent, if placed fairly near the subject.) The angle of a mirror-finish reflector is critical. When the light shines directly at its surface, the maximum effect is obtained. However, as the surface is angled toward the light, the reflected beam, which covers only a restricted angle anyway, narrows and becomes less effective. In a long shot, its limited coverage is seen as a localized patch of light.

If the reflector has a matte-white surface, it will produce a soft diffused light, which spreads over a wide angle. But this soft reflected light is much weaker, and will travel only a relatively short distance, which depends on the intensity and distance of the original light source.

Reflectors can be easily made from a board covered with aluminum foil (smooth or crumpled and flattened) or matte-white painted, according to the type of light reflection required (a board with a different surface on each side can be very useful). These "boards" can be made of wood, foam core (which is extremely lightweight), or cardboard. The bigger the reflector, the more light will be reflected over a broad area. Even a large cloth can be used. However, cloth reflectors of this size can be cumbersome to hang and are

---

REFLECTORS IN USE

A. The sun provides the key light and the reflector is the fill light.

B. A reflector is being used to reflect additional light into a building through its window.

C. Two reflectors are being used to increase the illumination. One of the reflectors is silver (the key light) and the other is white (the fill light).

How effective a reflector is depends on its surface and on its angle to the sun or other light sources. If a reflector is used beside the camera, and reflects a source directly ahead of the camera, the intensity and coverage of the reflected light is at a maximum. As the reflector is angled to the source, its output and its coverage fall.

Source: Photos by Josh Taber (A) and Nathan Waggoner (B).

FIGURE 11.38 *Reflectors can be used in many situations, using different techniques.*

likely to blow in the wind if outdoors. However, since the only alternative is to use powerful lamps, or lights close in to the subject, it is certainly worth trying reflectors, when the sun's direction is appropriate.

Indoors, reflectors can be used to redirect light from windows or spotlights into shadowy corners. When using backlight, a low reflector can be used near the camera to reduce the shadows under people's chins and eyebrows (Figure 11.38).

LIGHT SUPPORTS

Grip Clamps

There are a number of different clamps or grips available on the market in order to hold lights on location or in the studio. All of them include "mounting spuds" where lights can be attached. These clamps clip a light to any firmly based object, such as a door, table, chair, rail, post, window, or ladder. In the studio, they can also be clamped to a light stand and set flat. These clamps can be a very useful compact device to secure lamps in out-of-the-way places, especially when space is restricted (Figure 11.39).

Light Stands

Light stands come in all different sizes and shapes and are generally telescopic three-legged stands. They can be collapsed, folded, and/or dismantled into sections for transport. The size of the light will determine how sturdy the light stand needs to be. If the stand is too flimsy, it will be top-heavy and easily upset, even by the weight of the light's cable. With more robust types of stands, two or more lights can be attached to a stand when necessary.

Light stands have the disadvantages of occupying valuable floor space (perhaps impeding camera or sound-boom movement), casting shadows onto backgrounds, having trailing cables, and being vulnerable. But they do permit easily adjusted precision lighting (Figure 11.40).

In fact, most television lighting is suspended, in order to leave floor space uncluttered by lamps or cables.

Studio Ceiling Supports

Smaller studios and temporary sets in larger studios frequently use pipes to hang the lights on in the studio. These pipes enable lamps to be clamped or suspended as required (Figure 11.41).

FIGURE 11.39 *"Furniture clamps" and "gator" clamps are used to attach a light to anything around the shooting location.*
Source: Photos courtesy of Mole-Richardson.

Large studios frequently use "battens," "bars," or "barrels" arranged in a parallel pattern over the studio area, and individual battens may be counterbalanced by wall weights or motor winches to allow lamps to be rigged. Battens are usually a type of hanging bar that allows the lights to be hung and then plugged directly into the batten.

Lighting grids or catwalks allow the lighting crew to walk around the lighting located at the ceiling (Figure 11.42).

Portable Light Kits

Portable light kits are available for the remote production crew. These kits usually come in two- to four-light packages, which include lighting instruments, light stands, power cables, barn doors, and so on. There are a wide variety of kits available from different vendors (Figure 11.43).

FIGURE 11.40
Light stands can be collapsed and folded into a compact size for storage and transport.

Source: Photo by Josh Taber.

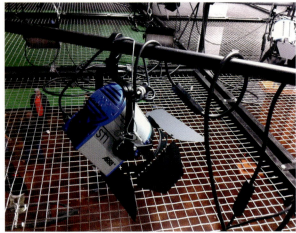

FIGURE 11.41 *Pipes are often used on sets to hang lights from.*

FIGURE 11.42 *Lighting grids allow lighting personnel to walk on the same level as the lighting instruments. Lights can be hung above or below the grid.*

FIGURE 11.43 *Many different types of portable light kits are available.*
Source: Photo courtesy of Mole-Richardson.

CONTROLLING THE LIGHTS

Lights can be controlled in a number of ways, including dimmer boards, cookies, filters, diffusers, barn doors, and flags. Each of these types of controllers can be used for different purposes, based on where the shoot is taking place, the equipment available, and the skills of the person doing the lighting.

- **Dimmer board**: There are many different styles of dimmer boards, from fairly inexpensive simple manual fader boards to highly technical computer-driven boards

(Figure 11.44). These systems are extremely adaptable and can be suitable for portability. Faders (control levers) and channel switchers are grouped on a lighting board (console), remotely controlling the lights. Intensity adjustments are smooth and proportional over their range, even for varying electrical loads. Generally, the lights are attached to dimmer units, which are controlled by the dimmer board.

- **Barn doors**: Barn doors have independently adjustable flaps (two or four) on a rotatable frame; these selectively cut off light beams. They are used to restrict light, shade walls, and prevent backlight from shining into the camera lens (causing lens flares). Barn doors are attached to the front of the lighting instrument and are most commonly used to control where the light falls on the scene (Figures 11.41 and 11.45).

• **Flags**: Flags are often used to control spill light from light sources. Flags are generally made of cloth that is stretched over a metal frame. The flags are usually mounted on a light stand placed in front of the light source (Figure 11.46). Flags can also be constructed by using any material that can block the light.

• **Light filters/gels**: Color gels (filters) can be placed over lights to enhance the color of the light. They are added to create special effects or control the type of light falling on the subject. Note the filter being used in Figure 11.8. Gels are usually placed into gel holders and then placed onto the light (Figure 11.47).

FIGURE 11.45 Barn doors are used to help mold the light.

FIGURE 11.44 Dimmer boards are available in a variety of styles, from simple manual fader boards to highly sophisticated computer-driven units.

Source: Photos courtesy of Strand.

FIGURE 11.46 Flags are available individually or in kits such as the one shown here. Generally, flags are mounted on light stands.

Source: Photos courtesy of Wescott.

FIGURE 11.47 *Color filters, also called gels, are used to create effects or balance the color of the light falling on the scene. This gel is placed in the metal frame and slid into a holder in the light.*

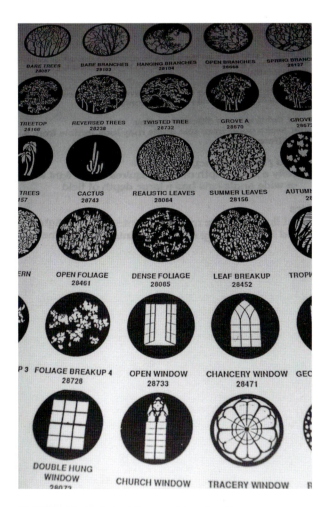

BARE TREES 28087 · BARE BRANCHES 28103 · HANGING BRANCHES 28104 · OPEN BRANCHES 28668 · SPRING BRANCH 28127

TREETOP 28166 · REVERSED TREES 28238 · TWISTED TREE 28732 · GROVE A 28670 · GROVE 2867

TREES 157 · CACTUS 28743 · REALISTIC LEAVES 28084 · SUMMER LEAVES 28156 · AUTUMN 28

ERN · OPEN FOLIAGE 28461 · DENSE FOLIAGE 28085 · LEAF BREAKUP 28452 · TROPIC

P 3 · FOLIAGE BREAKUP 4 28728 · OPEN WINDOW 28733 · CHANCERY WINDOW 28471 · GEO

DOUBLE HUNG WINDOW 28073 · CHURCH WINDOW · TRACERY WINDOW · R

FIGURE 11.48 *A sheet from a catalog of cookies.*

- **Diffuser**: Diffusers consist of some type of a translucent material (wire mesh, frosted plastic, or spun-glass sheet). They diffuse light (overall or locally) and reduce the intensity (Figure 11.10).

- **Cookies/gobos**: Cookies (or gobos) are perforated opaque or translucent sheets of glass or metal that create dappling, shadows, light breakup, or patterns that can be projected onto a set by a spotlight. Cookies can be inserted onto the front or into some spotlights (Figure 11.48).

BASIC LIGHTING PLAN

Before lighting any production, a number of preliminary questions should be answered. Whether you get these answers at a preproduction meeting with the director, set designer, or audio engineer, or from a chat in the studio while looking around the set, will depend on the size and type of the production.

What Is Going to Happen?

The subject: The main subjects in most programs are people. Even for a single person speaking to the camera, or a seated interview, you must know where people are going to be positioned and the directions they will be facing. You will need details of the action. The answers will decide where you place key lights. If keys are badly angled, it can considerably affect portraiture. All subjects tend to have optimum directions for the main light direction.

The cameras: Where are the cameras going to be located? Lighting must suit the camera's position. If the subject is to be shot from several positions, the lighting has to take this into account. However, that does not mean flooding light onto the set from all directions.

The surroundings: You will need to know about the general tones of the surroundings. Are they light-toned (then they could easily become overly bright) or dark-toned (in which case, more light may be needed to prevent lower tones from becoming detail-free shadows)? Will the subjects stand out from their background or tend to merge with it?

Atmosphere: Are you aiming at a specific atmospheric effect (upbeat, cozy evening interior, intriguing mystery, etc.)? The answer to this question will influence how the light and shade are distributed in the scene.

Production mechanics: These include such things as:

- Selecting *sound-boom positions* that avoid casting shadows that will be seen in the shot. Shot anticipation and coordination are necessary to prevent boom shadows from falling across people and backgrounds. The normal trick is to throw the inevitable shadow out of the shot by careful key light positioning. Obviously, difficulties arise when this lamp position is artistically incompatible (Figure 11.49).
- Determining *lighting cues* so that you know when lights need to be adjusted (such as someone apparently switching room lights on).
- Creating *lighting effects*, which include effects such as a fire flicker, lightning, moonlight, and so on.

There will be times when you will have to light a setting without knowing any of these details. In those circumstances, you can provide only a general pattern of lighting and check the images during the rehearsal to see where changes will be needed to improve the situation. Results under these conditions can be unpredictable.

The Lighting Plot

Most people find the basics of lighting easy enough to understand, but are very apprehensive when it comes to the actual lighting process. How do you begin? Creating a lighting plot will help you think through your lighting plan. An accurate lighting plot will enable you to immediately identify every lamp during rehearsal. Without a plot, you will be left wondering which lamp is causing the boom shadow, or why there is a hot spot on the set. Work systematically through the project step by step. Break the set into different areas, analyzing where the action is going to take place, and place the lights as needed (usually a type of triangle lighting) (Figure 11.50). Table 11.1 lists the types of questions that need to be considered when placing lights on the light plot.

Lamp Care and Safety

- Avoid moving hot lights when they are lit. Filaments are very fragile when hot. LED lights can be moved immediately since they do not get hot.
- Use gloves when handling hot lamp bulbs. Never touch quartz lamps or HMI bulbs with a bare hand (body acids destroy the glass bulb).
- Allow plenty of ventilation around lamps to avoid overheating (drapes, cloths, and other scenery can burn if they are too close to hot lamps). If you place diffuser material or color medium too close to a fresnel lens, you may crack the lens, so always use proper holders.
- Beware of overbalancing floor stands! Weigh the bottoms (weights, sandbags). Secure cables to prevent accidents.
- Always utilize safety chains or cables to secure all hanging lamps and accessories in case they fall. Note the safety cable shown on the light in Figure 11.41.
- Switch off lamps whenever possible to reduce heat, lengthen lamp life, and minimize power costs.

Mic boom
Shadow

FIGURE 11.49 *Sound boom shadows can be a challenge on a set.*

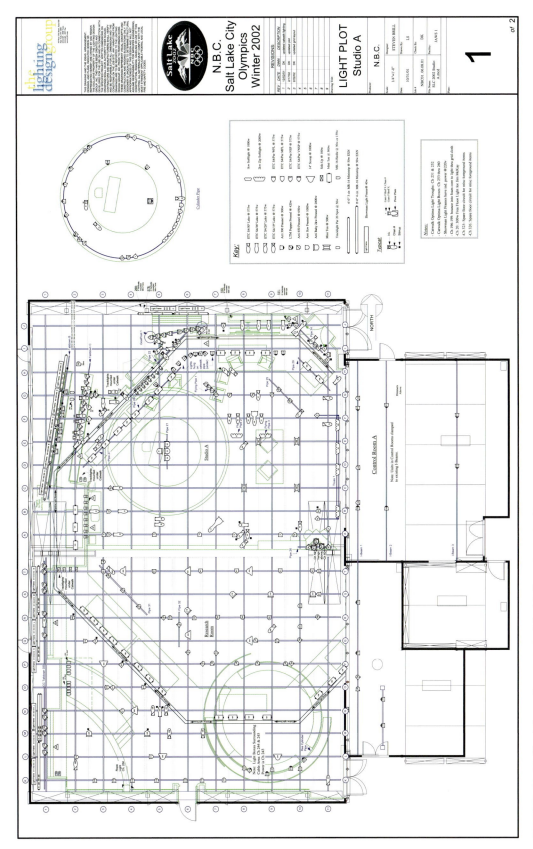

FIGURE 11.50 *This lighting plot was created for NBC's studio at the Olympics.*

Source: Courtesy of Lighting Design Group.

Lighting and Camera Rehearsal

The camera rehearsal is the moment of truth. Now you can see the results of your labors. Keep an eye on all preview monitors and look out for lighting defects, such as distracting shadows, hot spots, unsatisfactory contrast, wrong light direction, unattractive portraiture, boom shadows, and the like. If possible, readjust dimmers to improve the lighting balance directly as you see inaccuracies; rebalancing could affect all subsequent shots.

The director will also be arranging and readjusting shots. Many shots will be as expected; some may be quite different and could require total relighting. Whether you correct lighting problems as they arise or list them to be corrected during a break depends on the circumstances. It is reasonable to quietly adjust a barn door with a lighting pole during rehearsal, but diffusers and gels or hanging lamps cannot be changed unobtrusively.

Lighting on Location

When you are shooting away from the studio, you will encounter a wide range of lighting conditions:

- *Day exteriors*: These vary from overcast skies to strong blinding sunlight with deep shadows.
- *Night exteriors*: Anything from pitch-black night to strong moonlight; from the odd street lamp to "bright-as-day" surroundings.
- *Day interiors*: These can range from locations where sunlight through windows embarrassingly overwhelms any interior lighting to those where you need to provide a high-power lamp to simulate sunlight on a dull day!
- *Night interiors*: Conditions here can vary considerably, from total darkness to an extensively lit environment. Sometimes the interior lighting is quite unsuitable for the camera and must be switched off.

In Table 11.2, you will find a useful summary of typical equipment and techniques that you can use when lighting on location.

INTERVIEW WITH A PROFESSIONAL: TOMMY BROWN

Why do you like working with lighting? I am passionate about lighting because it is fun for me and I enjoy the challenges of it. When you see what is being accomplished with your light, you feel you are really contributing to the project. I must confess that sometimes it's a little like Christmas morning getting to plug things in and have them work! It's not that I'm surprised when a light comes on; it's that I take great pride in doing the job right.

What makes good lighting? Experience! Experience! Experience! Knowing what works and what doesn't will save you when there's limited time for lighting! Especially when someone needs that light yesterday! Preproduction is essential. Sitting down with the DP in advance to get his or her vision makes all the difference in the world. Like acting, lighting can make or break your production. You can have the best script in the world, but if your actors aren't doing their job, good luck getting the project to see the light of day. The same goes for lighting. You can have a vast array of lights at your fingertips, but if you don't know how to use them you are never going to move beyond the student film level.

FIGURE 11.51
Tommy Brown, lighting.

What are some of the challenges of lighting?

- Many times the biggest challenge is getting power from one location to another.

- It can be very difficult to stay attentive, and not get distracted, so that you know what is needed.

- Lighting instruments change. It is tough to quickly learn new lights.

Tommy Brown has worked on the lighting of corporate videos, network television, and feature films.

REVIEW QUESTIONS

1. What are three of the goals of lighting?
2. What is the advantage of a zebra system?
3. How are spotlights used in productions?
4. Describe how to set up three-point lighting.
5. How can a portable reflector be used in lighting a person?
6. What are two of the issues that need to be considered when doing a basic lighting plan?

TABLE 11.1 *Determining the Location for Your Lights*

The subject	• Scrutinize the subject and consider the specific features that you want to emphasize or suppress.
	• Will there be any obvious problems with that subject (such as a person with deep-set eyes, who needs a lower key light)?
	• Anticipate troublesome light reflections (e.g., on an oil painting). Will a carefully positioned key avoid bad light flare or is dulling spray needed?
	• How critical will the shots be? If revealingly close, will there be sufficient depth of field? Will there be possible camera shadow problems?
	• Is the subject stationary, or is it moving around?
	• Is it being shot from several angles? If so, from which directions?
	• Are shadows likely to fall on the subject from people nearby, or parts of the scenery (hanging chandeliers, arches, tree branches)? Lighting angles may need to be adjusted to suit the situation.
The key light	• Arrange a key light, suitable for the camera's position(s), subject direction(s), and action.
	• Does the key light cover the subject? If the subject moves, will it still be suitable?
	• Does the key need to cover more than one subject—an area, perhaps?
	• Does the light coverage need restricting? Is there any unwanted spill light onto nearby areas (people, settings)? Use barn doors or flags.
	• Does the key light suit the various camera angles?
	• Is the key light likely to cause boom shadows?
	• Is the key light position too steep (i.e., too high)? Look for dark eye sockets and long nose and neck shadows.
	• Is the key light position too shallow (i.e., too low)? It can blind talent, reflect in glasses, cause background hot spots, throw subject shadows onto the background, or cast camera shadows onto people.
The fill light	• Generally avoid high-intensity overall fill light (base light). It flattens modeling, reduces contrast unduly, and over-lights backgrounds.
	• Position the fill to illuminate the shadows, not to add illumination to the key light level.
	• Fill light must be diffused. If necessary, place diffuser over light sources.
	• Typical fill light intensity is around half of the key light level.
The backlight	• Avoid a steep backlight. It becomes an ugly "top light"—flattening the head and hitting the nose tip. If someone is close to a background and cannot be backlit properly, omit it. Do not use top light or side light instead.
	• Avoid very shallow backlight. Lamps come into the shot. Lens flares may develop.
	• Avoid excess backlight. Intensity is roughly the same as the key light level. Excess light creates unnatural hot borders to subjects and over-lights hair.
Background lighting	• Is the background associated with a specific style, atmosphere, or mood that needs to be carried through in the lighting treatment?
	• Is there any danger that the subject might blend into the background tones?
	• Plan to light the subject before lighting the background. Any subject lighting falling on the background may make extra background lighting there unnecessary. However, keep the subject and background lighting separate, as they usually require quite different treatment.
	• Whenever possible, relate the light direction to the environment, such as visible windows and light fixtures.

TABLE 11.2 *Lighting Treatment on Location*

EXTERIOR SHOTS	Day: Sunlight	• Avoid shooting into the sun or having the talent look into the sun. • Reflect sunlight with a hard reflector (metallic-coated) as key light, or soft reflector (white-coated) for fill light. • In closer shots, color-corrected lamps can fill shadows. • Bright backgrounds are best avoided, especially when using auto-iris.
	Dull day, failing light	• Color-corrected light can provide key light or fill shadows.
	Night	• Quartz lighting can provide key, fill, and backlighting for most situations. For larger areas, higher-powered sources are necessary, such as HMI lights. • Large lens apertures are often unavoidable (which produces shallow depth of field). Extra video gain may emphasize picture noise.
INTERIOR SHOTS	Day	• The camera must be balanced to the color temperature of either the daylight or the interior lighting. • When strong sunlight enters windows, pull shades or blinds, put filter material over the windows, perhaps adding neutral-density media to reduce sunlight strength. Another option is to avoid shooting windows. Alternatively, filter your lamps to match daylight. • A camera light flattens modeling.
	Night	• Either supplement any existing lighting or replace it with more suitably angled and balanced lighting.

CHAPTER 12

BACKGROUNDS AND SETS

"What we are looking for in a dramatic set is an imaginative substitute. Keep in mind, directors produce illusions. However basic the set materials really are, the end result can appear to the audience as the real thing."

—Gerald Millerson, Former Director, BBC

TERMS

Chromakey: Utilizing a production switcher, with this technique, the director can replace a specific color (usually green or blue) with another image source (still image, live video, graphics, prerecorded material, etc.).

Cyclorama (cyc): Serves as a general-purpose detail-less background. Cycloramas generally have a continuous, seamless background that can be projected on or keyed out to create unique backgrounds.

Flat: A panel usually created by building a frame and then either attaching a piece of wood or stretching cloth over it to create a wall for a set in the television studio.

Floor plan (staging plan, set plan): A rough plan of the staging layout usually begins with drawing potential scale outlines of settings, including their main features: windows, doors, and stairways. Ensure that there is enough room for cameras, sound booms, and lighting.

Hand properties (props): Any items that are touched and handled by the talent during the production. These could include a pen, dishes, a cell phone, silverware, and so on.

Modular set units: Prebuilt set components that can be compactly stored and then quickly assembled and disassembled.

Stage props: The furniture on the set. These would include news desks, chairs, couches, tables, and the like.

Studio plan: The basis for much of the organization, showing the studio's permanent staging area with such features and facilities as exits, technical supplies, cycloramas, and service and storage areas.

Virtual set: Uses a blue or green seamless background, chromakeying the computer-generated set into the scene. Most virtual sets employ sophisticated computer software that monitors the camera's movements so that as the camera zooms, tilts, pans, or moves in any other way, the background moves in a corresponding way.

The background is much more than "whatever happens to appear behind the subject." It directly affects the success of the program and needs to be carefully designed and controlled. Scenery can range in practice from a simple backdrop to extensive construction, but effective design is important to the success of any show (Figure 12.1).

THE INFLUENCE OF THE TELEVISION BACKGROUND

Television shows are shot in a wide range of locations: people's homes, offices, factories, rooms, public buildings, studios, streets, and wide-open spaces. Where we shoot the program may be vital (contextually relevant) to what we want

FIGURE 12.1

This television set was designed by Australia's Network 10 to show off the context of being in London, with recognizable buildings in the background (Parliament and Big Ben). This background added credibility to the broadcast, allowing viewers to see where the anchors were located.

FIGURE 12.2

The set pieces, the dressings and the props give an authenticity to the time period this story takes place in.

Source: Photo by Doug Smart.

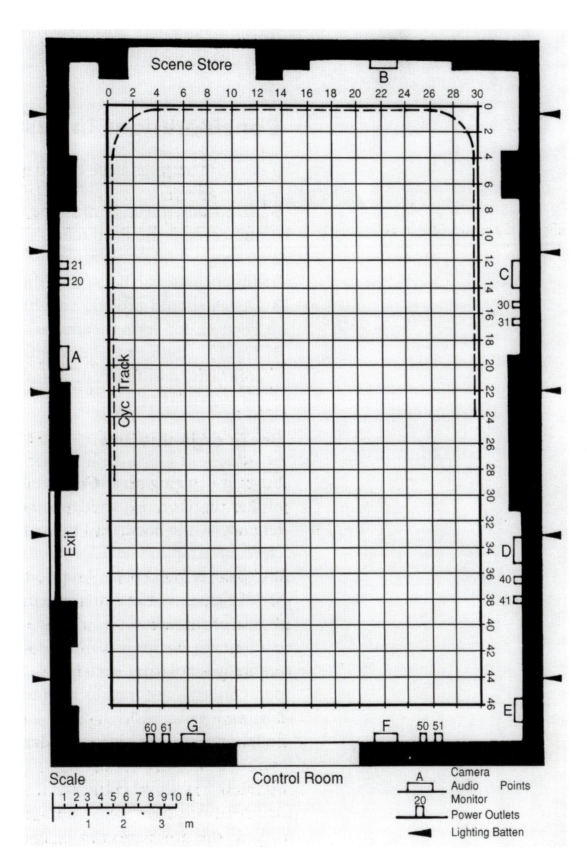

FIGURE 12.3 *Sample studio plan.*

to tell our audience, or it can be merely incidental. To some extent, the importance of the set depends on the director, the way the subject is approached, and the chosen style and form. Effective backgrounds or sets are more a matter of making wise choices than having a big budget.

Surroundings have a considerable influence on how we feel about what we are seeing and hearing. It is not just a matter of choosing a background that looks appropriate or attractive, but determining whether its impact on the audience is right for the specific points being made in the program (Figure 12.2).

The background we choose for our action, and the way we shoot it, can affect how persuasively points are communicated to our audience. It is one thing to see a person standing on a street corner, recommending a type of medicine, and another when we see that same person wearing a white lab coat in a laboratory. The surroundings have swayed our reactions, yet they have nothing to do with the true quality of the product.

The camera cannot avoid being selective. For example, if a video camera is taken to an offshore oil rig, depending on which parts of the structure are shot, a very different view of

life on the rig can be expressed. The final emphasis could be on its huge geometric structure or the isolation of this group of workers in treacherous seas, or it might appear as a scene of endless noise and tense activity. In the end, it is the shot selection and editing that will portray the concept of life on a rig to the audience. The result may be a fair cross-section of life there or it could be overly selective. Much depends on the point of view the director adopts.

BASIC ORGANIZATION

Staging begins with the demands of the script and the aspirations of the director. Much depends on how effectively these can be related to the facilities, time, and budget available. As with all craftsmanship in television, optimum results come from a blend of imaginative perception and practical planning. Television set designers achieve minor miracles in making ingenious use and reuse of materials.

Planning begins with discussions between the director and the set designer. Using sketches, scale plans, and elevations, production concepts are transformed into the practicalities of man-hours, cost, and materials. For larger productions,

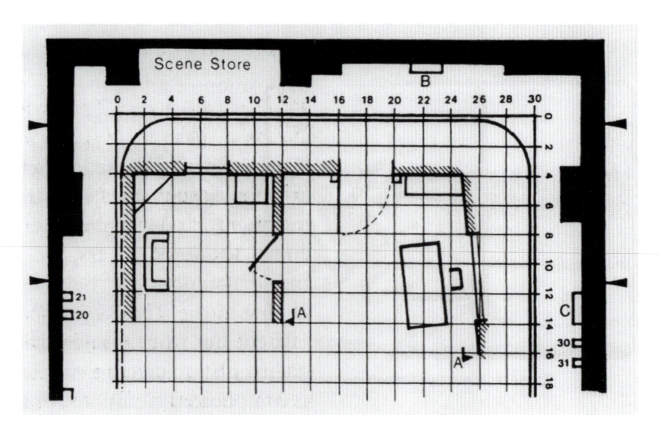

FIGURE 12.4 *Sample production floor plan.*

FIGURE 12.5 *The set design must provide ample shot opportunities.*
Source: Illustration courtesy of Fifteenhundred (www.fifteenhundred.com).

there is close collaboration with various specialists, who consider shot opportunities for cameras, talent action, and moves, and the various lighting, audio, camera treatment, costumes, makeup, and technical requirements. This type of project requires teamwork.

The Studio Plan

The basis for much of the organization is the standard printed *studio plan*, which shows the studio's permanent staging area with such features and facilities as exits, technical supplies, cycloramas, service and storage areas, and the like (Figure 12.3).

The Floor Plan

Also known as the *staging plan*, *ground plan*, or *set plan*, the *floor plan* is a rough plan of the staging layout that usually begins with drawing potential scale outlines of settings, including their main features—windows, doors, and stairways. Ensure that there is enough room for cameras, sound booms, and lighting (Figure 12.4).

Lighting Plot

The lighting director designs the lighting plot, showing the battens, or lighting grid, and each of the lighting instruments on the staging plan.

Design Considerations

Television settings must satisfy several requirements (Figure 12.5):

- Artistically, settings must be appropriate to the occasion, the subject, and the production's purpose.
- The staging must fit the production needs: its dimensions, facilities, and the production budget.
- The design should provide room for camera movement and shots, sound and lighting.

Set Design for 16:9

HD and its 16:9 format created some challenges. Sets that were acceptable on standard definition (4:3) were no longer acceptable due to the high resolution of HD. Scratches and dents are much more apparent.

The 16:9 format has also changed the design of sets, and made it more complex, affecting also the locations of the talent. Until the 16:9 format became the standard, viewers saw more header elements (high sections of the set), because camera operators needed to frame for 4:3 and 16:9. This was especially true in wide shots. However, once 16:9 framing is done exclusively, the header elements are seen much less. If your program is still shot in the 4:3 format, it's important to create clean, visually interesting header elements, while not dedicating too much of the budget to elements that may only be seen for a short period (Figure 12.7).

SETS ARE A MATTER OF TASTE

Television show set designs differ greatly from country to country, and even network to network. The following photographs are a study of some of the design choices that were made by two different countries' networks who covered the Olympics in Vancouver. These sets were photographed in the main broadcast center during the Games.

FIGURE 12.6 *Australian and Japanese sets. Set designs can be incredibly varied.*

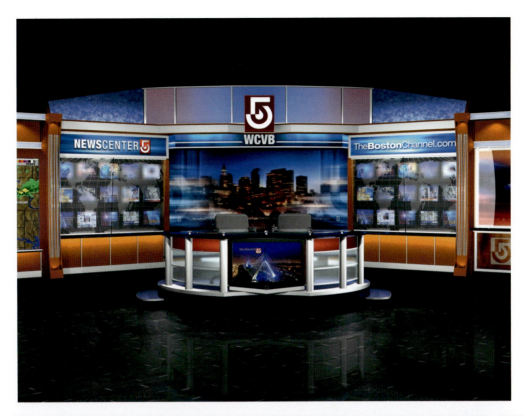

FIGURE 12.7 *Today, most sets are created with the 16:9 format in mind. However, it may be necessary to design for the 4:3 standard definition as well. It is important to remember that 4:3 viewers won't see the areas to the far right and left areas of the screen, so it's important not to put vital scenic elements or graphics in this space.*

Source: Photo courtesy of FX Group (www.fxgroup.tv).

SET DESIGN SOFTWARE

A number of companies have created software that can assist a set designer in the process of creating and visualizing how the final set is going to look. One of the most popular of the 3D modeling software, SketchUp, offers a free version for non-commercial use. Once the set is designed, you can look at it from different angles to see what the camera would see. You can even place modeled actors or commentators into the set to see if it works for your production (Figure 12.8).

FIGURE 12.8 *Modeling software such as SketchUp assists set designers in creating a set. The 3D model of the set allows them to view the set from various camera angles.*

Source: Design by Sarah Owens.

Real and Unreal Backgrounds

Most audiences are not concerned about whether the background is real or an illusion. They usually don't care if it is a real location or computer-generated. It is the effect that counts. However, it is worth remembering that backgrounds can be derived in a number of ways:

- *Use of actual place*: The action is really shot in the Sahara Desert.
- *Use of substitute*: Shoot the action in a convenient location that looks sufficiently like part of the Sahara Desert (Figure 12.9).
- *Use of a set*: Build a set that resembles the real thing in a studio (Figure 12.10).

- *Suggested location*: The camera shows location shots of the Eiffel Tower (part of a still photo) intercut with shots of someone standing against a brick wall. Thanks to sound effects of traffic and so on, the viewer assumes that this is shot in Paris.
- *Virtual set*: It is possible, with various electronic equipment, to insert the person standing in front of the camera into a separate background picture. With care, it can be done absolutely convincingly.

The Neutral Background

There are times when we want the background to provide totally neutral surroundings for the action. In the extreme, this background could be just a blank white (limbo) or black

FIGURE 12.9
*This "airplane accident"
was actually a scene
created for filming.*

FIGURE 12.10
*(Below) Golf Channel's
studio shown uses 4K
video monitors; there are
no live windows. This
creates a flexibility that
was unavailable in the
past.*

Source: Photo courtesy of
Golf Channel/NBC.

(cameo) area, because we are concentrating on the performers. However, we usually want something rather more interesting to look at than a plain area of tone, and television solved this problem by creating the neutral setting: a background that is visually attractive, without actually reminding us of any specific style, period, or place. You will see this sort of setting in broadcast talk programs, studio interviews, and discussions, or, in more elaborate versions, for game shows.

ECONOMICAL SETS

People working on a tight budget and with limited storage facilities will have little opportunity to build much scenery. But that does not need to be a major limitation since it is possible to develop very attractive sets, simply and economically, by

using just a few multipurpose set units in front of a cyclorama or a background wall (Figure 12.11).

- Lighting alone can significantly change the appearance of a background, whether it is a plain wall or a cyc. The set can be lit evenly, have shadows or patterns projected on it, or be used with plain or blended color areas.

- An *open set* can be created by carefully grouping a few pieces of furniture in front of the wall. Even as little as a couch, low table, table lamp, potted plants, screen, chair, and stand lamp can suggest a complete room.

Support frames can be constructed from lengths of aluminum or wood. They can be dismantled or folded, are easily transported, and require little storage space. Various materials can be stretched across these support frames to make flats; these are taped, nailed, or stapled on. Many

FIGURE 12.11 *Economical sets can be created by using just a few elements.*

FIGURE 12.12 *A simple flat constructed with 1 3/4 inch wood and a sheet of 1/4 inch plywood. This flat is a type of support frame that would require at least one stage weight to keep it from falling down.*

materials can be used, including mesh, trelliswork, scrim, netting, cardboard, wall coverings, translucent plastic sheeting, and so on (Figure 12.12).

Modular units can be constructed out of many different materials, from wood products to plastic sheeting and aluminium. The modules can also be purchased commercially in a variety of configurations. The advantage to modular systems is that they can be assembled and disassembled quickly, are generally designed for minimal storage size, and can look quite professional (Figure 12.13).

Semipermanent Sets

Set design has become more and more complex over the years. Sets incorporate technology, special lighting, monitors, and areas for keying graphics (Figure 12.14). Dramatic sets are being built with such detail that they have become incredibly complex. Most complex sets are installed semipermanently. They are complicated enough that they are not worth installing and uninstalling on a regular basis, at least until they need to be updated or the show is cancelled. This means that they are built into the studio, bolted to the floor, and probably connected to the ceiling (Figure 12.15). When studio space is available, it saves a lot of time to have a set sitting waiting to be used. A regular show can be shot quickly, without all of the setup time.

A permanent or semipermanent set has the advantage that the majority of the set is assembled and ready to be used whenever needed, the set can be dressed and left in place (various props and furnishings), and the lights usually have already been hung and adjusted and then left in position.

When a studio regularly produces a specific program, it may have a permanent set installed, such as a kitchen, a laboratory, an office, a lounge, or a news desk layout, all designed to fit the production.

Pictorial Backgrounds

Because in a 2D television picture it is difficult to distinguish between distant objects that are flat and that are solid, you can simulate 3D effects convincingly using a flat pictorial background. Ideally, it would need to be free from blemishes, evenly lit, and show no shadows; its perspective, proportions, and tones would match the foreground scene; and it would be shot straight on. In practice, you will find that even quite blatant discrepancies can still be very convincing on camera.

Painted Cloths (Backdrops, Backcloths, Scenic Cloths, Canvas Drops)

Ranging from pure vaudeville to near-photographic masterpieces of scenic art, these large painted sheets are used primarily as window backings. Painted cloths are normally hung on battens, pipes, or wooden frames (Figure 12.16).

FIGURE 12.13 *Modular units provide a quick and easy way to build a set. They can be constructed or purchased commercially.*

Source: Photos courtesy of Uni-Set.

FIGURE 12.14 *Sets have become more and more complex over the years, loaded with technology and multiple surfaces.*

Source: First photo by Jon Greenhoe.

FIGURE 12.15 *A semipermanent dramatic set being built into the studio. The back of the "plaster" walls are wooden flats, shown in Figure 12.27.*

FIGURE 12.16 *Painted backdrops, or canvas drops, can range from designs to near-photographic masterpieces of scenic art.*

Source: Background painted by John Coakley/Courtesy of JC Backings.

Photographic Enlargements (Photomurals, Photo Blowups)

Although expensive, enlarged photographs represent the ultimate realism obtainable from studio pictorial backgrounds. Enlargements are made on sections of photosensitized material that can be front-lit or backlit on the set (Figures 12.14 and 12.17).

Television Monitors

Television monitors are increasingly being used as pictorial backgrounds, from a single monitor, as shown in Figures 12.10 and 12.14, to a monitor wall that takes up a large part of the set's wall.

FIGURE 12.17 *Photographic enlargements printed on vinyl can be created so that they are night and day backgrounds. When lit from the front, they look like a daylight backing; when lit from the front and back, it can look like the golden hour; and when just lit from behind, the same backing can look like nighttime.*

Source: Photo courtesy of JC Backings.

FIGURE 12.18 *This television news station uses a green chromakey background for its weather reports. Note the final combined image on the monitor behind the talent. The foot pedal in front of the talent is for changing background images. The talent is looking at another monitor (not shown in the photo) to know where to point.*

Source: Photo by Josh Taber.

Cyclorama

Even the smallest studio can make full use of a cyclorama as a general-purpose background. The cyc (pronounced "sike") serves as a general-purpose detail-free background. Cycloramas generally have a continuous, seamless background that can be used for chromakey (see the next section) or projected on to create unique backgrounds. The cyc provides an extremely useful general-purpose background surface for studios of all sizes. The cyclorama can be the basis of a wide range of program backgrounds from the mundane to the spectacular. It can be built to fit the project and can range from a few feet long to a complete wall around the studio (Figure 12.18).

Cycloramas are usually illuminated by a row of cyc lights (Figures 11.33 and 11.34). It can be neutral, colored with lights, or include no light (black); also, video can be inserted into a blue cyclorama, providing a virtual set. Note that it curves between the wall and the floor. As corners are difficult to light and still look like corners, its corners are usually rounded so that the cyc can be effectively used as a virtual set.

The cyc is available in two primary forms: *soft cyc* and *hard cyc*. A soft cyc can be made of cloth, paper, or canvas. A cloth cyc is usually stretched taut by tubular piping along its bottom edge, where a totally wrinkle-free surface is required (Figure 12.18). Soft cycs are sometimes hung on a straight or curved track (cyc rail), allowing them to be repositioned, changed, and moved out of the way when not required. A hard cyc is created out of drywall, metal, or wood to provide a hard, continuous surface (Figure 12.19).

FIGURE 12.19 *A hard cyclorama was used to allow Korea's KBS network the flexibility of quickly changing sets.*

Chromakey/Matting

The easiest way to understand the matting processes is to imagine cutting a hole in one video image and inserting an area from another picture that corresponds exactly. Within the area covered by the matte, the studio switcher or postproduction software switches from the main picture channel to a second picture source. The edges of this insert may be sharply defined (hard-edged) or diffused (soft-edged) so that they blend unobtrusively into the composite.

Matting/keying techniques have endless potential as background and as special effects. In television, chromakeying is used extensively to create backgrounds and is based on a very simple principle. Wherever a chosen keying color (usually blue or green) appears in the on-air shot, it is possible to insert a second source (the background). Chromakey replaces the blue or green area (determined by the user) with the corresponding section of the second source (Figure 12.18). This keying technique can be created by using a production switcher in a multi-camera production or with a nonlinear editing system equipped with the appropriate software.

Basic one-camera chromakey techniques have the major disadvantage that unless attached to a computer tracking program, the camera needs to hold a steady shot—it's unable to pan, tilt, zoom, or move, because this would immediately destroy the realism of the scene. However, even simpler chromakey systems can be extremely effective.

Chromakey is probably most often used to give the illusion that a person is standing in front of a real location such as a castle, standing in a field, standing out on the seashore, or standing in a town square—merely at the press of a button. If done well, this technique can be very convincing and effective with an audience.

When utilizing the chromakey technique, the entire background does not need to be keyed out. Instead, any part of the background or foreground may be chromakeyed, as long as the appropriate keying color is used. Because the image can be as large as the key color, it offers a very economical method of providing an impressive giant display screen in a shot (Figure 12.20).

Virtual Sets

The use of virtual sets continues to grow. This sophisticated type of chromakey is changing the way sets are designed in many studios. Although in the beginning the cost of setting up a virtual set system integrated with cameras can be quite significant, the savings of not having to quickly change many

FIGURE 12.20 *The first image shows the set with a section that can be keyed out in blue, the second image is the external video shot, and the third image shows what the viewers would see. Any appropriately colored portion of the set can be chromakeyed.*

Source: Photos by Tyler Young.

FIGURE 12.21 *The women sitting on the blue cyc are being shot by a camera that is connected to the virtual set computer located in the foreground of this photo. Notice on the middle monitor that the computer-generated set can be seen. In the monitor on the right, you can see the combined virtual set and keyed talent.*

FIGURE 12.22 *This large outside blue wall is used for television and filmmaking at Universal Studios. Note the water tank area in front of the wall so that boats can be filmed, keying in unique backgrounds.*

FIGURE 12.23 *This wall, located at Northshore Studios in Vancouver, is being used for a stunt motorcycle jump. Again, the virtual wall can be used to insert video from anywhere.*

different kinds of sets can pay for itself in the long run. Studio space requirements and construction times are reduced with the use of these sets. Virtual sets use a blue or green seamless background, chromakeying the computer-generated set into the scene. Most virtual sets employ sophisticated tracking computer software that monitors the camera's movements so that as the cameras zoom, tilt, pan, or move in any other way, the background moves in a corresponding way. This system automatically adjusts the background with each shot change, changing the background size and angle to simulate a real set (Figures 12.21–12.23).

Outside/Backlot Sets

Building an outdoor backlot set requires quite a bit of financial commitment in terms of both building the set and maintaining it. It is far beyond the means of a small-budget project. However, these sets can be rented and do offer a lot of flexibility. The nice thing about them is that, if designed effectively, they can be reused multiple times. The outdoor set shown in Figure 12.24 was built decades ago and has been used in many different films and television productions.

These can be decorated to match the time era; most viewers do not realize that it is the same location that was used for earlier shows.

The Location as a Background

"One of the most important decisions to be made is where the studio set will be located at a sports venue. The people at home watching on TV need to feel like they are at the venue. That is always our main goal when it comes to the set location. They need to see the stadium, the players, and the fans at the event."

—Jennifer Pransky, Coordinating Producer, Fox Sports

Location backgrounds bring context to the production. They make the production look real and genuine in a way that is hard to imitate in any studio. It usually brings a credibility and urgency to the production. However, any time the production is moved out of the studio, a little control is lost in audio, lighting, camera placement, and so on.

FIGURE 12.24 *(Left) The buildings and street are part of a set used in a network television production. (Right) The back of the same set. Notice that the back has been designed so that actors can appear in windows or move in and out of doors.*

FIGURE 12.25 *The yellow-orange building, the red awning, and the "Guido's" sign is real on this outdoor backlot building. The bricks, windows, air conditioners, and fire escapes are all just painted on a flat wall.*

However, unless it is a very famous place, all the audience knows about the location background is what is shown to them. It is possible to go to an exotic place, and shoot someone leaning against a tree that looks just like one back home. If you're shooting on location, make good use of it. Ensure that there are sufficient visual clues for people to benefit from the specific atmosphere of the place (Figures 11.3, 11.4, and 12.1).

In the "busyness" of shooting a production, it is so easy to overlook things in the background that can become a distraction in the final image. On the spot, they are just accepted as part of the scene. In the final production, they distract the audience's attention. Even major films often have a microphone sneaking into the shot at the top of the picture, or a shadow of the camera crew, or prominent lighting cables, in spite of all their vigilance.

Sometimes these odd things are puzzling or disturbing to the audience; at other times, they look funny, such as someone standing with a flagpole "growing out of their head," or a circular ceiling light hovering like a halo. Some of these distractions are impossible to avoid, but they are worth looking out for. Following are some reminders of typical things that can spoil the image.

Windows can be an embarrassment when shooting interiors. A large patch of sky in the shot can create problems. Even if the interior is exposed properly, this bright blank area still grabs the audience's attention. Although corrective filters can be used to compensate for the high color temperature of the daylight, its intensity can easily overwhelm the interior illumination and prevent the camera from getting a good tonal balance.

In addition to all that, if the audience has a good view of what is going on outside the window, there is always the chance that they will find this more intriguing than the real subject. The simple remedy is to keep the window out of the shot, or close the shades.

Reflective surfaces in the background are difficult to avoid. But glass, plastic, and even highly polished furniture can be very troublesome. These surfaces can even reflect the camera and its crew. So instead of admiring the gleaming new automobile, the audience watches the interesting reflections in its door panel.

Worse still is that shiny surfaces reflect lamps. If a camera light is being used, its beam will bounce straight back into the lens. When the camera is moved, the blob of light will move along with the camera.

Low-intensity reflections give sparkle and life to a surface. Strong light reflections are a pain, both technically and artistically. Apart from avoiding shooting straight on at these surfaces, or keeping them out of the shot, the quick solutions are to change the camera's location, cover them up (position something or someone so that the highlight is not reflected), or angle the surface.

Any strong lights directly visible in the background of the shot can be similarly troublesome. But unless their intensities can be controlled, or kept out of the shot, the director will probably have to accept the results.

Flashing signs, prominent posters, direction signs, and billboards are among the visual diversions that can easily ruin a shot. They are all part of the scene, but if a dramatic situation is taking place anywhere near an animated advertising sign, do not be surprised if part of the audience's attention is elsewhere.

Even if shooting in a busy spot, it is often possible to find a quiet corner in which there are not too many interruptions. Try to avoid including a door or busy corridor in the background, or similar areas with a continually changing stream of people. People staring at the camera and bystanders watching (particularly the hand-waving types) are a regular problem, and there is little one can do, except try as much as possible to keep them out of the shot.

Miniatures

There are times when it just is not feasible to build the set you need to fit a specific goal. This may be due to limited budgets or time limits. One option that may be worth considering is the use of miniatures. This is especially helpful when creating a time period set, night scenes, and explosions. Miniature models are used on a regular basis in television and film (Figure 12.26).

SET COMPONENTS

There are many different types of set components used for production. Here are some of the most common:

- *Standard set unit*: Used instead of interior or exterior walls. A flat is a good example of a standard set unit (Figure 12.27).
- *Hanging units*: Basically, any background that is supported by hanging on a wall, a lighting grid, or another overhead support. These include curtains, rolls of background paper, and canvas (Figures 12.16–12.18 and 12.28).
- *Platforms*: Used to elevate the talent or set (Figure 12.29).

FIGURE 12.26 *These miniature model sets were built using train board supplies available at a hobby shop.*

- *Set pieces*: Usually three-dimensional objects used on a set. These would include modular set systems, steps, pillars, and so on (Figure 12.13).
- *Floor treatment*: Includes rugs, wood, rubber tiles, paint, and so on.
- *Stage props*: The furniture on the set. These would include news desks (Figure 12.29), chairs, couches, and tables.
- *Set dressings*: Set decorations are used to create the character of the set. They can establish the mood and style of the production. The dressings can include fireplaces, lamps, plants, pictures, or draperies (Figures 12.30 and 12.37).
- *Hand properties (props)*: Any items that are touched and handled by the talent during the production. These could include a pen, dishes, a cell phone, or silverware (Figures 12.31 and 12.32).

FIGURE 12.27 *These flats, or standard set units, are the back of the walls seen in Figure 12.15. This photo reveals how the walls were constructed.*

FIGURE 12.28 *(Left) This photo shows a cityscape hanging unit (far right of photo) hanging behind the actual set. (Right) This photo was shot from inside the set, showing the cityscape.*

FIGURE 12.29 *In this situation, a platform is used to bring the talent up to eye level with the camera.*

FIGURE 12.30 *These set dressings will be used to establish the character of the set.*

FIGURE 12.31 *Prop room showing the various props available for use in a production.*

FIGURE 12.32 *This prop was used during a production of The X-Files. The actor bent the "pipes" and climbed through the "concrete." In reality, the pipes were bendable rubber and the concrete was foam. The "metal" edges and bolts were actually plastic.*

SET CONSIDERATIONS

Camera Height

The camera's height has a significant effect on how much of the scene is visible in the shot. From a lower viewpoint, less of the middle ground is seen, which reduces the feeling of space and distance in the picture. Things nearer the camera become more prominent—perhaps overly so. Even very small foreground objects nearby can obscure the shot. But raise the camera just a little, and not only will it shoot over them, but the audience will not even realize that they are there (Figure 12.33).

As the camera's height is increased, more of the middle ground comes into view, and the audience gets a greater impression of space and distance. However, if the scene is shot from a very high angle, or overhead shots are used, the audience will no longer feel that they are within a location, but will find themselves looking down, inspecting it instead. Of course, the audience is also affected by the speed of the camera move and the content of the shot.

Foreground Pieces

Objects can also be deliberately positioned in the foreground of an image to improve its composition, to increase the impression of distance, or simply to hide something in the scene. Many exterior shots have foliage hanging into the top of the frame. It is almost a visual cliché. But the camera operator has done this because the picture looks more complete, and it gives a better sense of scale than if there were just a blank open sky. With this "frame," the picture tends to look more balanced and no longer bottom-heavy. When there does not happen to be an overhanging tree to shoot past, a piece of a tree branch can always be held above the lens. If this positively affects the look of the picture, do it—and your audience need never know.

Although the television's picture itself usually has a fixed horizontal aspect ratio, a foreground window, an arch, or a similar opening can be used to provide a border that alters the apparent shape of the picture.

By carefully framing foreground objects, it is possible to hide things in the background that would be distracting to the audience. They might ruin the shot in some other way. For example, if an historical drama is being re-enacted, it is very convenient (to say the least) if a carefully positioned gatepost, bush, or even a person in the foreground hides the modern signs, power lines, and other elements (Figure 9.6).

Creating Depth

Foreground pieces can add depth to a limited background. Depth is created by shooting through things, such as a bookshelf, a fence, or flowers. Usually the foreground is kept slightly out of focus, so that the attention is drawn to the primary subject. However, the foreground can also help bring context to the image, by using something in the foreground that adds information to the scene (Figure 12.34).

Versions of "Reality"

Obviously, the camera does not "tell the truth." It interprets. Each aspect of the picture and the sound influences how the audience responds to what they see and hear. A slight change in the camera position can entirely alter the picture's impact.

FIGURE 12.33 *By moving the camera to a high angle on a jib, the audience becomes curious, inspecting the scene.*

Source: Photo courtesy of Sony.

FIGURE 12.34 *Foreground subjects add depth to the images.*

If the sun comes out, what was a drab threatening block of building can become transformed into an attractive, interesting piece of architecture. In the winter, we see a dull-looking street planted with stark, leafless trees. In spring, it becomes a charming avenue, where sidewalks are dappled with shade.

A location can be shot so that the audience envies its inhabitants, or pities them for having to live there. It can appear like a fine place, or an eyesore. It's all a matter of what the director chooses to include, and omit; what is emphasized, and what is suppressed.

As the camera moves around the scene, it can dwell on busy purposeful bustle, as people go to work, or it can linger on those who appear to be lounging around with nothing to do. (In reality, they might be waiting for a bus, but at the moment the camera captures them, they are "aimlessly inactive.") The director can suggest spaciousness by shooting with a wide-angle lens. Use a telephoto (narrow-angle) lens instead, and the same streets can look congested.

In most cities, one can find litter, decay, and graffiti; conversely, there will be signs of prosperity—attractive buildings, green spaces, fountains, wildlife, and things that are amusing and others that are touching. How the images are selected and related will significantly influence how the viewing audience will interpret the scene.

What Can We Do about the Background?

If the director is shooting in a studio and the background is unsatisfactory for any reason, he or she can usually improve it in some way or other. But what can be done when on location if the background proves to be unworkable?

When a guest in someone's home, the answer may be disappointingly little. So much depends on the people involved, and the director's diplomacy. If the hosts are not accustomed to appearing on camera, they will probably be disturbed when it is suggested that things need to be moved around to any extent.

They may even feel uncomfortable if they are not sitting in their customary chair. There is little point in doing things that will jeopardize the interview. However, there are various little things that can be done unobtrusively to improve matters:

- Natural lighting can be used, rather than introducing lights. The person being interviewed will probably feel more at ease. The problem is that additional lighting cannot be avoided in most interiors in order to get good images.
- Although a room's tones cannot necessarily be changed, it may be possible to shade your lamps off a light-toned surface to prevent them from appearing too bright (using a barn door, flag, or partial diffuser). A little illumination may be able to be added to dark corners.
- If there are reflections in glass-fronted pictures or cabinets, and the lights and camera cannot be moved in order to avoid them, slightly angling the frame or furniture may resolve the dilemma. A wall picture or mirror can often be tilted up or down a bit by wedging something behind it. To avoid seeing the camera in a glass bookcase behind the talent, it may be possible to slightly open its doors.
- Closing or partly closing the room's curtains may help to adjust the lighting balance in a room.
- If shooting in a corridor or hallway, it can help if doors in a sidewall are opened enough to let extra light in.
- Even if shooting in daylight, it may provide more interesting images if the table lamps or other lights in the room are turned on.

- It may even be possible to conceal low-powered lamps behind furniture or wall angles in order to illuminate distant parts of the room, but be careful that they do not over-light or even burn nearby surfaces.

Rearranging the Background

Most of the time, the director will be able to alter the background to achieve the best possible scene. Again, this has to be done diplomatically, but if the host's confidence is gained, and the director seems to know what he or she is doing, there should be no difficulties. The simplest changes usually involve moving around what is already there to avoid any unnecessary distractions or unwanted effects (glares) in the picture.

It is important to look at the background of any location to make sure that nothing is apparently growing from the talent's head, or balancing on it, and that no vertical or horizontal lines cut through the center of the head or across the shoulders. These visual "accidents" can make the picture look contrived or comic (Figure 12.35).

Altering the Background

As mentioned earlier, there are times when the background must be altered to improve the appearance of the room for television. There are a number of quick, inexpensive, and simple things that can be done to adjust things for the camera:

- Rearrange the furniture.
- Replace furniture with other pieces from nearby rooms.
- Add or remove rugs.
- Hide a doorway with a folding screen.
- Attach display posters to the walls.
- Position indoor plants (such as ferns) to break up the background.
- Introduce notices and signs on walls, doors, or elsewhere.

When shooting outside, there are relatively few things that can be done cheaply and easily to change the background.

Partial Settings

This is a strategy for convincing your audience that a very modest setting is not only the real thing, but even that it is much more extensive than it actually is. Yet the cost and effort involved are minimal (Figure 12.36).

If the camera does not move, it can only show a limited amount of the scene in a medium or close-up shot. With partial settings, it is important to concentrate on building up a section of the scene, just large enough to fit the camera's shot, and no more. Within the scene, enough features are included to allow the audience to interpret where the action is supposed to be taking place.

FIGURE 12.35

People's homes are designed to be comfortable and to be lived in—they are not designed for television. When going into a home with a camera, it is common to have to move some furniture, things hanging on the wall, rugs, and knickknacks in order to create an effect that works on television.

FIGURE 12.36 *By putting the crowd in just the right place (about 30 people), repositioning the camera so that the crowd is directly behind the batter and catcher, and zooming in the lens to a close-up or medium shot, the viewing audience will assume that the baseball stadium was filled with a cheering audience.*

Do not underrate this idea. It has been used successfully in film and television for many years. The result does not need to look amateurish. Add the associated sound effects and the combined image can be indistinguishable from the real thing.

Typical Examples of Partial Settings

Sets that impact the audience can often be simply designed without building huge complex backgrounds.

- An "instant store" can be created by putting the appropriate type of merchandise on a foreground table (the "counter"), an advertisement or two on the back wall, and a shelf behind the salesperson, holding some more products.
- Sometimes even a single feature in the picture can suggest an environment. A stained-glass window and organ music become a church interior (the window could even be projected).
- A convincing "room" can be created in a studio with just a couple of flats or screens and a chair. Add an extra item or two, such as a side table with a potted plant and a

picture, and it begins to take on its own character. If a curtain is hung on one of the walls, a window is assumed to be there. Whether it is interpreted as being someone's home, or a waiting room, for instance, largely depends on the way people in the scene behave. If they're lying back casually dressed, reading a paper, it is obviously their home. In outdoor attire, sitting upright and anxious, they are waiting.

- Replace the plant with a computer, and the picture with a framed certificate—magically, the setting becomes an office.

On location, the same concept is still useful: restrict the shot and "doctor" the background. Suppose that a nineteenth-century drama is being shot, in which somebody visits his or her lawyer. A house exterior of about the right period is found, but the rest of the street is obviously busy and modern. Fortunately, all that is needed is a shot of the house doorway with the appropriate business sign attached to it, and the picture explains itself to the viewing audience. Have the actor go up to the door, or pretend to leave the house, and the audience will immediately accept the location as the lawyer's office. It needs only the sounds of horse-drawn vehicles to replace modern traffic noise, and the period illusion is complete. With a little care and imagination, locations can be created from a minimum of scenery and work.

It is incredible how seemingly trivial techniques can give a totally convincing effect on camera:

- The camera rhythmically tilting up and down sells the illusion of a ship at sea.
- The wafting breeze may really be the result of an assistant waving a piece of board.
- The shuddering camera accompanied by things falling to the ground (pulled by an unseen fishing line) implies an explosion or an earthquake. A hanging lamp swings alarmingly—tugged by an out-of-shot line.
- The flickering flames of a nearby fire come from a stick of cloth strips waved in front of a ground lamp.

These are just a few examples of how a little ingenuity can apparently achieve the impossible, and create a strong impression in your audience's minds.

Facing Reality

It is one thing to have dreams about creating a program, but it is quite different to turn them into reality. Among the problems facing all directors are the inevitable limitations of budget and facilities. Some of the things that are needed may not even be available.

When faced with such problems, it is tempting to think small: to cut back on the ambitious goals, and to do a simpler version. Do not immediately abandon your ideas! Instead, ask yourself whether there is another way of tackling the situation to get virtually the same effect. How can you get around the difficulty?

What we are looking for are imaginative substitutes. Keep in mind that directors produce illusions. However basic the materials really are, the end result can appear to be the real thing (Figure 12.37).

As an example, let's look at an actual scene that was used on the air. The scene was the banquet hall of an ancient castle. The king sat on a throne at one end of a long table, eating from golden dishes. That was the illusion.

FIGURE 12.37 *Sometimes ingenuity needs to be used instead of reality. A fireplace was desired for this NBC set, but it was impossible to have a working fireplace in the building in this specific situation, and the heat would have been a problem. So an HD monitor (with a video of fire) was used inside the fireplace as a substitute.*

Source: Photo courtesy of LPG/NBC.

What was reality? Two small foreground flags on wooden floor stands masked the edges of the shot, so that no one could see the rest of the small studio. The "wooden table" was created from painted boards placed on sawhorses. The far "stone wall" was photographic wallpaper attached to a flat (and was slightly sprayed black in order to "age" it). The "throne" was an old wooden armchair with a red drape thrown over it. The "gold dishes" were sprayed plastic plates. A "window" was painted, black on white, and stuck to the "stone wall." But no one in the audience could recognize all of this in the long shot. Of course the scene would not have worked for close-up shots, but under patchy lighting, it was seen as it was intended to be–the banquet hall of an ancient castle.

Set Problems

Here are some of the more common problems that you might have to address when working with sets:

- Check the set for distracting features. This could include things such as bright surfaces or reflections. Check to see whether the lighting can be adjusted to correct the problem. Other options may be using dulling spray, repainting the area with a different color, covering it, or removing that item from the set.
- Colors or tones may be unsuitable (subject merges with background). Modify the background, lighting, or subject (e.g., change the clothing).

- Background blemishes (dirty marks, tears, scrapes, wrinkles, etc.). HD is very unforgiving and seems to show every mark. Correcting the problem may require refurbishing the area, covering it, or possibly changing the lighting.
- Distracting shadows may appear on the background. This usually means that the talent or scenery is too close to the background. This often requires moving the scenery/people or relighting the shot.

What Can You Shoot?

Intellectual property (IP) is everywhere. Unless you have permission from the respective copyright and trademark owners, you can't include logos on your sets or in your production. IP-protected material includes things such as posters, product labels, paintings, billboards, books, toys, storefront signs, clothing insignias, and sculptures. If you plan to distribute your production, all IP material needs to be cleared. If you can't get permission to use the logos, it would be best not to use them. Otherwise, you jeopardize the marketability of your project.

Just because you have a valid release from a location owner does not mean that you have the right to shoot any IP-protected material at the location. Location releases do not cover IP rights.

There are exceptions to the rules, including some educational uses, news reporting, research, teaching, criticism, and parody.

INTERVIEW WITH A PROFESSIONAL: JOHN DECUIR

Why is design important? I would like to think that what separates the design professional from other ways of life is that we are constantly on the lookout for the new idea; a better way to say, package, or construct our work. For better or worse, we seem to be dedicated to a life of rolling over those "idea rocks" never too sure what just might crawl out.

What advice do you have for the beginning designer? Develop a working set of professional tools that have proven value. Here are a few of the tools you might wish to consider investing your time in developing:

FIGURE 12.38
John DeCuir, set designer.

- **Master the technology of your craft**. It is safe to assume that the technical knowledge we employ can make or break a project. In judging the value of a new idea, always consider the skill level of the person who has created it.
- **Apply storytelling techniques to your project**. Understanding the narrative of your project will ensure you are able to communicate its value both to the client, the world, and most importantly to yourself. The professional simply can't design effectively unless he or she can communicate a project's narrative, tell its backstory, and define the theme.

continued

- **Understand and respect the project environment**. You can't design well unless you first understand, then respect, then influence, then reshape the environmental issues surrounding a project. As an example: in a program scene, everything from the religious beliefs of the next-door neighbor, to the objects on the dresser, to the snow falling outside create a contiguous "design biosphere." This biosphere or "story ecosystem" must be allowed to impact each and every choice the designer makes. Essential in this process is how the interaction of two elements create a third, new idea, and how the new idea and its component parts all need to exist in a harmonious framework.

What are the challenges that you have to deal with as a designer?

- The most important challenge you will face is to **maintain a high state of energy and enthusiasm** for your work.
- **The challenge of maintaining a balanced attack**. To create a quality solution, the designer *must* balance issues of *budget, time and quality*. As money becomes less available, time may have to expand. If a super emphasis is placed on quality, then perhaps both time and budget will need to expand. When deadlines are shortened, budgets may have to be increased and perhaps quality (at some acceptable level) will need to be sacrificed. Balancing these three factors at every step in the development process is one of the young professional's great challenges.
- **Managing the client designer relationship**. The designer must earn the client's respect so that he or she can operate in a professional atmosphere.

John DeCuir designs sets for network television series and major motion pictures.

REVIEW QUESTIONS

1. How do the various format ratios (4:3 and 16:9) affect the set design?
2. What are some of the advantages of a modular set?
3. Why would someone want to use a cyclorama?
4. Explain the relationship between a virtual set and chromakey.
5. What is a contextual set?
6. Name three types of props and explain how they are used.
7. How is depth created on a set?

CHAPTER 13

MAKEUP AND COSTUMES

"For the makeup artist, the soft veil of film has given way to the uncompromising clarity of high definition. Images so sharp that even the most beautiful or handsome talent can have subtle imperfections visible for all to see."

—Bradley M. Look, Emmy-Winning Makeup Artist

TERMS

Character makeup: The emphasis is on the specific character or type that the actor is playing. By facial reshaping, remodeling, changes in hair, and similar effects, the subject can even be totally transformed.

Corrective makeup: Reduces less-pleasing facial characteristics while enhancing more attractive points.

Straight makeup: A basic compensatory treatment affecting the talent's appearance to a minimum extent.

FIGURE 13.1 *One of the hosts on NBC's Today Show receives a hair and makeup check before going on the air.*

The television camera is a critical tool and facial characteristics that pass unnoticed in daily life can appear surprisingly exaggerated or distracting on the television screen. Most of us can benefit from the enhancement that a skilled makeup artist provides. Whether this needs to be slight or elaborate depends on the type of production and the role of the talent (Figure 13.1).

FORMS OF MAKEUP

Television makeup treatment follows three general forms: straight, corrective, and character makeup.

Straight Makeup

Straight makeup is a basic compensatory treatment, affecting the talent's appearance to a minimum extent:

* *Skin-tone adjustment*: This provides a good tonal balance in the picture; it involves darkening pale faces and lightening dark complexions.
* *Routine improvements*: These subdue blotchy skin tones, shiny foreheads, strengthen lips and eyebrows, remove beard lines, lighten deep-set eyes, and lighten bags under the eyes.

For many television productions, performers require little or no makeup, with minimum correction and brief last-minute improvements often being sufficient (Figure 13.2). Regular talent may do their own makeup.

Corrective Makeup

Corrective makeup seeks to reduce less-pleasing facial characteristics while enhancing more attractive points. Actual treatment can range from slight modifications of lips, eyes, and nose to concealing baldness.

The general aim is to treat the person without them appearing "made up." Skin blemishes and unattractive natural color must normally be covered, preferably by using several thin applications of increased pigmentation, rather than trying to obscure these with heavy mask-like coatings. Arms, hands, necks, and ears may need blending to an even tone (with body makeup).

A person's skin quality will modify the makeup materials used. Coarser skin textures provide more definite modeling; finer complexions tend to reveal veining or blotches that the camera may accentuate (Figure 13.3).

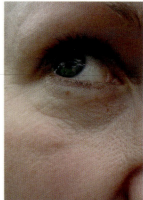

FIGURE 13.2 *Straight makeup is the basic type of makeup treatment.*

Source: Photo courtesy of Sennheiser.

FIGURE 13.3 *Before (left) and after (right) photos of corrective makeup.*

Source: Photos courtesy of Jessica Goodall (www.jessicag.tv).

Character Makeup

With character makeup, the emphasis is on the specific character or type that the actor is playing. By facial reshaping, remodeling, changes in hair, and making other such changes, the subject can even be totally transformed; for example, the actor can be aged dramatically. But most character makeup involves less spectacular and more subtle changes (Figures 13.4 and 13.5).

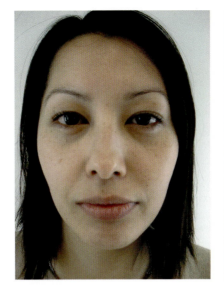
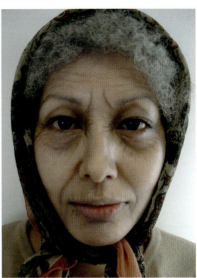

FIGURE 13.4 *Character makeup can dramatically age a person.*

Source: Photos courtesy of Jessica Goodall (www.jessicag.tv).

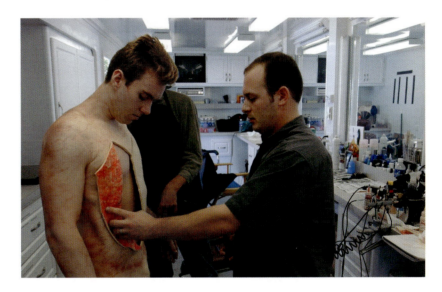

FIGURE 13.5 *A special effects artist applies character makeup and prosthetics to an actor on the set of the CBS television series NCIS.*

Source: Photo courtesy of CBS/Landov.

CONDITIONS OF TELEVISION MAKEUP

The principles and practices of television makeup are almost identical with those of motion pictures, except that in television the continuous performance often prevents the shot-by-shot changes that are possible in film.

A long shot ideally requires more defined, prominent treatment than a close-up. However, such refinements may not be possible under typical television production conditions. You may not even be able to do anything about such distractions as perspiration or disheveled hair when the actor is on camera for long periods, except to correct them for any retakes, when time permits.

For the very demands of television drama, careful planning and presentation are essential. At a preliminary meeting with the program's director, the makeup artist will discuss such details as character interpretation, hair styling, special treatments, and any transformations during the program (such as aging). Actors who need fitted wigs or trial makeup are then contacted.

Camera Rehearsal

For the camera rehearsal, the most common practice is to apply the makeup to the talent before the rehearsal. Then, while watching the camera rehearsals on a picture monitor, the makeup artist can note the changes that will be required. This process also allows the lighting director to assess tonal balance, contrast, and exposure of the talent while in makeup.

MAKEUP TREATMENT

Generally speaking, a straight makeup for men may take around 3 to 10 minutes; women require 6 to 20 minutes, on average. Elaborate needs can double or even triple these times.

After a few hours, cosmetics tend to become partly absorbed or dispersed through body heat and perspiration. Surface finish, texture, and tones will have lost their original definition, and fresh makeup or refurbishing becomes necessary.

There will always be problematic occasions. Some performers cannot have makeup, due to allergies or other situations. There are also times when the makeup has to be done immediately before airtime, without any opportunity to see the performer on camera beforehand—a situation that the wise director avoids.

PRINCIPLES OF MAKEUP

The broad aims of facial makeup and lighting are actually complementary. Makeup can sometimes compensate for lighting problems, such as lightening eye sockets to anticipate shadowing cast by very high-angled lights. However, whereas the effect of lighting changes as the subject moves, that of makeup remains constant.

Localized highlighting by slight color accents will increase the apparent size and prominence of an area; darkening reduces its effective size and causes it to recede. By selective highlighting and shading, you can vary the impression of proportions considerably. However, you must take care to prevent shading from looking like grime!

You can reduce or emphasize existing modeling or suggest modeling where none exists; remember, though, that the deceit may not withstand close scrutiny.

A base or foundation tone covers any blotchiness in natural skin coloring, blemishes, beard shadows, and the like. This can be extended, where necessary, to block out the normal lips, eyebrows, or hairline before drawing in another different formation.

Selected regions can be treated with media a few tones lighter or darker than the main foundation and worked into adjacent areas with fingertips, a brush, or a sponge. After this highlighting and shading, any detailed drawing is done using special lining pencils and brushes.

HAIR

Hair may be treated and arranged by the makeup artist or by a separate specialist. Hair work may include changes to the talent's own hair, the addition of supplementary hairpieces, and complete wigs covering existing hair.

Hair Alteration

In television, a certain amount of restyling, resetting, or waving may be carried out on the talent's own hair, but when extensive alterations such as cutting or shaving are needed complete wigs are more popular. Hair color is readily changed by sprays, rinses, or bleaches. Hair whitener suffices for both localized and overall graying. Overly light hair can be darkened to provide better modeling on-camera; dark hair may need gold or silver dust, or brilliantine, to give it life.

HD AND 4K MAKEUP

- Makeup artists need to be cautious that they use only enough makeup to conceal a defect, so as to remain undetectable on television.
- Brushes and sponges can easily leave "brush strokes" that are detectable by high-resolution televisions.
- Airbrushed makeup has become very common when shooting on HD and 4K.
- Facial hair can be extremely apparent on close-ups with 1080+ resolution.
- Many makeup artists have replaced the traditional opaque crème foundation (which is sometimes obvious) with a sheer liquid foundation that is available in department stores.

COSTUMES (WARDROBE)

In larger television organizations, the talent's clothing (costume or wardrobe) is the responsibility of a specialist (Figures 13.6–13.8). But for many productions, people wear their own clothing. Diplomatic guidance may be required to ensure that unsuitable attire is avoided.

Be sensitive to the talent's feelings and taste when suggesting changes, particularly when you want them to wear an item from stock (such as an off-white shirt, or a different necktie) to replace their own. Experienced talent may bring along alternative garments for selection on-camera.

A costume that looks attractive in a long shot (full-length) may be less successful when seen as a medium shot (head and shoulders) behind an anchor's desk. Color matching that looks good to the eye can reproduce quite differently under various lights. Some shades of color may look great in the long shot and terrible in the close-up (Table 13.1).

Costumes that fit what the director wants are not always available. The baseball costumes shown in Figure 12.36 had to be created in order to fit the time period.

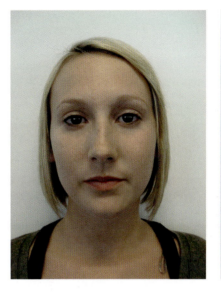

FIGURE 13.6 *A combination of makeup, hair, and costume can radically change the impression that the actor makes on the viewing audience.*

Source: Photos courtesy of Jessica Goodall (www.jessicag.tv).

FIGURE 13.7
The costume shop at a production facility.

FIGURE 13.8
A costume department creates the costumes needed in productions.

INTERVIEW WITH A PROFESSIONAL: RUTH HANEY

FIGURE 13.9
Ruth Haney, makeup artist.

Briefly define your job. My job is to apply makeup to all actors in a film, television, video or commercial. There are many variables in the application, which depends on the person's skin, vision of the director, producers, and writers. There are also variables with the needs of the actor and the character. Knowledge of lighting and what cameras we are using for the production are critical to the makeup application. After the application, there is maintenance during the filming and removal at the end of the day.

What do you like about your job? My job fits me perfectly. Every day is different; every job is exciting and fresh.

What are the types of challenges that you face in your position? There are challenges with working with new people and talent. Being a people pleaser is very important. Knowledge of psychology has been a huge asset, when one might have an actor or artist with an ego that must be dealt with during a production. Staying calm and thinking on your feet is an important attribute. I consider myself a chameleon when it comes to work. I have some abilities in makeup that others do not have; this often brings me more work because I am able to help with many facets of makeup during a production. One of my friends refers to me as the problem-solver.

Are there specific things that you do to prepare to work on a program? Preparation for a production is very important! I read the script usually twice, and on the second read I make notes: tears, blood, injuries, etc. Then I do a character breakdown, making a list of questions for the production meeting. During the production meeting, the script is read out loud. We can ask questions and receive ideas. The visions are set in place by the director

continued

Interview with a Professional: Ruth Haney—*continued*

and the writers. Many times there is research for a certain look or if there is a creature, etc. Many times I will do drawings or Photoshop pictures to gain a better idea from the director and writer.

What suggestions or advice do you have for someone interested in a position such as yours? I let people know that this is not an easy job. It is so important to work well with others. One must be able to get up very early in the morning, drive all over the city, or even fly to other parts of the world. Work no less than 12 hours a day and keep an upbeat personality at all times. If you still want to work in this industry as an artist, I recommend that you have a background in art and psychology. Attend a beauty school and get a license in makeup and then practice on everyone you can find who will let you apply makeup to their face. Get as much experience for a portfolio as you can. Don't let anyone discourage your dream; it is achievable.

Ruth Haney is a makeup artist who has worked on close to 40 different television shows and films, including 24, CSI, Boston Legal, Star Trek: Enterprise, *and* Star Trek: Nemesis.

REVIEW QUESTIONS

1. What is the difference between straight makeup and corrective makeup?
2. How is character makeup used in television production?
3. How is makeup different for HD and 4K?
4. How can costumes be obtained for a character in a television production?

TABLE 13.1 *Costume Problems*

White shirts, blouses, and so on	White shirts can be so bright that they lose all detail when the camera has to adjust the iris to obtain detail in the talent's face.
Glossy materials, such as satin	High sheen, especially from shoulders, may become too white or even reflect incident light.
Light tones	These emphasize size, but, if loosely cut, light garments can appear formless.
Dark tones	These minimize size, but all detail is easily lost in reproduction, particularly with dark velvets.
Strong, vibrant colors	Usually appear over-saturated and can reflect onto the neck and chin.
Fine stripes, checks, or herringbone patterns on clothing	Patterns can "vibrate," causing a localized flicker. Color detail is liable to be unsharp and lost in longer shots.
Shiny, sequined, or metallic finishes	Incredibly bright highlights can distract the viewers and can reflect onto the face or nearby surfaces.
Noisy jewelry or decorations such as multi-string beads	Microphones can pick up extraneous clinks, rattles, or rustles.
Rhinestones and other highly reflective jewelry	Reflects bright spots of light onto chin, neck, and/or face and flashes obtrusively.
Very low necklines	In close shots, can create a topless look.

CHAPTER 14

AUDIO FOR TELEVISION

"Just as you should preplan camera angles for a shoot, you should preplan mic types and positioning. Carefully plan cable runs and connections, and scout every location to determine if it will cause undesirable noise from air conditioners, trains, traffic, or other uncontrollable sounds that will end up being recorded."

—Douglas Spotted Eagle,
Grammy Award-Winning Producer

"Who the hell wants to hear actors talk?"

—Harry Warner, President, Warner Bros.,
in the silent film era, 1927

TERMS

Acoustics: High-frequency sound waves travel in straight paths and are easily deflected and reflected by hard surfaces. They are also easily absorbed by porous fibrous materials. Lower-frequency sound waves (below 100 Hz) spread widely, so they are not impeded by obstacles and are less readily absorbed. As sound waves meet nearby materials, they are selectively absorbed and reflected; the reflected sound's quality is modified according to the surfaces' nature, structures, and shapes.

Audio mixer: The audio mixer is needed whenever there are a number of sound sources to select, blend together, and control (such as microphones, CD, VCR audio output, etc.). The output of this unit is fed to the recorder.

Audio sweetening: The process of working on the program sound after the video portion is completed; also known as a dubbing session or track laying.

Condenser microphone: This microphone produces very high audio quality and is ideal for musical pickup. A significant advantage to the condenser is that it can be very small, making it the logical choice for a shotgun, lavalier mic, and other miniature microphones.

Directional microphone: The directional (or cardioid) mic pickup pattern. This broad heart-shaped pickup pattern is insensitive on its rear side.

Dynamic microphone: Dynamic microphones are the most rugged, provide good-quality sound, and are not easily distorted by loud sounds such as nearby drums.

Dynamic range: The range between the weakest and loudest sounds that can be effectively recorded by a recording device.

Foley: Creating sounds in a studio that can replace the original sounds.

Line level: The audio signal generated by a non-microphone device, such as a CD player.

Mic level: The audio level of a signal that is generated by a microphone.

continued

Monaural sound: Also known as *mono*, this single track of audio is limited, because its only clue to distance is loudness, and direction cannot be conveyed at all.

Omnidirectional microphone: The omnidirectional pickup pattern is equally sensitive in all directions and cannot distinguish between direct and reflected sounds.

Perambulator: A large microphone boom on wheels.

Super-cardioid microphone: A super-cardioid (or highly directional) pickup pattern is used wherever you want extremely selective pickup, to avoid environmental noises, or for distance sources.

Stereo sound: Two audio tracks create an illusion of space and dimension. Stereo gives the viewer a limited ability to localize the direction of the sound.

Surround sound: Can provide a sense of envelopment when mixed correctly. Instead of the one channel for mono or the two channels for stereo, 5.1 surround has six discrete (distinct, individual) channels.

Wild track: General background noise.

Historically, audio has been slighted in the world of television. Most manufacturers and producers cared more about the image, relegating audio to an inexpensive, poor-sounding little speaker on televisions. However, if you really want to find out how important audio is, just turn off the audio on a video and try to follow the story. You will soon get lost. Look away from the screen, with the audio turned up, and you can still follow the story. Audio is as important to television as the video image. Audio gives images a convincing realism. It helps the audience feel involved in what they are seeing. Dennis Baxter, Emmy-winning sound designer for television, believes that "audio, in partnership with video, delivers a holistic experience with all of the intense emotion and interesting nuances to the viewer."

The valuable contribution that sound makes to television cannot be underestimated. In a good production, sound is never just a casual afterthought. It is an essential part of its appeal. Soundtrack composer Marc Fantini says, "Music cues are the unseen actors in the scene. Music in a production's goals are the same as the visual, to move the story along."

People often think of television as "pictures accompanied by sound." Yet, when the best television productions are analyzed, people are usually surprised that most of the time it is the sound that is conveying the information and stimulating the audience's imagination, while the image itself may be the accompaniment. Audio has the power to help the audience conjure mental images that enhance what is being seen.

Sounds are very evocative. For example, consider an image of a couple of people leaning against a wall, shown with the open sky as a background. If noises of waves breaking and the shrill cries of birds are heard, we quickly assume that they are near the seashore. Add the sound of children at play, and now we are sure that our subjects are near the beach. Replace all those sounds with the noise of a battle, explosions, and passing tanks, and they are immediately transported to a war situation. They might even appear particularly brave and unfazed as they remain so calm in the middle of this tumult.

In fact, all we really have here is a shot of two people leaning on a wall. The wall itself might have been anywhere—up a mountain, in a desert, or in a studio. The location and the mood of the occasion have been conjured up by the sound and our imagination.

Successful audio is a blend of two things:

- Appropriate *techniques*—the way the equipment is used to capture the audio.
- Appropriate *artistic choices*—how the sounds are selected and mixed.

Both are largely a matter of technical know-how, combined with experience.

THE AUDIO SYSTEM

The *dynamic range* (volume range) that any audio system can handle is limited. When too loud, sounds will cause overload distortion, producing a deteriorated sound signal. If too quiet, wanted sounds will become merged with background noise of comparable level (volume), such as hum and ventilation. So, to avoid exceeding these limits, it is essential that you do not overload the microphone itself (too near a loud source), or over-amplify the signal (over-modulation). Conversely, you must prevent the audio signal from becoming too weak (under-modulation) by placing the microphone close enough and using sufficient amplification. But at the same time, as you will see later, you must not destroy an impression of the dynamics of the original sound source.

ACOUSTICS

You have only to compare sound in an empty room with the difference when that same room is furnished or filled with people to realize how acoustics alter sound quality. If the basics of acoustics are understood, many of the audio problems can be avoided during the production.

A certain amount of reverberation enriches and strengthens sounds, conveying an impression of vitality and spaciousness. Therefore, most television and audio studios have quite carefully chosen the acoustics so that they are neither too live or too dead.

In practice, you will find that the amount of sound absorption or reflection within an environment can change considerably as the conditions alter. Sound quality may be dampened (dull) or brightened (well defined) as furnishings are added or removed. The difference in a theater's acoustics with and without an audience can be quite remarkable. Moving a large scenic flat can alter local sound quality, making it harsh, hollow, or boxy—particularly if there is an extensive ceiling to the setting.

When a sound wave hits a hard surface (plastic, glass, tile, stone walls, metal), little is absorbed, so the reflected sound is almost as loud as the original. In fact, when its higher frequencies have actually been reinforced by this reflection, the sound bouncing off the surface can actually sound brighter and sharper.

When a sound wave hits a soft surface (curtains, couches, rugs), some of its energy is absorbed within the material. Higher notes are the most absorbed, so the sound reflected from this sort of surface is not only quieter than the original sound wave, but lacks the higher frequencies. Its quality is more mellow, less resonant—even dull and muted. Certain soft materials absorb the sound so well that virtually none is reflected (Figure 14.1).

Where there are a lot of hard surfaces around (as in a bathroom, a large hall, or a church), a place can become extremely reverberant, or live. Sound waves rebound from one surface to another so easily that the original and the reflected versions completely intermixed are heard. This can cause considerable changes in the overall sound quality and significantly degrade its clarity.

When surroundings are very reverberant, reflections are often heard seconds after the sound itself has stopped—in extreme cases, as a repeated echo. Whether reverberations

FIGURE 14.1
The variety of angles of the set walls, curtains, carpet, furniture, and people help dampen the live sound, but be careful that it is not dampened too much.

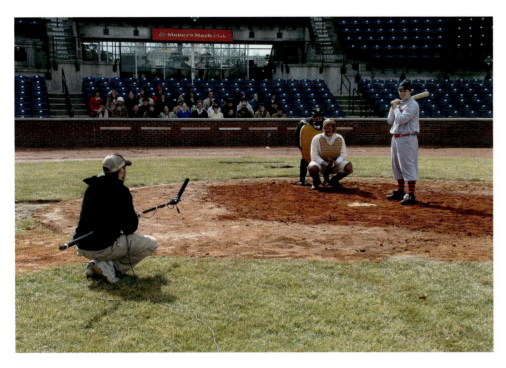

FIGURE 14.2
During the shooting of a dramatic program, the boom operator got the microphone as close as possible, while still being off-camera, because open-air sound does not usually travel far.

DEALING WITH ACOUSTICS

When surroundings are too live, to reduce acoustic reflections:

- Move the microphone closer to the sound source.
- Pull curtains if available.
- Add thick rugs.
- Add cushions.
- Use upholstered furniture.
- Drape blankets on frames or over chairs.
- Add acoustic panels (Figure 14.3).

When surroundings are too dead, to increase acoustic reflections:

- Move the microphone further away.
- Open curtains to increase hard surface space.
- Remove rugs.
- Remove cushions.
- Remove upholstered furniture.
- Add board or plastic surfaced panels.
- Add floor panels (wood, fiberboard).
- Add artificial reverberation.

FIGURE 14.3 *Acoustic panels were placed on the walls of this audio room in order to reduce the "liveness" of the room.*

add richness to the original sound or simply confuse is determined by the design of the space, the position of the sound source, the pitch and quality of the sound, and the position of the mic.

If, on the other hand, the sound is made in a place with many absorbent surfaces, both the original sound and any reflections can be significantly muffled. Under these dead conditions, the direct sound can be heard with few reflections from the surroundings. Even a loud noise such as a handclap or a gunshot will not carry very far and dies away quickly. When outside, in an open area, sound can be very dead. This is due to the air quickly absorbing the sound, as there are few reflecting surfaces (Figure 14.2).

We all know how dead sound seems when outside in the open. Open-air sound is very weak and does not travel far, because the air quickly absorbs it and there are few reflecting surfaces. Microphones often have to get much closer to a subject than normal to pick up sufficient sound, especially if a person is speaking quietly.

Open-air sound has a characteristic quality that can be immediately recognized; it has an absence of reflected sounds, combined with a lack of top and bass. This effect can be very difficult to imitate convincingly in the studio, even when the subject is completely surrounded with highly absorbent acoustic panels.

Acoustics often influence where the microphone is positioned. To avoid unwanted reflections in live surroundings, the mic needs to be placed relatively close to the subject. If working in dead surroundings, a close mic is necessary, because the sound does not carry well. When the surroundings are noisy, a close mic helps the voice (or other sound) to be heard clearly above the unwanted sounds.

However, there can be problems if a mic is placed too close to the source. Sound quality is generally coarsened and the bass can be overemphasized. The audience can become very aware of the noise of breathing; sibilants (the letter S); blasting from explosive letters P, B, and T; and even clicks from the subject's teeth striking together. Placed close to an instrument, a mic can reveal various mechanical noises such as key clicks, bow scrapes, and so on.

MONO SOUND

In everyday life, the audience is used to listening with two ears. As their brains compare these two separate sound images of the external world, they build a three-dimensional impression from which the direction and distance of sound is estimated (see the next section).

ROOM ACOUSTICS

Live Surroundings

When a room contains predominantly hard surfaces, the sound is strongly reflected. Many of these reflections are picked up by the microphone, reinforcing and coloring the direct sound pickup.

Dead Surroundings

When surfaces in a room are very sound-absorbent, the direct sound waves strike walls, floor, ceiling, and furnishings, and are largely lost. Only a few weak reflections may be picked up by the microphone.

Non-stereo television sound is not as sophisticated as this. It presents a *monaural (mono)* representation of sound in space. The only clue to distance is loudness; direction cannot be conveyed at all. Listening to mono reproduction, we are not able to distinguish between direct and reflected sounds, as we can when listening in stereo. Instead, they become intermixed, so that the sound is often "muddy" and less distinct. In mono sound, we become much more aware of the effects of reverberation.

Because the audience cannot easily distinguish direction and distance, the mono microphone needs to be carefully positioned. Audio personnel need to be careful that:

- Too many sound reflections are not picked up.
- Louder sounds do not mask quieter sounds (particularly in an orchestra).
- Extraneous sounds do not interfere with the ones we want to hear.

STEREO SOUND

Stereo sound creates an illusion of space and dimension. It enhances clarity. Stereo gives the viewer the ability to localize the direction of the sound. This localization gives the audience a sense of depth—a spatial awareness of the visual image and the sound. However, because the speakers in television receivers are quite close together, the effect can be somewhat limited. Sound quality and realism are enhanced, but our impressions of direction and depth are less obvious.

To simplify sound pickup, many practitioners mix central mono speech, with stereo effects and music. When a stereo microphone is used, care must be taken to maintain direction (such as mic left to camera left), and to hold the mic still; otherwise, the stereo image will move around. In a stereo system, reverberation even appears more pronounced and extraneous noises such as wind, ventilation, and footsteps are more prominent, because they have direction, rather than merging with the overall background.

SURROUND SOUND

Surround sound can provide a sense of envelopment when mixed correctly. Instead of the one channel for mono or the two channels for stereo, 5.1 surround has six discrete (distinct, individual) channels: left front and right front (sometimes called stereo left and right), center, a subwoofer for low-frequency effects (LFEs), and left rear and right rear speakers (sometimes called surround left and right). The feeling of depth, direction, and realism is obtained by the audio personnel panning between the five main channels and routing effects to the LFE channel (Figure 14.4).

Although 5.1 surround sound is currently the most popular type of surround, it is not the only type available. There are currently 6.1 and 7.1 surround systems, and Japan's NHK is currently marketing a 4K video that includes a 22.2 surround sound system for theaters (Figure 14.5).

MICROPHONE CHARACTERISTICS

The microphone characteristics that are most important to you will depend largely on the type of sound pickup involved and operating conditions. For example, ruggedness may be at the expense of fidelity. The main things that you need to know about microphones are:

- *Physical features*: Although size may be unimportant for some situations, it can matter where the microphone is to appear in the shot or to be held by the talent. Appearance also counts. Ruggedness is a consideration where rough or inexperienced usage is likely.
- *Audio quality*: Ideally, a microphone should cover the entire audio spectrum evenly. Its transient response to brief sharp sounds should be impeccable. Audio should be accurately reproduced without coloration or distortion. Fortunately, such parameters are less critical in many situations.
- *Sensitivity and directionality*: A microphone's sensitivity determines how large an audio signal it produces for a given sound volume, although audio amplifiers can compensate for even the least sensitive microphones. The problem with this is that excessive amplification can add hiss and hum to the audio signal. All microphones normally have to work closer to quiet sounds than louder ones, but less sensitive microphones must be positioned even closer.

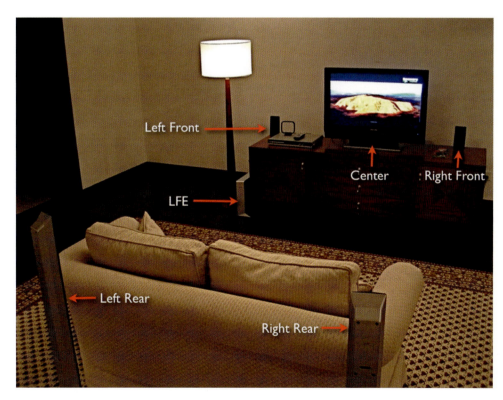

FIGURE 14.4
A home surround sound setup.

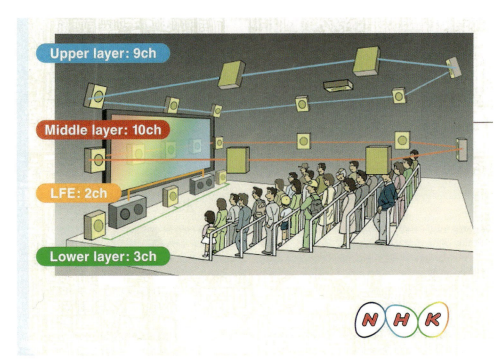

FIGURE 14.5
The NHK network in Japan has designed a 22.2 surround sound system aimed primarily at theaters.

Source: Illustration courtesy of NHK.

However, they are less liable to be overloaded or damaged by loud sounds, so that in certain applications (percussion) they may be preferable. The directional properties of the microphone are determined by its sensitivity pattern.

• *Choice of microphone/installation suitability*: All audio personnel have prejudices about the right microphone for the job and exactly where to place it; no two situations are identical. While many audio personnel may agree on specific types of mics in many settings, the positioning of the mic is very subjective.

Microphone Care

Though most people regard the video camera with a certain apprehension, there are those who tend to dismiss the microphone (or mic) all too casually. They clip it onto a guest's jacket with an air of "that's all we have to do for audio," instead of treating the mic as a delicate tool. If the microphone is damaged or poorly positioned, the program sound will suffer. No amount of postproduction work with the audio can compensate for doing it right from the beginning. Program sound all begins with the microphone.

Although most microphones are reasonably robust, they do need careful handling if they are to remain reliable and perform up to specification. It is asking for trouble to drop them, knock them, or get liquid on them.

ON-THE-JOB REPAIRS

Audio cables are pulled and walked over so much that they are often the weak link in the audio system. That means that audio personnel must make occasional repairs on microphones. These repairs could mean re-soldering cables to the connectors and repairing cables that get severed (Figure 14.6).

FIGURE 14.6 *Repairing cables usually means using wire cutters, a knife, a soldering iron, and possibly a vice grip of some type.*

Directional Features of Microphones

Microphones do not all behave in the same way. Some are designed to be *omnidirectional*—they can hear equally well in all directions. Others are *directional* (also known as *cardioid*)—they can hear sounds directly in front of them clearly, but are comparatively deaf to sounds in all other directions.

The advantage of an omnidirectional mic (Figure 14.7) is that it can pick up sound equally well over a wide area. It is great when covering a group of people, or someone who is moving around. The disadvantage is that it cannot discriminate between the sound you want to hear and unwanted sounds such as reflections from walls, noises from nearby people or equipment, ventilation noise, footsteps, and so on. The more reverberant the surroundings, the worse the problem. The mic must be positioned so that it is closer to the wanted sounds than to the extraneous noises. This mic is great for picking up ambient or natural (NAT) sounds.

When a directional mic (Figure 14.8) is pointed at the desired sound, it will tend to ignore sounds from other directions, providing a much cleaner result. On the other hand, the directional mic needs to be aimed very carefully. It is also important to make sure that the audio source does not move out of the main pickup zone, otherwise the source will be "off-mic." The off-mic sound becomes weaker, will probably include high-note losses, and may cause the audience to hear what it is being pointed at, instead of the desired source.

There are several different forms of unidirectional pickup patterns. The cardioid (see Figure 14.8) or heart-shaped pattern is broad enough for general use, but not overly selective, and the *super-* or *hyper-cardioid* (Figure 14.9) response also has a limited pickup area at its rear to receive reflected sounds.

Microphone Pickup Methods

There are two predominant methods for converting sound energy to an electrical-equivalent signal: dynamic and condenser.

Dynamic microphones are the most rugged, provide good-quality sound, and are not easily distorted by loud sounds

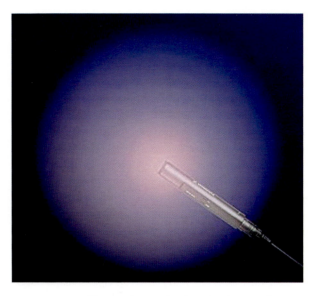

FIGURE 14.7 *The omnidirectional pickup pattern is equally sensitive in all directions, generally rugged, and not too susceptible to impact shock. This mic cannot distinguish between direct and reflected sounds, so it must be placed close to the sound source.*
Source: Image courtesy of Sennheiser.

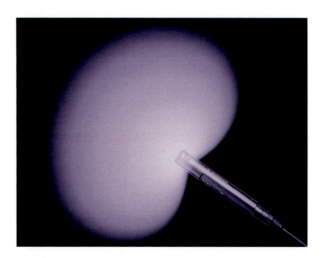

FIGURE 14.8 *The directional (or cardioid) mic pickup pattern. This broad, heart-shaped pickup pattern (roughly 160 degrees) is insensitive on its rear side.*
Source: Image courtesy of Sennheiser.

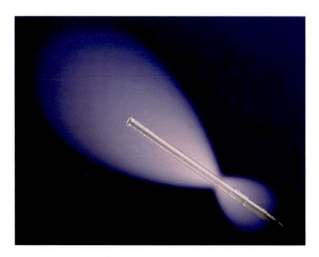

FIGURE 14.9 *A super-cardioid (or highly directional) pickup pattern is used wherever you want extremely selective pickup, to avoid environmental noises, or for distance sources.*
Source: Image courtesy of Sennheiser.

such as nearby drums. These mics need little or no regular maintenance. They can be handheld without causing unwanted "handling noise" and used with all types of microphone mountings. These mics generally cannot be as small as a condenser mic and some are not of as high quality. However, they can be just as high quality as the condenser microphone.

The *condenser* microphone produces very high audio quality and is ideal for musical pickup. A significant advantage to the condenser is that it can be very small, making it the logical choice for a shotgun, lavalier mic, and other miniature microphones. The condenser mic is generally powered by an inboard battery, phantom-powered (power sent from the mixer) audio board, or a special power supply. The electret condenser microphone has a permanent charge applied when it is manufactured, which remains for the life of the microphone and does not need to be externally powered.

TYPES OF MICROPHONES

In most television production situations, any number of a variety of microphones can be used to record the audio. One audio person may select one type of mic and another may choose a radically different mic. Each person is looking for the best mic that will provide the sound that he or she is looking for. The following microphones are just some of the audio tools that are available. As the audio plan is created, the mic that is right for you must be chosen.

Camera Microphones

If the camera is fitted with a microphone, the theory is that when it is aimed at the subject to capture the video, the mic will pick up quality audio. However, a lot depends on the situation, the camera/mic placement, and the type of sound involved. Nothing beats a separate, high-quality microphone placed in exactly the right place. However, single-camera operators, working by themselves and moving around to various shooting positions, may have to use a camera microphone.

Some of the less professional mics are known to pick up sound from all around the camera, including noise from the camera zoom lens and camera operator sounds. With care, though, this basic microphone is useful for general atmospheric background sounds (traffic, crowds), and occasionally has good enough pickup to capture a voice quality, if the camera is close to the subject.

FIGURE 14.10 *Shotgun mics are the most popular type of camera microphones.*

The most popular type of camera microphone is the shotgun mic, attached to the top of the camera (Figure 14.10). Plugged into the camera's external mic socket, this mic will give the best quality pickup from the subject. As always with directional mics, these must be aimed accurately.

Camera microphones do have their drawbacks, and should only be used for voice when better options are unavailable. Distance is the biggest problem with a camera microphone:

- If the camera mic is more than 4–6 feet away from the talent, it may result in unacceptably high levels of background noise and/or acoustical reflections.
- Distance is the same for all shots. The camera may zoom in to a close-up shot or take a wide-angle shot, but the sound level remains the same.
- The microphone is often too far away from the subject for the best sound. Its position is determined by the camera's shot, not by the optimum place for the microphone.
- The camera microphone cannot follow somebody if he or she turns away from a frontal position, such as to point to a nearby wall map. The sound's volume and quality will fall off as he or she moves off-mic.

Handheld Microphone

The handheld mic (or stick mic) is a familiar sight on television, used by interviewers, singers, and commentators. It is a very simple, convenient method of sound pickup, provided that it is used properly. Otherwise, results can be erratic. The handheld mic is best held just below shoulder height, pointed slightly toward the person speaking. Make it as unobtrusive as possible (Figure 14.11).

FIGURE 14.11 *The handheld microphone is widely used for interviews, commentaries, and stage work. If the mic has a cardioid directional response, extraneous noise pickup is lower. If it is omnidirectional, the mic may need to be held closer to the subject to reduce atmosphere sounds. The mic is normally held just below shoulder height.*

To reduce the low rumbling noises of wind on the microphone and explosive breath-pops when it is held too close to the mouth, it is advisable to attach a foam windshield to the microphone. Note the blue foam windshield in the commentator photo in Figure 14.11. Whenever possible, talk across the microphone rather than directly into it. This will provide the optimal audio quality.

Some people attempt to hold the mic around waist height to prevent it from being visible in the picture; however, this generally results in weak pickup, poor quality, and more intrusive background noise. Handheld microphones with cardioid patterns help reduce the amount of extraneous sound overheard, so it can be used about 1–1.5 feet (0.30–0.45_m) from the person speaking. If an omnidirectional handheld mic is used, it should normally be held much closer—around 9 inches (22 cm) for optimal sound quality.

Shotgun Microphone

The shotgun microphone (hyper-cardioid) consists of a slotted tube containing an electret microphone at one end. This microphone is designed to pick up sound within quite a narrow angle, while remaining much less sensitive to sounds from other directions. It is great at isolating a subject within a crowd or excluding nearby noises (Figure 14.12).

Unfortunately, the shotgun microphone is not good at maintaining these directional properties throughout the audio range. At lower frequencies, it loses its ability to discriminate. The narrow forward-pointing pickup pattern then becomes increasingly broader.

When shooting in very "live" (reverberant) surroundings, a shotgun microphone has advantages, as it will pick up the subject's sound successfully while reducing unwanted reflections, although how effectively it does so depends on the pitch or coloration of the reflected sounds.

The shotgun microphone is quite adaptable, and is regularly used as:

- A handheld microphone supported by some type of shock mount (Figures 14.12 and 14.13).
- A mic connected to the end of a boom pole or fishpole (Figures 14.13 and 14.14).
- A mic in the swiveled cradle support of a regular sound boom or perambulator boom (Figure 14.1).
- As a camera microphone, fitted to the top of the camera head (Figure 14.10).

Most people working in the field fit a shotgun microphone with some type of a windshield (also called a windjammer or wind muffler). The most effective types at suppressing obtrusive wind noises are a furry overcoat with "hairs" or a plastic/fabric tube (Figure 14.15). An alternative design of the

FIGURE 14.12 *Shotgun microphones are among the most commonly used mics in television. They are very susceptible to handling noise and must be held or connected to a pole or stand with a shock mount.*

FIGURE 14.13 *The shotgun microphone is not attached directly to the pole. Instead, a shock mount, such as the one shown here, must be used to prevent the rumbles of handling noise traveling along the pole and being picked up by the microphone.*

Source: Photo courtesy of Audio-Technica.

FIGURE 14.14 *A mic boom (or fishpole) is a regular method of mounting the shotgun microphone, particularly in the field. It allows the operator to stand several feet away from the subject, reaching over any foreground obstacles, and to place the microphone at an optimum angle. This position can be tiring if it has to be maintained for a long period of time. However, it may be the only solution when people are standing and/or walking about.*

Source: Photo by Naomi Friedman.

FIGURE 14.15 *Different types of windshields are used to protect a shotgun mic from wind noise.*

wind filter is a tubular plastic sponge (Figure 14.12). Although much lighter, this design may prove inadequate except in the lightest breeze.

Using the Shotgun Microphone

Selecting the best position for shotgun microphones takes some advance planning. Audio personnel need to know how the action is going to develop. They may get this from a briefing beforehand or find it out from a camera rehearsal:

- Will the shotgun be used for long takes or for brief shots? It is one thing to stand in a fixed position for someone talking straight to the camera and another to have to follow action around as people and cameras move through a sequence.
- Will audio personnel have an uninterrupted view of the action?
- Is anyone or anything going to get in their way or are they going to get in anyone else's way?

The Shotgun and the Boom Pole (Fishpole)

The boom pole (or fishpole) has become the most popular choice for sound pickup on location and in many smaller studios. This adjustable lightweight aluminium pole is usually about 6–9 feet long, carrying a microphone at its far end. The sound cable is either designed inside the pole or is taped securely along the pole (Figures 14.2 and 14.14).

Lavalier (Lapel or Clip-On) Microphone

The lavalier microphone, also known as a lav, lapel, mini-mic, or a clip-on mic, has become the favorite mic in productions where it is unimportant whether the viewer sees a mic attached to someone's clothing. These microphones are compact, unobtrusive, and provide high sound quality. This microphone is usually clipped to outside clothing (such as a tie, lapel, shirt, or blouse) so that noises from rubbing on clothing will be kept to a minimum. If lavalier mics are tucked beneath a heavy sweater or coat, understandably the sound becomes muffled and less distinct (Figure 14.17). The incredibly small lav mics, as shown in one of the photos in Figure 14.17, can also be used as an "earset" or "headworn" mic, as shown in Figure 14.18.

One of the challenges when using a lavalier mic is that the volume and clarity of the sound can change as a wearer turns his or her head left and right, or toward and away from the microphone.

A lavalier mic can be relied on to effectively pick up only the sound of the person wearing it. When two or three people are speaking, each will need to wear his or her own microphone. However, that does not mean that the mics won't pick up the sound from others—it just won't be the same quality and will not be the same level. When working in noisy surroundings, a small foam windshield over the end of the microphone will reduce the rumble of wind noise. The mic's cable can be concealed beneath clothing.

OPERATING THE FISHPOLE OR BOOM POLE

Until audio personnel get used to it, the boom pole can be a very unwieldy, unbalanced instrument when fully extended, as the weight is all at its far end. There are many different ways that a boom pole can be held (Figures 14.2, 14.14, and 14.16), such as:

- Above the head, with arms fully extended along the pole to balance it.
- Set across the shoulders for added stability, as shown here in the left figure.

The goal is to get the microphone close enough, without showing up at the bottom or top of the camera's shot.

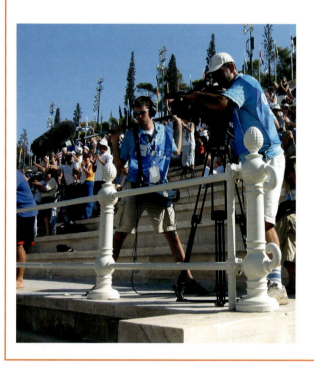

FIGURE 14.16

FIGURE 14.17 *Lavalier microphones come in many different sizes and shapes. Generally, a lavalier mic is clipped to a necktie, lapel, or shirt. Sometimes a "dual redundancy" pair is used, whenever a standby mic is desired.*

Source: Photos courtesy of Audio-Technica and Countryman Associates.

Lavalier microphones can also be used to record subjects other than people. They are used effectively in sports productions (mounted in places such as the nets at a soccer/football field) and they can be used to pick up the sound of some musical instruments. The clip-on mic in Figure 14.19 is actually a type of lavalier mic.

Boundary or PZM Microphone

The boundary microphone and pressure zone microphone (PZM) are low-profile mics that can be used to capture audio from talent that is six or more feet away without the "hollow" sound of a hanging handheld mic. Although the pickup technology is very different, these two mics are used similarly. These microphones are especially good for dramatic productions where microphones should not be seen (they can

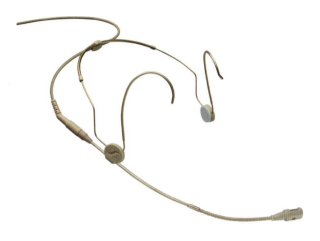

FIGURE 14.18 *The talent is using an "earset" or "headworn" microphone that utilizes the lavalier microphone. It is mounted on a tiny mic "boom" and attached to the ear. It is available in a flesh color and is almost invisible to the viewing audience. In this situation, it was used by hosts of ESPN's X Games.*

Source: Photos courtesy of Dennis Baxter and Sennheiser.

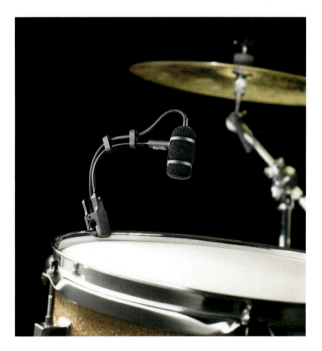

FIGURE 14.19 *These photos show a variation of lavalier microphones often used to record musical instruments or any other subject that requires close microphones.*

Source: Photo courtesy of Audio-Technica.

be attached to the back of set pieces). They are also good for stage performances of large groups. They can be hung from the ceiling, set on a floor, or attached to furniture. The pickup distance can be increased by mounting these mics on a hard surface (Figure 14.20).

Hanging Microphone

Hanging microphones are especially designed for high-quality sound reinforcement of dramatic productions, orchestras, and choirs. The mics are suspended over the performance area. Their small size is ideal since they will probably be visible to the viewing audience (Figure 14.21).

FIGURE 14.20 *The boundary microphone is a low-profile mic that can pick up accurate sounds from six or more feet away.*

FIGURE 14.21 *Hanging microphone.*
Source: Photo courtesy of Audio-Technica.

Surround Sound Microphone

Surround sound microphones can capture 5.1 to 7.1 channels of discrete (separate) audio with the multidirectional pickup pattern. Using the small microphone shown in Figure 14.22, the audio can be recorded directly onto the camera's internal media along with video images. The small microphone has five microphone elements (left, right, center, left side, right side) and a dedicated LFE (this counts as the ".1" in the channel count) microphone. The higher-end version is much larger and provides 7.1 surround sound. The smaller system, designed specifically for a camcorder, utilizes an internal Dolby(r) Pro Logic II encoded line-level stereo output for connection directly to the camera on a single 3.5 mm stereo female mini-plug jack. Some nonlinear editing systems have a Dolby logic decoder built in, allowing the channels from the stereo input to be split into the five surround channels, which allows a user to record programs in surround sound without having a full surround mixing board (Figures 14.22 and 14.23).

Surround sound microphones must be positioned carefully. They should not just be mounted on top of a camera if the camera will be panning and tilting around quite a bit. Moving the microphone around with the camera can really spatially disorient the audience. Generally, these microphones are mounted on a separate stand or clamped to something stationary in order to pick up a quality ambient sound (Figure 14.24).

FIGURE 14.22 *This small surround sound microphone includes a Dolby(r) Pro Logic II encoder with a line-level stereo output designed for stereo inputs on camcorders.*
Source: Photo courtesy of Holophone.

Suggestions for Using a Surround Sound Microphone

- Use the surround mic to provide the "base" ambient surround sound for the audio mix.
- For a concert situation with arena-style seating, the sound mic should be placed a little higher than the orchestra, tilting the nose down toward the performers.
- When panning and tilting, mount the mic on a stationary stand, not on the camera.
- In most situations, try to position the surround mic as close to "front row center" as possible, rather than near the back of the room.
- When shooting sports events, it is best to place the surround mic either near the center of the field or near a main camera position. Always keep in mind the perspective of the television viewer. Mounting the surround mic on the side of a field or rink opposite to the main camera angle would seem backwards and unnatural.
- For surround recording of acoustic instruments, including drum kits, pianos, and voice at close range, try placing the mic near or above the instrument that is being recorded. For vocals or choirs, position the singers around the mic and monitor in surround to hear the results.

FIGURE 14.23 *There are a variety of surround sound microphones available. Audio personnel must select the one that best fits their specific situation.*

Source: Photos courtesy of Holophone and Core Sound.

FIGURE 14.24 *ABC Television's Extreme Home Makeover uses a professional surround sound microphone, separate from the camera, to capture the audio.*

Souce: Photo courtesy of Holophone.

Wireless Microphones

The most commonly used wireless microphones (or radio microphones) are the lavalier mic and the handheld mics. Both of these types of mics can be purchased with the wireless transmitter built into the mic (or belt pack) and include a matching receiver. Lavalier mics are very popular because they allow the talent to have generally unrestricted movement while moving around the location. They are used in the studio with interview shows, on referees to hear their calls, and hidden on actors to catch their words (Figures 14.25–14.27).

Wireless microphones generally work on a radio frequency (RF) and many are frequency-programmable, allowing the audio personnel to select the best frequency for a specific location. Care must be taken to make sure that legal frequencies are being used.

FIGURE 14.26 *Wireless receivers can be located on a camera. In this situation, the interviewer is using a handheld wireless.*

Source: Photo courtesy of Sennheiser.

FIGURE 14.25 *A wireless (radio) belt pack transmitter and receiver. A lavalier microphone can be plugged into the transmitter.*

Source: Photo courtesy of Audio-Technica.

FIGURE 14.27 *Any type of microphone can become a wireless microphone if some type of wireless plug-on transmitter is used. This transmitter converts a dynamic or condenser microphone to wireless, transmitting the signal back to a receiver.*

Source: Photo courtesy of Audio-Technica.

FIGURE 14.28 *It is important to ensure that the batteries in the transmitter, and possibly the receiver, are at full capacity at the beginning of a program. The transmitters are usually clipped to the back of the talent.*

Source: Photo courtesy of Sennheiser.

There can be a number of challenges when working with wireless microphones:

- These mics work off of batteries. The battery life is roughly 4–6 hours. When working in freezing temperatures, battery life is usually cut in half. New batteries should be placed in the mic before each new session. Do not leave it to chance, assuming that there is enough capacity left from the last time (Figure 14.28).
- If two or more wireless microphones are being used in an area, they must be set on different RF channels to avoid interference.
- When working near large metal structures, there can be difficulties with RF dead spots, fading, distortion, or interference. Diversity of reception—using multiple antennas—can improve this situation, but it is still cause for some concern.

Hidden Mics

When other methods of sound pickup are difficult, a hidden microphone may be the best solution to the problem. Mics can be concealed among a bunch of flowers on a table, behind props, in a piece of furniture, and so on.

However, hidden mics do have limitations. Although the mic can be hidden, the cable must not be seen and/or the transmitter must be out of sight. Sound quality may also be affected by nearby reflecting or absorbing surfaces. Because the microphone covers a fixed localized area, the talent has to be relied on to play to the mic and not speak off-mic.

MICROPHONE STANDS AND MOUNTS

Microphone stands are very useful in situations where the director does not mind the microphone possibly being seen in the shot. It is especially useful for stage announcements, singing groups, and for mic'ing musical instruments. It does have some disadvantages. If people move around much, they can easily walk out of the mic's range. Directors have to rely on the talent to get to the right place and keep the right distance from the mic. It is a good idea to give talent taped marks on the floor to guide them. And, of course, there is always the danger that he or she will kick the stand or trip over a cable (Figures 14.29–14.31).

FIGURE 14.29 *The most common microphone stand for musicians.*

FIGURE 14.30 *An audio person adjusts a microphone stand for an on-location interview project.*

CONTROLLING DYNAMICS

Dynamic Range

Everyday sounds can cover a considerable volume range. Fortunately, our ears are able to readjust to an astonishing extent to cope with these variations. But audio systems do not have this ability. If audio signals are larger than the system can accept, they will overload it and become badly distorted. If, on the other hand, sounds are too weak, they get lost in general background noise. In order to reproduce audio clearly, with fidelity, it must be kept within the system's limits.

A lot of sounds pose no problems at all. They don't get particularly soft or loud; that is, they do not have a wide dynamic range. When recording sounds of this type, there is little need to alter the gain (amplification) of the system once it has been set to an appropriate "average" position.

It can be very difficult to capture the wide range of audio between a whisper and an ear-shattering blast. Because the blast will exceed the system's handling capacity, the audio person must compensate in some way. The most obvious thing to do is to turn down the system's audio gain so that the loud parts never reach the upper limit. But then the quiet passages may be so faint that they are inaudible. So somehow or other, most of the time, the audio levels need to be controlled.

Automatic Control for Audio

Cameras generally allow the operator to set the camera's audio manually or automatically. To avoid loud sounds overloading the audio system and causing distortion, most audio and video recording equipment includes automatic gain control. When the sound signal exceeds a certain level, the auto gain control automatically reduces the audio input.

FIGURE 14.31 *There are many different types of microphone stands and mounts, from bottom-weighted telescopic stands to small versions with thin flexible or curved tubing intended for lavalier or miniature mics.*

Source: Photos by Paul Dupree and Dennis Baxter.

A completely automatic gain system amplifies all incoming sounds to a specific preset level. It "irons out" sound dynamics by preventing over- or under-amplification. Quiet sounds are increased in volume and loud sounds are held back. There are no adjustments to make and the camera operator must accept the results.

This can be an effective way of coping with occasional overloud noises, but if the sounds happen to be so loud that they are continually "hitting the limiter," the results can be very distracting from an unpleasant strangled effect as sound peaks are "pulled back" to moments when quiet background sounds are over-amplified and surge in persistently whenever there is a pause.

Some auto gain systems do have manual adjustments. The idea is to ensure that the gain control is set high enough to amplify the quietest passages without running into over-amplification of the loudest sounds. The auto gain control circuitry limits sound peaks only as an occasional safety measure, and depends on the gain adjustment. It is generally best to manually control the audio.

Manual Control

Manual control means that you continuously monitor the program while watching an audio level meter (Figure 14.32). The audio person is responsible for readjusting the audio system's gain (amplification) whenever necessary to obtain a quality audio signal. That does not mean that the dynamics should be "ironed out" by making all the quiet sections loud, and holding back all the loud passages. "Riding the gain" in this way can ruin the sound of the program. Instead, when sounds are going to be weak, anticipate by gradually increasing the gain, and conversely, slowly move the gain back before loud passages. Then the listeners will be unaware that changes are being made to accommodate the audio system's limitations.

How quiet the softest sound is allowed to be will depend on the purpose of the program. If, for example, the recording will be used in noisy surroundings, or shown in the open air, it may be best to keep the gain up to prevent the quietest sounds from falling below 215 or 220_dB. If a piano performance is being shot to be heard in a controlled location indoors, take care not to over-control the music's dynamics, and use the system's full volume range.

Unlike automatic control circuits, audio personnel are able to anticipate and make artistic judgments, which can make the final audio far superior than it would be otherwise. The drawback to manual control is that the audio personnel need to be vigilant all the time, ready to make any necessary

FIGURE 14.32 *It is the audio person's responsibility to readjust the audio system's gain (amplification) whenever necessary to obtain a quality audio signal.*

Source: Photo courtesy of Sennheiser.

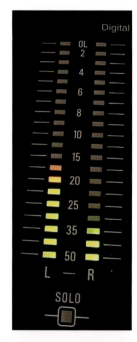

FIGURE 14.33 *VU meters and bar graphs are used to monitor the audio signal.*

readjustments. If they are not careful, the resulting audio may be less satisfactory than the auto circuits would have produced.

There are several types of volume indicators, but the most common on video equipment take the form of visual displays using bar graphs or some type of VU meter (explained shortly).

A bar graph (Figure 14.33) has a strip made up of tiny segments. This varies in length with the audio signal's strength. Calibrations vary, but it might have a decibel scale from 250 to 110_dB, with an upper working limit of about 12 dB. Adjust the audio gain control so that the sound peaks reach this upper mark. Twin bar graphs are used to monitor the left and right channels. The sound generally distorts if the audio signal goes into the red area of the bar graph.

The VU meter (Figure 14.33) is a widely used volume indicator. It has two scales: a "volume unit" scale marked in decibels, and another showing "percentage modulation." Although accurate for steady tones, the VU meter gives deceptively low readings for brief loud sounds or transients such as percussion.

The maximum signal coincides with 100 percent modulation at 0_dB. Above that, in the red area, sounds will distort, although occasional peaks are acceptable. The normal range used is 220 to 0_dB, typically peaking between 22 to 0 dB.

In summary, if the camera operator needs the audio system to look after itself, because he or she is preoccupied with shooting the scene, or is coping with very unpredictable sounds, then the automatic gain control has its merits—it will prevent loud sounds from overloading the system. However, if an assistant is available who can monitor the sound as it is being recorded, and adjust the gain for optimum results, then this has significant artistic advantages.

There are special electronic devices called "limiters" or "compressors" that automatically adjust the dynamic range of the audio signal, but these are found only in more sophisticated systems.

Monitoring the Audio

Monitoring sound for a video program involves:

- *Watching*: Checking the volume indicator and watching a video monitor to make sure that the microphone does not pop into the shot inadvertently. Watching the video monitor also allows you to anticipate what microphone will be needed next.

- *Listening*: Checking sound quality and balance on high-grade earphones or a loudspeaker to detect any unwanted background noises.

Ideally, the audio level can be adjusted during a rehearsal. However, if a performance is going to be recorded without a rehearsal, such as in an interview, ask the talent to speak a few lines so that the audio level can be accurately adjusted. It is best if the talent can chat with the host in a normal voice for a few minutes or he or she can read from a script or book long enough to adjust the level. Do not have him or her count or say "test." Both of these can give inaccurate readings, as they do not necessarily reflect normal speaking levels.

It is also important to monitor the sound to get an impression of the dynamic range while watching the volume level. If the results are not satisfactory, the talent may need to be asked to speak a little louder or more quietly, or to reposition the mic.

When shooting a program alone, it is a little more complicated to control the audio manually. Most professional cameras include an audio meter in the eyepiece of the camera, which allows the camera operator to monitor the audio signal. In recent years, camera manufacturers have improved the ability for camera operators to adjust the audio on the fly. Usually, if the level is set in advance, it is not difficult to capture good audio levels. However, in difficult situations, the automatic gain control can be used. It is also essential to monitor the audio with an earpiece or the small built-in speaker on the side of some cameras.

The Audio Mixer

An *audio mixer* is needed whenever there are a number of sound sources to select, blend together, and control (such as a couple of microphones, CD, VCR audio output, and so on). The output of this unit is fed to the recorder (Figures 14.34 and 14.35).

On the front panel of the audio mixer are a series of knobs or sliders. Each of these "pots" (*potentiometers*) or faders (Figure 14.36) can adjust its channel's volume from full audio to fade-out (silence). In some designs, the channel can be switched on or off on cue with a "mute" button. When sources are plugged into the patch panel (connector strip), generally on the back of the switcher, the audio person can select the appropriate channel.

On a large audio mixing panel, there may be group faders (group masters, sub-masters). Each of these group faders

FIGURE 14.34 *The portable audio mixer is used in the field to mix up to three mics; the overall output is controlled by a master fader. A VU meter provides the volume indicator. Some mixers include a limiter to prevent audio overload.*

Source: Left photo courtesy of Shure Incorporated and right photo by Courtney LeMay.

FIGURE 14.36 *Faders (potentiometers) on a large audio mixer.*

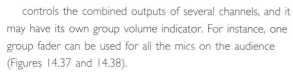

FIGURE 14.35 *Some field mixers include a hard drive that can store the audio program directly on the mixer.*

controls the combined outputs of several channels, and it may have its own group volume indicator. For instance, one group fader can be used for all the mics on the audience (Figures 14.37 and 14.38).

Finally, there is a master fader that controls the overall audio strength being sent to line (such as on the recorder). This can be used to fade the complete mix in or out. A master volume indicator shows the combined strength of the mixed audio.

Larger audio mixers include a *cue circuit* (also called *audition*) that enables audio personnel to listen "privately" on earphones or a loudspeaker to the output of any individual channel, even when its pot is faded out. That way, the source, such as a CD, can be set up at exactly the right spot, ready to be started on cue, without this being overheard on the air.

FIGURE 14.37 *A large surround sound audio mixer for television productions.*

Using the Audio Mixer

Located at the audio mixer (board, desk, audio control console), the audio control engineer (sound mixer, sound supervisor) selects and blends the various program sound sources. His or her attention is divided variously between:

- Selecting and controlling the outputs of various audio sources (microphones, discs, hard drives, etc.).
- Keeping the volume indicators within system limits by adjusting appropriate channel faders (amplifier gains).

- Following the program audio and pictures and the director's intercom.
- Checking the audio quality on high-grade loudspeakers.
- Watching the video monitors that show the program and preview shots, to check sound perspective, and warn against microphones or shadows coming into the shot.
- Guiding (and cueing) operators on mic booms, audio disk, and recording playback.
- Possibly operating audio recording equipment.
- Liaison with other production team members.

Audio mixing can be as simple as fading up a microphone or two and controlling the sound levels, or it can be a complex process involving edge-of-seat decisions, depending on the type of production. For example:

- A "live" show usually involves rapid decisions. When a production is being recorded scene by scene, not live, there is time to set up complicated audio treatments. Anything that goes wrong can usually be corrected and improved.
- If there are a number of sound sources that need to be cued in at precisely the right moment, it can be stressful. This type of program poses a very different situation than a less-complicated program such as an interview.

These are just a few of the issues that decide how complex the audio mix needs to be. Let's look at typical operations in some detail:

FIGURE 14.38
Some television audio mixers prefer to mix live events, such as a concert, awards show, or similar performance, from inside the performance hall.
Source: Photo courtesy of Dennis Baxter.

AUDIO CONNECTORS

The most common audio connector used by audio professionals is the XLR, a three-pin connector. The 3.5 mini plug is used on most amateur cameras. Remote productions that have a lot of microphones often use the DT system, which allows multiple XLR plugs to be cabled (using one cable) from the studio or remote to the audio control room (Figure 14.39).

XLR Jack   XLR Plug   1/4" Plug   3.5mm Mini Plug   RCA Plug

DT-12 Plug   DT-12 Box with DT-12 jack and multiple XLR jacks

FIGURE 14.39 *There are a variety of audio connectors used in television.*

- Sound sources should be faded in just before they begin (to the appropriate fader level), and faded out when finished.
- Source channels should not be left "open" (live) when not in use. Apart from accidentally recording overheard remarks ("Was that all right?") and other unwanted sounds, it may pick up someone who is on another mic.
- It is important that the right source is selected and faded up/down at the right moment. Individual or group faders can be used. Here is an example:

Imagine a scene for a drama production is being shot. The audience sees the interior of a home, where the radio (actually from an iPod located near the mixer) plays quietly. An actor is talking (on live mic 1) to another person (on live mic 2) who is not shown in the shot. A nearby telephone rings (a fade-up using a special effect from a hard drive of a phone ringing). The actor turns down the radio (we fade down the iPod) and picks up the

MIC LEVEL VERSUS LINE LEVEL

- A **mic level** is the audio level (or voltage) of a signal that is generated by a microphone. The mic level, typically around 2 millivolts, is much weaker than a line-level signal.
- A **line level** is the audio signal generated by a non-microphone device such as a CD player, amplifier, video playback unit, MIDI, or line mixer output. The two normal line levels are 0.316 µvolts and 1.23 volts.

phone (we stop the special effects). Continuous background noises of a storm (from another special effect from the hard drive) can be heard at a low level throughout the scene. All these fades must be completed on individual channel faders. The different audio sources must be operated by the person at the mixer or by an assistant.

- When combining several sound sources, all of them should not be faded up to their full level. They should be blended for a specific overall effect. For example, if a single microphone were used to pick up the sound of a music group, chances are that the one microphone would pick up certain instruments much better than others. Loud instruments would dominate and quiet ones would be lost. The overall balance would be poor. Instead, use several microphones, devoted to different parts of the group. Then the volume of the weaker instruments, such as a flute, can be increased and the volume of the louder ones, such as drums, can be decreased. With care, the result would sound perfectly natural and have a clearer overall balance.

- Sometimes the relative volume of a sound will need to be adjusted in order to create an illusion of distance. If the sound of a telephone ringing is loud, we assume it is nearby; if it is faint, it must be some distance away.

- Sounds may need to be deliberately amplified. For example, you can readjust the fader controlling the crowd noise to make it louder at an exciting moment and give it a more dramatic impact.

- Bottom line, the final audio mix needs to fit the mood of the overall production.

Natural Sound

Most video productions are made up of a series of shots, taken in whatever order is most convenient, and edited together later to form the final program. This approach has both advantages and drawbacks. As far as the program sound is concerned, there are several complications.

First of all, although the various shots in a sequence have been taken at different times, it is important that their quality and volume match when they are edited together. Otherwise, there will be sudden jumps whenever the shot changes. If, for instance, a shot is taken of a person walking down a hallway, using a close mic, and then a side view of the same action using a more distant mic, the difference in the sound, when cutting from one shot to the other, could be quite noticeable. The overall effect would draw attention to the editing.

AUDIO SAFETY

It is always important to take the time to tape down all audio cables (Figure 14.40) that are in pedestrian walkways. This is not only for the safety of other people, but also protects your cables and equipment from excessive damage.

FIGURE 14.40 *Taping cables.*
Source: Photo courtesy of Dennis Baxter.

When editing together a sequence of images shot at different times, the background sounds may not match. In the time it takes for shooting one shot, repositioning the camera, adjusting the light, and then retaking the shot, the background noises often have significantly changed. Because the crew members are busy doing their jobs, they may not notice that the background sounds are quite different. Sounds that we become accustomed to while shooting the scene, such as overhead aircraft, farm equipment, hammering, or typing, have a nasty habit of instantly disappearing and reappearing when the shots are edited together.

Anticipation

Anticipation comes with experience. When things go wrong, hopefully you will learn from it and be better prepared next time. There are a number of ways to anticipate audio challenges.

Preparation

- Check through the script or preplanning paperwork and then pull together the appropriate equipment so that every audio situation in the production can be covered.
- Prerecorded audio inserts should be checked before the show. Make sure that they are appropriate. Is the duration too long or too short? Is the quality satisfactory? Will an insert require equalization? Is it damaged in any way (e.g., surface scratches on a disk)? Is the insert material arranged in the order in which it is to be used?
- Check all of the equipment to make sure that it is working correctly. Don't rely on the fact that "it was OK yesterday." If additional plug-in equipment is being used, such as a portable audio mixer, have someone fade-up each source (microphones or CD) to ensure that each one is working.
- Go to each microphone in turn, scratch its housing (an easy way to test the microphone), and state its location to make sure that the microphone is working and plugged into the correct input (this is "Boom A").

FIGURE 14.41 *This lavalier clip is designed to hold two microphones, providing a backup microphone in case one of them fails.*

- Have a backup microphone ready in case the main microphone fails. If it is a one-time-only occasion and a lavalier mic is being used, it may be advisable to add a second "dual redundancy" lavalier mic, too (Figure 14.41).
- Is the microphone cable long enough to allow the boom pole to move around freely?

Of course, these suggestions are all a matter of common sense, but it is surprising how often the obvious and the familiar get overlooked. These are just reminders of what should become a regular routine.

Anticipating Sound Editing

When shooting a scene, it is important to overcome the challenges of sound editing by anticipating the types of problems that will occur:

- *Continuity*: Try to ensure that the quality and level of successive shots in the same scene match as much as possible.
- *Natural/atmosphere sounds*: Record some general natural sound (atmosphere) and typical background sounds (wild track) in case they are needed during postproduction.
- *Questions*: When shooting an interview, concentrating on the guest, the questions of the interviewer may not be audible. Make sure that the host has his or her own microphone or go back and have him or her ask the same questions after the interview so that he or she is recorded.

Filtered Sound

Significant changes can be made to the quality of sound by introducing an audio filter into the system. This can be adjusted to increase or decrease the chosen part of the audio spectrum, to exaggerate or suppress the higher notes or the bass or middle register, depending on the type of filter system used and how it is adjusted.

The simplest "tone control" progressively reduces higher notes during reproduction. A more flexible type of audio filter is called an equalizer. This filter can boost or reduce any segment(s) of the audio spectrum by changing the slider pots.

Here are typical ways in which filtering can enhance the subjective effect of the sound:

- Cutting low bass can reduce rumble or hum, improve speech clarity, lessen the boomy or hollow sound of some small rooms, and weaken background noise from ventilation, passing traffic, and so on. Overdone, the result sounds thin and lacking body.

- Cutting higher notes can make hiss, whistles, sibilant speech, and other high-pitched sounds less prominent. However, if you cut them too much, the sound will lack clarity and intelligibility.
- If the bass and top notes are cut, the sound will have a much more "open air" quality—a useful cheat when shooting an exterior scene in a studio.
- By slightly increasing bass, the impression of size and grandeur of a large interior can be increased.
- The clarity and "presence" of many sounds can be improved by making them appear closer, by boosting the middle part of the audio spectrum (such as 2–6 kHz).
- Filtering can make the quality of sound recorded in different places more similar (such as shots of someone inside and outside a building). It can help to match the sound quality of different microphones.

Reverberation

As mentioned earlier, most of the everyday sounds we hear are a mixture of direct sound from the source itself, together with "colored" versions reflected from nearby surfaces. The quality of that reflected sound is affected by the nature of those surfaces. Some surfaces will absorb the higher notes (curtains, cushions, carpeting). This reflected sound may even be muffled. Conversely, where surroundings reflect higher notes more readily, these hard reflections will add harshness (also called "edginess") to the final sound.

Where there are few sound-absorbing materials around, there will be noticeable reverberation as sound rebounds from the walls, ceiling, and floor. If the time intervals between these reflections are considerably different, a distinct echo will be heard.

This is a reminder that the room tone will depend on its size, shape, carpeting, drapes, easy chairs, and other furnishings. Where there are no reflections—as in open spaces away from buildings or other hard surfaces—the resulting sound will seem dead. The only way we can simulate dead surroundings within a building is to use carefully positioned sound-absorbing materials. On the other hand, if there is too much reverberation, the result is a confused mixture of sound that reduces its clarity.

In practice, the appeal of many sounds can be enhanced by adding a certain amount of real or simulated reverberation to them. Today, the most commonly used method of adding

some "liveness" is to use a reverberation unit that digitally stores the sound and is selectively reread over and over to give the impression of reflected sounds.

BUILDING THE SOUNDTRACK

Most television programs contain not only speech, but also music and sound effects. These vary with the type of program. A talk show is likely to have music only at the start and end of the production, to give it a "packaged" feel; a dramatic presentation may be strewn with a variety of carefully selected atmospheric sound effects and musical passages (bridges, mood music, etc.).

Some sounds are prominently in the foreground, and others are carefully controlled to provide an appropriate background to the action. Some may creep in, and are barely audible, yet add an atmospheric quality to a scene (such as the quiet tick-tock of a grandfather clock). Others may be deafening, even drowning out the dialogue.

Some sound effects are continuous, and others rely on split-second cueing to exactly match live action.

Types of Program Sound

When a production is running smoothly, it's easy to overlook the complexities that lie behind it. This is particularly true of the sound component of television. The person controlling the program sound often has to work simultaneously with a diversity of sound sources. Some of these are live and therefore liable to vary unpredictably; others may have been prerecorded specifically for the production, or selected from stock libraries.

Contributory sounds can include:

- *Dialogue*: Direct pickup of the voices of people in the picture.
- *Off-camera voices*: This could include an unseen bystander, a radio playing in the background, or a public announcement in a train station.
- *Voiceover (VO)*: The voice of a commentator or announcer (with introductory or explanatory information).
- *Sound effects*: Sound coinciding with action in the picture.
- *Background or environmental effects*: General atmospheric sounds, such as the wind, ocean, or traffic.
- *Foreground music*: Someone playing an instrument in the shot.
- *Background music*: Atmospheric mood music (usually recorded).
- *Special effects sounds*: Sounds that enhance the scene.

Program Music

The role of music in television programs is so established that we don't need to dwell on it here. Musical themes often remain in the memory long after the program itself has faded from the mind.

Music can have various purposes. For example:

- *Identifying*: Music associated with a specific show, person, and/or country.
- *Atmospheric*: Melodies intended to induce a certain mood, such as excitement.
- *Associative*: Music reminiscent of, for example, the American West or the Orient.
- *Imitative*: Music that directly imitates, such as a birdsong. Music with a rhythm or melody copying the subject's features; for example, the jog-trot accompaniment to a horse and wagon.
- *Environmental*: Music heard at a specific place, such as a ballroom.

Sound Effects

Sound effects add depth and realism to a video production. They significantly affect the audience's experience. Interestingly, if a production is shot in a real location, yet is missing the everyday sounds that occur there, the audience will perceive that it is a contrived location. However, if the same scene is shot in a well-designed television studio setting, but accompanied by the appropriate sound effects, the audience can easily be convinced that it was shot on location. The barely heard sounds of a clock ticking, wind whistling through trees, birdsong, passing traffic, and the barking of a distant dog (or whatever other noises are appropriate) can bring the scene to life.

Sound effects can come from a number of sources:

- *The original sounds recorded during a scene*: For example, a person's own footsteps accompanying the picture, which may be filtered, have reverberation added, and so on.
- *Reused original sounds*: An example would include the sounds of wind, traffic, or children at play, recorded during a scene, then copied and mixed with that same scene's soundtrack to reinforce the overall effect.

MICRO FOLEY: CREATING SOUNDS WITH VERY SMALL-SIZED TOOLS

By Brant Falk

When creating projects, it is quite common to add sounds that either may have been missed during shooting or are needed to create a more interesting moment. The term micro foley was created to refer to situations where small tools are used to get the sounds you need. Most low-budget directors don't have the recording space large enough to accommodate all the tools a regular foley artist might use. It's amazing how limitations can often lead to creativity. A great strength about foley is how you can satisfy the ear's belief with audio that is not directly related to that object. This allows a lot of room for creativity. Ideally, you should have equipment that has these basic components when creating foley: pitch change, reversing (playing the audio backwards), reverb, delay, and an equalizer.

I once was creating a haunted house style project that required a lot of creepy effects. I needed ghosts and goblins to be floating around here and there, heard but not seen. One of the ways we created the creepy effects was with a paintbrush. I would slowly drag the brush over different material, from wood blocks to the back of a frying pan. Pitch shifting them down added the mass I was looking for.

Below is a small list of just a few of the micro foley tools you can use to get lots of interesting sounds. Keep in mind that with pitch changes and compositing sounds together, you can get plenty of sound effects for your project.

- **Thumb tacks and wood**: Sounds like shoes walking.
- **Retractable pen**: Sounds like a clock ticking.
- **Rubik's Cube**: When pitched down, they sound like machinery.
- **Paintbrush and foam**: Sounds like someone combing their hair.
- **Duffle bag, 5 pound weight, and clothing**: When dropped, it sounds like a body falling.
- **Zippo lighting**: Sounds like you've flipped open the hatch of a nuclear sub.

- *Foley*: Creating sounds in a studio that can replace the original sounds. For example, introducing sounds of your own footsteps for the original ones, keeping in time with those in the picture.
- *Sound effects library*: Effects from a commercial audio effects library are generally available to download online.
- *Digital processing or sound sampling*: Computer software offers a plethora of options for creating and manipulating sounds. Connected to a keyboard, these effects can be repeated and changed in an endless variety of ways.

Anticipating Sound Editing

When shooting on location, you can make eventual editing and audio sweetening a lot easier if you habitually follow certain practices:

- *Level continuity*: Aim to keep the level and quality of successive shots in the same scene reasonably similar. Particularly in location interiors, you may find that longer shots have lower sound levels and strong reverberation, while closer shots are louder and relatively dead acoustically.
- *Wild track*: It is good practice to make supplementary recordings of general "atmosphere" (background noise) from time to time. This is often referred to as NAT (natural) sound. Even in the studio, there is always a low-level background sound of air conditioning/ventilation (room tone). On location, this atmosphere may include wind in trees, passing traffic, and so on. When editing the program, there will be occasions where the soundtrack has been cut, or a sequence is muted, and these cover sounds can be introduced to avoid distracting lapses into silence and to give a feeling of continuity to the edited shots.

When recording on location, keep an alert ear for any potential sound effects that arise when you are not actually shooting, yet might be integrated into the final soundtrack. Even unwelcome intrusive noises, such as a passing fire truck, might prove useful in your sound library for another occasion.

Extra audio recording is useful, too, when shooting interviews on location. Typically, the camera and microphone concentrate on the guest, and separate shots of the interviewer are cut in during editing.

Audio Sweetening

The process of working on the program sound after the production is called audio sweetening (or a dubbing session or

track laying). Although it can be a lot cheaper to record a show "live to tape" and have a complete production ready for use at the end of the session, audio sweetening is both necessary and preferable where extensive video editing is involved.

Audio sweetening can be carried out at various levels:

- *Additional material*: Adding extra material to the finished edited program (playing title music, special effects sounds, adding a commentary, etc.).
- *Corrections*: Improving the sound within a scene. You might, for example, readjust varying levels between speakers' voices. Careful filtering could reduce hum, rumble, ventilation noise, and other issues.
- *Enhancement*: Modifying sound quality to improve realism or to achieve a dramatic effect (adding reverberation or changing its equalization). Adjusting the relative strengths of effects and music tracks to suit the dialogue and action.
- *Continuity*: Ensuring that the sound levels, balance, etc. are consistent from one shot to the next when various shots are edited together.
- *Bridging*: Adding bridging effects or music that will run between shots. An overlay track can be played throughout an entire sequence to ensure that the same background sound levels continue without level jumps or changes in quality. It may be kept down behind dialogue and made more prominent during action.
- *Overdubbing*: Replacing unsatisfactory sections of the soundtrack that were spoiled, for instance, by passing aircraft or other extraneous noise.

As you can see, there are very practical reasons to rework the soundtrack after production. Now that television makes increasing use of short takes and sophisticated video editing, it has become a regular part of the production process on complex shows.

Copyright

Whenever material prepared and created by other people is used—a piece of music, a sound recording, video recording, film, a picture in a book, a photograph, and so on—the producer/director are required to pay a fee to the copyright holders or an appropriate organization operating on their behalf for copyright clearance.

Copyright law is complex and varies between different countries, but basically it protects the originators from having their work copied without permission. You cannot, for example, prepare a video program with music dubbed from a commercial recording, with inserts from television programs,

magazine photographs, advertisements, and other sources without the permission of the respective copyright owners. The owners will probably require the payment of use fees, and these fees will depend on the purpose and use of the program. Some of the exceptions to this policy occur when the program is only to be seen within the home or used in a class assignment that will not be seen by the public. In most cases, the copyright can be traced through the source of the material needed for the production (e.g., the publisher of a book or photograph).

Agreements take various forms and may be limited in scope. A license and/or a fee may need to be paid for using the material. For music and sound effects, directors are usually required to pay a royalty fee per use; it may be possible to buy unlimited rights.

The largest organizations concerned with performance rights for music (copyright clearance for use of recorded music or to perform music) include ASCAP (American Society of Composers, Authors, and Publishers), SESAC (Society of European Stage Authors and Composers), and BMI (Broadcast Music, Inc.). When clearing copyright for music, both the record company and the music publishers may need to be involved.

Music in public domain is not subject to copyright, but any arrangement or performance may be covered by copyright. Music and lyrics published in 1922 or earlier are in the public domain in the United States. Anyone can use a public domain song in a production—no one can "own" a public domain song. Sound recordings, however, are protected separately from musical compositions. If you need to use an existing sound recording—even a recording of a public domain song—you usually have to record it yourself or license the recording.

INTERVIEW WITH A PROFESSIONAL: BRYANT FALK

Briefly define your job. I handle all aspects of audio for productions, from writing and recording voice-over to music and sound design.

What do you like about your job? I like the variety. Last week I was handling a corporate video piece and this week I'm dealing with a documentary.

What are the types of challenges that you face in your position? Surprisingly, one issue these days is technology. Because it is moving so fast, there is constantly something on the market that I need to be aware of. Also, the formats are constantly changing. It's important to stay on top of all the audio formats available.

FIGURE 14.42
Bryant Falk, audio mixer engineer.

How do you prepare for a production? I try to get as much information on the project as possible. It's more difficult than you think. The director is so busy with other parts of the project he or she barely has time to talk. Also the "Flux Factor," at least that's what I like to call it. It's the continual changes that are occurring while the production is in motion. From, say, deleting a scene and then putting it back in, to adding a voice-over that never existed before, etc. Next, I immediately start an SFX folder and begin collecting audio. If it's a horror project, I hunt down creaky doors and record them. I'll go to a friend's house and sit in the backyard to record the crickets. I like to use as much fresh sound as possible. I try to only use sound effects libraries when I'm in a pinch. If I'm given a script, I will look through all the scene locations and begin to collect the SFX.

What suggestions or advice do you have for someone interested in a position such as yours? If you really want to get into audio production, make sure to understand all the other aspects of production as well. For example, be as knowledgeable as possible about the editors' software; this will make it easier to give them what they need. Also intern with people that are doing what you like to do so you get a firsthand glimpse as to what you are in for.

Bryant Falk has been a producer and audio engineer for video projects for clients such as MTV, Coca-Cola, and Sports Illustrated. He is the owner of Abacus Audio, located in New York City.

REVIEW QUESTIONS

1. How are acoustic reflections (or echoes) reduced?
2. How is the speaker setup different between stereo and surround sound (5.1)? How many speakers are involved?
3. What are three microphone characteristics, and how do each of these affect the audio recording process?
4. How is a super-cardioid microphone used for television production?
5. Review the different types of microphones and specify the type that you would use to obtain a quality recording of a piano. Explain your choice.
6. What are some of the different situations where a type of lavalier mic can be used?
7. What are the advantages of a wireless mic?
8. How is a soundtrack designed (built)?
9. When do you need to be concerned about copyright?

CHAPTER 15

RECORDING THE STORY

"There is absolutely no need or reason at this time to believe
that 8K is for home television. With 4K, we have reached the
limit of display resolution that can reasonably be observed in a
home television environment."

—Tore B. Nordahl, Television Consultant

"Don't become too emotionally attached to a medium for its
own sake—there's no future in it."

—Leonard Guercio, Producer, Director, and Educator

TERMS

Codec: A device or software that enables compression or decompression of digital video.

Digital recording: The digital system regularly samples the waveforms and converts them into numerical (binary) data. This allows many generations of copies to be made without affecting the quality of the image.

Flash memory: Flash cards can store large amounts of digital data without having any moving parts. This makes them durable and able to work in a variety of temperatures, and data can be easily transferred into a nonlinear editor.

HDD: Hard disk drives can be used for recording digital video images and can be built into the camera, attached to the outside of the camera, or an entirely separate recorder.

Interlaced scanning: The television's electrons scan the odd-numbered lines first and then go back and scan in the remaining even-numbered lines. The "i" in 1080i stands for interlaced.

Progressive scanning: Also called sequential scanning; uses an electron beam that scans or paints all lines at once. Note that the "p" in 720p/1080p stands for progressive.

What is the best video recording format? That can be a complicated question to answer. While there are high-level formats available, such as 4K and 8K, it is not realistic unless you are shooting for the large screen. Some formats may have an incredible quality but the cameras are too large for a one-person investigative reporter. The bottom line is that the best format varies for different people and situations. You have to weigh the costs involved, size of the equipment, portability, size of the crew, amount of data storage needed, and the situation that you will be covering, and then go for the highest-quality format that works with your specific requirements and situation. Sometimes, it is not an easy decision to make.

VIDEO RESOLUTIONS

Fortunately, there is continual development in the design and format of video and audio recording systems. Some are mainly used for acquisition (shooting original material); others are for postproduction editing and archiving (storage) work. Recordings can be made on flash memory cards, hard drives, discs, and some are still recorded on videotape. Some cameras can record on flash memory cards, hard disks, and tape—all in one device. Traditionally, videotape was the most popular medium. However, with the advent of low-cost, high-capacity, and non-mechanical flash memory cards, tape is rapidly disappearing.

Standard-Definition Television

Standard-definition television (SD or SDTV) is what was used around the world before HD came onto the scene. However, most of the world is now in the process of moving from SDTV to some type of HD format. SD generally refers to analog and digital broadcasts that scan the image at 480i (NTSC) or 576i (PAL). Although most SD signals are broadcast in a 4:3 aspect ratio, it is also possible to transmit them with a 16:9 aspect ratio as well.

High-Definition Television

"Today's television is so much sharper that more attention must be given to the small details. Since the flaws in the background, or even makeup, can hold the audience's attention, directors may not need as many close-up shots, which previously was the only way details could be shown."

—Brian Douglas, Olympics Producer
and Director

High-definition television, known as HD or HDTV, is the most popular format in the world. Although some areas of the world are still working with standard definition, HD is the most common type of television. The advantage it has over the larger standards is that most homes have an HD television set, production equipment is readily available, and it costs less to transmit the HD image since it takes much less bandwidth. There are two types of HD scanning systems currently being used:

- **Interlaced**: Interlaced scanning means that the television's electron scans the odd-numbered lines first and then goes back and scans in the remaining even-numbered lines. The "i" in 1080i stands for *interlaced*.
- **Progressive**: Also called sequential scanning, progressive scanning uses an electron beam that scans or paints all lines at once. The HDTV 720p and 1080p systems use progressive scanning. Note that the "p" in 720p/1080p stands for *progressive*. The progressive image displays the total picture.

Although there are a number of different formats within HD (720p, 1080p, 1080i, etc.), it is very difficult for the untrained eye to see the difference between them. There were fairly significant differences when the formats first came out (such as motion blur and flicker), but today, as technology has improved, the majority of the bugs have been fixed. All of the HD formats include a 16:9 aspect ratio. There are also multiple frame rates to choose from that give a "film look," a "video look," or something in between. Here are the current HD formats with some data that can be used for comparison:

- **720p**: This is a progressive scanning format that has a pixel aspect ratio of 1,280 x 720 and 921,600 pixels per frame. The progressive scanning gives it a little bit of a film look.
- **1080i**: This format utilizes interlaced scanning. It has a pixel aspect ratio of 1,920 x 1,080 and 1,555,200 pixels per frame. 1080i gives a very high-quality video look.
- **1080p**: This is the newest progressive scanning format and has a pixel aspect ratio of 1,920 x 1,080 with 2,073,600 pixels per frame. It was designed to compete with film cameras.

The *high definition* of HDTV is currently highly dubious. Depending on the content delivery system, it can look amazing or pitiful. An HD network on some cable systems looks astounding, because network executives cut carriage deals that dictate a minimum bit rate. Many networks don't have the leverage to negotiate the same bit rate."

—Deborah McAdams, Senior Editor,
Television Broadcast

FIGURE 15.1 *This photo is a single frame from the digital video 4K film Crossing the Line by director Peter Jackson. Note the incredible image clarity. The CMOS sensors in the RED camera were 12-megapixel.*

Source: Photo courtesy of Red Digital Cinema.

4K Television

4K, also known as "Ultra High Definition," "Ultra HD," or SHV-1, is defined as at least 3,840 horizontal and at least 2,160 vertical pixels, with an aspect ratio of at least 16 × 9. 4K may be the next home format for your television. However, while 4K's imagery is stunning and manufacturers are heavily pushing the format, it will take a while for broadcasters to decide if and when they will make the required technological changes, which will be expensive (Figure 15.1). Broadcasters are also watching to see if consumers care enough to purchase 4K televisions for their homes. At this point, some broadcasters are "future-proofing" their productions by shooting the events in 4K so that they can archive their productions in a high-quality format for future use as technology improves. Today's 4K productions are often then converted to HD for use on today's popular systems.

At this point, no one seems to know if home television will stop at 4K. The BBC's head of technology, Andy Quested, believes "4K, at best, will only be an interim standard on the way to 8K."

8K Television

Super Hi-Vision, also known as SHV-2 or 8K television, was created by the Japanese broadcaster NHK. It is 16 times more detailed than HD, with 33 million pixels compared to HD's

4K MOBILE PHONES

4K video devices are rapidly becoming available for consumers. This Samsung phone can shoot in 4K and has a 2K viewing screen. Manufacturers hope that consumer devices such as this will help build the demand for 4K programming on television.

FIGURE 15.2 *Mobile phone with a 4K camera.*

Source: Photo courtesy of Samsung.

2 million (Figure 15.3). The design concept is that it has higher quality than the human eye, so that no pixels are visible. It also has a frame rate of 120 frames/second, so that movement is very smooth and realistic. Scenes in low light are also incredibly well-rendered, while the surround sound follows motion around the screen and will make you look over your shoulder to see what's going on.

The Super Hi-Vision system has three-dimensional 22.2 surround sound, with three vertical layers of front speakers, two layers of surround speakers, and two subwoofers for bass frequencies.

Many broadcasters are very hesitant about moving toward 8K. The resolution is so high that the minimum screen size is considered to be around 84 inches in order to see that quality. Ideally, home televisions would be 100–200 inch screens, with the audience sitting 3–6 feet away from the television. This ensures that the audience's peripheral vision is filled with the 8K image. While it has incredible imagery for the theater, the thought is that very few homes will have 8K televisions.

A comparison of the various recording formats can be found in Figure 15.4.

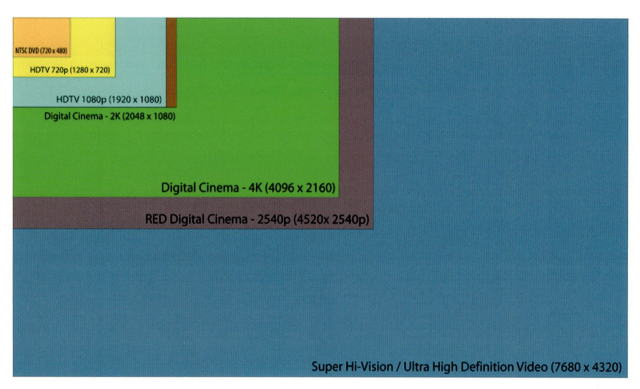

FIGURE 15.3 *This image compares the quality of the various video formats. As the number of scan lines is increased, the image quality increases proportionally.*

Comparison of HD, 4K and 8K Formats

	HD	4K	8K
Resolution (pixels)	1920 wide x 1080 high	3,840 wide x 2,160 high	7,680 wide x 4,320 high
Total pixels	2 million	8 million	33 million
Frame rate	50 or 60 frames/second	50 or 60 frames/second	120 frames/second
Audio channels	5.1	5.1	22.2

FIGURE 15.4 *A comparison of the HD, 4K, and 8K video formats.*

8K AT THE LONDON OLYMPICS

"All the Hi-Vision equipment is now reliable enough to use in the everyday environment and this has been achieved in 10 years as opposed to the 36 years it took to develop HDTV."

—Dr. Keiichi Kubota, Director-General, Science & Technical Research Laboratories, NHK.

NHK, BBC, and Olympic Broadcast Services teamed up for a test and demonstration of 8K television (Super Hi-Vision) during the 2012 London Olympics. The transmissions were without commentary and virtually devoid of graphics as the partners worked toward building an immersive experience.

FIGURE 15.5 *NHK's (Japan) 8K production truck covering the London Olympics.*

3D TELEVISION

At this point, 3D has not been successful with television. It has been very difficult for networks around the world, who have committed extensive amounts of money to experiment and attempt to make it work. Producers and directors have struggled with determining the right depth of field, and they seem to overload the viewer with detail. Compression technologies have also added to the difficulties; broadcasting 3D over HD networks dropped the resolution. Besides the production difficulties and the increased cost of production, the biggest problem has been the required use of 3D glasses. The audience just does not like them. Another difficulty has been the lack of 3D content.

3D for most companies has gone into a type of hibernation. Most believe that it is not gone for good. Some broadcast networks believe that 3D will be successful when 4K 3D is available without requiring 3D glasses. Television manufacturers continue to explore new ways to provide glassless 3D television. While there are a number of types being experimented with, none of them have been produced with both the expected quality and affordability.

FUTURE TELEVISION?

No one quite knows for sure where television is going. The general consensus is that there is no reason to go to a higher quality than 8K because no one would be able to tell the difference. Some manufacturers are pursuing 4K and 8K 3D without glasses. Others are pursing the possibility of transmitting full-size holograms. Atos Singapore spokesman Gregoire Gillingham told *CIO Australia* that from a technical perspective, holographic projection technology is developing rapidly: "We predict that it will be possible to show holograms in a stadium within 10 to 15 years and the concept of a 'live' event being projected via holograms into other stadiums filled with spectators to be a realistic prediction."

Wherever it goes in the future, it will be interesting to see.

VIDEO RECORDING MEDIA

With the advent of digital recording, the divisions between consumer and professional formats significantly blurred. Productions shot on the consumer formats began showing up on television networks and film festivals and winning top awards. The following is a summary of today's most popular format.

Videotape

While some professionals are still using tape, it is quickly disappearing. As equipment is being replaced, consumers and professionals are buying tapeless cameras. Videotape maintained its popularity due to its ease of availability, relatively low cost, the sheer capacity of tapes for recording and storage, and the infrastructure that was built around it. For example, many companies built, over the years, a tape-based infrastructure that will take time to move away from (Figure 15.6). There are still a significant number of companies who archive their productions on videotape.

One of the disadvantages of tape is the sheer number of incompatible videotape formats. Tapes can be recorded and reproduced only on equipment that uses identical standards. Design features vary considerably between video recorders. Tapes come in all different widths, the cassettes have different sizes, some systems are digital and others are analog, and there is an incredibly wide range of image quality between the various formats.

Consumer tape-based cameras have all but disappeared. There are still a few being sold but they are only usually available for someone who wants to stay with an old standard.

Professional Digital Tape Formats

Manufacturers are building fewer and fewer of these professional tape formats. However, many are still being used around the world. Tape is also still used as a backup recording media. The following videotape formats are still currently being used by some professionals: Digital Betacam, HDCam, and DVCProHD. All of these formats use 1/2" videotape (Figure 15.7).

Flash Memory

Flash memory has quickly become the most popular recording medium for video. A significant advantage of the flash memory card is that it is very easy to transfer files from the card to a nonlinear editor. The small size of the card allows for very compact camcorders. Cameras using flash memory as their medium generally do not have moving parts and should thus require less maintenance.

One of the advantages of flash memory card cameras is that they can have multiple slots, which are "hot-swappable." This means that while one is being recorded onto, an already full card can be removed during the recording process and replaced with a blank card. This feature allows uninterrupted recording. There are a wide variety of flash memory cards. Figures 15.8–15.12 show some of these flash memory systems.

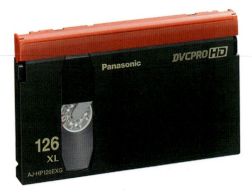

FIGURE 15.6 *Videotapes are still in use by some professional broadcasters.*

Source: Photos courtesy of Panasonic.

FIGURE 15.7 *Many companies invested heavily in a tape-based infrastructure. Moving toward tapeless production takes time.*

Source: Photo by Jon Greenhoe.

FIGURE 15.9 *A CompactFlash (CF) memory card.*

FIGURE 15.8 *SD flash memory card.*

FIGURE 15.10 *Some Sony and JVC devices use an SxS memory card.*

Source: Photo courtesy of JVC.

FIGURE 15.11 *This P2 card player/recorder can be used to record, edit, play back, and play in slow motion. Note that there are six hot-swappable slots.*

Source: Photo courtesy of Panasonic.

FLASH MEMORY DEVICES

Figure 15.12 shows a comparison of some of the most popular flash memory devices. There are variations of some of the cards. For example, there may be micro size, extended capacity, or writing speeds for each device listed below. When purchasing, you must look at each specific manufacturer's specifications.

FIGURE 15.12

Flash memory media comparison.

Flash Memory Media

Memory Device	Storage Capacity*	Write Speeds**
SD Card	Up to 512GB	Up to 250 Mb/s
CompactFlash (CF) Card	Up to 512GB	Up to 155 Mb/s
XQD Card	Up to 128GB	Up to 350 Mb/s
SxS Card	Up to 128GB	Up to 2.8 Gb/s
P2 Card	Up to 64 GB	Up to 1.2 Gb/s
SSD	Up to 1TB	Up to 520 Mb/s
Pro-Duo	32GB	Up to 50 Mb/s

*Card size currently available. These sizes can theoretically increase significantly.
**Specific card formats are often available in a variety of transfer speeds and storage capacities.

Hard Disk Drive/Internal Memory

Hard disk drive (HDD) cameras record directly to a hard drive (or sometimes flash memory) built into the camera (Figure 15.13). Roughly 4 GB of disk space is required for each hour of video. Some of these compact HDD cameras have as much as 160 GB of hard disk storage. Many of the HDD cameras also include an SD slot for video recording to a transportable medium, although this feature is not required to transfer footage. The card makes it extremely easy to transfer the data to a nonlinear system.

External Camera Hard Drives

External camera hard drives can now be attached to most digital cameras. Some of the drives have an internal memory while others utilize flash memory cards. These drives can provide extremely long recording times depending on the amount of internal memory or the type of card they utilize. The drives can connect directly to nonlinear editing systems, allowing the editor to begin editing the program immediately. Most of the drives attach to the camera via an HDMI port. Audio, time code, and video and control information are passed directly through the connector (Figure 15.14).

Hard Disk Recorders

Stand-alone hard disk recorders are now used to record video at a very high quality. They utilize a variety of recording media, from flash memory cards to servers. Some of these real-time recorders/players include smooth fast-motion and slow-motion playback. They are replacing tape decks (Figures 15.15 and 15.16).

FIGURE 15.15 *Hard disk recorders have gained popularity in the professional video field. The recorders often include color monitors and multiple inputs/outputs.*

Source: Photo courtesy of BlackMagic Design.

FIGURE 15.13 *This high-end consumer-targeted camcorder includes an internal HDD.*

Source: Photo courtesy of JVC.

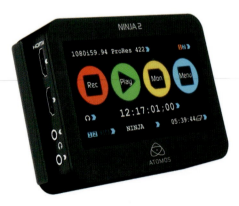

FIGURE 15.14 *External drives can be attached to many digital video cameras. Some of these drives include high-resolution monitors, dual recording media slots, and even waveform monitors.*

Source: Photo courtesy of Atomos.

FIGURE 15.16 *A professional server-based EVS recording/playback system. These systems offer slow-motion replay and at the same time can record as many as six separate channels of video.*

Recording Media Care

It is important to care for recording media. Here are some suggestions for prolonging the life of the various types of recording and storage media:

- The optimum storage temperature is around 65 degrees. Above 100° F and below 14° F can cause problems with some media.
- Avoid rapid temperature/humidity changes (such as moving from a cold exterior to a warm interior) and allow both media and equipment to become climatized before use.
- Before loading a medium, confirm that it isn't protected against recording (i.e., reposition the safety switch). Make sure that you're not recording over something that should be retained!
- Store media in their protective boxes to avoid damage and dust.
- Make sure that each recorded medium is clearly identified on the label (name, contact information, shot/scene numbers, etc.).

Video Recording Suggestions

Here are some suggestions to consider when recording to video media:

- Use the highest data rate possible on digital media other than tape. Although this will not allow you to record as many minutes on the medium, it will give you a higher-quality image.
- Watch the elapsed time on the camera to make sure that you know how much recordable media is left and that you know the state of the battery.
- Review the end of the takes to check whether the recording is satisfactory.
- When the medium is taken out of the camera, make sure that the protection device is implemented so that someone does not accidentally record over your original footage.
- Clearly label all media and the media container.

INTERVIEW WITH A PROFESSIONAL: ROBIN BROOMFIELD

FIGURE 15.17
Robin Broomfield, technical manager, Sky Sports.

Briefly define your job. I work with new technologies, processes, and workflows for Sky. I helped develop Sky's 3D channel and am now testing and evaluating 4K for Sky.

What do you like about your job? I have really enjoyed being involved with pioneering new and very different areas (3D and 4K) for the future television audience—it is a privilege to be involved with such an exciting project. TV is a fantastic industry to work in—it gives you access to events, places, and people that you wouldn't have in many other careers—not just in sport events, but in entertainment too.

What are the types of challenges that you face in your position? Starting, developing, and implementing new technologies such as 3D and 4K for a network is a huge challenge because it had never been done before. Equipment and systems were new, often under development, or in some cases just not there at all.

How do you prepare for a production? A site survey is necessary, especially as different equipment, camera positions, and production requirements are often employed. Suitable time must be put aside for rigging, testing, and rehearsing; this will guarantee that a good production is achieved.

What suggestions or advice do you have for someone interested in a position such as yours? To work in television and on big projects needs a dedication and enthusiasm that will drive you through many long hours, sometimes in unpleasant weather and surroundings. "Television is a lifestyle, not just a job." Working hard to get the best results and giving time to fine detail in planning and production is essential to achieve good results. Look at what you do critically and find ways to improve, be innovative in what you do; if you think that something is "good enough" then it probably isn't!

Robin Broomfield is technical manager for Sky Sports.

INTERVIEW WITH A PROFESSIONAL: RYAN HAMMER

How do you decide what video format to use when working on a production? Many people look entirely at cost of camera rentals and media cost. I look at it more from a postproduction perspective; how much is it going to cost to digitize/ingest, are there editing problems with certain video formats, and what issues will we have in the online process?

Do you still use videotape? Definitely. There's something about having a piece of tape to hold onto that helps me sleep at night. Plus, "tapeless" hasn't been ironed out (in the post process), so there are still issues that just *go away* when you use tape instead.

Where do you think editing is headed in the future? With the competition between Avid, Premiere and Final Cut Pro, prices are dropping. I personally think that people will be owning their own systems. There will be buildings where bays, storage, and recorders are kept and maintained by assistant editors. Editors will primarily work from home—either off their own systems, or remotely logging into the edit systems at the facility. Cuts will be streamable to executives for notes, and when they're supposed to be output, they will be emailed to the assistant editor at the facility; at least, that's what I'm working on.

Ryan Hammer is a partner in a major postproduction house in Los Angeles.

FIGURE 15.18
Ryan Hammer, Atlas Digital.

REVIEW QUESTIONS

1. What are three differences between HD and 4K television?
2. What are two of the advantages of flash memory?
3. Why has 3D struggled to become sustainable on television?
4. Give two ways that you can care for recording media?

CHAPTER 16

EDITING THE PRODUCTION

"The three great editing tools are context, contrast, and rhythm."
—Jay Ankeney, Editor

TERMS

Clip: A video segment.

Cut/take: An instantaneous switch from one shot to another.

Dissolve: An effect produced by fading out one picture while fading in another.

Fade (a type of dissolve): A gradual change between black and a video image. For example, at the end of a program there is usually a "fade to black" or, if there is a "fade-up," it means that the director is transitioning from black to a video image. A slow fade suggests the peaceful end of action. A fast fade is rather like a "gentle cut," used to conclude a scene.

Linear editing: The copying, or dubbing, of segments from the master tape to another tape in sequential order.

Logging: Loggers view the footage and write down the scene/take numbers, the length of each shot, time code, and descriptions of each shot.

Nonlinear editing: The process where the recorded video is digitized (copied) onto a computer. The footage can then be arranged and rearranged, special effects can be added, and the audio and graphics can be adjusted using editing software.

Postproduction: Taking the video that was previously shot and assembling a program shot by shot, generally with a computer-based editing system.

Timeline: Usually includes multiple tracks of video, audio, and graphics in a nonlinear editing system.

Wipe: A novel transition that can have many different shapes.

Selecting the right image and skillful editing will make a vital contribution to a production's impact. The way the shots are interrelated will not only affect their visual flow, but will also directly influence how the audience reacts to what they are seeing: their interpretation and their emotional responses. Poor editing can leave them confused. Proficient editing can create interest and tension or build up excitement that keeps the audience on the edges of their seats. At worst, poor image selection can degenerate into casual switching between shots. At best, editing is a sophisticated persuasive art form.

EDITING TECHNIQUES IN TELEVISION

There are roughly two broad categories of editing:

- Live editing: The director, using live cameras and other video sources, "live edits" (directs) a production using a video switcher (Figure 16.1).
- Postproduction: Taking the imagery that was previously shot and assembling a program shot by shot, generally with a computer-based editing system (Figure 16.2).

FIGURE 16.2 *Postproduction editing.*

One style involves a director with cameras and the other style involves a director with prerecorded programming. However you do it, it is still editing—you are deciding which shot the audience will see.

FIGURE 16.1 *Live editing occurs when a director is cutting a show together from multiple cameras.*
Source: Photo by Morgan Irish.

EDITING BASICS

Editing Decisions

During the editing process, a series of decisions need to be made:

- *Which of the available shots do you want to use?* When directing (editing) a live show, choices are irrevocable. You can select only from the shots being presented at each moment by cameras, prerecorded material, graphics, and so on. When you are editing in postproduction, there is time to ponder, to select, to reconsider, and to redo if needed.
- *What is the final shot sequence?* The relative durations of shots will affect their visual impact.
- *At exactly which moment in the action do you want to change from one shot to the next?*
- *How will each shot transition to the next shot?* Transitions include cut, dissolve, wipe, fade, and others.
- *How fast or slow will this transition be?*
- *Is there good continuity between pictures and sound that shows continuous action?* These elements may not have been shot at the same time or places.

Each of these decisions involves you making both a mechanical operation and an artistic choice. Even the simplest treatment (a cut from one image to the next) can create a very different effect, based on the point at which you decide to edit the action. Let's look at an example:

- You can show the entire action, from start to end:

    The intruder reaches into a pocket, pulls out a pistol, and fires it. The victim falls. (The action is obvious.)

- You can interrupt an action, so that we do not know what is going to happen:

    The hand reaches into the pocket/CUT/to the second person's face. (What is the intruder reaching for?) Or the hand reaches into a pocket, and pulls out a pistol/CUT/to the second person's face. (Is the intruder threatening, or actually going to fire it?)

- You can show the entire action, but hold the audience in suspense about its consequences:

    We see the pistol drawn and fired/CUT/but did the shot miss?

Editing Possibilities

Let's take a look at how editing techniques can contribute to the success of your production:

- You can join together a series of separately recorded takes or sequences to create a continuous smooth-flowing storyline—even where none originally existed.
- Through editing, you can remove action that would be irrelevant or distracting.
- You can seamlessly cut in retakes to replace unsatisfactory material—to correct or improve performance, to overcome camera, lighting, or sound problems, or to improve ineffective production treatment.
- You can increase or reduce the overall duration of the program—by adjusting the length of sequences, introducing cutaway shots, altering playing speed, or repeating strategic parts of an action sequence.
- Stock shots can be inserted and blended with the program material—to establish location, for effects, or to introduce illustrations.
- When a subject is just about to move out of shot, you can cut to a new viewpoint and show the action continuing, apparently uninterrupted. (Even where it is possible to shoot action in one continuous take, the director may want to change the camera viewpoint or interrupt the flow of the action for dramatic impact.)
- By cutting between shots recorded at different times or places, you can imply relationships that did not exist.
- Editing allows you to instantly shift the audience's center of interest, redirecting their attention to another aspect of the subject or the scene.
- You can use editing to emphasize or to conceal information.
- You can adjust the duration of shots in a sequence to influence its overall pace.
- Editing can change the entire significance of an action in an instant—to create tension, humor, horror, and so on.
- By altering the order in which the audience sees events, you can change how they interpret and react to them.
- Audio sweetening (adding additional sounds to the recording) can be done.
- Graphics can be added.
- Special effects can be added.

THE MECHANICS OF EDITING

The actual process you use can have an important influence on the ease and accuracy with which you can edit and on the finesse that is possible. There are several systems, described in the following sections.

Editing In-Camera

It is possible to edit in-camera. To do so, most cameras require that you shoot the action in the final running order, which may not be practicable or convenient. There are even cameras that will allow limited in-camera editing, such as trimming a scene or changing the order of the clips. Some of the latest smartphones and tablets actually have the ability to shoot and edit HD and 4K videos. However, cameras have varying abilities, and all of them have more limitations than editing on an actual editor (Figure 16.3).

Production Switcher (Vision Mixer)

The live-edit method of using a production switcher, to combine (cut, dissolve, wipe, fade, etc.) video sources such as cameras, video players, or graphics is used in many productions. During a production, conditions in most production control rooms are generally a world apart from the relative calm of an editing suite (Figure 16.4).

The director needs to watch the monitor wall, which is made up of small monitors showing images from cameras, video feeds, video players, graphics, and any other sources. There are also two larger monitors, a "preview" monitor and a "program" or "on-air" monitor. The director uses the preview monitor to review any video source before going to it. The program monitor shows the final program output (see Chapter 3 for more details about the control room).

While all of this is going on, the director's attention is divided between the current shot in the program monitor and upcoming shots—guiding the production crew by instructing,

FIGURE 16.3 *Some mobile phones have the ability to both shoot and edit video. Although there are limitations to the editing functions, it can still be a valuable resource for doing a rough cut of the video.*

correcting, selecting, and coordinating their work. Of course, this is all done while the director is also checking the talent's performance, keeping production on schedule, and dealing with issues as they arise. It is no surprise that under these conditions, "editing" with a production switcher can degenerate into a mechanical process (Figure 16.1).

Linear Editing

Linear editing involves "dubbing" or copying the master tape to another tape in a sequential order. This process worked well for editors until the director or client wanted significant changes to be made in the middle of a tape. With a linear tape, that usually meant that the whole project had to be entirely re-edited; this was incredibly time-consuming and frustrating. Analog linear editing also did not work well if multiple generations (copies of copies) of the tape had to be made, as each generation deteriorated a little more. Linear systems are generally made up of a "player" and a "recorder" along with a control console. The original footage is placed into the player and then is edited to the recorder (Figure 16.5). Although some very limited segments of the television industry are still using linear editing, the majority of television programming today is edited on a nonlinear editor.

Nonlinear Editing

Today, almost all video and television programs, as well as films, are edited on a *nonlinear editor*. Nonlinear editing is the process in which the recorded video is stored on a computer's hard drive. The footage can then be arranged, rearranged, special effects added, and the audio and graphics can be adjusted using editing software. Nonlinear editing systems make it very easy to make changes such as moving video and audio segments around until the director or client is happy. Hard disk and memory card cameras have allowed editors to begin editing much more quickly, as they do not need to digitize all of the footage. Nonlinear systems cost a fraction of what a professional linear editing system does. Once the edited project is complete, it can be output to whatever medium is desired.

An Overview of the Nonlinear Process

- *Step 1*: Store the footage on flash memory or a hard drive that can be accessed by the editing computer.
- *Step 2*: Each video segment or *clip* is then *trimmed* (cleaned up) by deleting unwanted video frames.

FIGURE 16.4 *A small "laptop" video switcher can be used with four-camera productions.*

FIGURE 16.5 *Linear editing, or copying the contents of one tape to another tape, one clip after another linearly, is still used on a limited basis. Although the use of linear editors has significantly reduced, segments of the industry, such as news, still use them.*
Source: Photo by Jon Greenhoe.

- *Step 3*: The clips are placed into the *timeline*. The timeline usually includes multiple tracks of video, audio, and graphics. This timeline allows the editor to view the production and arrange the segments to fit the script (Figure 16.6).
- *Step 4*: Video special effects and transitions are added. Nonlinear edit systems allow all kinds of effects, such as ripple, slow/fast motion, color correction, and others. Transitions include dissolves, cuts, and a variety of wipes.

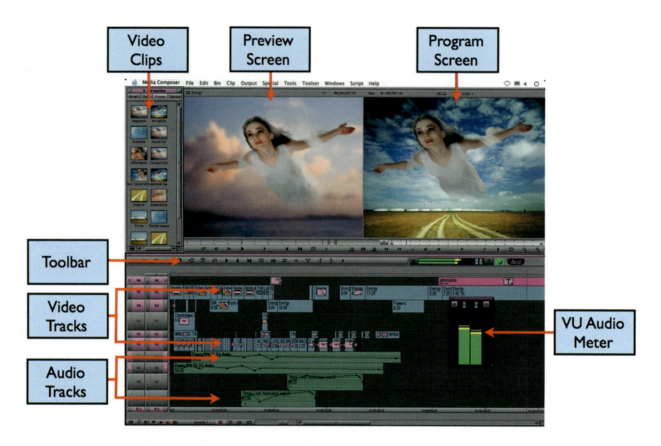

FIGURE 16.6 *Screenshot showing the composition page of a nonlinear editor. The video clip bin is where video clips that are to be used in the program are stored. The preview monitor allows the editor to preview the video clips before moving them to the audio and video timeline. The program monitor allows the editor to see the audio and video in the timeline.*

Source: Photo courtesy of Avid.

FIGURE 16.7 *A nonlinear editing suite with an adjoining sound booth for voice-overs.*

- *Step 5*: Additional audio may or may not be added at this point. Audio effects may be used to "sweeten" the sound. Music and/or voice-overs may be added at different points in the project (Figures 16.7 and 16.8).
- *Step 6*: The final program is output to the distribution medium.

Nonlinear Editing Equipment

Editing equipment has drastically changed over the last decade. Where once a minimal editing system required two editing decks, two monitors, and an edit controller, today the equipment can be as simple as a camcorder and a computer with an editing software package installed.

Higher-level editing suites may contain multiple types of input devices using a variety of different connectors to transport the data at a much faster speed. They may also include multiple edit screens, speakers, an audio mixer, and other tools (Figure 16.9).

Video can be imported into the computer from a camcorder, deck, or a memory storage device. One cable can move large amounts of data, as well as control signals, between the camera and computer.

FIGURE 16.8 *The talent is doing a voice-over in an editing suite for a news story at a local news station.*

Source: Photo by Jon Greenhoe.

FIGURE 16.9 *A large nonlinear editing system.*

HABITS OF A HIGHLY EFFECTIVE POSTPRODUCTION EDITOR

1. **Schedule enough time to make a good edit.** Quality editing takes time. Be realistic, and then pad your schedule with a little extra time. It is always better to be done a little early than late. Plus, you can always use a little more time to refine the edit.
2. **Get a little distance from the project occasionally.** It is easy to become emotionally involved with a specific element of a project. Take a break from it; when you come back, your perspective may have changed. Ask others for their opinion—there is a good chance that they will see things that you didn't see.
3. **List the issues before fixing them one by one.** It is good to come up with an organized plan for editing the project. Although it takes time to think it through, it is worth it.
4. **Know the priority of your editing elements.** The most important editing elements are the emotion and story. If you lose those two elements, you lose the production.
5. **Keep a copy of each edited version.** Each time you make changes to the project, keep the original (or previous) version. That way, you have something to go back to if you run into problems.
6. **Focus on the shots that you have.** By the time you sit down to edit, it is time to get the best project that you can possibly get from the footage you have recorded. You may even have to forget about the script.

(Adapted from Mark Kerin's *Six Habits of Highly Effective Editors*)

POSTPRODUCTION LOGGING

An often-neglected important aspect of the postproduction process is logging the recorded material. Logging saves time, which also translates into budget, during the actual editing process, as it can be completed before the editing session. After logging the footage, the editor can then move the specific clips that will be used in the program instead of taking time on the editor to search through all of the clips. By moving specific clips instead of all of the footage, it also saves hard drive space. Generally, some type of log sheet is used on which notes can be written that include the time code (the address where the footage is located), scene/take numbers, and the length of each shot. The notes may also include a description of the shot and other comments such as "very good," "blurry," and so on. Logging can be simple notes on a piece of paper or can be done with logging software. An advantage to some of the logging software is that it can work with the editing software, importing the edit decisions automatically into the computer (Figures 16.10 and 16.11).

Shots can be identified for the log a number of different ways:

- **Visually**: "the one where he gets into the car."
- **By shooting a *slate*** (*clapboard*) before each shot, which contains the shot number and details (or an inverted board, at the end of shots) (Figure 16.12).
- **By *time code***: a special continuous time signal throughout the tape, showing the precise moment of recording.

The Art and Techniques of Editing: Multiple Cameras and Postproduction

Directors edit by:

1. Selecting the appropriate shots (camera shots or prerecorded video).
2. Deciding on the order and duration of each shot.
3. Deciding on the cutting point (when one shot is to end and the next to begin).
4. Deciding on the type of transition between shots.
5. Creating good continuity.

Let's look at these points in more detail.

Selecting the Appropriate Shots

Multi-Camera Editing

The director needs to review the available shots on the monitors and determine which one works best to tell the story (Figure 16.13).

Digital Video Log Sheet

Title:	
Overview:	

Tape ID #:		Location:	
Date shot:		Camera/gear:	
Filmed by:		Job #:	

Page # _____ of Total Pages _____

Scene #	Description	In hr/min/sec	Out hr/min/sec	Quality A–F	Notes

DV Log © Copyright The Avanti Group, Inc but is FREE to Use and Distribute as is

FIGURE 16.10 *Sample of a log sheet.*

Source: Courtesy of the Avanti Group.

FIGURE 16.11

Logging can be done on paper or via software. Here, a camera is connected directly into the computer to capture still frames from each clip and automatically import time code ins and outs. The screenshot shows the stored thumbnail frame, duration, and description.

Source: Photos courtesy of Imagine Products.

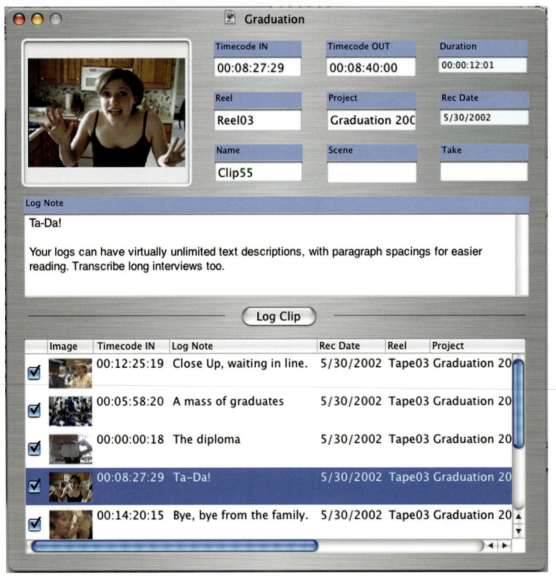

FIGURE 16.12
Slates, or clapboards, are often used to identify each shot taken. The numbers on the slate are transferred to the logging sheet.

Postproduction Editing

It is a normal practice to shoot much more material than can be used on the final video. As the video can be immediately checked for quality, the director knows when the needed material has been captured. When the shooting is finally complete, it is time to review the footage. Generally, the following shots are found:

FIGURE 16.13 *This sitcom director, sitting out on the set, is deciding which of the cameras best communicates the story.*

- Good shots that can easily be used.
- Shots that cannot be used due to defects or errors of various types.
- Repeated shots (retakes to achieve the best version).
- Redundant shots (too similar to others to use).

So the first stage of editing is to determine which of the available video should be used. Once the shots are chosen, the next step is to decide on the order in which they will be presented.

The Order of Shots

To edit successfully, the editor must imagine being in the position of the audience. He or she is seeing a succession of shots, one after another, for the first time. As each shot appears, the editor must interpret it and relate it to previous shots, progressively building up ideas about what he or she is seeing.

In most cases, the shots will be shown in chronological order. If the shots jump around in time or place, the result can be extremely confusing. (Even the familiar idea of "flashbacks" works only as long as the audience understands what is going on.)

When a series of brief shots are cut together, the fast pace of the program will be exciting, urgent, and sometimes confusing. A slow cutting rhythm using shots of longer duration is more gentle, restful, thoughtful, and/or sad.

In most circumstances, you will find that the order in which the series of shots are presented will influence your audience's interpretation of them. Even a simple example shows the nuances that easily arise: a burning building, a violent explosion, men running toward an automobile. Altering the order of these shots can modify what seems to be happening:

- Fire—automobile—explosion: Men killed while trying to escape from fire.
- Fire—explosion—automobile: Running from fire, men escaped despite explosion.
- Automobile—explosion—fire: Running men caused explosion, burning the building.

Not only is the imagination stimulated more effectively by implication rather than direct statements, but indirect techniques overcome many practical difficulties.

Suppose you join two shots: a boy looking upward, and a tree falling toward the camera. One's impression is that a boy is watching a tree being felled. Reverse the shots and the viewer could assume that the tree is falling toward the boy, who, sensing danger, looks up. The actual images might be totally unrelated—they're just a couple of shots from a stock library.

Where Should the Edits Be Made?

The *moment* chosen for a cut affects the visual flow of the program.

Directors usually transition at the following points:

- At the completion of a sentence, or even a thought.
- When the talent takes a breath.
- Whenever a reaction or clarifying shot is needed.
- About a third of the way into an action, such as standing up (this is a rule of thumb that can be broken).

If the first shot shows a man walking up to a door to open it, and the second shot is a close-up of him grasping the handle, the editor usually has to make sure that there is:

- No missing time (his arm hasn't moved yet, but his hand is on the handle in the close-up).
- No duplicated time (his hand takes hold of the handle in the first shot, then reaches out and grasps it again in the close-up).
- No overextended time (his hand takes the handle in the first shot, and holds it, and in the second shot is still seen holding it, waiting to turn it).

SPECIAL EFFECTS

Most nonlinear editing systems include a number of special effects that can be used to enhance the project. However, directors must be careful to use them appropriately. Overuse of special effects is the sign of an amateur production. Here is a brief list of typical effects:

- **Freeze-frame**: Stopping movement in the picture, and holding a still frame.
- **Strobe**: Displaying the action as a series of still images flashed onto the screen at a variable rate.
- **Reverse action**: Running the action in reverse.
- **Fast or slow motion**: Running the action at a faster or slower speed than normal.
- **Picture in picture**: A miniature picture inserted into the main shot.
- **Mosaic**: The picture is reduced to a pattern of small single-colored squares of adjustable size.
- **Posterizing**: Reduces tonal gradation in image.
- **Mirror**: Flipping the picture from left to right, or providing a symmetrical split screen.
- **Time lapse**: Still frames shot at regular intervals. When played back at normal speed, the effect is of greatly speeded-up motion.

There are occasions when editors deliberately "lose time" by omitting part of the action. For instance, a woman gets out of a car, and a moment later we see her coming into a room. We have not watched her through all the irrelevant action of going into the house and climbing the stairs. This technique tightens up the pace of the production and leaves out potentially boring bits during which audience interest could wane. Provided that the audience knows what to expect, and understands what is going on, this technique is an effective way of getting on with the story without wasting time.

Similarly, it is possible to "extend time," creating a dramatic impact. We see someone light the fuse of a stick of dynamite—cut to people in the next room—cut to the villain's expression—cut to the street outside—cut to him or her looking around—cut to the fuse, and so on, building up tension in a much longer time than it would really have taken for the fuse to burn down and explode the dynamite.

What Transition Should Be Used?

Transitions play a significant role in the audience's understanding of what is going on in a scene.

Cut

The *cut* or *take* is the most common, general-purpose transition. It is an instantaneous switch from one shot to another—a powerful dynamic transition that is the easiest to make.

Dissolve

The dissolve is an effect produced by fading out one picture while fading in another; a quiet, restful transition. A quick dissolve tends to imply that the action in the two scenes is happening at the same time. A slow dissolve suggests the passing of time or a different location. If a dissolve is stopped halfway, the result is a *superimposition* (Figure 16.14).

Wipe

The *wipe* is a novel transition that can have many different shapes. While it is occasionally effective, it can be easily overused and quickly become the sign of an amateur (Figures 3.30 and 16.15).

Fade

A *fade* is a gradual change (dissolve) between black and a video image. For example, at the end of a program there is usually a "fade to black," or if there is a "fade-up," it means that the director is transitioning from black to a video image. A slow fade suggests the peaceful end of action. A fast fade is rather like a "gentle cut" used to conclude a scene.

FIGURE 16.14 *A dissolve is an effect produced by fading out one picture while fading in another.*
Source: Photos by Josh Taber.

FIGURE 16.15 *A wipe is a novel transition than can take many different shapes. The wipe shown is a transition of one image wiping onto another, covering the original image with the new image.*

How to Use Transitions

The Cut

The cut is the simplest transition. It is dynamic, instantly associating two situations. Sudden change has a more powerful audience impact than a gradual one, and that is the strength of the cut.

Cutting, like all production treatment, should be purposeful. An unmotivated cut interrupts continuity and can create false relationships between shots. Cutting is not the same as repositioning the eyes as we glance around a scene, because we move our eyes with a full knowledge of our surroundings and always remain correctly oriented. On the screen, we know only what the camera shows us, although guesses or previous knowledge may fill out the environment in our minds.

The Fade

Fade-in

A fade-in provides a quiet introduction to action. A slow fade-in suggests the forming of an idea. A fast fade-in has less vitality and shock value than the cut.

Fade-out

A quick fade-out has rather less finality and suspense than a cut-out. A slow fade-out is a peaceful cessation of action.

Cross-Fade or Fade-Out/Fade-In

Linking two sequences, the cross-fade introduces a pause in the flow of action. Mood and pace vary with their relative speeds and the pause time between them. This transition can be used to connect slow-tempo sequences in which a change in time or place is involved. Between two fast-moving scenes, it may act as a momentary pause, emphasizing the activity of the second shot.

The Dissolve

As mentioned earlier, a dissolve is produced by fading out one picture while fading in the next. The two images are momentarily superimposed; the first gradually disappears, being replaced by the second.

Dissolving between shots provides a smooth restful transition, with minimum interruption of the visual flow (except when a confusing intermixture is used). A quick dissolve usually implies that their action is concurrent (parallel action). A slow dissolve suggests differences in time or place.

Dissolves are often comparative:

- Pointing out similarities or differences.
- Comparing time (especially time passing).
- Comparing space or position (dissolving a series of shots showing progress).
- Helping to relate areas visually (when transferring attention from the whole subject to a localized part).

Dissolves often show a change of time or location and are widely used as "soft" cuts, to provide an unobtrusive transition for slow-tempo occasions in which the instant nature of a cut would be disruptive. Unfortunately, they are also used to hide an absence of motivation when changing to a new shot!

A very slow dissolve produces sustained intermingled images that can be tediously confusing or boring.

The Wipe

The wipe is a novel visual transition that is often used to provide a change of time, change of location, or is just used for decorative transitions.

Although the wipe can add a novelty to transitions, it can easily be overused. The audience can easily pay more attention to the wipe effect than the storyline that it is intended to be moving forward. The wipe can also draw attention to the flat nature of the screen, destroying the three-dimensional illusion.

Wipes have many geometric forms with a variety of applications. For example, a rectangular wipe may be used as a transition between close-up detail (entertainers) and an extreme long shot of the venue.

The Wipe's Split Screen

If a wipe is stopped before it is complete, the screen remains divided, showing part of both shots. In this way, you can produce an inset, revealing a small part of a second shot or, where the proportions are more comparable, a split screen (Figure 16.16).

The split screen can show us things simultaneously:

- Events taking place at the same time.
- The interaction of events in separate locations (such as a satellite feed).
- A comparison of appearance and/or behavior of two or more subjects.
- A before-and-after comparison (developments, growth, etc.).

FIGURE 16.16 *Wipes can be used in many different situations to enhance the viewing experience. In this situation, it was used as a picture in picture.*

Good Continuity

Let's say we are watching a dramatic television show. As the director switches from one camera to the next, we notice that in the close-up, the talent's hair is askew, but in the second camera's medium shot, the talent's hair is perfect. Cutting between the two shots in the editing room exposes a continuity error. If we see a series of shots that are supposed to show the same action from different angles, we do not expect to see radical changes in the appearance of things in the various images. In other words, we expect continuity.

If a glass is full in one shot and empty in the next, we can accept this—if something has happened between the two shots. But if someone in a storm scene appears wet in all the long shots, but dry in the close-ups, something is wrong. If they are standing smiling, with an arm on a chair in one shot but with a hand in a pocket and unsmiling when seen from another angle in the next shot, the sudden change during the cut can be very obvious. The sun may be shining in one shot and not in the next. There may be aircraft noises in one but

silence in the next. Somebody may be wearing a blue suit in one shot and a gray one in the next. These are all very obvious—but they happen. In fact, they are liable to happen whenever action that is to appear continuous in the edited program stops and restarts.

There is an opportunity for a continuity error when the crew:

- Stops shooting, moves the camera to another position, and then continues the shoot.
- Repeats part of an action (a retake); it may be slightly different the second time, so you cannot edit unobtrusively with the original sequence.
- Shoots action over a period of time; part of it one day, and the rest of the scene on the next day.
- Alters how they shoot a scene, after part of it was already shot.

The only way to achieve good continuity is to pay attention to detail. Sometimes a continuity error will be much more obvious on the screen than it was during shooting. It is so easy to overlook differences when concentrating on the action, and the 101 other things that arise during production. If there are any doubts, there is a lot be said for reviewing the recording to see previous shots of the scene before continuing shooting.

Cause-Effect Relationships

Sometimes images convey practically the same idea, whichever way they are combined: a woman screaming and a lion leaping. But there is usually some distinction, especially where any cause-effect relationship is suggestible.

Cause-effect or effect-cause relationships are a common link between successive shots. Someone turns his or her head—the director cuts to show the reason. The viewer has become accustomed to this concept. Occasionally, you may deliberately show an unexpected outcome:

1. Two men are walking along a street.
2. Close-up shot of one of the men who is intently telling a story and eventually turns to his companion.
3. Cut to shot of his companion far behind, window-gazing.

The result here is a bit of a surprise and a little amusing. However, sometimes the viewer expects an outcome that does not develop and then feels frustrated or mystified, having jumped to the wrong conclusions:

1. Shot of a lecturer in long shot beside a large wall map.
2. Cut to a close-up of map.
3. Cut back to the lecturer, who is now in an entirely different setting.

The director used the close-up of the map to relocate the speaker for the next sequence, but the viewer expected to find the lecturer beside the map, and became disorientated.

Even more disturbing are situations where there is no visual continuity, although action has implied one:

1. Hearing a knock at the door, long shot of the girl as she turns.
2. Cut to a shot of a train speeding through the night.

The director thought that this would create tension by withholding the identity of the person who was outside the door. However, this inadvertently created a false relationship instead. Even where dialogue or action explains the second shot, this is usually an unsuitable transition. A mix or fade-out/fade-in would have prevented the confusion.

Montage

In a montage, a series of images are presented that combine to produce an associative effect. These images can be displayed sequentially or as a multiple-image screen (Figure 16.17).

Sequential Montage
One brief shot follows another in rapid succession, usually to convey a relationship or an abstract concept.

Multiple-Image Montage
Several images can be shown at the same time by dividing the screen into two or four segments. Although more segments can be used, images become so small that they can lose their impact. These images may be of the same subject, or of several different subjects. They can be stills (showing various stages as an athlete completes a pole vault) or moving pictures (showing different people talking from different locations).

Multiple-image montages can be used for many different purposes—to show steps in a process, to compare, to combine different viewpoints, to show action taking place at different places, to demonstrate different applications of a tool, to show variety, and so on.

Duration of Shots

If a shot is too brief, the viewer will have insufficient time to appreciate its intended information; if it's held too long, his or her attention wanders. Thoughts possibly begin to dwell on the sound and then eventually to channel switching. The limit for most subjects is roughly 15 seconds, depending on the complexity of the shot. A static shot has a much shorter time limit!

The "correct" duration for a shot depends on its purpose. We may show a hand holding a coin for half a minute as its features are described by a lecturer, whereas in a drama a one-second shot can tell us that the thief has successfully stolen it from the owner's pocket.

Many factors influence how long a shot can be held:

• The amount of information you want the viewer to assimilate (general impression, minute detail).
• How obvious and easily discernable the information is.
• Subject familiarity (its appearance, viewpoint, associations, etc.).
• How much action, change, or movement the shot contains.

FIGURE 16.17 *Montages can be created a number of ways, including: (1) a rapid succession of related images; and (2) juxtaposed images, such as this quad split.*

Source: Photos in bottom image courtesy of the U.S. Department of Defense.

FIGURE 16.18 *Tension can be increased by quicker cutting. Here, an increasing cutting rate is combined with closer and closer shots.*

Source: Photos by Tyler Young.

- Picture quality (detail and strong composition hold most interest).

During an exciting scene, for example, when the duration of shots is made shorter and shorter as the tension grows, the audience is conscious only of growing agitation, and fast-moving action (Figure 16.18).

Audience attention is normally keyed to production pace. A short flash of information during a slow-tempo sequence may pass unnoticed, yet in a fast-moving sequence it would have been fully comprehended.

Priority: Video or Sound?

It is worth remembering during the editing phase that either the pictures or the audio may be given *priority*. For example, the dialogue has priority when shooting an important speech. Although the camera should focus on the speaker, a single unchanging shot would become visually boring, even with changes in shot size. To make it more interesting, a number of "cutaway shots" are usually used of the audience, special guests, reactions, and so on. But the dialogue is continuous and unbroken—even when editing the image.

If the speech was too long, it may need to be edited in postproduction in order to hold the audience's attention.

Generally, the most important passages are then edited together. In this situation, visually it would be easy for the audience to see that segments had been removed; therefore, shots of the audience may need to be placed over the edits.

There are occasional scenes in which two people are supposed to be speaking to each other, although they were actually shot separately. For instance, all the shots of a boy stranded on a cliff would be taken at the same time (with dialogue). All the shots and comments of his rescuer at the top of the cliff would be shot at another time. During editing, the shots with their respective speech would be cut together to provide a continuous conversation.

So there are times when the images have priority, and the sound must be closely related to what we are seeing. Other times, the sound will be the priority, and everything has to be edited to support that sound.

Good Directing/Editing Techniques

If editing is done well, the audience does not notice it, but is absorbed in its *effect*. There are certain established principles in the way one edits, and although like all "rules" they may be occasionally disregarded, they have been created out of experience. Here are a few of the most common:

- Avoid cutting between shots of extremely different sizes of the same subject (close-up to long shot). It is a bit jolting for the audience.
- Do not cut between two shots of the same size (close-up to close-up) of the same subject. It produces a *jump cut* (Figure 16.19).
- If two subjects are going in the same direction (chasing, following), have them both going across the screen in the same direction. If their screen directions are opposite, it suggests that they are meeting or parting.
- Avoid cutting between still (static) shots and moving images (panning, tilting, zooming, etc.), except for a specific purpose.
- If you have to break the continuity of action (deliberately or unavoidably), introduce a cutaway shot. But try to ensure that this relates meaningfully to the main action. During a boxing match, a cutaway to an excited spectator helps the tension. A cutaway to a bored spectator (just because you happen to have the unused shot) would be meaningless, although it can be used as a comment on the main action.
- Avoid cutting to shots that make a person or object jump from one side of the screen to the other.

Anticipating Editing

It does not matter how good the video images are; if they are inappropriate, they may not be able to be used. As you plan your shots, keep in mind that the transition from one image to another has to work smoothly. Following are some of the issues to think about when shooting.

Multi-Camera and Postproduction Editing

- Edits should be motivated. There should be a reason for the edit.
- Avoid *reverse-angle shots* (shots from the other side of the axis of action) unless needed for a specific reason (such as slow-motion shots of a sports event), or if it is unavoidable (such as when crossing the road to shoot a parade from the other side). Include *head-on shots* (frontal shots) of the same action. These shots can work as transitional shots.
- Keep "cute shots" to a minimum, unless they can really be integrated into the program. These include such subjects as reflections, silhouettes against the sunset, animals or children at play, footsteps in the sand, and so on. They take up valuable time and may have minimal use. However, there are times when beauty shots have their place, such as an establishing shot.
- Where possible, include features in shots that will help provide the audience with the context of the event. This helps them identify the specific location (such as landmarks).
- Always check what is happening in the background behind the talent or subject. Distractions, such as people waving, trash cans, and signs, can take the audience's attention away from the main subject. When shooting multiple takes of a scene, watch the background for significant changes that will make editing the takes together difficult.

Postproduction Editing

- Include *cover shots* (long shots) of action wherever possible to show the overall view of the action.
- Always leave several seconds of *run-in* and *run-out* (sometimes called *heads* and *tails*) at the start and finish of each shot. Do not begin recording just as the action is beginning or the talent is about to speak or stop immediately when action or speech finishes. Spare footage at the beginning and end of each shot will allow more flexible editing.
- Include potential *cutaway shots* that can be used to cover edits when any sequence is shortened or lengthened.

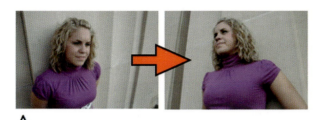

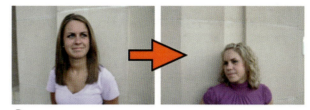

FIGURE 16.19 *Matching cuts: people.*

Source: Photos by Josh Taber.

When cutting between images of people, avoid the following distracting effects: A. Mismatched camera angles. B. Changes in headroom. C. Jump cuts: avoid cutting between shots that are only slightly different in size. The subject suddenly appears to jump, shrink, or grow.

These could include crowd shots, long shots, and people walking by.

- Try to anticipate continuity. If there are only a few shots taken in daylight and others at night, it may not be practical to edit them together to provide a continuous sequence.
- Where there is going to be commentary over the video (voice-over), allow for this in the length and pace of takes. For example, avoid inappropriately choppy editing due to shots being too brief (editors sometimes have to slow-motion or still-frame a very short shot to make it usable).
- Plan to include long shots and close-up shots of action, to provide additional editing options. For example, where the action shows people crossing a bridge, a variety of angles can make a mundane subject visually interesting. For example, an *LS*—walking away from camera toward bridge; *MS*—walking on the bridge, looking over; *XLS*—shooting up at the bridge from the river below; *LS*—walking from the bridge to the camera on the far side; and so on.
- Remember that environmental noises can provide valuable bridging sound between shots when editing. They can be recorded as a *wild track* (unsynced sound).
- Wherever possible, use an identifying board or slate at the start of each shot. Otherwise, the talent or camera operator can state the shot number so that the editor knows where it goes in the final production.

Directing/Editing Ethics

Editing is a powerful tool. And we cannot forget that the way a sequence is directed or edited can and should strongly influence an audience's interpretations of what is happening. Editing can manipulate—sometimes unwittingly—and, particularly in factual programs (newscasts, documentaries), one needs to be aware of the underlying ethics of certain treatment. A sequence of pictures can be selective, misleading the audience:

- One could deliberately avoid showing significant reactions by cutting out enthusiastic applause or heckling during a speech.
- Omitting important action or dialogue: When a person rises, turns to another, bows reverently, and slowly leaves the room, this action could be edited so that we see the person rise, open the door, and exit—apparently departing abruptly and unceremoniously, and giving a very different impression of events.
- Introduce misleading or ambiguous action: During a speech, cutting between shots of people leaving or of a person in the audience yawning or sleeping.
- Introducing false material: Showing enthusiastic applause that is actually associated with a speech different from the one we have been watching.

INTERVIEW WITH A PROFESSIONAL: SCOTT POWELL

FIGURE 16.20
Scott Powell, editor.

Briefly define your job. My Job is to take images and sounds and mold them into a coherent story.

What do you like about your job? One of the things that I like about my job is that I get to treat people to the fruits of their labor. After many battles have been fought over the script, the casting, the production design, and all of the many frustrations of production, I get to (hopefully) show them that it all turned out well.

What are the types of challenges that you face in your position? The types of challenges that face me have to do with taking whatever is dropped in my lap and weaving it into something that stirs the senses. Whether it's exciting, emotional, suspenseful, or sad, our job is to make the most out of each moment. We shape an actor's performance, one line, sometimes one word, at a time.

Are there specific things that you do to specifically prepare to work on a production? What I do to prepare is read the script and then just dig in. As the scenes come in, I cut them. I use the lined script as a roadmap to my material and I let the material guide what I do.

What suggestions or advice do you have for someone interested in a position such as yours? My best advice is to go with your gut. Use your own instincts to find the emotional rhythm of a scene. Learn the rules of editing and make them second nature before you start breaking them.

Scott Powell is an editor who has worked on almost 30 different television shows, including: Hawaii Five-O, 24, The Lot, *and* Lost City Raiders.

REVIEW QUESTIONS

1. Describe the two main types of editing (live and postproduction) and explain how they are alike.
2. How does the act of editing two clips together affect the audience?
3. What is the difference between linear and nonlinear editing?
4. Explain the nonlinear editing process.
5. Why log video footage?
6. How do you determine the order of shots?
7. What are the basic switcher transitions and when are they used?
8. Why is ethics an issue when editing?

CHAPTER 17

PRODUCTION PRACTICES

"If something has gone wrong, a director who panics and shouts merely spreads the panic to all those on the intercom and will create an air of uncertainty. Speaking with a calm voice, even if you are inwardly in turmoil, will help to solve the problems quickly and efficiently."

—Simon Staffurth, Director, *Britain's Got Talent*

TERMS

Action line/axis of action line/eyeline: The imaginary line along the direction of the action in the scene. Cameras should shoot from only one side of this line.

Crash start: Takes us straight into the program, which appears to have begun already: an automobile screams to a stop outside a store, a figure throws a bomb that explodes, an alarm bell shrills, a police siren wails—the show has begun and titles roll over the chasing vehicles.

Cutaway shot: These shots are used to cover edits when any sequence is shortened or lengthened. Generally, it is a shot of something outside of the current frame.

Filmic space: Filmic space intercuts action that is concurrent at different places: as a soldier dies, his son is born back home.

Filmic time: Omits intermediate action, condensing time and sharpening the pace. We cut from the automobile stopping to the driver entering an apartment.

Pace: The rate of emotional progression. A slow pace suggests dignity, solemnity, contemplation, and deep emotion; a fast pace conveys energy, excitement, confusion, brashness, and so on.

Teaser: Showing dramatic, provocative, intriguing highlights from the production, before the opening titles. The goal is to convince the audience to stay for the entire production.

Television production companies all over the world follow surprisingly similar practices. This is partially due to the nature of the medium and partially due to the result of sharing knowledge and experience.

PRODUCTION PRESSURES

Organizing and coordinating the production often leaves most directors little time to meditate on the medium's aesthetics. Rehearsal time is usually limited. The camera and sound crews are meeting the director's brainchild for the first time and need to understand the vision. If the treatment is elaborate, there is greater opportunity for problems that require immediate decisions. If the production is shot out of sequence, it becomes that much more difficult to ensure that each segment is coherent and will provide good continuity when edited together. In this type of situation, it is often tempting to just experiment instead of follow the normally accepted production process; there has to be a balance.

SHOOTING STYLES

There are a number of different approaches to shooting, ranging from unscripted to fully scripted.

In unscripted shooting, the camera tends to record available events, perhaps working from a rough outline (shooting plan). Material is selected during postproduction and compiled into a program format with added commentary. This technique is often used when shooting documentaries and news stories.

Fully scripted productions, on the other hand, are first broken down into their individual shots or scenes, and then the techniques involved for each shot or group of shots are assessed. The staged action is then methodically shot out of story sequence (according to a shooting schedule/plan) for maximum economy and efficiency. For example, if the storyline requires a series of intercut shots between the heroine at the cliff top and the hero below, all the sequences involved are shot at one camera setup, then the camera is moved to the other angle for the next shots.

This out-of-sequence shooting prevents having to constantly move the camera back and forth. It saves time (which is money) with fewer camera setups, lighting, and audio, and uses actors more economically.

SINGLE-CAMERA TECHNIQUES

Single-camera shooting is the traditional method of filmmaking. Two or more cameras are used only for situations in which repeat action would be impracticable or costly, or simultaneous angles are required, such as when derailing a train.

Shooting out of sequence often requires the actors to repeat their action for each change in camera viewpoint or shot length, which can introduce continuity problems.

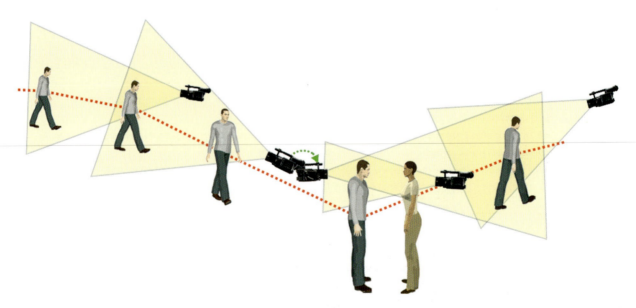

FIGURE 17.1 *Single-camera shooting: in order to achieve smooth-flowing action with one camera, a series of camera setups are often necessary.*

Action, gestures, expressions, costume, lighting, and so on must be consistent (this is continuity) within the various intercut shots and relate to the story development, although they may have been filmed at different times.

Single-Camera Setups

After a sequence of shots is carefully edited, the continuity can appear so natural that viewers will not think about the fact that it was originally shot as a disjointed series of individual setups.

In a typical sequence, we may find ourselves close to a man as he walks—yet an instant later see him from a distant viewpoint. The camera is often "left behind" as he moves away—yet is at his destination as he arrives (Figure 17.1). The overall effect here is smooth-flowing and provides a variety of angles. To do the obvious and shoot this entire sequence as continuous action would have required several cameras, spread over a significant distance—an approach that might not be practical anyway. Usually, it would not provide the same effect.

Shooting Uncontrolled Action

When using a single camera to shoot broad action over which you have no control, you have the option of:

- Remaining at a fixed viewpoint and relying on the zoom lens to provide a variety of shots.

- Moving the camera to a series of different vantage points, changing the angle and shot size to capture the main features of the occasion.

Clearly, if you are shooting a public event such as a parade with a single camera, the first option would be very limiting. However, if you move around or stop shooting, you will probably lose coverage of some of the action, unless you can time your moves to fit in with a lull in the proceedings (Figure 17.2).

In order to simplify the editing process, it is important to shoot plenty of general material that can be used as cutaway shots. Cutaways tend to be of two types: active and passive. The *active cutaway* includes subjects that can be edited into the program only at a specific moment, such as shots of a town clock showing that the parade is about to start or the crowd's responses to a specific event. *Passive cutaways* can be used at almost any time, and might include reflections, flags, sun through branches, and so on.

With news events, you make the most of whatever opportunities are available. You may be able to plan exactly what you are going to do, checking out the route and your angles in advance. You may even be able to influence the events a little, arranging for the beauty queen in the procession to turn and smile in the direction of the camera as she passes. But, generally, you will have little control over the entire situation and there are usually no repeats.

FIGURE 17.2
When shooting with a single camera at a public event, such as this parade, the camera operator needs to decide whether to stay in one place and get whatever comes his or her way, or move around to different angles to obtain better shots, but possibly miss some of the action.

"My ambition was to be one of the people who made a difference in this world. My hope is to leave the world a little better for having been there."

—Jim Henson, Director, *Sesame Street*

Shooting Controlled Action

When shooting situations that you can control, such as interviews, drama sequences, and the like, the situation is very different from shooting uncontrolled action. Now you can arrange the action, the camera setups, and sometimes the lighting and staging (background and props) to fit each individual shot.

If you want to repeat the same action so that you have long shots, medium shots, and close-up shots of it to ultimately cut together, it can be planned out in advance. Instead of having to select from whatever is happening at the location, you have the option to change things to improve the storyline. This could include having the person lean over the bridge to improve the composition, waiting for the sun to come out, or pausing while a loud aircraft passes overhead.

You generally take the shots in the most convenient, rational order. You can then edit them together afterwards to fit the storyline sequence. If, for instance, you have a series of shots of people on either side of a river, it is obviously more sensible, and efficient to shoot all of the action on one side and then move to the other, rather than moving back and forth, shooting in the script's running order.

When shooting out of order, you have to take care that there are no obvious discrepancies that spoil the continuity between the shots when they are cut together. If in our example there happened to be bright sunshine when shooting on one bank of the river but pouring rain when shooting on the other, the intercut shots would look pretty unconvincing (Figure 17.3).

FIGURE 17.3
If the editor wanted to cut between these two shots, continuity would be a problem. Note that the man's shoes are a different color in each shot.

Segmented Shooting

Shooting controlled action with a single camera normally requires the director to break down the action into a series of brief camera setups or action segments. If part of the action adds nothing to the situation, it may be deliberately omitted altogether. At other times, the action may be emphasized. Let's imagine some options with a simple situation:

Someone is walking along the street.

- If we are only really interested in what the person does when he or she reaches a destination, the walk can be omitted completely.
- Perhaps we want to emphasize that the person has hurt a leg, and every step is taken in agony. In that case, we might follow him or her closely—the heavy footsteps, the hesitant walking stick, and the uneven sidewalk—in a series of shots taken from different angles.
- Suppose the story's aim is to intrigue us. We wonder what is going to happen; is he or she going to be attacked by a thief? In this situation, we would probably shoot the sequence quite differently. The entire walk might be filmed on a stationary camera, using a continuous panning shot. We show the person walking past the camera into the distance—and out of the shot. We hold the empty frame, and their pursuer moves into the foreground of the shot.

Clearly, how the action is broken up will depend on the dramatic purpose of the image sequence.

As an exercise, record a long action sequence from a motion picture or documentary. It might be showing someone rock climbing, sailing a boat, or in a car chase. Analyze the shots that make up this action sequence. Time them. Think carefully about the variety of angles.

In a car chase, you might find that at one moment, the camera is in the car beside the driver; in the next, we see a frontal shot of the car showing the driver alone in the car. Cut, and we are behind the driver's shoulder looking through the windshield; and now we are looking down from a high building as the car maneuvers a corner. It moves away into the distance and—cut—we see it coming up toward the camera.

When watching this treatment, we never question the rapidly changing angles, or why the camera is suddenly located on a helicopter. When shooting in the desert, we see the car vanish into the distance—without wondering whether it is going to come back to pick up the camera operator!

Many types of action, in reality, can take too long and become somewhat boring to watch. Shoot them continuously, and the audience's interest will wander. Instead, break sustained action into continually varying angles, and you can create a lively pace and maintain the viewer's interest.

One specific function of editing is that we can create an illusion of continuity between shots or situations where in fact none exists. Editing enables us to develop action sequences that in reality we could not shoot.

Imagine a scene showing a bird on a branch; a close shot as it dives into the stream, an underwater shot of fish, a fish is caught by the bird, the bird sits beside the river eating its catch. These are probably shots of different birds, and of different fish, shot at different times. The separate shots are edited together to form a complete effective action sequence.

We have looked at these examples in some detail, because they epitomize the thinking and the techniques that underlie single-camera shooting. As you will learn in the next section, they are noticeably different from multi-camera treatments.

MULTI-CAMERA TECHNIQUES

"Cutting has sometimes become too frenetic. It is important to let the action in the frame tell the story instead of fast cutting."

—John Nienaber, Producer

Using two or more cameras gives you additional flexibility to handle any situation. Without missing any of the action, or needing action repeated, you can cut between different shots of the same subject, alter the camera's angle, and move to another area of the set (Figure 17.4).

You can instantly change the significance of a shot. You can provide new information, alter emphasis, point out new detail, show reactions, shift the audience's attention, compare relationships, and introduce visual variety. And all while the action continues.

Using several cameras, you can arrange to have one covering the action and another ready to catch the audience's reactions (Figure 17.5). But if you are shooting with a single camera, you are unlikely to turn away from the main event to see whether someone in the crowd is responding. You are more likely to shoot some reaction shots later (probably relating to an entirely different action), and edit them in for effect during postproduction.

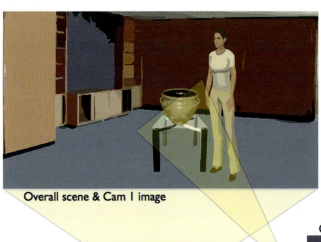

Overall scene & Cam 1 image

Cam 1

Cam 2 image

Cam 2

FIGURE 17.4
*Two-camera treatment: shots
are divided between two
cameras. Cam 1
concentrates on long shots
and Cam 2 takes close-ups
that provide detailed
information.*

Visual Variety

Static shots easily become boring. So audience interest can be encouraged by introducing movement and change—although, if overdone, the images can become distracting to watch. Rapid cutting demands continual viewer concentration, which is not easily sustained. Continual action and camera movement can wear out the audience. As always, the aim is a well-balanced blend with variations in pace, tempo, and emphasis.

Successful visual variety can be achieved by a number of ways, as described in the following subsections.

Performer Movement

Most visual variety stems from action within the scene, as people alter positions and regroup. Quite often, more meaningful changes are achieved by talent movement than by camera moves or editing.

In a close-up shot, the talent dominates. As he or she moves away from the camera, the audience becomes more aware of their relationship to the surroundings. In long shots, these surroundings may dominate. So by changing the length of the shot, you change emphasis and create visual variation.

Changes by Grouping

Where people are seated (panels, talk shows), you can achieve visual variety by isolation, selectively shooting individuals, two-shots, and subgroups (Figure 17.6).

Shooting Static Subjects

Visual variety can be introduced into your treatment of non-moving subjects (statuary, pottery, paintings, flowers) by the way you shoot and light them. Variety can be created with viewpoint-altering camera movements: selectively panning over the subject, interrelating its various parts, changing the lighting to isolate or emphasize sections or alter the subject's appearance, or handling smaller objects, turning them to show the different features.

Variety by Décor

The presentation of certain subjects may be restricted (piano playing, singers). They can be relatively static and meaningful shot variations can be limited. You can prevent sameness between productions by introducing variations in the décor. This can happen with background or scenic changes or by lighting. The pictures can look significantly different, even if the shots and angles follow a recognizable format (Figure 17.7).

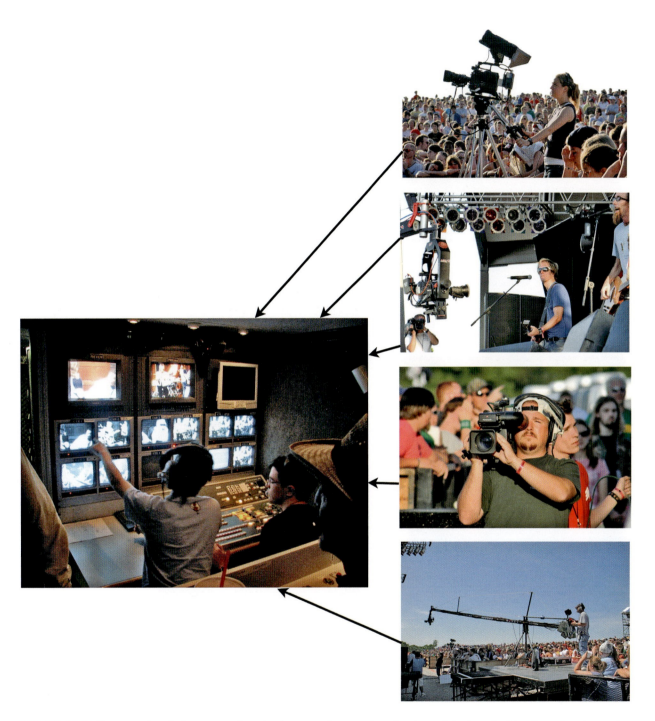

FIGURE 17.5 *Shooting with multiple cameras allows the director to show a variety of angles, assisting in maintaining the audience's attention.*
Source: Photos by Josh Taber and Paul Dupree.

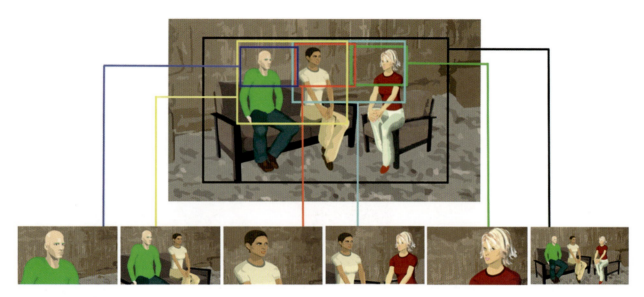

FIGURE 17.6 *Shooting groups: static groups can be broken into a series of smaller individual shots. For single-camera shooting, this involves repositioning the camera (or repeated action). With multi-camera shooting, the director can cut between the various angles.*

FIGURE 17.7
The décor, background, or lighting can add variety to a production.

Crossing the Axis of Action Line

Camera viewpoints can easily confuse the audience's sense of direction and their impression of spatial relationships if care is not taken when selecting camera positions. For example, during a basketball game, if cameras are placed on both sides of the court, it is confusing to see a player running toward the left side of the screen and then, when the director cuts to the camera on the other side of the court, seeing the same player running toward the right side of the court if he or she hasn't actually changed direction.

To avoid this, draw an imaginary line along the direction of the action (*action line*, *axis of action*, *eyeline*, or *center line*). Then be careful that cameras shoot from only one side of this line—generally, it is not crossed. It is possible to dolly across the line, or shoot along it, or change its direction by regrouping people, but cutting between cameras on both sides of this imaginary line produces a reverse cut or jump cut (Figures 9.36, 9.37, 17.8, and 17.9).

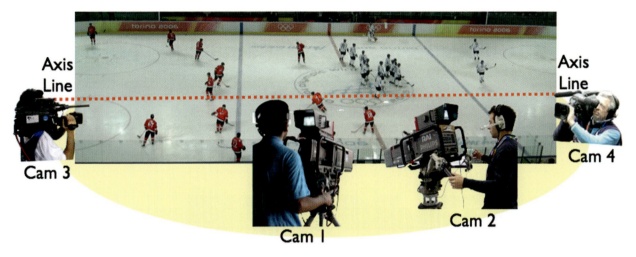

FIGURE 17.8 *The red line represents the axis of action at a sporting event. All "live" cameras must stay on one side of that line. The yellow area represents where the cameras can be placed on one side of the field of play. Sometimes additional cameras are placed on the other side for slow-motion replay, but not live.*

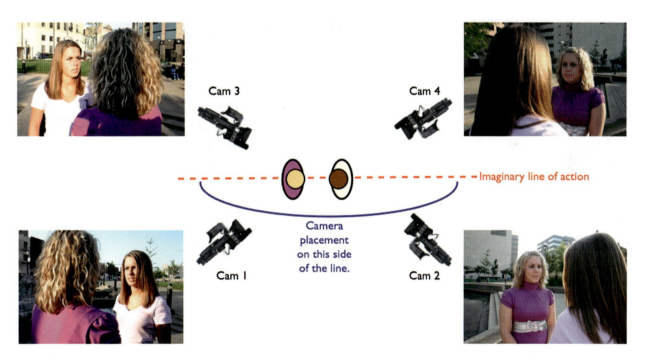

FIGURE 17.9 *The line of action: shots can be cut between cameras located on the same side of the imaginary line of action: between 1 and 2, or 3 and 4. Cutting between cameras on opposite sides of this line could cause a jump cut (1 and 3, 1 and 4, 2 and 3, 2 and 4).*

Source: Photos by Josh Taber.

Organizing the Angles

It is important for directors to create an organized approach to covering the story. Obviously, you could just place the cameras around a location and use the images that happen to be captured. However, even if the cameras captured some great images, there may be little relationship between the images, there would probably be some nearly identical shots and some useless shots, and the coverage would be erratic. Shots must be appropriately chosen according to a coordinated plan, and only the director is in a position to do this (Figure 17.10).

Using this technique, you begin by analyzing where the action will take place (where people move, what they do there, and so on) and arrange strategic camera angles that are able to capture the needed images. Each camera position provides a series of shot opportunities; you select from these available shots as needed (Figures 17.11 and 17.12, and Table 17.1).

FIGURE 17.10 *Directors create an organized approach to covering the story, selecting camera positions that provide angles that are effective in showing the subject.*

TABLE 17.1 *Arranging the Shot*

Broad objectives	In arranging a shot, you should be able to answer such questions as: • What is the purpose of the shot? What is it aiming to show? • Is the shot to emphasize a specific point or feature? • Which is the main subject? • Are we primarily concerned with the subject or its relationship to another or to its background/environment?
The actual image	• Is the shot too close or distant for its purpose? • Is the attention reasonably localized—or split or diffused? • Is the composition arrangement appropriate? • Is the subject suitably framed headroom, offset, edge cut off, overcrowded frame? • Are subjects clearly seen sharp, not obscured, good background contrast? • Is there any ambiguity or distraction in the shot?
The action	• Are we aiming to show what a person is doing—clearly, forcefully, incidentally, not at all? • Does action (movement, gestures) pass outside the frame? • Are any important features or actions accidentally excluded?
Specific objectives	• Is the presentation to be straightforward or dramatic? • Do we want to indicate subject strength or weakness? • Do we aim to reveal, conceal, mislead, puzzle? • What effect, mood, atmosphere are we seeking: businesslike, clinical, romantic, sinister? • Does the shot relate successfully to the previous and subsequent shots?

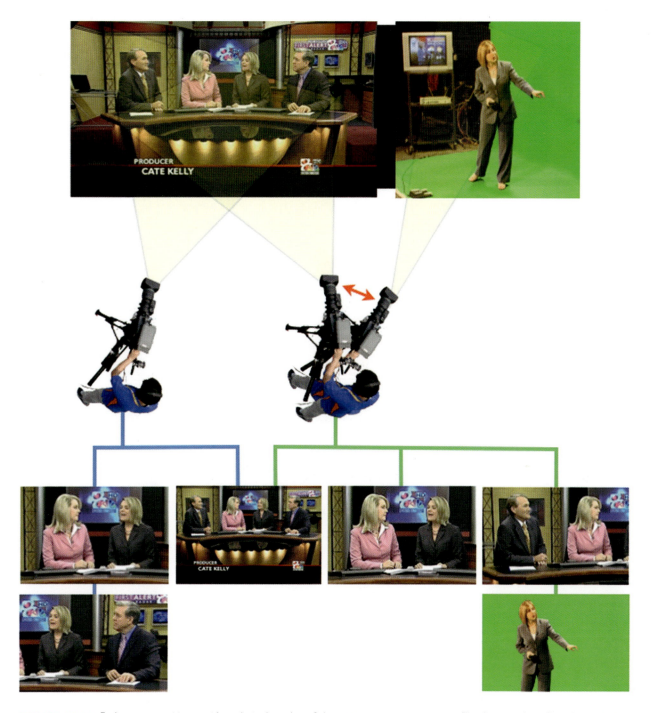

FIGURE 17.11 *Each camera position provides a limited number of shot opportunities on a news set. The director selects from the available shots as required. Note that the second camera moves from the main set to the chromakey set.*

Source: Photos courtesy of KOMU-TV, WLEX, and Josh Taber.

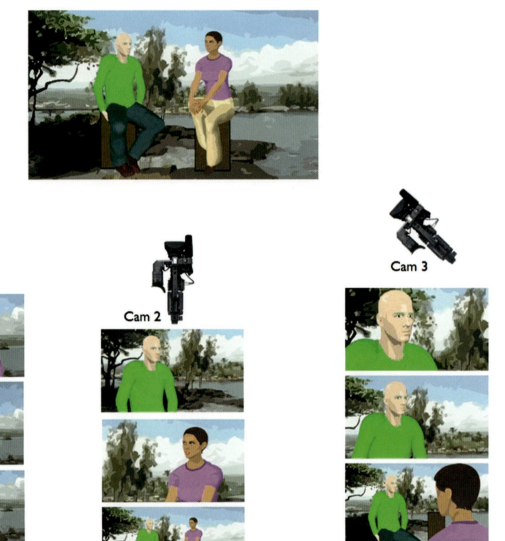

FIGURE 17.12 *In a formal interview, there are relatively few effective shots, although more are available on a news set.*

At sports events, large concerts, and other large-scale events, this is the only practical approach. Cameras are often widely dispersed, camera movement is often restricted, and the director often has to rely on camera operators to use their initiative to find appropriate shots. The director's knowledge of actual shot opportunities is mostly derived from what the cameras reveal and on-the-spot assistants (usually called spotters). The director guides selection, adjusts shot sizes, suggests desirable shots, and chooses from the available material.

You can use a similar strategy in studio production for certain types of shows with a regular format. For example, you can assign specific shots to each camera:

- *Camera 1*: Long shots and extreme long shots.
- *Camera 2*: Primarily close-up shots.
- *Camera 3*: Medium shots and close-ups.

This approach is especially useful for unscripted "live" productions. However, it is also valuable for talent so that they know which camera to look at for close-ups at any time during the production.

CAMERA MOVES

FIGURE 17.13 *When shooting talent sitting down, be prepared to either follow him or her with the camera or have a second camera ready that is already zoomed out for when he or she stands. Otherwise, his or her head will be cut off. Similarly, be ready for the talent to sit, if he or she is standing.*

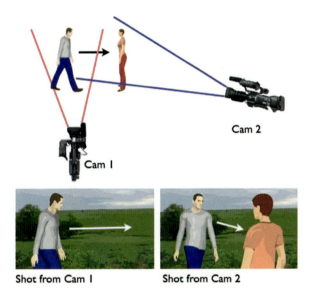

Cam 2

Cam 1

Shot from Cam 1   Shot from Cam 2

FIGURE 17.14 *Widely spaced moves: if two people are quite a distance apart, you can cut between them, pan between them, or just shoot the action from a different angle.*

Program Opening

Establishing shots, usually a long or extreme long shot, introduce the scene and the action, setting the mood and influencing the audience's frame of mind toward what you have to say and show. There are many different types of openings.

The Formal Start

A "Good evening" or "Hello" introduces the show and moves into the content. Whether the presenter appears casual,

reverent, indifferent, or enthusiastic can create an ambience directly influencing the audience's attitude.

The Teaser

We are shown dramatic, provocative, intriguing highlights from the production before the opening titles. The goal is to convince the audience to stay for the entire production.

The Crash Start

We are taken straight into the program, which appears to have begun already. An automobile screams to a stop outside a store, a figure throws a bomb that explodes, an alarm bell shrills, a police siren wails—the show has begun and titles roll over the chasing vehicles.

The Character Introduction

A montage of symbols or shots of the hero in various predicaments provides an introduction to the characters that we are to meet.

The Eavesdropping Start

The camera peers through a house window, sees a group of people sitting around a table talking quietly, and moves in to join them.

The Slow Buildup

The camera pans slowly round, arousing curiosity or suspense until we reach a climax point (the bloodstained knife)! It is essential to avoid diminishing interest or anticlimax.

The Atmosphere Introduction

A series of strongly associative symbols establish the place, period, mood, or personality. On a shelf: a brass telescope, a ship in a bottle, and a well-worn uniform cap—the old sea captain is introduced long before we see him.

Focusing Audience Attention

Directors can hold the audience's attention by taking care with how they arrange and present subjects.

Audience concentration easily lapses, so you need to continually direct and redirect their attention to hold interest along specific lines. Such redirection implies change. But excess or uncontrolled change can lead to confusion or irritation. Pictorial change must be clearly motivated, to allow the viewer to readjust easily to each new situation.

Some of the most common ways of focusing the audience's attention on a specific area are (Table 17.2):

- **Brightness**: Eyes are always drawn to the brightest areas on the screen. This could include lighting, set design, or costumes (Figure 17.15).
- **Focus**: Eyes are attracted to the sharpest areas of the image (Figure 17.16).
- **Motion**: Movement can attract attention, such as having a person stand up within a seated group or moving the camera around the subject (Figure 17.17).
- **Dominant figure**: Placing the dominant person or subject in a position that makes him or her stand out (Figure 17.18).

TABLE 17.2 *Methods of Focusing Attention*

Exclusion	Taking close-up shots. Excluding unwanted subjects. Using neutral backgrounds.
Visual indication	Indication with a finger or graphic of some type (arrow, circle).
Aural indication	A verbal clue or instruction: "Look at the black box."
Color	Using prominent contrasting hues against neutrals or muted (pastel) colors. The area of interest might have the lightest tones or maximum contrast in the picture.
Camera viewpoint	Avoid weak angles (side or rear shots) or weakening angles (high or long shots).
Composition	Use convergent lines or patterns, picture balance, isolation, and prominence through scale.
Contrasting the subject with its surroundings	Through differences in relative size, shape, proportions, scale, type of line, movement, or association.
By movement	Movement attracts according to its speed, strength, and direction. Change direction during motion, or interrupt and resume, rather than maintaining sustained action. According to how it is introduced, a movement can attract attention to itself (a moving hand), the subject (person whose hand it is), or the purpose of the movement (what the hand is pointing at). Remember, when the camera moves (or zooms), it can virtually create the illusion of the subject movement. Synchronizing a movement with dialogue, music, or effects (especially the subject's own) gives it strength and draws attention to it. Having a person move on his or her own dialogue emphasizes both action and speech.
By subject attitude	Having performers use strong movements—upward or diagonal, stand up, play to camera, move in front of others, or scenic elements.

FIGURE 17.15 *Attention can be drawn by lighting or even clothing. Eyes are always drawn to the brightest area of the screen.*

Source: Photos by Lee Peters and David Clement.

FIGURE 17.16 *Eyes are always drawn to the sharpest parts of the image.*

FIGURE 17.17
Movement can draw attention. This includes talent movement as well as camera movement (such as a moving dolly, as shown here).

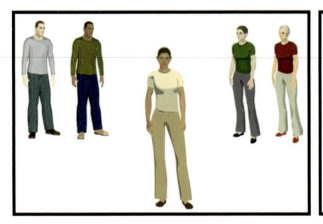 

FIGURE 17.18 *By placing a person or an object in a dominant position, viewers know where to look.*

Shifting Visual Interest

It is just as necessary to be able to shift the viewer's attention to another aspect of the picture as it is to localize his or her attention originally. You can do this by readjusting any of the influences that we use to originally direct his or her attention. Here are a few of the alternatives:

- Transfer the original emphasis (contrast, isolation, etc.) to the new subject.
- Use linking action; first person looks to camera right—cut to new subject on screen right.
- Weaken the original subjects by reducing their size within the picture or alter the camera height when shooting.
- Adjust the focus, pulling it from the old subject to the new.
- Change the sound source; a new person speaks instead.

Creating Tension

Tension in a dramatic situation derives partly from the dialogue, storyline, and interaction between characters, but it can also be considerably influenced by the way in which the subject is presented:

- Using progressively more powerful shots (cutting to closer and closer shots; lower viewpoints, gradual canting).
- Using suspenseful music and effects.
- Presenting ambiguous information—is the nocturnal shadow a bush, or a prowler?
- Presenting insufficient information—the audience sees the figure in the doorway—is it the bad guy (Figure 17.19)?

- The audience knows something a character does not— the victim runs to escape, but we know the route is blocked.
- A character is suddenly confronted with an insurmountable problem—the stairway collapses upon reaching it.

The borderline between creating tension and creating a ludicrous situation can be narrow. An intended climax too easily becomes an anticlimax. So you have to take care not to emphasize the trivial, or unintentionally to permit an emotional letdown after an emotional peak.

Pace

We might define pace as the rate of emotional progression. A slow pace suggests dignity, solemnity, contemplation, and deep emotion; a fast pace conveys energy, excitement, confusion, and brashness.

A well-balanced show continually readjusts its pace. A constant rapid tempo is exhausting; a slow sustained one becomes dull. Pace comes from an accumulation of factors:

- **The script**: Scene length, speech durations, phrasing, word lengths. Sharp witty exchanges produce a faster pace than lengthy monologue.
- **The delivery**: Fast, high-pitched, interrupted sounds provide a rapid pace compared with slow, low-pitched ones.
- **Production treatment**: The rate of camera movement, switching, performer movements.

The eye can maintain a quicker pace than the ear. The eye can assess, classify, and evaluate almost immediately, but the ear has to piece together consecutive sounds to interpret their overall meaning.

FIGURE 17.19
Tension is often created by presenting insufficient information. The audience sees the man watching from a shadowy area and wonders: bad guy or good guy?

TABLE 17.3 *Selecting Camera Treatment*

	Production Purpose	Artistic Purpose
Why pan or tilt?	• To follow action • To exclude unwanted subjects • To show an area that is too large to be contained in a static shot	• To join separate items • To show spatial relationships • To show cause/effect • To transfer attention • To build up anticipation or suspense • As an introductory move
Why elevate the camera (camera up)?	• To see over foreground objects • To see overall action	• To reduce prominence of foreground • To look down onto objects • To reduce subject strength
Why drop the camera down (ped down)?	• To frame picture with foreground objects • To obscure distant action • To obtain level shots of low subjects	• To increase their prominence • To emphasize subject strength
Why dolly in?	• To see or emphasize detail or action • To exclude nearby subjects • To recompose a shot after one of the subjects has been repositioned or exited • To change emphasis to another subject • To emphasize an advancing subject	• To underline an action or reaction • To emphasize subject importance • To localize attention To reveal new information • To follow a receding subject • To provide spatial awareness • To increase tension • To create a subjective effect
Why dolly back (out)?	• To extend field of view • To include more of subject(s) • To include widespread action • To reveal new information • To include entry of a new subject • To accommodate advancing action • To withdraw from action	• To reduce emphasis • To show relationship or scale of previous shot to the wider view (whole of subject, other subjects) • To increase tension, as more significance is gradually revealed • To create surprise (e.g., reveal unsuspected onlooker) • To provide spatial awareness • To create a subjective effect
Why truck (also referred to as a crab or track)?	• To follow subject moving across the scene • To reveal the extent of a subject/scene, section by section • To examine a long subject or series of subjects	• To emphasize depth
Why follow a moving subject?	• In close-up shots—to keep it in frame while showing reactions or detail information • In long shots—to show subject progress through an environment, or its relationship to other subjects	• To spatially interrelate subjects • To avoid transitions or viewpoint changes; to maintain continuity

Although a fast visual pace is readily assimilated, it is usually at the expense of less attention to the accompanying sound—unless visual changes are so rapid that the viewer ignores the information and just listens! When emphasis is to be on the sound, visual pace generally needs to be relaxed.

Timing

There are two different kinds of timing in television. One kind deals with a clock or stopwatch. This type of timing refers to keeping a television program on schedule—when to play commercials, how long a story can be on the news, and so on. Because television is run on a very specific time schedule, the clock is an important piece of production equipment.

The second type of timing is artistic timing. Artistic timing is choosing the right moment and duration for an action—exactly when to cut, the speed of a transition, the pause duration between a comment and a response.

Inept timing may mistakenly communicate the wrong emphasis and can disrupt continuity.

VISUAL CLARITY

If an image is to effectively communicate its message quickly and unambiguously, the audience must be able to see relevant details easily and clearly. The goal is to avoid confusion, ambiguity, obscured information, restricted visibility, distractions, and similar visual confusion (Table 17.3).

Viewing Angle

A poor viewing angle can make even the commonest subjects look unfamiliar. Occasionally, this may be done deliberately. However, the goal is to provide a clear, unambiguous presentation, even if, for instance, it becomes necessary for a demonstrator to handle or place items in an unaccustomed way to improve shot clarity.

Distractions

Poor lighting treatment can distract and confuse the audience, hide the subject's contours, and cast misleading or unmotivated shadows (especially from unseen people). It can also unnecessarily create hot spots, lens flares, and specular reflections.

Strong tonal contrasts can distract, as can strongly marked detail. When someone stands in front of a very busy background, the audience may begin to look at the background instead of the subject (Figure 17.20).

Slightly defocused details, particularly lettering that we cannot read easily, can sidetrack the audience's attention. Directors should also be careful about including strongly colored defocused objects in the background. These objects should be excluded from shots whenever possible because they become distractions.

Confusing and Frustrating Subject Treatment

How often, when watching regular television productions, do you experience total frustration or antagonism at the way the show is being handled? It is important for directors to be very

FIGURE 17.20 *Background contrast is often difficult to deal with. Although strong tonal contrasts can be too distracting, similar tones can often prove distracting as well.*

sensitive to these situations. Here is a list of frequent annoyances:

- Someone points to a detail—but it is too small or fuzzy for the audience to see.
- Titling or graphics are shown—but too briefly for us to read or examine.
- The interviewer who asks questions (quoting from notes)—but is not really listening to the guest's replies.
- The interviewer who asks the guest questions that require only "Yes" or "No" replies.
- A too-brief glimpse of a subject—followed by a shot of the commentator talking about it instead.
- The shot that leaves us wondering what we are supposed to be looking at or where we are now.
- Too much information, such as statistics, is covered, resulting in the audience being overwhelmed.

The Visual Problem

Directors often have to deal with subjects in their programs for which directly appropriate visuals are not obvious. These may include:

- Abstract subjects—philosophical, spiritual, social concepts.
- Imaginary events—hypothetical, fantasy.
- Historical events—before photography, or unphotographed events.
- Events that will be occurring in the future.
- Shooting is not possible—prohibited or subject inaccessible.
- Shooting is impracticable—too dangerous, meaningful shots impossible.
- Appropriate visuals are too costly—would involve distant travel, copyright problems.

Possible Solutions

When directors cannot show the actual subject being discussed, they often have to provide suitable alternative images—a kind of visual padding or screen filler (also known as "wallpaper shots").

The most economical solution, and the least compelling, is to have a commentator tell us about what we cannot see, as he or she stands at the now-empty location (historical battlefield, site of the crime, or outside the conference hall). Inserts in the form of photographs, video clips, paintings, or drawings (typically used for courtroom reports) can all provide illustrative material. Occasionally, a dramatic re-enactment may work.

When discussing future events, stills or stock library shots of a previous occasion may be used to suggest the atmosphere or to show what the event may look like (celebration days, processions) (Figure 17.21). A substitute subject (not the animal that escaped but one just like it) may also be used, letting the audience know that they are not seeing the actual subject.

Occasionally, the camera can even show the absence of the subject: the frame of the stolen painting or where the castle once stood.

Associated subjects are frequently used. We visit the poet's birthplace, often using stock tourist location shots, statues or create models of the event. However, they may not be strictly applicable (wrong period), or irrelevant (not architecture, but social conditions influenced his poetry). But apart from family album snapshots or newspaper cuttings, nothing else may be available (Figure 17.22)!

When using library (stock) shots or stills, there is always the danger that available shots will become overly familiar through repeated use. This issue is particularly likely with historic material, or unexpected tragedy (assassination, air crash).

Some forms of visual padding suit a variety of occasions. The same waving wheat field can epitomize food crops, daily bread, prosperity, agriculture, the war on insect pests, and so on.

FIGURE 17.21 *Often, stock footage can be used when discussing abstract ideas. For example, this stock airport could be shown when discussing future growth of the airport.*

FIGURE 17.22 *Statues or historic models may be available to illustrate an historic concept.*

FIGURE 17.23 *Abstract subjects can be difficult to visualize. This type of photo is often used by people attempting to visualize the concept of "heaven."*

Abstracts can be pressed into service at almost any time. Atmospheric shots of rippling water, shadows, light reflections, into-the-sun flares, and defocused images are all used regularly (Figure 17.23).

The Illusion of Time

Motion picture editing has long accustomed us to the concepts of filmic space and time:

- *Filmic space* intercuts action that is concurrent at different places: As a soldier dies, his son is born back home.
- *Filmic time* omits intermediate action, condensing time and sharpening the pace: We cut from the automobile stopping to the driver entering an apartment.

When all the intervening action has no plot relevance, the viewer can often get frustrated with a slow pace just to state the obvious.

Time Lapses

There are a number of techniques that can be used to indicate the passage of time. Explanatory titles are direct and unambiguous, but other and subtler techniques are generally preferable. For short time lapses:

- Slow fade-out, new scene slowly fades in.
- Cutting away from a scene, we assume that time has elapsed when returning to it.
- A time indicator (clock, sundial, burning candle) shows passage of time.
- Lighting changes with passing time (a sunlit room gradually darkens).
- Dissolve or wipe between before/after shots of a meal, fireplace, and similar images.
- Transition between sounds with time association— nocturnal frogs and owls to early-morning roosters and birdsong.
- Defocus shot, cut, dissolve, or wipe to another defocused shot and then refocus.

For longer time lapses:

- A calendar changes pages or changes date.
- Seasonal changes—from winter snow to spring flowers.
- Changes in personal appearance—beard growth, aging, fashion changes.
- New to old—dissolving from a fresh newspaper to a yellow, crumpled, discarded version.

Flashbacks

A familiar device, the flashback turns back time to see events before the present action as a reminder or an explanation, or for comparison purposes. Typical methods include reversing the previously discussed time-lapse techniques (such as the old becoming new again), a special effects wipe, a defocus, or even a brief flash cut-in (1/2–2 seconds long) can convey recognition or moments of memory recall.

Cutaway Shots (Insert Shots)

By cutting from the main action to a secondary activity or associated subjects (such as spectator reactions), you can:

- Remove unwanted, unsuccessful, dull, or excess material.
- Suggest a time lapse to compress or expand time.
- Show additional explanatory information (detail shots).

- Reveal the action's environment.
- Show who a person is speaking to; how another person is responding (reaction shots).
- Show what the speaker is seeing, talking about, or thinking about.
- Create tension to give dramatic emphasis.
- Make comment on a situation (cutting from a diner to a pig at trough).

Interviews can be shot using a continuous multi-camera setup or as a one-camera treatment. A separate series of cutaway shots (cutaways, nod shots) are often recorded afterwards, in which the interviewer and interviewee are seen in singles or over-the-shoulder shots, smiling, nodding, reacting, or looking interested. When edited in (to disguise cuts, continuity breaks, or to add visual variety), the added shots appear quite natural. Without these cutaways, any continuously held shot would jump frames when edited (although dissolves may improve the disruption).

Reaction Shots/Partials/Cut-In Shots

By skillfully concealing information, you can prime the audience's imagination and arouse their curiosity. Instead of showing the event, you can demonstrate its effect:

- *Reaction shot*: The door opens—but we see the victim's horror-stricken face, not the intruder.
- *Partial shot*: A switchblade opens, is moved out of frame; we hear the victim's cry, then silence.

- *Cut-in shot*: We watch the victim's cat drinking milk to sounds of a fight and a body falling; the victim's hand comes into frame, upsetting the milk dish.

This technique can provide maximum impact with minimum facilities, conveying information by implication rather than direct statement. It aims to intrigue and tantalize.

The Recorded Insert

Occasionally, during a studio production, we may cut away from studio cameras' pictures to show prerecorded material. There are several reasons why we might want to do this:

- To illustrate a lecture, demonstration, or talk, such as footage of a trip down the Amazon.
- To imply that the studio setting is in a specific location, showing stock shots of the outside of a building or cityscapes.
- To authenticate a setting. We see a small boarded room in the studio. A stock shot of a ship is inserted so that we accept that this is a ship's cabin.
- To show environments that could not be recreated effectively in the studio, such as a typhoon.
- To extend action. You can have a person walk through a door in the studio set (an apartment) and then out into the street (prerecorded).

- Once-only action. Prerecording is essential when action might prove unsuccessful in the studio (e.g., involving animals) or take too much time during the recording period (an elaborate makeup change), or is too dangerous (fire) or unrepeatable (an explosion) or very critical (an accurately thrown knife).
- Visual effects. To produce time-lapse effects, reversed action, transformations, and the like.
- Animation sequences. These may be cartoon or animated still life.
- To include performers otherwise unavailable—filming persons who could not attend the taping session, such as overseas guests.

Stock Shots (Library Shots)

These are short film, still, or video sequences of illustrative material, held in an organization's archives or rented from specialist libraries. Stock shots are inserted into a program where it would be impracticable or uneconomic to shoot new material. These short clips cover a very wide field, including news events, location shots, manufacturing processes, natural history, personalities, and stunts. They are widely used to illustrate talks, demonstrations, and newscasts, as well as to provide atmospheric and environmental shots for drama (Figure 17.21).

INTERVIEW WITH A PROFESSIONAL: DAVID NIXON

Briefly define your job. I am a producer/director for my own production company, which means I am the salesman, the account executive, the creative, I hire and fire, I put together the shoots, I direct the talent, I supervise the edit, and in the end I am the bill collector! As a producer/director, you have to do it all. It's not just about the glamorous stuff of directing on set. Usually, I have to go out and sell the job or raise the funds to make it, hire a writer to write the script, then put together the crew, the talent, the schedule, and direct the shoot, either supervise the edit or edit myself, deal with the client's demands, and make sure I get paid so I can survive to "fight another day!"

What do you like about your job? I love that I'm creating something out of nothing . . . an idea in someone's head that's never been seen before. It's incredible to see the idea come to life and then see the smile on the client's face. It's never the same. It's different every day and every job. That's refreshing! And if you can inspire someone, or touch someone's heart, or bring a tear to an eye, you've really done something. What we do is incredibly powerful. People empathize with what they see on the screen, and we have the power to change hearts and minds. That's incredible!

FIGURE 17.24
David Nixon, producer/director, DNP Studios.

continued

Interview with a Professional: David Nixon—*continued*

What are the types of challenges that you face in your position? It's really difficult to take words on a page and end up with something on the screen that is remotely the same or even better. It's easy to see it in your head, but translating that to the camera is very difficult. It's never easy. The technology has limitations, the locations have limitations, the talent have limitations, the weather plays havoc with you, the audio is always difficult to get "clean," the edit always takes longer, but when everything is aligned and you work out all the bugs, it's magic! As a director, the biggest challenge is creating a tone on set that allows the technicians to give you the best that they can, and the actors to be as real as they can. You're kind of like the coach of the football team. You can't carry the ball, but you have to inspire the talented ones to, and if you can get them to stretch beyond the usual, and give you something extraordinary, you'll have magic on camera. Don't ever compromise, especially when the technology or the situation becomes difficult. Never give up; never accept mediocrity! You never know that the next take might be spectacular!

How do you prepare for a production? Pre-pro is *king*! Always spend three times more time preparing than shooting. It will come back to you in spades. If you haven't thought through everything before you get on set, you'll be behind the eight ball. Don't be in a hurry to get on set and start shooting. If you don't have an ironclad plan, then all hell will break loose! There's always problems that arise on the shoot that you hadn't anticipated, so make sure you have all your ducks in a row before you shoot. Scout the location with your key personnel, think through every shot, discuss the project with every member of the team beforehand, including the editors, the music composers, the animators, etc., not just those that will be on set. You never know what great ideas they might come up with. If you can get really good people around you, people that are smarter than you, they will make you look good. Leave your ego at the door. You shouldn't be the loudest voice in the room. Listen to good counsel. Every member of your team is better than you at their specific expertise, such as camera or sound or lighting, so let them do their job and make the project better than you see in your head. The sum of the parts is greater than the whole!

What suggestions or advice do you have for someone interested in a position such as yours? Learn your craft! Practice your craft! You should have working knowledge of every aspect of the production: camera, acting, sound, lighting, editing, music, graphics, etc. You don't have to be good at all those things, just well versed enough that you can talk their language, know when they're getting good stuff (or not), are able to direct them well and efficiently, and inspire them to give you their best. And the most important thing is that you should know when you've "got it!" The most frustrating thing for technicians and actors is the director that does take after take trying different things and not knowing whether he or she's "got it" or not. You need to have such an intimate knowledge of your craft and the project that you know when you've "got it" on set and you know when you've "got it" in the edit. And when it turns out better than in your head, you've really "got it!"

David Nixon is a producer/director at DNP Studios in Orlando, Florida. His clients have included SeaWorld Parks, Nickelodeon, Subway Restaurants, Nestlé, Walt Disney World, and the Ritz-Carlton Hotel & Resorts.

REVIEW QUESTIONS

1. How do you determine whether to shoot with a single camera or multiple cameras?
2. Describe a continuity error and explain how it could occur.
3. What are the advantages and disadvantages to multi-camera production?
4. Why do cameras need to stay on one side of the axis of action?
5. What are some of the issues that you need to think about as you arrange a shot?
6. How do you focus the audience's attention on a specific subject on the screen?
7. How does pacing affect a production?

CHAPTER 18

THE STUDIO PRODUCTION

"Always get more sleep than your actors."

—Orson Welles, Director

"There is a sense of exhilaration that I feel every time I step into a TV studio that has not lessened over the years. For me, the studio is like my very own giant sandbox, where I can build castles and get dirt under my fingernails."

—Doug Smart, Director

TERMS

Camera blocking (stumble-through): The initial camera rehearsal, coordinating all technical operations and discovering and correcting problems.

Cue card: Large card-stock can be used by a floor crew member to show brief notes, an outline, or other information.

Dress rehearsal (dress run): The goal of the dress rehearsal is to time the wardrobe and makeup changes.

Dry blocking (walk-through): Actors perform, familiarizing themselves with the studio settings and surroundings while the studio crew watch, learning the format, action, and production treatment.

ISO (isolated camera): A camera of which the signal is not only being sent to a switcher, but is being recorded on its own recorder or camcorder.

Teleprompter: Shows the script on a two-way mirror that is positioned directly over the camera's lens.

Although there are many well-established methods of preparing and recording television productions, the method that you should choose depends not only on the type of show, but whether it is to be live or recorded, the facilities, and the amount of time and the size of the budget available for the production. There are differences between the way local productions and network productions work in the studio. However, in this chapter, the information is condensed to provide an overview of the subject.

UNREHEARSED FORMATS

Every production benefits from a rehearsal before being recorded or going out live. But what do you do when the talent is going to arrive at the last minute, or even while you are live? If it is a live show with long prerecorded and edited packages, such as a magazine program, it may be possible to quickly review what will be happening. Otherwise, you need to have rehearsed it in your mind, know what you want, do the best that can be done without a rehearsal, and accept the result as part of shooting a live production.

Fortunately, many television productions fit into familiar routine formats. Consequently, even when it is not possible to rehearse the action beforehand, you can still prepare a setup that will work successfully when the talent does arrive. Interviews, for example, have regular plans so that you can quickly line up the appropriate chair positions and move the cameras into their positions. Crew members can be used as stand-ins while lighting and sound are being checked. When the talent does arrive, you can quickly review the camera shots and adjust voice levels, makeup, and lighting.

When the unrehearsed action is less defined—such as a late-arriving band—you have to rely on cameras arranged strategically in front and cross-shooting positions. Instead of cameras grabbing shots of whatever is near them, you can allocate cover shots (long shots) to one camera, and have another concentrate on close-ups of the instruments, while another shoots close-ups of individuals or small groups. Before the production begins, always explain to the performers the floor area limits within which they must work or their action may uncontrollably spread into areas that cannot be covered by the lighting or cameras. Production treatment is largely a matter of recognizing effective shots as they are offered by the cameras—taking care to dwell on any special features, such as action detail of hands playing a piano or grouping shots of a chorus.

ADVANCE REHEARSALS

If you are renting cameras and/or even a studio, time is expensive, leaving few opportunities to experiment, try out variations, or work out half-formed ideas. It is imperative that you practice as many elements of the production as possible before the actual camera rehearsal. The director and talent can discuss the various production options, making sure that the ideas work. This can be done in any room that has enough space to work through the material.

For large complex productions, preparations need to be completed long before the camera rehearsals. It is essential to practice dialogue and action, coordinate performances, and discuss the camera angles. Drama and comedy shows are often rehearsed a week or two before the shoot date. Another rehearsal may even take place hours before the actual shoot time. This practice reduces the cost of the production by avoiding the need for a rental space.

A pre-rehearsal for a typical drama production usually begins with a read-through (also known as a *briefing* or *line rehearsal*). The director goes over the script, indicating specific points about style and presentation that will help familiarize the cast with their parts. They read their lines from the script, becoming more accustomed to the dialogue, the other actors, and their characterizations.

The rehearsal room's floor can even be tape-outlined with a full-size layout of the studio set. Doors, windows, stairways, and so on can be outlined. Stock rehearsal furniture can be substituted for the actual studio items, and action props (telephone, tableware, etc.) can be provided. This way, actors can become accustomed to the scale and features of their surroundings.

The director arranges the action, the actors' positions, and their groupings to suit the camera production plan. Rehearsing a scene at a time, the cast is able to learn their lines and practice their performance until it flows naturally and the show runs smoothly, finally ready for the actual camera rehearsal. The durations of segments are checked and adjusted. (In calculating the overall timing, allowances are made for the time taken by later inserts such as prerecorded sequences.)

TABLE 18.1 *The Effective Studio Rehearsal*

Unrehearsed or briefly rehearsed studio production	• The director's assistant confirms that available production information is distributed (breakdown sheet/running order) and that contributory graphics or prerecorded inserts are correct. The assistant has details of all word cues for inserts, announcements, and their timings.
	• The director, with the floor manager, technical director, lighting director, cameras, and sound crew, arranges basic talent/performer and camera positions (usually the floor is marked). Even a basic plan needs coordination. The director outlines action or moves and camera coverage. Lighting is set.
	• If the talent is available, line up the shots and explain to them any shot restrictions or special care needed in demonstrating items and so on. Otherwise, use stand-ins for shot line-up and then brief talent on arrival.
	• Check that performers and crew, including the teleprompter operator, are ready to start.
	• If a full rehearsal is impractical, carry out the basic production checks: rehearse beginnings (open) and ends (close) of each segment, and check any complicated action. Fix any errors or problems.
Intercom and the director	• Remember that the production team has to depend on each other in order to do a good job. A quiet, methodical, patient approach is best with the crew. Be firm but friendly. Avoid critical comments on the intercom.
	• Generally, cameras are called by their numbers (such as "Cam 1" or "Cam 2"), guiding all camera moves (and zooms) during the rehearsal, warning the crew of upcoming action and movements. However, it is not uncommon for a director who works with a regular crew to call them by name instead of their camera number.
	• Examine each shot. As necessary, modify positions, action, movement, and composition.
	• Consider shot continuity. It is important to remember that altering shots may affect earlier shots.
	• Remember, the crew and performers are learning what you want them to do from your production paperwork and your intercom instructions.
	• Do not be vague; be concise. Make sure that your intentions are understood. In correcting errors, explain what was wrong and what you want. Not "Move him left a bit," but "He is shadowing the desk, move him stage left." Do not assume that performers will see and correct problems.
	• Keep in mind that there is a difference between stage directions ("Actors, move stage left") and camera directions ("Cam 1, truck left").
	• If a camera operator offers alternative shots (to overcome a problem), briefly indicate whether you accept or disagree. If you have time, explain why.
	• Where practicable, at the end of each sequence, ask whether there are any problems, and whether anyone needs to rehearse that section again.
	• Remember that various staging and lighting defects may be unavoidable in early rehearsals. Certain details (set dressing, light effects) take time to complete. Some aspects need to be seen on camera before they can be corrected or get final adjustment (overly bright lights, lens flares, etc.). Shot readjustments during rehearsal often necessitate lighting changes.
	• At the end of the rehearsal, check timings, and let the performers and crew know if there are any errors to be corrected, changes needed, or problems to be solved. Check whether they had difficulties that need your attention.
	• It is important to get at least one complete uninterrupted rehearsal.

Studio Rehearsal

Before the studio rehearsal, the stage crew, supervised by the set designer, erects and dresses the set. Lamps are rigged and adjusted under the guidance of the lighting director. Camera and sound equipment are then positioned. The performers arrive, seeing the set possibly for the first time. The studio rehearsal is ready to begin (Figure 18.1 and Table 18.1).

Directors organize their studio rehearsals in several ways, according to the complexity of the production, available time,

and the performers' experience. Following are some of the options (Table 18.2).

Walk-Through (Dry Blocking)

Actors perform, familiarizing themselves with the studio settings, and so on, while the studio crew watch, learning the format, action, and production treatment. The director is usually in the studio. The camera crew usually leave their cameras alone and just look at their script.

FIGURE 18.1

The set of CBS's The King of Queens sitcom is shown here, ready for the next rehearsal. The set has been put in place and the lights have been rigged.

Source: Photo courtesy of CBS/Robert Voets/Landov.

Camera Blocking (Stumble-Through)

This is the initial camera rehearsal, coordinating all technical operations, discovering and correcting problems. The goal is to make sure that the corrections worked and that the timing is appropriate (Figure 18.2).

Dress Rehearsal (Dress Run)

The goal of the dress rehearsal is to time the wardrobe and makeup changes. Notes about issues are taken and then shared with everyone at the end of the rehearsal.

FIGURE 18.2 *Photographs are used as stand-ins for the cameras to rehearse with at the Academy of Country Music Awards. The photos allow the camera operators to learn where each nominee or performer will be sitting during the show. This helps the operators know where to point their cameras when an award is announced.*

Source: Photo courtesy of Monty Brinton/CBS/Landov.

TABLE 18.2 *The Director's Instructions during Rehearsals and Productions*

To cameras	Opening shots please	• Cameras to provide initial shots in the show (or scene).
	Stand by 2	• Stand by cue for Cam 2.
	Give me a one shot (single), two-shot; three-shot/group shot	• Isolate in the shot the one, two, or three person(s) specified. A group shot includes the whole group of people.
	Zoom in (or out)	• Narrow (or widen) the zoom lens angle.
	Tighten your shot	• Slightly closer shot (slight zoom-in).
	Get a wider shot	• Get the next wider standard shot, such as from a CU to an MS.
	More (or less) headroom	• Adjust space between top of head and top of frame.
	Center (frame-up)	• Arrange subject in the center of the image.
	Pan left (or right)	• Horizontal pivoting of camera head.
	Tilt up (or down)	• Vertical pivoting of camera head.
	Dolly back/out	• Pull camera mounting away from the subject.
	Dolly in	• Push camera mounting toward the subject.
	Truck left (or right)/crab left (or right)	• Move camera and mount to the left (or right).
	Arc left (or right)	• Move camera and mount around the subject in an arc.
	Ped up (or down)	• Raise (or lower) the column of the pedestal mount.
	Crane up (or down)/jib up (or down)	• Raise (or lower) crane arm or jib (boom).
	Lose focus on the building	• Defocus the building, remain sharp on other subject(s).
	Follow focus on the horse	• Maintain focus on the moving horse.
	Focus 2, you're soft	• Criticism that Cam 2's shot is not sharply focused.
To the floor manager	Opening positions please	• Talent (and equipment) in position for start of show (or scene).
	Standby	• Alert host to get ready to start the show.
	Cue action/cur	• Give sign for action to begin (perhaps name person or character).
	Back to the top/take it from the top	• Begin again at the start of the scene; repeat the rehearsal.
	Pick it up from shot 20	• Start rehearsal again from script shot 20.
	Clear 2's shot	• Something or someone is obscuring Cam 2's shot.
	Tighten them up	• Move them closer together.
	Show Maria 3/she's on 3	• Indicate to her Cam 3, which is shooting her.
	Tell host to stretch/keep talking	• Tell host to improvise until the next item is ready.
	Give host 2 minutes . . . 1 minute . . . 30 seconds	• The host has 2 minutes left (followed by countdown on fingers).
	Kill it/cut it	• Stop immediately.
	He is clear	• We have left him. He is free to move away (or relax).

TABLE 18.2 *continued*

To audio (sound mixer)	Fade-up sound	• Fade-up from zero to full audio.
	Stand by music	• Warning before music cue.
	Cue music/go music	• Start music.
	Fade-in music	• Begin very quietly, gradually increasing the sound.
	Fade-out music	• Reduce volume (level) of audio.
	Music under/music to background	• Keep music volume low, relative to other audio sources.
	Start music	• Begin music, at full volume.
	Sound up (or down)	• General instruction to increase (or decrease) overall volume.
	Kill (cut) the music	• Stop the music.
	Cross-fade; mix	• Fade-out present source(s), while fading in the next.

TABLE 18.3 *Pre-Rehearsal Blocking Suggestions*

Timing	• A preliminary read-through gives only a rough timing estimate. Time needs to be allowed for prerecorded video or graphic inserts. Anticipate potential script cuts if an overrun is evident. Many productions include sequences that can be dropped, reduced, or expanded to obtain the correct timing. At this point, only the dialogue and blocking moves are timed (not laughter).
Briefing the talent	• Ensure that all performers have a clear idea of the program format, their part in it, and their relationships with others. • Ensure that performers have a good understanding of the setting: what it represents, where things are.
Props	• Provide reasonable substitutes when the real props are not available. Sometimes only the actual item will suffice.
Directing performers	• Maintain a serious attitude toward punctuality, inattention, and background chatter during rehearsals in order to avoid frustration and wasting time. • Avoid excessive revisions—of action, grouping, line cuts, and so on. Wrong versions can get remembered and new ones forgotten.
Shot arrangement	• A monitor may be helpful to arrange shots. • Always think in terms of shots, not of theatrical-styled groupings, entrances, and exits. • When setting up shots in a rehearsal facility, do not overlook the scenic background that will be present in the studio. Check shot coverage with plans and elevations. • Consider depth of field limitations in close shots or deep shots (close and distant people framed together). • Think in terms of practical camera operations. Although you may be able to rapidly reposition your handheld monitor, the move may be impossible for a studio camera.
Audio and lighting	• Be sensitive to audio and lighting problems when arranging talent positions and action. For example, when people are widely spaced, a mic boom may need time to swing between them, or have to be supplemented.

Rehearsal Procedures

In many types of productions, directors prepare and rehearse the entire action and treatment until it really works, then record the polished performance. In other productions, each small section is rehearsed and recorded before going on to the next segment.

Camera Blocking

Camera blocking—also known as the *first run* or *stumble-through*—is usually when the director controls the production from the production control room, only "going to the floor" when on-the-spot discussion is essential (Table 18.3). Otherwise, his or her eyes and ears are focused on the camera monitors and the speakers. All communication with the crew is through the intercom system, with the floor manager cueing and instructing the talent, receiving the director's guidance through the intercom (Figure 18.3).

Many directors use the *whole method*, going continuously through a segment (sequence, scene) until a problem arises that requires stopping and correction. The director discusses problems, solutions, and revisions, then reruns the segment with corrections. This method gives a good idea of continuity, timing, transitions, and operational difficulties. But by skimming over the various shortcomings before the breakdown point, quite a list of necessary minor corrections may develop.

Other directors, using the *stopping method*, stop action and correct faults as they arise—almost shot by shot. This precludes error adding to error, ensuring that everyone knows exactly what is required throughout. For certain situations (chromakey treatment), this may be the only rational approach. However, this piecemeal method gives the impression of slow progress, and can feel tedious for everyone involved. The continual stopping makes checks on continuity and timing much more difficult. Later corrections are given as notes after the run-through.

Floor Blocking

Using this method, the director works out on the studio floor, viewing the camera shots on studio monitors. Guiding and correcting performers and crew from within the studio, the director uses the intercom to give instructions to the AD or technical director for the switcher (Figure 18.3). During the actual recording, the show may be shot as segments, scenes, or continuous action.

THE FLOOR MANAGER

The floor manager (FM) plays a major role in studio production and is usually brought into the project from the beginning of production.

During the studio preparations for the rehearsal, the FM checks to make sure that the studio and crew are ready

FIGURE 18.3

Floor blocking allows the director to sit in the studio, reviewing the camera shots from a monitor. Note the quad-split monitor in the lower-right corner of the photo. The director can see the shots from all four cameras being used on this production.

to go at the scheduled time. Progress checks ensure that there are no staging hang-ups. This may include making sure that all action props work, the scenery and furniture are in the planned positions, the doors do not stick, and similar checks.

As the performers arrive, the FM welcomes them and makes sure that they are taken care of. This usually includes telling the talent when and where they are required. The goal is to ensure punctuality, which is an important aspect of the FM's duties, whether for rehearsal starts, turnarounds at the end of a rehearsal (returning equipment, props, sets to opening positions), or studio breaks (meals).

Rehearsal

When the director is in the control room, the FM is the director's contact on the studio floor. Wearing a headset, the FM listens to the director's instructions on the intercom system and then passes this information on to the talent and crew. The FM anticipates problems, rearranges action and grouping, adjusts furniture positions, and performs other tasks as instructed by the director, marking the floor where necessary and supervising staging and prop changes.

Normally, the FM is the only person in the studio who will cue and stop all action based on the director's instructions. Most directors rarely use the studio loudspeaker system to talk directly to the talent, as this can be quite disrupting. The FM talks back to the director over the intercom. Being on the spot, the FM can often correct problems not evident to the director in the control room (Figure 18.4).

A good FM combines calmness, discipline, and firmness with diplomacy and friendliness—making sure to put talent at ease, and always available, yet never in the shot. The FM maintains a quiet studio, yet understands that various last-minute corrective actions may be needed.

FIGURE 18.4 *The floor manager, or stage manager, is the director's representative in the studio. The floor manager is the person in the middle of the bottom of the photo.*

TABLE 18.4 *Methods of Recording Productions*

	Advantages	Disadvantages
Live-to-tape Utilizes a single camera or multiple cameras.	• Program is ready for transmission immediately after taping. • No lengthy postproduction editing is needed (saves cost, equipment, staff, time). • Recording period needed is only slightly longer than the length of the show. • No loss of pace or interaction, which can occur when action is continually interrupted. • Performers and team are all pumped up to give their best performances.	• Most shots and transitions are normally unchangeable. • There are no opportunities for second thoughts or alternative treatment. • The production may include some errors in performance and treatment (camerawork, sound, lighting). • All costume, makeup changes, scene changes, and other changes have to be made in real time during the show. • Any editing is usually nominal, unless insert shots are recorded afterward (cutaway shots or brief retakes inserted).
Shooting in segments/scenes/sections Utilizes a single camera or multiple cameras.	• Opportunities for corrective retakes and alternative treatments are available. • Editing within the sequence can be made with the switcher. When necessary, further editing decisions can be left to editing. • The talent and crew can concentrate on each segment and get it right before going on to the next. • Recording breaks can be arranged to allow costume, makeup changes, scene changes, and so on. • Scenes can be shot in the most convenient time-saving order: in sequence or out of sequence. • Performers who are not required for later scenes can be released (actors, extras, musicians). This can save costs and reduce congestion in the studio. • If a setting is used at various times throughout a show (e.g., a classroom), all scenes with action there may be shot consecutively, and the scenery struck to make room for a new set. This technique virtually extends the size of the studio, and makes optimum use of available space. • Where a set appears in several episodes of a series, all sequences there can be shot in succession to avoid the need to store and rebuild the set over and over. • Similarly, when a program consists of a series of brief episodes (a story told in daily parts), these can all be recorded consecutively during the same session.	• Much more time is needed to record the show. Time lost during breaks can become greater than anticipated. • Retakes and revisions can take so much time that the session falls behind schedule (an overrun). This can result in economic and administrative problems. There is a continuing pressure to save time and to press on. • Extensive postproduction editing is essential. At the end of the recording session, you have no show until the separate segments are sorted and edited together. • All extended transitions between segments (mixes, wipes, video effects) must wait until postproduction. Similarly, music/effects/dialogue carrying over between segments must be added during audio sweetening. • In a complicated production, recording out of sequence can lead to confusion and to continuity errors in a fast-moving taping session.
Multi-camera with ISO camera All cameras are fed to the switcher and then recorded by a master recorder. The director may choose one camera of which the video signal will be recorded by an ISO recorder.	• The ISO recorder can be used to continuously record the output of any chosen camera. • Provides slow-motion replay for inserts into the program. • Provides extra shots for postproduction editing.	• The arrangement ties up a second recorder. • Most of the material on the ISO recording is not used.

TABLE 18.4 *continued*

	Advantages	Disadvantages
Multi-camera with dedicated recorders/ camcorders Action is shot live-to-tape or in segments/scenes/ sections.	• Every camera has its own recorder. • No shots are missed. • Alternative shots are available throughout the show. • No irrevocable editing is carried out during the performance. • Unexpected action or unrehearsed action is recorded and available. • Postproduction editing can be adjusted to suit the final performance. • Camera treatment needs little planning. • Most nonlinear editors allow multiple recordings to be synced for switching.	• The entire show requires total postproduction editing. • There is a separate recording for each camera, creating a large amount of material. • The arrangement can lead to poor production treatment; off-the-cuff shooting. Concentration on action rather than visual development. • When camcorders are used, an effective program requires the director to give advance instructions to each camera operator.
Single camera Single camcorder taking brief shots or continuous sequences.	• An extremely flexible, mobile method. • One camera is easier to direct than a multi-camera crew. • There are advantages in working with a small unit. • The director can concentrate on the action and treatment. • No switching is involved. • No cueing of preproduced inserts.	• All the strain of camera work falls on one operator. • The production process is much slower than multi-camera treatment. • There is a temptation to shoot retakes for better performance, or extra "just-in-case" shots that might be used. These extra shots can absorb allotted shooting time. • All material requires sorting and detailed editing. • Production flow is continually interrupted (unlike multi-camera treatment). • All preproduced inserts in the program must be introduced during postproduction.

Recording/Transmission

When the director is ready to record or transmit the program, it is the FM's responsibility to confirm that all studio access doors are shut, talent and crew are standing by in their opening positions, and the studio monitors are showing the correct images. The FM will cue the talent or announcer.

During a studio production, the FM is responsible for cueing talent, being aware of everything happening in the studio, listening to intercom guidance, anticipating problems, and generally smoothing proceedings (Figure 18.5). At the end of taping, the FM holds talent and crew until the director has had a chance to determine if retakes are needed and "releases the studio" (talent and crew). The FM checks on safe storage of any valuable or special props or equipment and ends duties with a logged account of the production.

Guiding the Talent

How much guidance the talent needs varies with their experience and the complexity of the production. However, you must always ensure that the talent understands exactly what you want them to do and where to do it. A preliminary word may be enough, or a painstaking rehearsal may be essential.

Inexperienced Talent

Having welcomed talent, put them at ease and explain what is needed of them. They are best supported by an experienced host who guides and reassures them. Have them talk to the host rather than the camera, as the host can steer the guest by questions. Unfamiliar conditions make most people uncomfortable. However, it is important that the talent feels self-confident—that confidence is essential for a good performance. Keep problems to a minimum, with only essential instructions to the talent. Avoid elaborate action, discourage improvisation, and have minimal rearrangements. Even slight distractions can worry inexperienced talent. Sometimes a small cue card or a list of points held beside the camera may strengthen their confidence. However, few inexperienced people can read a script or teleprompter naturally—they usually give a stilted, uncomfortable delivery.

FIGURE 18.5
During a production, the floor manager is responsible for cueing the talent. In this photo, a floor manager is cueing the talent on NBC's Today Show set.

FIGURE 18.6
Familiar with the production process, professional talent such as Disney actor Dean Jones can respond to the most complicated instructions from a director or floor manager.

The balance between insufficient and excessive rehearsal is more crucial with inexperienced talent. Uncertainty or overfamiliarity can lead them to omit sections during the final program. Sometimes the solution lies in either taping the production in sections or compiling from several takes.

Professional Talent

You will meet a very wide range of experienced professional performers in television production—actors, presenters, hosts, commentators, demonstrators, and anchors. Each has a specific part to play in a production. Familiar with the studio process, they can respond to the most complicated instructions from the FM, or an earpiece intercom without the blink of an eye, even under the most difficult conditions—and yet maintain cool command of the situation. Comments, timing and continuity changes, item cuts, and ad-libbing and padding are taken in stride.

The professional can usually make full use of a teleprompter displaying the script—this requires that cameras must be within reading distance. The talent can play to specific cameras for specific shots, making allowances for lighting problems (e.g., shadowing) or camera moves (Figures 18.5 and 18.6).

CUEING

To ensure that action begins and ends at the instant the director wants, precise cueing is essential. If you cut to performers and then cue them, you will see them waiting to begin or watch action spring into life. Cue them too early before cutting and action has already begun. Wrong cueing leaves talent bewildered. If they have finished their contribution and you have not cut away to the next shot, they may stand there, not knowing what to do, wondering whether to ad lib or just smile.

Methods of Cueing

Hand Cues

Performed by the floor manager, these are the standard methods of cueing action to start or stop. Remember, some companies have their own variations on the signals, and it is advisable to use whatever is given to you by your supervisor. However, make sure that your signals can be understood by the talent. Explain the basic cues to them if necessary. Do not assume that they know what you are doing—especially with inexperienced talent (Figure 18.7).

Word Cues

An agreed-upon word or phrase during a dialogue, commentary, or discussion may be used to cue action or to switch to a prerecorded package (insert). It is also important to note out-cues (the last spoken words) at the end of a preproduced television package to ensure that the talent knows when to begin again.

Monitor Cues

Commentators and demonstrators often watch the preproduced package on a nearby television monitor, taking their cues from seeing the package end.

Tally Light Cues

Performers can take a cue from the camera's tally (cue) light that lights up when the switcher selects a specific camera. Announce booths sometimes use a small portable cue light to signal the commentators to begin (Figure 18.8).

Intercom Cues

These cues are given directly to a performer (newscaster, commentator) wearing an earpiece—it is usually called an IFB, which stands for interrupted (interruptible) feedback. It can also be called a program interrupter (Figure 18.9).

Clock Cues

The talent may be told to begin at an exact time.

Prompting the Talent

Rarely can talent be expected to memorize a script accurately and deliver it easily. Even the most experienced performers are liable to deviate from the written script and forget or drop lines. Although this is not a big issue with some types of shows that are spontaneous, it can be serious with programs such as news shows.

There are a number of different ways that the talent can get help with delivery:

- *Notes:* 4–6 inch index cards are probably the most common note size. They are also popular, because the card-stock does not crinkle. Most talent prefer not to use standard sheets of paper on an interview show, because they can distract the audience and can crinkle, making unwanted noises. Standard-sized paper is usually used for news scripts on a news desk or table.

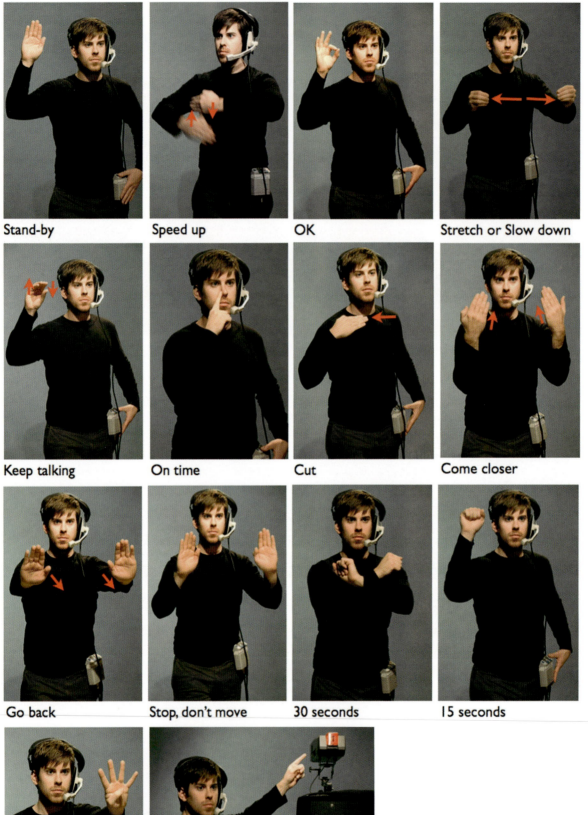

Stand-by

Speed up

OK

Stretch or Slow down

Keep talking

On time

Cut

Come closer

Go back

Stop, don't move

30 seconds

15 seconds

4 minute countdown

You are on this camera

FIGURE 18.7

Floor manager signals or cues.

Source: Photos by Austin Brooks.

FIGURE 18.8 *Camera tally lights—note the red light on the studio viewfinder—are often used to cue the talent when to begin their presentation.*

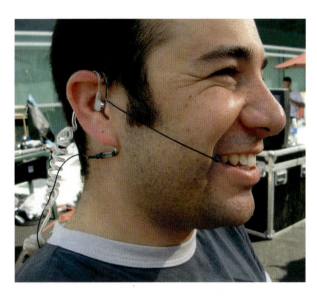

FIGURE 18.9 *This ESPN talent uses an IFB (intercom) earpiece to hear cues from the producer. He also has a small boom microphone attached to his ear.*

Source: Photo by Dennis Baxter.

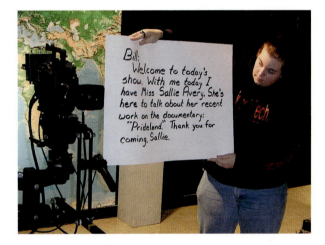

FIGURE 18.10 *Cue cards should be held so that the crew member holding the card can see the wording. This allows him or her to move the card up, adjusting the card so that the line being read is even with the lens.*

FIGURE 18.11 *Mobile telephones can be used as field teleprompters. One of the major benefits is that the latest script update can be emailed directly to the prompter.*

Source: Photo courtesy of Bodelin Technologies.

- *Cue card*: Large card-stock can be used by a floor crew member to show brief notes, an outline, or other information. These cards are usually held so that the line being read is even with the lens (Figure 18.10).
- *Teleprompter*: Teleprompters generally show the script on a two-way mirror that is positioned directly over the camera's lens. Other devices can also be used, such as a lightweight flat-screen monitor, a tablet computer, or a mobile phone. A teleprompter operator can vary the speed of the text on the screen so that it is comfortable for the talent. Cameras equipped with a teleprompter must be close enough to the talent for comfortable reading (Figures 18.11 and 18.12).

FIGURE 18.12 (Left) The camera is equipped with a teleprompter. (Right) What the text looks like from the talent's point of view.
Source: Left photo courtesy of JVC.

Production Timing

Broadcast television productions run by the clock. They start and stop at specific times, because they must fit their allocated time slot exactly. If a live show goes over the time limit, it may just be cut off or will cause problems with other programs. Shows that are live-to-tape will get edited so that they are the appropriate length.

Scripts may be roughly timed by reading aloud and allowing time for any mute action, commercials, or preproduced material. Most scripts run at roughly a minute per full page. During the rehearsal, each segment or scene is timed, then adjusted to fit into the allocated time slot.

RECORDING THE PRODUCTION

Although "live" productions were how television began, most television programs today are prerecorded. Recording a program allows the directors more freedom. They can edit out the mistakes and enhance the show with special effects that may not have been possible in a "live" situation. So it is not surprising that today, most shows are recorded whenever possible. Let's take a look at the variety of recording methods that are used, including their advantages and limitations (Table 18.4).

The most common recording methods are:

- Live-to-tape.
- Basic retakes.
- Shooting out of order.
- Isolated camera (ISO)—each camera has its own recorder.

Live-to-Tape

Here, the program is recorded continuously in its entirety, as if it were live. Most editing is completed on the production switcher during the performance. While the production may be done at the conclusion of the show, most of the time additional postproduction will be required.

Basic Retakes

Here, the production is recorded continuously, the same as for a live production. However, any errors (performers, production, technical) are rerecorded and the corrected sections substituted. Duration trimming may require some editing cuts—often covered by introducing cutaways/reaction shots.

Shooting out of Order

Many productions are shot out of order, stopping and starting as needed. For example, if one set is needed both for the beginning and the end of a production, both segments can be shot one after another. This method allows the set to be removed and another brought in for the next scene, which saves lighting and set-building time.

Isolated Camera (ISO)

In this approach, each camera's output is separately recorded on its own dedicated recorder or camcorder. Images from the ISO camera can be selected on the switcher whenever they are needed, but all its shots (on-air and off-air) are recorded on the isolated recorder. This arrangement provides several useful opportunities. ISO shots can be used a number of different ways:

- *Instant replay shots*: The ISO recorder can be used to provide instant replay shots in slow motion during a production. At any chosen moment (such as after a goal has been scored in a game), the ISO recorder is used to play its recorded footage into the program.
- *Cover shots*: The ISO camera can be used to concentrate on one specific aspect of the scene (such as the goal area).

- *Standby shot*: When a production switcher is used on a multi-camera shoot, the resulting program recording contains all the shots of the event that are available. So if you subsequently need to modify the final recording, the ISO camera recording can be edited into the program wherever necessary as cutaways, to allow changes or corrections to be made.

Single-Camera Recording

In this situation, a single portable camcorder shoots the action. As the main production camera, the material is recorded in any convenient order. Because no switching is involved, the director has the opportunity to concentrate on the action on the set.

INTERVIEW WITH A PROFESSIONAL: FIONA CATHERWOOD

Briefly define your job at BBC. Compiling and overseeing scripts, camera cards, running show formats and schedules, setting up technical planning meetings, and supervising and distributing the technical requirements. Responsibilities included assembling scripts, timing, continuity, and sometimes logging. I worked closely with the director, talent, and the crew as a team. My work also included postproduction paperwork, research, and clearances for music.

What did you like about your job? The "buzz" of live television, working with large OB remote crews. The programs I have worked on have all demanded the highest levels of organization, multitasking, and initiative, and have given me the opportunity to develop excellent relationships with Buckingham Palace, members of the armed forces, the Palace of Westminster, the Church, and government departments.

FIGURE 18.13
Fiona Catherwood, former script supervisor and production assistant, BBC.

What are the types of challenges that you faced in your position? Quick program turnarounds, responding to "live" situations resulting in changes in running orders and consequent timing changes, being able to add up time and respond quickly to situations as they happen. Always to remain calm.

Are there things that you did to specifically prepare to work on a production? Research as much as possible into the history and background of programs that are coming up. Get as much experience and training as possible.

What suggestions or advice do you have for someone interested in a position such as that? Confidence, good training, and be prepared to work your way up from the bottom. I joined the BBC as a production secretary and initially followed and watched people doing the role I wanted to eventually do.

Fiona Catherwood's experience of over 20 years at the BBC included events such as the Royal Wedding of HRH the Prince of Wales, State Opening of Parliament, Rugby World Cup Victory Parade, and the Queen's Jubilee Address.

INTERVIEW WITH A PROFESSIONAL: LAUREN CLAUS

FIGURE 18.14
Lauren Claus, director.

Briefly define your job. Basically, what I do is block and direct a half-hour newscast once a day. I also often create graphics for the show that day.

What do you like about your job? I love that my job is different every day and challenges me to constantly improve. Since no two shows are ever alike, I always have to be on my game. A part of me also enjoys the adrenaline rush of being in the midst of a live show where anything could happen.

What are the types of challenges that you face in your position? I think the toughest challenge is really being able to think on my feet and always having a contingency plan for anything that could go wrong. In one live show, a myriad of things can go wrong, but the viewer won't notice if the director is doing his or her job correctly. I am still learning how to be able to put out "fires" so the show is still clean to air.

What suggestions or advice do you have for someone interested in a position such as yours? Persistence and a go-getter attitude. Do everything you can to improve at your position. Ask a lot of questions, be early, all the time. Lastly, love what you do; the minute you stop loving it, stop doing it.

Lauren Claus is a director at a local television station.

REVIEW QUESTIONS

1. What are some of the situations that can happen when an unrehearsed studio production occurs?
2. What are the goals of a pre-rehearsal?
3. What are the differences between dry blocking and camera blocking?
4. Describe the role of the floor manager in the studio.
5. What are some of the cueing methods in the studio?
6. What are some of the devices used to prompt the talent?
7. How do you help build the confidence of inexperienced talent?

CHAPTER 19

DIRECTING TALENT

"Many directors expect actors to be like Apple computers: plug-and-play, performing perfectly right out of the box. Writers work for months on a script; directors take weeks to create their vision. Actors are often cast only a few days before the shoot and have little—if any—time to prepare. Actors need time to catch up to everyone else on the crew. They are not machines; they are fallible humans desperate to do their best. Actors must feel safe and supported on their creative journey."

—John Badham, Director

FIGURE 19.1 *"Talent" generally refers to anyone who appears in front of the camera. That can include a wide variety of people, from anchors to actors to athletes to the man on the street doing an interview.*

The talent you choose can make or break a production. The face and the body language of a person communicate faster to an audience than any other method. Showing a close-up shot of the face quickly communicates the story as the actor expresses disgust, joy, sorrow, and seriousness. Many video productions totally rely on how well people perform in front of the camera. It is up to the director to get the best performance from the talent. The director must be able to communicate his or her vision for the program and keep the talent motivated and informed. Coaxing the best performance can be tough, but it is well worth it in the end.

TALENT

Although the broad term "talent" is widely used to refer to those appearing in front of the camera, we must always remember how varied this talent really is. "Talent" covers a remarkable spread of experience and temperaments, from the

FIGURE 19.2 *News talent records voiceovers for story packages.*
Source: Photo by Jon Greenhoe.

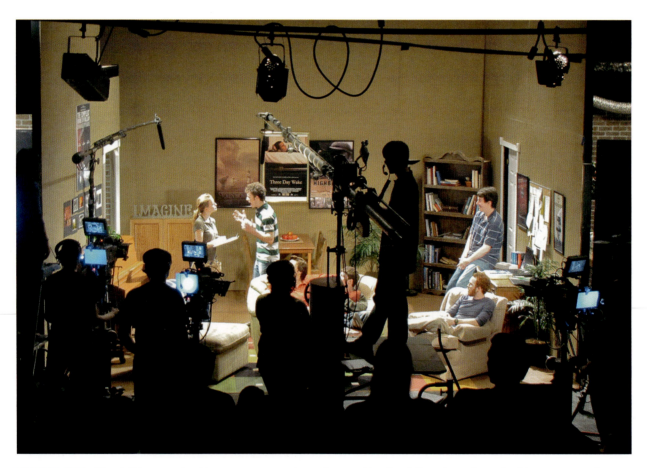

FIGURE 19.3 *(Above) Talent must be able to concentrate on the work, even with a lot of distractions going on involving cameras, audio personnel, lighting, and other crew members.*

FIGURE 19.4 *Note that while these two sports commentators rehearse, audio and production personnel are working around them, testing their mics and adjusting the talent's headset on his neck.*

Source: Photo by Josh Taber.

professional actor working to a script and playing a part, to the impromptu, unrehearsed interview with a passerby on the street (Figure 19.1).

When the director invites someone to appear in front of the camera, that is only the beginning of the story. Since the person's performance will have an influence on the program's success, it is important that the director helps in any way needed, to make the talent's contribution as effective as possible. The television camera can be an unwavering critic. Under its scrutiny, the audience weighs arrogance, attitude, and credibility, while sending sympathy to those who are shy and ill at ease.

For practical purposes, we can divide talent into the *professional performers*, who are used to appearing in front of the camera, and the *inexperienced*, for whom the program is likely to be a new, strange, exciting yet worrying event.

Professional performers usually like to work through a prepared format. Some, such as an actor in a play, will learn

FIGURE 19.5 *A "standup" is when talent are placed in front of something that provides context for the story being covered.*

their lines, their moves, and the mechanics of the production. Others work from an abbreviated cue sheet near the camera or read running text from the screen of a teleprompter. Some will extemporize from guide notes; others will read from a printed script (Figure 19.2).

The best professionals can be relied on to repeat, during the actual taping, the dialogue, the moves, the pace, and the timing that they gave in rehearsal. The talent needs to know what he or she is going to say and that he or she can modify the delivery of the piece to fit the situation, showing enthusiasm, vigor, calm detachment, patience, or reverence. Talent can take guidance and instructions and follow this through without being confused. Good talent can improvise when things go wrong and remain calm if the unexpected happens (Figures 19.3–19.5).

IMPORTANCE OF PEOPLE IN THE SCENE

Suppose you are documenting the red carpet at a major awards ceremony. You are trying to describe what it is like to be on the red carpet, or even standing next to the red carpet. You can shoot footage of the extravagant decorations, the glowing lights, and you can capture the sounds. If you begin shooting early enough, you can get uncluttered views of the carpet. Obviously, none of these shots really convey the atmosphere as the stars descend upon the carpet.

On the other hand, if you begin shooting as the limousines arrive and the stars begin to descend upon the red carpet to the delight of an adoring crowd, you can capture the mood of the event. While it may be more difficult to capture the footage you want, being jostled by other people, the place now has exactly the right feel to it. This is a reminder that it is worth the battle, in order to help the viewing audience get a feel for what it was really like to be there (Figure 19.7).

Finally, do not overlook how important people are in giving life to a scene. Rather than pan along an empty sidewalk, it is more interesting to follow someone who is passing (although there is the possibility that the audience may now wonder who this newcomer is, speculating on the person instead of on the real subject).

A shot looking along an empty bridge is likely to bore your audience within a few seconds. But have someone walk across that same bridge, and the scene immediately becomes more interesting. Now you can hold the shot much longer while a voiceover tells us about the bridge's history.

Selecting the right talent to fit the script and then helping the talent understand what you need from him or her to make a successful production are two of the director's greatest challenges. However, when the talent gets it right, with the right script and with the right director, the results can be incredible.

THE ACTOR'S CRAFT

"The best advice I've ever received about directing came from a veteran motion picture cameraman who had successfully switched to feature film directing in mid-career. The advice was this: no matter which medium, video, film, theater, etc., and no matter what kind of production—sales training tapes to news segments to documentaries to feature films—as long as people are involved, a good director must have a knowledge of the craft of acting.

A knowledge of the actor's craft will fundamentally change your approach to directing any kind of program in which people appear on camera. I took his advice and can now declare: he was right!"

—Frank Beacham, Director

FIGURE 19.6 *Actors as seen through a camera monitor.*
Source: Photo by Tyler Hoff.

FIGURE 19.7 *People bring life to scenes, making the image more interesting to the audience. This image captured the atmosphere of the red carpet at the British Academy Awards.*

PRESENTING THE INFORMATION

Just because the talent may be an expert about a specific subject does not mean that he or she can present it convincingly to the camera. Ironically, there are times when a talent working from a teleprompter script may appear more authentic, even though the words may have been written by someone else.

How the action is shot is influenced not only by the nature of the subject, but by the talent involved. The camera can behave like a bystander who moves around to get a good view of whatever is going on. This happens in street interviews. At best, one hopes that whomever is being interviewed will be preoccupied with the questioner and barely notice the camera and microphone. However, it can sometimes be difficult to relax when one is being besieged by the curiosity of bystanders.

SELECTING TALENT

It is important to find the right person for your project. Sometimes you just have to go with your gut feelings. This may be one of the most difficult decisions that you have to make in the whole production. Talk with the talent, get to know them a little. Building a bit of a relationship will help you see what they are like, and if they will work well for your production.

New directors often have to deal with limited budgets, and one of the areas they often are tempted to cut is the budget line for professional talent. There are basically three types of talent for productions:

- *Professional talent.* Professionals cost more money up front. However, their quality usually facilitates a better final project, and they usually can complete the project much faster than the two other types of talent discussed next (Figure 19.8).

- *University/college theater performance students.* Directors can usually "hire" students for very little money or even for free. The advantage is that these students have usually received training in acting and are often willing to work hard just for an experience they can put on their resume. So these students cost less than professionals but may take a bit more time to create the final project, including more retakes.

- *Amateur talent.* It is always tempting to grab a relative or friend to be the talent on a project. The problem is that amateurs do not have the experience needed to complete the project in a timely fashion or to master the pronunciation.

So, although they may be free, amateur talent may actually cost more money than professional talent if equipment or studio space is being rented for the production. The cost you

save on talent may be spent on rental costs, because the amateur can take much more time to complete a script. Not only can you save money by using professional talent, but the overall quality will probably be better.

THE CHALLENGES OF WORKING WITH TALENT

Like every other aspect of production, every person involved, talent in this situation, will have their own ideas about how things will work. As the director, it is your responsibility to guide the project along. However, you need to give them enough space to let them do the job well. Director Kristin Ross Lauderbach once commented that "one of the best compliments I received from an actor was that I gave him a box to play in and he had total freedom to experiment inside it, but I let him know if he went outside that box." You have to keep in mind that, as director, you know the big picture. It is your responsibility to help the talent have the confidence and understanding to fulfill the vision that you have for the project.

When working with talent, the following is important:

- Put them at ease. Explain what will be happening.
- Give them clear instructions, such as where to look and the direction you want the interview to go.
- Keep the area clear of distractions.
- Place items on a prearranged mark.
- Be aware of the time limits.

FIGURE 19.9 *Kristin Ross Lauderbach, director.*

What suggestions do you have for working with talent? Take the time to rehearse and talk scenes out with your talent before you ever get to set. Your entire cast and crew is looking to you for vision and guidance; make the time to help your talent develop their characters so you are not holding up the entire crew on shoot day. With that said, be willing to take a break on set if your actor is struggling with a scene. It might feel like you don't have time to do it, but you will thank yourself when you get back to the cutting room.

Inexperienced Talent

Many people who appear to be "natural performers" in the final program are really the product of an understanding, yet creative director. Typical ways of handling inexperienced talent include the following:

- You can put your performers at ease by making them feel welcome, letting them know what they need to do, making them feel that their contribution is an important and interesting part of the program, and assuring them that if anything goes wrong, there is nothing to worry about because the scene can be recorded again (unless it is a live program).
- Let them do it their way, and then edit the result.
- Let them reminisce, for example, then select the most interesting or the most relevant parts.
- Interview them in a situation that is natural for them (such as their workshop) rather than in a formal studio setting.
- Whenever possible, avoid showing the talent just standing, waiting for their part. They will feel more at ease (particularly in their own surroundings) if they have something to do. For example, a cook might be cutting up some vegetables before speaking.
- Give them a few instructions, such as telling them when, where, or how to hold the items in front of the camera.
- Use plenty of cutaway shots, so that the program does not concentrate on the talent all of the time. As much as possible, look at what they are talking about. During the editing phase, the cutaways can be cut into the program.
- Some directors keep the tension down by shooting when the talent does not know they are being recorded.

In a scripted program, it may be wise to record the dress rehearsal and, if possible, to run through the material several times, recording everything, so that the director can select and combine the best parts of each version.

If sections do not edit well, the interviewer can record a question or a statement later that will bridge what would otherwise be disjointed material. With care, this can be done quite well, even though it was not part of the original interview.

The Host

Many of the problems of handling inexperienced talent can be reduced if someone who is familiar with the production routine offers support. The person who serves as the host will probably meet the guest beforehand, help the person to feel at ease, and explain what will be happening during the program (Figure 19.10). The host can also gently guide the guest through the interview by posing the questions within the right context: "Earlier, you were telling me about . . ." or "I wonder if we could look at . . ." or "Isn't the construction of this piece interesting?" The host can move the program smoothly from one topic to the next. Especially when shooting intermittently or out of order—both very confusing to the novice talent—the host can be very effective at guiding the guest through the production.

In situations where the talent will be discussing an object that must appear on camera, the host can help the guest know how to handle the object. Instead of the nervous guest holding the prop in a way that does not work for television (partly hiding it, reflecting the lights, or moving it around in a close-up shot), the host can tactfully take it from the guest to look at it and then hold it in a way that facilitates a good camera shot. At times, the host can also look in a nearby monitor to check the shot.

The Off-Camera Host

Talent do not always appear on camera. Some hosts are required to sit in a booth and commentate while watching the event on monitors. As pay-per-view Internet extended versions of broadcast programs continue to grow in popularity, this practice will increase. The problem with off-camera broadcasting is that the talent needs to keep the excitement in their voice and act as though they are there. This can be difficult at times when the talent may not even be in the area of the event (Figure 19.12).

When There Are Problems

There are times when the invited guests simply do not live up to expectations. Perhaps he is too stiff. Or maybe she is so hesitant and nervous that she does herself little justice. Then there are the guests who overcompensate and become loud and obnoxious. In a live show, there may be little opportunity to actually replace someone. If there is a rehearsal and it is obvious that the guest is not going to work, it is possible to cut that individual from the show. However, doing so can create a difficult situation. It may be possible to tape the segment instead of going live, in the hope you will be able to use at least some of the person's contribution. If a shaky performance becomes a serious issue, however, it can be cut in postproduction.

FIGURE 19.10 *One of the hosts from NBC's Today Show, sitting in the light shirt, is prepping his guest for the next segment of the program.*

THE ACTOR

"A jolt of panic rises in even the most experienced actor when he starts to perform in front of other people, even if it is just a rehearsal. He is putting all his talent and self-image on the line and is probably terrified that his audience may not buy into what he is doing. What if the crew hates it? It would be a personal rejection of the actor. The more he uses his own psyche to build the character, the more vulnerable he becomes. There is one key principle about working with actors: actors must feel they are in a place where they are totally safe."

—John Badham, Director

FIGURE 19.11 *Actors getting ready for a scene.*
Source: Photo by Tyler Hoff.

FIGURE 19.12
This reporter is in a trailer at an event, commenting on a sporting event webcast worldwide on NBC.com. Keeping your energy going as a commentator, when you are not actually at the event, can be difficult.

Other ways to deal with a problematic guest are to simplify or to shorten the person's segment. You can use multiple cutaways, such as close-ups of details, to cover the edits during postproduction. If necessary, these cutaways can be shot after the main show is recorded.

Another approach is to have the host present the items in the program while continually referring to the guest. For example, the host could pick up an item, hold it to the camera, and remind the guest of the information he or she discussed before or during rehearsal. At worst, the guest has to do little more than say yes and no.

As a last resort, you can use a voice-over track for the guest's segment and use video images of the subject matter.

RESEARCH

For decades, television reporters, commentators, and anchors, along with newspaper reporters, were the main source of information for the public. Today, with a large number of competing cable channels, blogs, websites, Facebook, and Twitter, commentators have the difficult task of being innovative every time they are on the air. Whatever event you are covering, success is usually based on research. Talent has to be able to comment on the issues of the day and be

able to describe and interpret the event to the listeners in a knowledgeable way. If your talent does not understand the subject, the audience probably will not understand it.

WHAT SHOULD TALENT WEAR?

Talent often ask the question, "What do I wear to this shoot?" Good question. Here are some basic guidelines:

- *Appropriateness is key.* The clothing needs to fit the situation. If it is a studio interview, you want the talent to wear nice clothes and polished shoes. If they are on location at car race, you won't want them in a suit. If they are on a dramatic set, they need to dress to fit the character, which may mean dirty shoes, T-shirt, and jeans. The clothing needs to fit the situation.
- *Wear low-contrast clothes.* Contrasting colors are a major problem. Video cameras see differently than the human eye. Cameras cannot handle the contrast between light and dark objects, when the camera shows detail in the light object and the dark becomes indistinguishable. The reverse is also true; when detail is shown in the dark, the light colors become washed out, losing all of their detail. Contrast is also a problem for clothing and skin tone.

Very light-skinned people should stay away from black or dark clothing when on television and dark-skinned people should avoid wearing white or very light colors such as pink and yellow. White clothing, including shirts or blouses, is discouraged for television.

- *Blue is universal on television.* The clothing color that seems to work best on television is a medium blue.
- *Watch the details.* Socks or stockings should be high enough so that they cover the leg when crossed. Collars should be positioned correctly. Hair should be in place.

- *Minimal jewelry is used on television.* Generally, talent wears small jewelry since large shiny objects cause glares from the lighting. Small chains or pearls are usually non-reflective. There are times that even watches can cause problems.
- *Street makeup is usually appropriate.* HD cameras are so sensitive that they no longer require the heavy makeup used in early television. Of course, special makeup may be required in order to fit a certain look or character.

INTERVIEW WITH A PROFESSIONAL: SARAH LECKIE

When shooting people, what do you think about? What do you look for? Most of the action I film is people sitting and talking, which can get really boring. Because the action is extremely boring, I have to liven up the shot with something interesting. So I have to look for interesting angles. I usually try to look for something I can juxtapose the action with (e.g., a guitar in the foreground or something). You can't just think about the action. You have to think about what surrounds the action and whether or not it's relevant to the shot or will make it more appealing or more understandable. If I am filming someone playing soccer, I have to think about how relevant the context of the action is to the image. For instance, if I am filming a soccer game at a stadium, people know what stadiums look like and don't need a lot of stadium shots. I can focus on close-ups of the action. However, if I am filming a soccer game inside a prison, I'll probably want to widen up my shots to shoot the context of the prison. It will make the shots more dramatic.

FIGURE 19.13
Sarah Leckie, director/videographer.

What are challenges to shooting people? Usually the challenge is to get people to not look at the camera; to get them to act like I am not there. Communicating exactly what I have pictured in my head and to get them to see that image can be extremely difficult. Also, getting the camera actions perfectly lined up with the people's actions is a challenge. For instance, if I am rack focusing or zooming or tracking while they look away and then at the camera, the timing has to be perfect! This is difficult.

How do you get people to do what you need them to do in your productions? I usually have to do a lot of coaching. I have to be able to not only tell them what to do in an interview, but often I have to take the time to show them exactly what I want them to do. It takes a lot of patience.

Sarah Leckie is a corporate television producer/director who creates international documentaries.

INTERVIEW WITH A PROFESSIONAL: MORGAN SCHUTTERS

What is your job? I report the news: daily pitch lead-able story ideas, shoot, edit, interview, set up and coordinate contacts/work my beat, report, go on camera live news, newsroom shots, out in field or in the studio reporting.

What do you like about your job? I love my job because I get to meet new people every day allowing me to develop new relationships. It is fast-paced; I get to use my energetic personality and talkative friendly nature. I really enjoy the fact that it is never the same old routine—it's always different and new and exciting, always fast deadlines and pressure—stressful, but I like the stress and pressure. I love talking and writing and telling people's stories.

FIGURE 19.14
Morgan Schutters, reporter.

What are the types of challenges that you face in your position? It's emotionally draining. I arrive at work already tired and leave work tired, and the same cycle of exhaustion—physical and mental—is constant. The stress is hard to handle sometimes—tight deadlines, you sometimes feel your work is never good enough, constantly criticized on your appearance, writing, editing, etc.

How do you prepare for a production? I focus, focus, focus, and I do not relax, stop to think, or even catch my breath until I am done with my live story or whatever my deadline is that day. Often there is barely enough time to even touch up my hair and makeup before I go live, so there isn't much preparing—but I always try to read my script through at least once before I say it and I always try to memorize my lead; it's more conversational that way.

What suggestions or advice do you have for someone interested in a position such as yours? If you want to be a reporter or in the news industry in any way, you better be tough, have a thick skin or be willing to develop one, be willing to work long hours for barely any money, and face constant criticism from people outside the industry and inside it. You can't let anything get you down. You have to keep your chin up, and if you don't love it, don't do it. You have to have 100 percent passion for this career to survive. There are thousands of others just like you who want the same job you have. You're easily replaceable. The more internships, the better, with this job. You can never ever have too much experience. Also, the more mentors and contacts within the industry who are willing to help you with a tape and guide you along your journey of improving, the better, because you need genuine support in this career.

Morgan Schutters is a television news reporter.

REVIEW QUESTIONS

1. What are three of the challenges of working with talent?
2. What colors work best on television when selecting clothing?
3. What are two ways to deal with problematic talent?
4. When is it sometimes less expensive to hire a professional talent than one of your friends?

CHAPTER 20

PRODUCTION STYLE

"Drill down into what it is that excites you about the project and
harness it. Your passion as a director is what will make the show
happen. Make the show in your head and think through
everything that could happen."

—Tony Gregory, Director, Fashion Rocks

TERMS

Actuality: A type of production that is very transparent, even willing to deliberately reveal all of the production
equipment and crew to help at "authenticity."

Ambience: Production ambience influences the audience's perception of the show. Some of the ambience factors may
include music, graphics, and the set.

Display: An unrealistic, decorative way of presenting the subject to your audience. Game shows would be an example
of a display type of production.

Treatment: The production method used to encourage the interest of the audience. There are many different styles
or treatments that can be chosen such as narrative, comedy, news, documentary, etc.

There is no "correct method" of presenting any subject. Directors have tried a variety of approaches over the years. Some of these have become standard; others were just a passing trend. Techniques that have been used adroitly by some (such as background music) have been overdone by others, and become distractingly intrusive. Certainly, if you choose an inappropriate technique, you are likely to find your audience becoming confused, distracted, or simply losing interest.

VISUAL STYLE

Appropriateness

So what is appropriateness? In reality, it is largely a matter of custom, fashion, and tradition:

- *Informal presentations* usually take the form of "natural" situations. We chat with the craftsperson in his or her workshop, at a fireside, or while on a country walk (Figures 20.1 and 20.2).
- *Formal presentations* often follow a very stylized artificial format. We see people in carefully positioned chairs, sitting on a raised area in front of a specially designed set (Figure 20.3).
- *Display* is an unrealistic, decorative way of presenting your subject. Emphasis is on effect. We see it in game shows, open-area treatment (music groups, dance), and in children's programs (Figure 20.4).
- *Simulated environments* aim to create a completely realistic illusion. Anything breaking that illusion, such as a camera coming into shot in a period drama, would destroy the effect.

FIGURE 20.1 *Crew talking onboard the ship for the television show Whale Wars. The show attempts to show the program's talent in "natural" situations on the ship.*

FIGURE 20.2
Informal chats with talent walking in a hall.

- *Actuality* is a more transparent style. We make it very clear to the audience that we are in a studio by deliberately revealing all of the production equipment and crew. On location, the unsteady handheld camera and microphone dipping into shots supposedly give an authenticity to the occasion.

Routines

Some production techniques have become so familiar that it would seem strange if we presented them in any other way—such as a newscaster presenting an entire newscast in shorts or on a set with flashing lights.

Certain styles have become so stereotyped that they enter the realms of cliché—routine methods for routine situations. A number of standard production formats have emerged for productions, such as newscasts, studio interviews, game shows, talk shows, and others. If we analyze these productions, we usually find that styles have evolved and become the most effective, economical, and reliable ways of handling their specific type of subjects (Figure 20.5).

If we regard these formats as "a container for the content," these routine treatments can free the audience to concentrate on the show. However, if we consider the treatment as an opportunity to encourage interest and heighten enjoyment, then any "routine" becomes unacceptable.

Clearly, a dramatic treatment would not work for many types of television productions. Instead, it is best to aim for a

FIGURE 20.3 *Formal presentations often follow a very stylized artificial format.*
Source: Photo courtesy of the U.S. Navy.

FIGURE 20.4 *A display style is an unrealistic, decorative way of presenting your subject.*

FIGURE 20.5
News shows have a specific routine to them. The news style has evolved and become an effective, economical, and reliable way of presenting the information.

variety of camera shots, coupled with clear, unambiguous visual statements that direct and concentrate the audience's attention. How many sensible meaningful shot variations can you take of people speaking to each other, or driving an automobile, or playing an instrument, or demonstrating an item? The range is small.

For certain subjects, the picture is virtually irrelevant. What a person has to say may be extremely important; what the talent looks like is immaterial to the message. It may even prove a distraction or create prejudicial bias. "Talking heads" appear in most television shows. However, unless the speaker is particularly animated and interesting, the viewers' visual interest is seldom sustained. Changing the shot viewpoint can help, but may be a distraction.

Ambience

From the moment a show begins, we are influencing our audience's attitude toward the production itself. Introductory music and graphic style can immediately convey a serious or upbeat tone about what is coming next. While hushed voices, quiet organ notes, and a slow visual pace can provide a reverential atmosphere, the difference between a regal or game show opening fanfare adjusts our expectations in other directions.

Surroundings can also directly affect how convincingly we convey information. Certain environments, for example, provide a context of authority or scholarship: classroom, laboratory, museum, or other setting. A plow shown at work on the farm is better understood than if it were just shown standing alone in a studio with someone trying to explain what it does.

> "Dialogue should simply be a sound among other sounds, just something that comes out of the mouths of people whose eyes tell the story in visual terms."
>
> —Alfred Hitchcock

The Illusion of Truth

As you've seen, even when trying to present events "exactly as they are," the camera's angles, lens angles, image composition, the choice and sequence of shots, and other factors will all influence how our audience interprets what they are seeing. Keep in mind that the way a program is directed and shot has a significant impact on the final show. Where we lay emphasis, what we leave out, even the weather conditions (gloomy, stormy, or sunlight) will all modify the production's impact.

In a documentary program, the audience usually assumes that they are seeing a fair and informative story. However, that will always depend on how the director tackles the subject:

- There is the hopeful approach—an "adventure" in which the director points the camera around, giving a "tourist's view" of the events. This invariably results in a set of disjointed and unrelated shots. Sometimes, by adding commentary, graphics, music, and effects, it is possible to develop a coherent program theme. However, it can also be a disaster.

- Usually the director begins by researching the subject and then making a plan of approach. As discussed earlier, there are great advantages in anticipating and advance planning. By finding out about potential locations and local experts, the director can develop a schedule, arrange transport and accommodations, obtain permits, and so on. However, there is also the danger that preconceived ideas will dominate, so that you develop a concept before you arrive on location, and then reject whatever does not seem to fit in with the concept when you actually get there.

- Today's "reality shows" are what could be called a contrived approach, in which the director has staged what we are seeing—arranged the action and edited selectively. Dressed in their best, the participants put on a show for the camera, yet the television audience assumes that they are looking at reality.

Leaving aside ethics, even these brief (but real) examples are a reminder of the power of the image, and the director's responsibilities regarding the way in which it is used.

Pictorial Function

Most television images are factual, showing subjects in a familiar way. However, by carefully arranging these same subjects, with careful composition and a selective viewpoint, you can modify the subject's entire impact and give it a quite different implied significance. You can interpret the scene for the audience. You can deliberately distort and select reality so that your presentation bears little direct relationship to the actual situation—and you might do this to create a dramatic illusion, or to produce an influential force (advertising, propaganda).

Abstracting further, you can stimulate emotions and ideas simply by the use of movement, line, and form, which the viewer personally interprets. There are occasions when we seek to stimulate the audience's imagination—to evoke ideas that are not conveyed directly by the camera and microphone. In this chapter, we will examine these concepts and how they can be used.

Picture Applications

Because so much television program material is explicit, it is easy to forget how powerful it can be. Images can be used for a number of different purposes:

- To convey information directly (normal conversation).
- To provide context (establishing the location), such as Big Ben, to imply that the location is London (Figure 20.6).
- To interpret a situation, conveying abstract concepts (ideas, thoughts, feelings) through associative visuals, such as plodding feet suggesting the weariness of a trail of refugees.
- To symbolize—we associate certain images with specific people, places, and events.
- To imitate—pictures appear to imitate a condition (the camera staggers as a drunk reels, defocusing to convey loss of consciousness).
- To identify, showing features such as icons, logos, or trademarks associated with specific organizations or events.
- To couple ideas—using pictures to link events or themes (panning from a boy's toy boat to a ship at sea, on which he becomes the captain).
- To create a visual montage—a succession of images interplay to convey an overall impression (epitomizing war).

Production Rhetoric

Rhetoric is the art of persuasive or impressive speech and writing. Unlike everyday conversation, it stimulates our imagination through style and technique, by inference and allusion, instead of just direct pronouncement. The rhetoric of the screen has similar roots that directors such as Alfred Hitchcock have explored over the years to great effect.

It is amazing what the camera can communicate: without a word of dialogue, it can convey the whole gamut of human responses. For example, a veteran performer ends his brave but pathetic vaudeville act amid heckling from the crowd. He bows, defeated. We hear hands clapping; the camera turns from the sad lone figure, past derisive faces, to where his aged wife sits applauding.

AUDIO STYLE

Imaginative Sound

The ear is generally more imaginative than the eye. We are more perceptive and discriminating toward what we see. Consequently, our ears accept the unfamiliar and unrealistic more readily than our eyes, and are more tolerant of repetition. A sound effects recording can be reused many times, but a costume or curtain quickly becomes too familiar after being seen just a couple of times.

FIGURE 20.6 *Providing context for the audience helps them understand what is being said by the speaker.*
Source: Photo courtesy of the U.S. Navy.

In many television shows, the audio is taken for granted, while attention is concentrated on the visual treatment. Yet without audio, the presentations can become meaningless (talks, discussions, interviews, newscasts, music, game shows), whereas without video the production can still communicate.

Audio can explain or support the image, enriching its impact or appeal. Music or effects can suggest locale (seashore sounds), or a situation (pursuing police siren heard), or conjure a mood (excitement, foreboding, comedy, horror).

A nonspecific picture can be given a definite significance through associated sound. Depending on accompanying music, a display of flowers may suggest springtime, a funeral, a wedding, or a ballroom.

Sound Elements

Voice

The most obvious sound element, the human voice, can be introduced into the production in several different ways:

- A single person addressing the camera, formally or informally.

- An off-screen commentator (voice-over) providing a narrative (documentary), or the spontaneous commentary used for a televised sports event.
- We may "hear the thoughts" of a character while watching the talent's silent face, or watching the subject of the thoughts.
- Dialogue—the informal natural talk between people, with all its pauses, interruptions, overlapping, and even formal discussion.

Effects

The characteristic sound image that conjures a specific place or atmosphere comes from a blend of stimuli: from action sounds (footsteps, gunfire), from environmental noises (wind, crowd, traffic), and from the subtle ways in which sound quality is modified by its surroundings (reverberation, distortion).

Music

Background music has become "required" for many programs. It can range from purely melodic accompaniment to music that imitates, or gives evocative or abstract support.

Silence

The powerful dramatic value of silence should never be underestimated. However, silence must be used with care, as the audience could easily perceive it to just be a loss of audio. Continued silence can suggest such diverse concepts as: desolation, despair, stillness, hope, peace, and extreme tension (we listen intently to hear whether pursuing footsteps can still be heard).

- Sudden silence after noise can be almost unbearable: A festival in an Alpine village, happy laughter and music, the tumultuous noise of an unexpected avalanche engulfing the holiday makers . . . then silence.
- Sudden noise during silence creates an immediate peak of tension: The silently escaping prisoner knocks over a chair and awakens the guards—or did they hear him after all?
- Silent streets at night . . . then a sudden scream. The explosive charge has been set, the detonator is switched —nothing happens . . . silence.

Sound Emphasis

You can manipulate the volumes of sounds for dramatic effect: emphasizing specific sources, cheating loudness to suit the situation. A whisper may be amplified to make it clearly audible, a loud sound held in check.

You may establish the background noise of a vehicle and then gradually reduce it, taking it under to improve audibility of conversation. Or you could deliberately increase its loudness so that the noise drowns the voices. Occasionally, you might take out all environmental sounds to provide a silent background, perhaps for a thought sequence.

You can modify the aesthetic appeal and significance of sound in a number of ways. For factual sound, you can use:

- Random natural sound pickup (overheard street conversations).
- Selective pickup of specific audio sources. For atmospheric sound:

- By choosing certain natural associated sounds, you can develop a realistic illusion. (A rooster crowing suggests that it is dawn.)
- By deliberately distorting reality, you create fantasy to stimulate the imagination. (A flute's sound can suggest the flight of a butterfly.)

Sound Applications

As with visual images, sound can be used for a number of different purposes:

- *To convey information directly*: Normal conversation.
- *To establish a location*: Traffic noises that imply a busy street scene nearby.
- *To interpret a situation*: Conveying abstract concepts (ideas, thoughts, feelings) through associative sounds (such as musical instruments used to convey feelings).
- *To symbolize*: Sounds that we associate with specific places, events, and moods (such as a fire truck's siren).
- *To identify*: Sounds associated with specific people or events (such as signature tunes).
- *To couple ideas*: Using music or sound effects to link events and themes, such as a musical bridge between scenes, or aircraft noise carried over between a series of images showing the plane's various stops along the route.
- *To create a sound montage*: A succession or mixture of sounds arranged for comic or dramatic effect. For example, separate sound effects of explosions, gunfire, aircraft, sirens, and whistles create the illusion of a battle scene.

Off-Screen Sound

When someone speaks or something makes a sound, it might seem logical to show the source. But it can be singularly dull if we do this repeatedly: She starts talking, so we cut and watch her.

You can use off-screen sound in many ways to enhance a program's impact:

- Once you have established a shot of someone talking, you might cut to see the person he or she is speaking to and watch their reactions, or cut to show what they are talking about. The original dialogue continues, even though we no longer see the speaker. As you can see, you can establish relationships, even where the two subjects have not been seen together in the same shot.
- Background sounds can help to establish location. Although a medium shot of two people might fill the screen, if the background sounds have been used appropriately the audience will interpret whether they are near the seashore, a highway, or a sawmill.
- Off-screen sounds may be chosen to arouse our curiosity.
- A background sound may introduce us to a subject before we actually see it, informing us about what is going on nearby or what is going to appear (the approaching sound of a diesel truck).

- The background sound may create audio continuity, although the shots switch rapidly. Two people walk through buildings, down a street—but their voices are heard clearly throughout at a constant level.

Substituted Sound

It is often difficult to find a source for a specific sound, requiring us to create a substitute. There are several possible reasons for this approach:

- No sound exists, as with sculpture, painting, architecture, inaudible insects, and prehistoric monsters.
- Sometimes the actual sounds are not available, not recorded (mute shooting), or not suitable. For example, the absence of bird sounds when shooting a country scene; location sounds were obtrusive, unimpressive, or inappropriate to use; or the camera used a telephoto lens to capture a close-up shot of a subject that is too distant for effective sound pickup.
- The sounds you introduce may be just replacements (using another lion's voice instead of the missing roar), or artificial substitutes in the form of effects or music.

Background music and effects should be added cautiously. They can easily become:

- Too loud or soft.
- Too familiar.
- Distracting.
- Inappropriate (have wrong or misleading associations).

Controlling Sound Treatment

Various working principles are generally accepted in sound treatment:

- The scale and quality of audio should match the picture (appropriate volumes, balance, audio perspective, acoustics, etc.).
- Where audio directly relates to picture action, it should be synchronized (such as movements, footsteps, hammering, other transient sounds).
- Video and audio should normally be switched together. No audio advance or hangover should occur on a cut.
- Video cutting should be on the beat of the music, rather than against it, and preferably at the end of a phrase. Continual cutting in time with music becomes tedious.
- Video and audio should usually begin together at the start of a show, finishing together at its conclusion, fading out as a musical phrase ends.

The Effect of Combining Sounds

When we hear two or more sounds together, we often find that they interrelate to provide an emotional effect that changes according to their relative loudness, speed, complexity, and so on:

- Overall harmony conveys unity, beauty, organization.
- Overall discord conveys imbalance, uncertainty, incompletion, unrest, ugliness, irritation.
- Marked differences in relative volume and rhythm create variety or complication.
- Marked similarities result in sameness, homogeneity, mass, or strength of effect.

Selective Sound

In recreating the atmosphere of a specific environment, the trick is to use sound selectively if you want the scene to carry authenticity, rather than try to include all typical background noises. You may deliberately emphasize, reduce, modify, or omit sounds that would normally be present, or introduce others to convey a convincing sense of location.

The selection and blend of environmental sounds can strongly influence the interpretation of a scene. Imagine, for example, the slow, even toll of a cathedral bell accompanied by the rapid footsteps of approaching churchgoers. In developing this scene, you could reproduce random typical sounds. Or, more persuasively, you might deliberately use audio emphasis:

- Loud busy footsteps with a quiet insignificant bell in the background.
- The bell's slow dignity contrasted with restless footsteps.
- The booming bell overwhelming all other sounds.

So you can use the same sounds to suggest hope, dignity, community, or domination—simply through selection, balance, and quality adjustment.

Instead of modifying a scene's natural sounds, you might augment them or replace them with entirely fresh ones.

AUDIO/VIDEO RELATIONSHIPS

The picture and its audio can interrelate in several distinct ways:

- The picture's impact may be due to its accompanying audio. A close shot of a man crossing a busy highway:
  - Cheerful music suggests that he is in a lighthearted mood.

- But automobile horns and squealing tires suggest that he is jaywalking dangerously.
- The audio impact may be due to the picture:
  - A long shot of a wagon bumping over a rough road; the accompanying sound is accepted as a natural audio effect.
  - But take continuous close-ups of a wheel, and every jolt suggests impending breakdown!

- The effect of picture and audio may be cumulative: A wave crashes against rocks along with a loud crescendo in the music.
- Sound and picture together may imply a further idea: Wind-blown daffodils along with birds singing and lambs bleating can suggest spring.

INTERVIEW WITH A PROFESSIONAL: DAVE GREIDER

Briefly define your job. I work mostly with short-form content such as commercials, promos, trailers, etc. I mostly edit, but also produce and shoot.

What do you like best about your job? No question: working with world-class people. I'd rather work on a public relations video with knowledgeable pros than a highly creative music video with people who have no clue what they're doing. I also love coming up with creative solutions for a client and then knocking it out of the park.

What are the types of challenges that you face in your position? It is difficult to know which projects to take on. Do you accept the highest paying or the most fulfilling? Do you take one for lower pay in a field you want to work in hoping it's going to open the door for something better? If you're excellent at what you do, finding work won't be a problem; deciding what to do will.

FIGURE 20.7
Dave Greider, freelance editor and producer.

How do you prepare for a production? I lay out all the styles, references, shot list for the cam op or DP long before we step foot on set so they know exactly what I'm looking for. Same for the producer—we should have already gone over all the details, the call sheet should already be sent, schedule finalized, etc. Then on set, if everyone does their job, you get to steer the ship as the director. There's nothing more beautiful when a plan comes together.

What suggestions or advice do you have for someone interested in a position such as yours? Be absolutely excellent at a specific function. I decided postproduction, so I learned everything I possibly could about editing (digital asset management, post-workflows, codecs, etc.). Then I surrounded myself with people way smarter than me and learned from them. I'm always learning, always reading, always working on something, which has led me beyond the world of post. It's all about building relationships. In summary, figure out what you want to do and learn as much as you possibly can, become a master at teaching yourself and learning from others.

Dave freelances in production and post for a variety of companies in Los Angeles. His main clients include Associated Press (Corporate Services dept), 20th Century Fox (New Media Promotion), Oprah Winfrey Network, Sony Pictures, and various other production companies, big and small. He works mainly with short-form content such as commercials, promos, trailers, etc. Dave usually edits, but also produces and shoots.

INTERVIEW WITH A PROFESSIONAL: JEREMY RAUCH

Briefly define your job. I report/shoot regional sports, put together stories, and anchor/produce shows on the nightly newscast.

What do you like about your job? I meet a lot of new people and encounter many unique stories covering daily events, then I get to tell those stories with my own flavor and writing, which makes the job very rewarding.

FIGURE 20.8
Jeremy Rauch, anchor/reporter.

What are the types of challenges that you face in your position? By far, what content to include in a show and what to leave out. You are daily faced with what stories/highlights take the precedent when fitting into a newscast. Inevitably, someone watching will be offended that his or her story wasn't featured.

How do you prepare for a production? That varies depending on the production, reporting, or anchoring. I'll give both sides. (1) Reporting on a story: learn as much as I can about the person(s) I am doing a story on, try to have a visual map of what shots I'm looking for, and determine what materials (camera, mic, lighting, tripod, tape, etc.) I will need to execute the story. (2) Anchoring the news: check all media outlets for up-to-date stories (and refresh, refresh, refresh), gather all of the stories I need to put into the newscast and write the show in the newscast rundown, edit video, print scripts, put on makeup, and hit the set.

What suggestions or advice do you have for someone interested in a position such as yours? Don't ever let anyone discourage you by saying things such as "The job market is tough," "This is a dying business," or "It's impossible to make it in this biz." Be confident but *never* cocky. Be prepared to work hard without having a sense of entitlement. Be persistent and positive (job searching). Have fun!

Jeremy is an anchor, reporter, and producer.

REVIEW QUESTIONS

1. How is "appropriateness" determined with visual style?
2. Describe three of the various sound elements.
3. What are some of the effects of combining sounds?
4. How can audio make the video more powerful?

CHAPTER 21

REMOTE PRODUCTION

"Live remote events are the core of television. They are the one thing television can do that no other medium can match. There are things movies can do better, there are things radio can do better, but no other medium can bring you a visual report of an event as it's happening."

—Tony Verna, Director

TERMS

Coordination meetings: These meetings provide a forum for all parties involved in the production to share ideas, communicate issues that may affect other areas, and ensure that all details are ready for the production.

Remote survey (recce): This site survey assesses the venue and determines how, where, how many, who, what, and how much.

Today's television production equipment is highly mobile, and able to access any location. There are times when the only way a show will be authentic is to get out to the event. Most remotes are live productions, with the director having little or no control over it, requiring the crew to cover whatever happens. However, other remote productions, such as dramas, have scripted (controlled) productions. Remote productions often utilize multiple cameras. As mentioned in earlier chapters, single-camera remote productions are generally referred to as ENG (electronic news production) or EFP (electronic field production) productions. In this chapter, we will use remote productions to refer to both multi-camera and single-camera productions.

WHAT IS A REMOTE PRODUCTION?

Any production that occurs outside of the studio is considered to be a remote production or an outside broadcast (OB). Remote productions include all kinds of events:

- News events.
- Sports events.
- Parades.
- Concerts.
- Award shows.
- Telethons.
- Talk or variety shows that are "on the road."

Remote Production versus Studio Production

Both of these types of productions have pros and cons. Studio productions provide the maximum amount of control over the subject. The lighting and audio can be minutely controlled, providing the perfect levels for the production. Studios provide a clean location that is usually impervious to weather conditions and has full climate control.

However, there are times when the crew has to be on location. Remote locations can provide context and an exciting atmosphere (e.g., cheering crowds) (Figure 21.1).

FIGURE 21.1 *NBC decided to interview Michael Phelps after he had won his eight Olympic gold medals, outside and with a crowd as the backdrop to provide an exciting atmosphere.*

While weather can disrupt or even cancel a remote production, when the weather is nice, natural lighting and outdoor scenery can provide stunning images. There are also times when it is actually less expensive to shoot in the field than to rent and schedule studio time.

Shooting on Location

Remote productions require anticipating what may happen. It is essential to assemble a team that can anticipate what is going to happen and know how to deal with it. The crew must be able to work well together and plan for contingencies in case something goes wrong.

The more familiar the crew is with an event—especially a news or sports event—the better they can cover it. Understanding the intricacies of the event allows the director and talent to clearly communicate what is happening.

THE SINGLE CAMERA ON LOCATION

The single-camera production has some important advantages. It is extremely mobile, and thus easily relocated. It can be surprisingly unobtrusive. It is largely independent of its surroundings, and it is economical (Figure 21.2).

Typical Setups

Single-camera crews come in all different sizes, depending on the goals and the size of the production budget:

- One-person single-camera crews are increasingly being used by many news stations, sports shows, and documentaries. These operators run the camera, a microphone, and even an on-camera light. Some may even act as the reporter (Figure 21.3).
- Two-person single-camera crews are often made up of a camera operator who is responsible for the camera and a second person who is the reporter/director and may be responsible for the audio.
- Three-person single-camera crews are usually made up of a camera operator, an audio person, and a director/reporter (Figure 21.4).

Power Supplies

Professional video cameras normally require a DC power supply, which can be obtained by using an AC power adapter or batteries, or even by plugging into a car's cigarette lighter or DC outlet. Batteries come in all different configurations. Some batteries fit on the camera, others fit in a compartment

FIGURE 21.2
ENG camera crews cover a press conference with a government official.
Source: Photo courtesy of the U.S. Department of Defense, by Cherie A. Thurlby.

FIGURE 21.4 *This three-person single-camera crew is made up of a camera operator, a boom mic operator/audio person, and a reporter/director.*

Source: Photo courtesy of Andy Peters.

FIGURE 21.3 *This one-person ENG crew writes the stories, does the reporting, and runs the camera, microphone, and on-camera light.*

under the camera, and some are designed to be worn on a belt in order to spread the weight around (Figure 21.5).

Batteries cannot be taken for granted. Carelessly used, they can become unreliable. Correctly used, they will give excellent service. Always carry spare batteries with you. How many depends on the nature and duration of your project and your opportunities to recharge exhausted cells.

SINGLE-CAMERA SHOOTING

Handling the Camera

If you are using the camera shoulder-mounted, make sure that it is comfortably balanced before you begin shooting. Try to use the camera as an extension of your body: turning with a pan, bending with a tilt. With your legs comfortably braced apart, turn to follow movement—preferably from a midway position between the start and end of the pan. Learn to

---

BATTERY CARE

There are a lot of dos and don'ts here, but remember: When batteries fail, the shooting may come to a halt:

- Batteries power your camera viewfinder, the recorder, on-camera lights, and anything else attached to the camera. So switch off (or use the standby mode) and conserve power whenever possible. If you are not careful, time taken reviewing recordings and lighting can leave you with low power for the take.
- Handle batteries carefully. Dropping can cause a battery breakdown.

- Always check a battery's voltage while it is actually working.
- Recharge batteries as soon as possible.
- Stored batteries tend to discharge themselves to a noticeable extent.
- In addition to the main battery that powers the camera, there may be "keep-alive" batteries for memory circuits within the camera that should be checked regularly.

FIGURE 21.5 *(Left) Most professional cameras use a battery that attaches to the back of the camera. (Right) The news photographer is using a battery belt to power his camera. Belts help distribute the weight of the battery around the body.*

Source: Photos courtesy of Matt Giblin and Thom Moynahan.

shoot with one eye looking through the viewfinder and the other *open*, seeing the general scene (with practice, it's not as difficult as it sounds). You stand a better chance of walking around without an accident this way than if your attention is glued to your viewfinder's picture alone.

Even though its image is magnified, you must look carefully at the viewfinder picture to detect exact exposure adjustments. You can overlook something intruding into the frame that will be very obvious on a large-screen television monitor. Distracting, brightly colored items can easily pass unnoticed in a black-and-white viewfinder.

Walking should be kept to a minimum while shooting. If you must move, slightly bent legs produce smoother results than normal walking. Practice to see what you find comfortable and effective—and critically examine the results (Figure 21.6).

FIGURE 21.6 *Camera covers can protect the cameras from dust, rain, and snow.*

Source: Photo by Josh Taber.

TABLE 21.1 *Be Prepared: System Checklist*

As with the studio camera, preliminary checks before shooting can help you anticipate and prevent problems later. If you do not have access to a color television monitor as you shoot, you may have to rely on a small black-and-white viewfinder for all image checks. So if, for example, you forget to white-balance the camera, it will not be apparent until the picture is reviewed at base!	
The camera	• *Viewfinder*. Check the controls and adjust for a good tonal range. • *Lens*. Lens clean? Lens firmly attached? • *Focus*. Check both manual and autofocus systems. • *Aperture (f-stop)*. Manually adjust the range of the aperture ring, checking the results in the viewfinder. Check auto-iris movements while panning over different tones. • *Zoom*. Operate manual zoom over full range, checking focus tracking. Operate power zoom in and out over full range. • *Microphone*. Test both camera microphone and hand microphone. • Check the audio output through an earpiece/headset. Record and then replay the recording to confirm that the audio recording was accurate. • *Subsidiary controls*. Looking at a color monitor, check filter wheel positions, white balance, black adjust video gain, and other controls.
Recorder	• If possible, examine playback in viewfinder and on a color monitor.
Power supplies	• Check all batteries. Confirm that all spare batteries are fully charged. Are AC cords (main cables), AC adapters, car battery adapters, and other accessories all set?
Cables	• Don't leave cables attached to equipment during storage. Never be casual about cables; they have a will of their own, and tangle easily. Coil them neatly, and secure them with a quick-release fastener or tie.
Spares	• Examine all recording media to make sure that you have the type you need—as well as extras. • Spare fuses may be needed for certain pieces of equipment. • Take a spare microphone. • Avoid field maintenance and cleaning unless you are experienced and the equipment is in need. Protective camera covers prevent problems (Figure 21.6).
Tools	• Small and medium screwdrivers and pliers; electrical insulating tape; masking tape; gaffer tape; cord. • Head-cleaning fluid (isopropyl alcohol), cotton-tipped cleaning sticks. • Compressed air can.
Accessories	• Flashlight (with spare bulb and batteries). • White material (or card) for checking white balance. • Small reflectors (white, silver foil). • Another matter for personal preferences. There are circumstances in which mosquito repellant, sunscreen preparations, or waterproofing material can be more essential than a comprehensive toolkit! Much depends on where you are working.

Table 21.1 is a checklist that is helpful when working with video equipment.

Lens Angles

Generally speaking, avoid extreme lens angles. A wide-angle lens makes camera work easier, and slight jolts are much less noticeable. However, everything looks so far away. Moving an extreme wide-angle lens closer may seriously distort the subject, which might actually be OK unless the subject is a person. With an extremely wide angle, you imagine that you have much more space to maneuver than you actually have, and are liable to trip or walk into things in the foreground.

Long telephoto lenses produce unsteady shots, and focusing can be difficult. On handheld cameras, they are often suitable only for brief stationary shots while holding your breath. A little wider than normal is probably the best compromise.

If you have a camera assistant, while walking around and shooting you can be guided by a hand on your shoulder, particularly when moving backwards. Walking backwards unguided is at best hazardous, and at worst foolish, unless you are on an open flat area without any obstructions. People can move toward you a lot faster than you can hope to walk backwards.

Automatic Controls

Automatic controls such as autofocus and auto-iris provide a safety net when you do not have the opportunity to control your camera accurately by hand adjustments. But remember that automatic controls are far from foolproof and should be used only in certain situations.

Audio

You can adjust the audio system manually, or switch it to *AGC (automatic gain control)*. In the *manual mode*, watch the volume indicator for sound peaks. If they are far below the upper level (100 percent modulation/0 VU), increase the audio gain until the peaks reach this limit. But remember, if anything louder comes along, it will probably distort. Switching to AGC instead will protect against unexpected overloads, but may bring up background sounds and smooth out the audio dynamics as mentioned in Chapter 14.

Storing the Gear

Before putting the camera away at the end of a shoot, camera care should be a priority:

1. Cover the lens using a lens cap or filter wheel, or just close the iris.
2. Make sure that all equipment is switched off.
3. Remove all camera accessories (light, mic, cables, etc.).
4. Remove all recording media and replace with fresh ones if appropriate.
5. Clean and check all items before storing.
6. Make sure that the camera is safely stored in its case.
7. Replace any worn or damaged items.
8. Check and recharge batteries.

MULTI-CAMERA REMOTE PRODUCTION

Although a single-camera production has its advantages, there are many production situations in which a single camera has little hope of capturing much more than a glimpse of the event, and multi-camera coverage is the only answer:

- Coverage from different viewpoints is to be continuous and comprehensive.
- Action is spread over a large area (a golf course).
- At an event where there is no time or opportunity to move cameras around to different viewpoints.

- There is to be a "one-time-only" event (demolition of a bridge).
- The location of action continually changes (sports field of play).
- Cameras could not move to new angles or locations (because of obstructions).
- Cameras must be concealed, or located in fixed places.
- You cannot accurately anticipate where the action is to take place.

Multi-Camera Planning and Preparation

Multi-camera productions have a number of aspects that make them quite different from studio productions or single-camera productions. Because they are larger productions, requiring more equipment and personnel, they need much more planning and preparation regarding the basics, such as whether there is power, how long the cables need to be, whether there is enough light—things you don't need to think about in the studio. Following are discussions of some of these unique issues.

Production Meetings

Production meetings, also known as coordination meetings, are essential to the planning phase of a multi-camera production. These meetings provide a forum for all parties involved in the production to share ideas, communicate issues that may affect other areas, and ensure that all details are ready for the production. Production meetings usually include event officials, venue management, and production personnel.

Remote Survey (Recce)

Once the production team has a good general idea of how the event will be covered, a survey team should visit the shoot location. This visit must assess the venue and determine how, where, how many, who, what, and how much. The answers to these questions will provide the foundation for the production's planning. The purposes of the remote survey are to:

- Determine the location for the production.
- Determine where all production equipment and personnel will be positioned.
- Determine whether all of the production's needs and requirements can be handled at the location.

Areas that must be determined and assessed include: contacts, location access, electrical power, location costs, catering/food, security, telephones/Internet access, parking, and lodging. Figure 21.7 shows a sample remote survey form.

Remote Survey Form

Client: _____   Date of Survey _____
Shoot Date _____   Time of Shooting _____
Program Name _____   Air Date(s) _____
Location _____
Director _____ Producer _____ TD _____

Location Contacts:
Primary Contact _____   Phone _____
Secondary Contact _____   Phone _____
Permits Needed _____   Phone _____
Truck Location _____
Other Parking _____
Credentials Contact _____

Cameras: (add sketch of camera locations at event)

Camera	Position/locations	Lens	Cable run
1			
2			
3			
4			
5			
6			

Audio: (add sketch of microphone locations at event)

Mic Type	Location		Mic Type	Location
1		6		
2		7		
3		8		
4		9		
5		10		

Lighting: (add lighting plot if needed)
Available Light _____
Talent Light _____
Special Instructions _____
_____

Power:
Location Electrician Contact _____
Program Requirements _____
AC Outlets

Location	Voltage	Connect. Type		Location	Voltage	Connect. Type
1			4			
2			5			
3			6			

Communications:

	Type/Style	Location(s)
Camera Headsets		
Intercoms (PL)		
Business Phone		
Wireless		

Location Sketch: (should include important dimensions, location of props and building, truck, power source and sun during time of telecast)

FIGURE 21.7 *Sample of a small event remote survey form.*

REMOTE PRODUCTION: THE ROYAL WEDDING

One of the largest non-sports remote productions in recent history was the wedding of Prince William and Catherine Middleton. Inside Westminster Abbey, where the wedding took place, there were over 40 broadcast cameras capturing every angle of the event. Outside, broadcasters had set up more than 35 television studios. Over 100 different broadcast organizations, with roughly 140 OB remote trucks, from around the world converged on London for the festivities.

FIGURE 21.8 *Over 40 cameras were set up in Westminster Abbey to cover the 2011 royal wedding. This high point-of-view camera gave a unique view of the ceremony.*

Cameras

There are many decisions that have to be made when it comes to cameras in remote locations. These can include:

- How many cameras are required to cover the event?
- What type of camera should be used (dolly, jib, handheld, POV)?
- Where are the best locations to place the cameras? Does anything obscure a camera's viewpoint?
- Are special camera mounts required (scaffolding, jibs, etc.)?
- Are special lenses required (such as long telephoto lenses)?
- Where can camera cables be run?

Audio

Although audio may be one of the least appreciated aspects of a television production, it is one of the most important areas of a production. Some of the issues for consideration include:

- What does the audience need to hear? How many mics are needed to cover the event?
- What type of microphone works best in each situation (handheld, lapel mic, shotgun, etc.)?
- Stereo or surround sound?
- Can microphones appear in the shot?
- Wired or wireless microphones?
- Is the natural sound of the location a problem (traffic, crowds, airplanes)?

REMOTE PRODUCTION VEHICLES

The facilities needed to cover a remote production will depend partly on the scale of coverage and partly on the nature of the event. Production trucks vary in size from very small to very large. Other vehicles could include cars, motorcycles, golf carts, boats, and helicopters. For example, marathon coverage might include a few stationary cameras, a couple of handheld cameras, a motorcycle with a camera to stay with the lead runner, and a helicopter or two to get the long-shot/"big picture" images that establish the scene. It always comes down to the event and what you want to accomplish.

Remote Production Truck/OB Van

The most common unit for remote productions is the remote production truck, otherwise known as an outside broadcast (OB) truck or van. These units vary in size from a small van with two or three cameras to a large truck with more than 20 cameras (Figures 21.10 and 21.11). They provide a full broadcast standard production control center with complete video and audio facilities. The trucks include everything needed to produce a television program: monitor wall, video production switcher, audio, recording and playback decks, graphics, intercom, and anything else you might need for a remote production. See Chapter 3 for more details about remote production trucks.

REMOTE PRODUCTION: THE OLYMPICS

The largest broadcast event in the world is a Summer Olympic Games. The video feed of the Olympics, created by Olympic Broadcast Services (OBS), is used by networks in over 200 countries around the world. OBS uses over 1,000 cameras, 60 OB remote trucks, and a crew of over 5,000.

FIGURE 21.9 *Over 1,000 cameras were used in the coverage of the Summer Olympic Games.*

The unit may be used in several ways:

- Parked within the action area (in a public square or at a sports venue) with cables extending to cameras at various vantage points.
- As a drive-in control room parked outside a permanent or temporary studio. This could include a public hall, a theater, or even a soundstage (usually used for film).
- The program may be recorded (and edited) on board the remote truck, or it may be relayed back to their base by a microwave link, data line, or some other transmission path.

LIVE TRANSMISSION

If an event is being produced for a live audience, the production's signal must be sent back to a location in order to be broadcast, cablecast, or cybercast. There are a variety of transmission methods that are discussed in Chapter 22.

Transmission trucks provide units that can be quickly relocated and provide broadcast-quality images and sound. Most systems provide a two-way communication link between the unit and base or studio (Figure 21.12).

Event Coverage

As in every other type of event, directors shooting remote productions must keep the axis of action in mind (Chapter 17), placing all cameras on one side of that line (Figures 17.8 and 21.13).

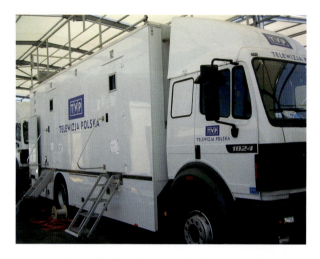

FIGURE 21.10 *Small production trucks are easier to drive into some locations but usually offer limited facilities.*

FIGURE 21.11 *Large remote trucks offer more space and additional facilities but are difficult to maneuver within some venues.*

FIGURE 21.12 *Transmission trucks come in all sizes, depending on what is needed. This unit is a combination satellite uplink and microwave van.*

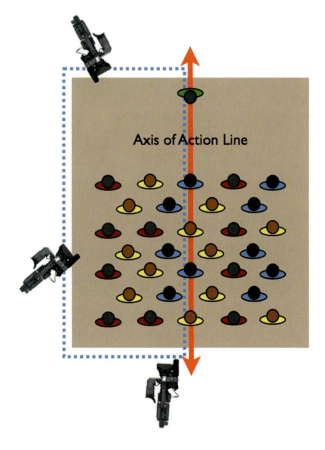

FIGURE 21.13 *In a meeting in which a speaker is speaking to a group of people, the axis of action is between the two. That means that all of the cameras must be on one side of the line.*

PREPARING TO COVER SPORTS

How to prepare to cover sports:

- Know the rules of the sport.
- Know the participants (athletes, coaches, officials).
- Know the venue/field of play.

Areas for which special decisions need to be made:

- Cameras/lenses.
- Camera mounts/platforms.
- Graphic design.
- Audio plan.
- Lighting.
- Award ceremonies.
- Star and finish protocols.

How the production plan is created:

- Production planning meetings with group who will be producing the event.
- Review previous recordings of events.
- Rehearsals.

(Adapted from Pedro Rozas, Television Producer)

Sports Action

Sports productions are a bit unique, because the participants can be going all over the venue. Some venues are large (a car racetrack or golf); other venues are very small (a wrestling match). Events here are categorized by different types of action: horizontal, vertical, and round.

Horizontal Action

Horizontal sports include basketball, soccer, and American football, among others. The cameras are placed on a long side of the venue, panning right to left to capture the athletes' action (Figure 21.15).

COVERING THE GAME

A five-camera production of football/soccer includes the following cameras:

- **Main camera**: The main camera (also known as the action or game camera) is a high camera that usually covers general action of the sport. Sometimes that may mean a long shot of the field, half of the field, or zooming in a little closer, as long as the majority of the athletes are in the shot. The responsibility of this camera is to allow the audience to see the ball and the relationship between the players.
- **Hero camera**: The hero camera (sometimes called the shag camera) is high in the stands, usually next to the main camera, and is usually as close a shot as you can get while keeping the person who just scored (hero) or a person who just fouled. The goal is to show the audience who this person is, as well as the expression on his or her face.
- **Center camera**: The center camera is low, very close to the field level, and is designed to give us field-level shots of the action, especially for replays.
- **Handheld cameras (2)**: Handheld shots can give the audience a close-up view of net action and field-level play action, and they can turn around and get shots of the coach, bench, cheerleaders, and audience.

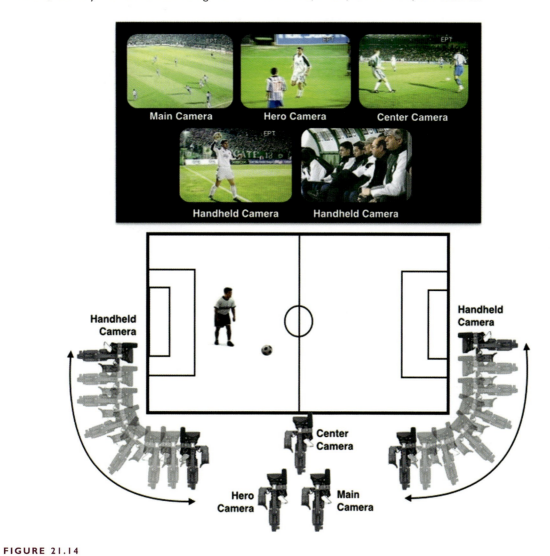

FIGURE 21.14

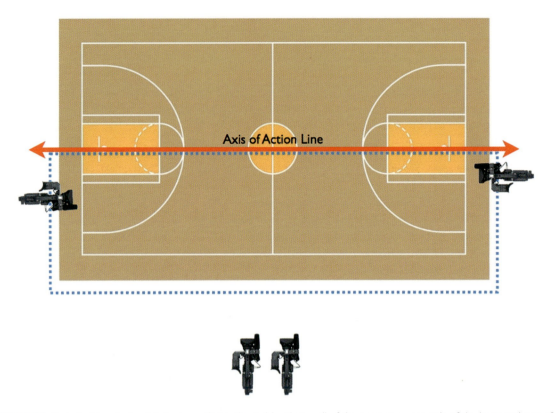

FIGURE 21.15 *Basketball is a horizontal sport, one that is directed by placing all of the cameras on one side of the horizontal axis of action (blue area) on the field of play.*

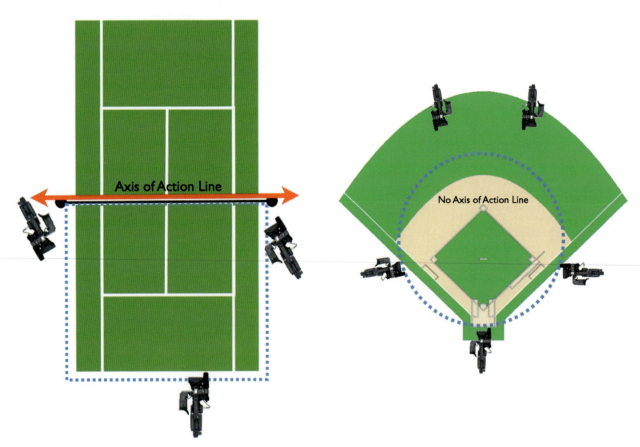

FIGURE 21.16 *Tennis is a vertical sport. In order to make the shots more interesting, the axis of action line runs across the court vertically. Keep all of the cameras on one side in the blue area.*

FIGURE 21.17 *Baseball is a round action sport. There is no axis of action line, allowing the cameras to be placed anywhere. However, the director often needs to go to a long shot, re-establishing the location for the audience.*

Vertical Action

One vertical sport is tennis. Although it is a sport that takes place on a rectangular venue, like basketball and soccer, the action is difficult to follow for the audience with two players hitting a small ball back and forth. Instead, the cameras are placed behind one of the athletes, looking over his or her shoulder at the other athlete. So the axis of action is located at the net (Figure 21.16).

Round Action

Round sports include auto racing and baseball. Because cameras are needed to cover the action the whole way around the circle or oval, an axis of action is not chosen. Instead, the director has to constantly re-establish the scene in order to avoid confusing the viewers. This means that if a camera has a close-up shot of a car as it drives around the track, every once in a while a long shot must be shown to establish the current location of the car and where it is in relation to the other cars (Figure 21.17).

INTERVIEW WITH A PROFESSIONAL: TOM CAVANAUGH

Briefly define your job. Responsible for creating content for Webstream Sports programming.

What do you like about your job? Creating and participating in a collaborative environment that promotes creativity, communication, and acknowledges the breadth of talent and experience of the people I work with. Focusing on the ability to fail and then recover rather than playing it safe or avoiding mistakes quite often delivers a very successful operation. In the broadcast business, the ultimate goal is to have that successful operation show up on the screen in the form of compelling content.

What are the types of challenges that you face in your position? Habits are the toughest thing to break, and they are even more difficult to change when people don't even realize they have formed a habit. So, change can be a big challenge, but if you focus on giving feedback and developing a collaborative environment substantial change will take care of itself.

How do you prepare for a production? It is important to make sure the obvious questions are asked and that typical assumptions are addressed. Stepping back and identifying the big picture storytelling elements helps to set up . . . everything.

- What is the storyline?
- Who are the stars?
- Who is the underdog?
- Where is the rooting interest coming from for the fans?

Productions really end up being quite simple because good stories are usually fairly simple, but we can get too "inside" and forget to tell the basic aspects of the story.

What suggestions or advice do you have for someone interested in a position such as yours? Follow your passion and let your ability develop because you will usually be surprised by the potential you have when you find yourself in "your element." You will know when you find it, and basically you should never stop looking for it.

Tom Cavanaugh is the vice president at Webstream Sports. He previously was vice president of production of NASCAR at Speed, vice president of production at Wheeler Television, coordinating producer at World Sports Enterprises, and a producer at ESPN.

INTERVIEW WITH A PROFESSIONAL: CINDY PENNINGTON

Briefly define your job. I coordinate all aspects of preproduction planning, technical crew management, and remote operations, including: mobile unit location/parking, uplink and generator orders, camera locations, cabling, power, phones and Internet, credentialing, talent pickups/escorts, shipping, budgets, and surveys and reports.

What do you like about your job? My job allows me to travel across the country and to be part of some of the biggest events in the world of sports. I have the opportunity to meet and work with some of the best folks in the industry.

FIGURE 21.18
Cindy Pennington, freelance operations producer.

What are the types of challenges that you face in your position? How much time do you have? That is a tricky question to answer because each sport and each venue present their own challenges. Whether it be difficult cable runs, not enough space for our mobile units, or simply a grumpy athletic director who is not happy that we are on site, there are always obstacles to overcome. That being said, it is my job to solve these issues, and it is very rewarding to walk away from a show that has been successful.

How do you prepare for a production? I start with the basics, which is determining what the specs are for the show that I am working. How many cameras, recorders, microphones, etc. Once I find out which mobile unit has been assigned for the show, I coordinate parking and power times with them. I call the venue and set a meeting with the broadcast and athletic directors to discuss logistics. I then build a technical survey that is distributed to the crew upon arrival on site.

What suggestions or advice do you have for someone interested in a position such as yours?

1. Start local! So many folks want to jump directly to the network, and I strongly encourage people to take a job at a local television station in a smaller market where you can truly learn all aspects of TV so that you have an overall understanding of how the entire process works. Internships are also fantastic!
2. Be persistent and network! This business is about who you know, not necessarily how much you know, which can be frustrating at times. ESPN, in particular, covers events all over the country and we are constantly looking for students to work as runner and utilities on our broadcast. I encourage students to work as many events as possible to make those contacts.
3. Work hard and be joyful!

Cindy Pennington is a freelance operations producer for ESPN.

REVIEW QUESTIONS

1. Define a remote production.
2. What are the advantages and disadvantages of studio productions and remote productions?
3. What are some of the suggestions for handling a single camera?
4. What are some of the challenges with extreme wide-angle and telephoto lenses?
5. What are some of the situations in which multi-camera productions are essential?
6. Describe the three primary types of sports action and give examples of each type.

CHAPTER 22

DISTRIBUTING YOUR PRODUCTION

"One cell phone can now simultaneously feed real-time video to the entire world. This, quite honestly, blows my mind. There is real opportunity out there and it's available now. Virtually all of the media playing fields have been leveled. That is now the reality for many programs. It has pay television operators, the networks, and other broadcasters running scared."

—Frank Beacham, Producer and Writer

"Just like the coming of digital to production, the coming of digital to distribution has empowered television and filmmakers to do things that weren't previously possible. In the old-school model, there were always people in between the producer/director and his or her audience. Now, for the first time, producers and directors have the chance to access or reach an audience directly, going around the traditional middlemen and gatekeepers."

—Peter Broderick, President, Paradigm Consulting

TERMS

Bonded cellular: The merging of multiple cellular channels in order to increase the quality of the transmitted video.

IPTV: Internet Protocol Television (IPTV) utilizes the Internet to provide programming to its audience instead of broadcast or cable television.

iTV: Interactive television, or iTV, refers to online programming that allows the viewer to make choices about how he or she watches an event.

Streaming: Programming that is shown live or transmitted from a video-sharing site to a computer.

Video-sharing sites: Online sites that enable producers to upload programming so that it can be seen by an audience.

Traditionally, distribution was not a part of the production personnel's problem. They created the content, and someone else worked to get it out to the audience. However, today distribution often becomes part of the role of the production personnel. Once the production has been completed, production personnel often need to burn it to post it online or stream it. Distribution depends on where you are sending it since the devices viewers watch their programming on are incredibly varied: traditional televisions, large screens, tablets, computers, and mobile phones.

Historically, it cost thousands of dollars to broadcast your program, through a local television station to your local community. If you wanted to transmit your program nationally, it would be very expensive. Transmitting it around the world was almost unheard of unless it was an international event. Transmitting it *live* locally, nationally, and internationally was even more expensive.

The equipment needed to create live broadcasts can be incredibly expensive. Microwave or satellite trucks are used to send the signal back to the station or network (Figure 21.12).

FIGURE 22.1 *YouTube and Vimeo are some of the largest video-sharing websites. The second image is of a private website that distributes videos to a specific audience. In this situation, it is distributing short films to university students. These video-sharing sites enable producers to distribute their material around the world.*

Welcome to Revival House

Revival House exists to empower passionate filmmakers with the ability to distribute their stories to greater audiences. For audiences, Revival House is a haven of excellent films — narratives and documentaries that contain high production values and relevant content, told by directors with a creative flair.

If you would like to distribute your films to the Revival House community, please register for an account to begin. If you already have an account, login to add new films.

Featured Film

Girlfriend In A Box
A fascinating new invention is purchased by a few kids.

Watch film

Most Popular Films

1   A Terrible Stampede

2   Girlfriend In A Box

3   Barry White

Newest Films

1   Barry White

2   Girlfriend In A Box

3   A Terrible Stampede

You are not currently logged in. Login or register to add your own films.

FIGURE 22.1 *continued*

Videotapes, DVDs, and Blu-ray have traditionally been used to distribute video programs to the home audience.

While these traditional means of distribution are still being used around the world on a daily basis, there are many other technologies that are proving to be more efficient, and many times provide a much lower distribution cost.

NON-LIVE DISTRIBUTION

Online Distribution

Online distribution allows programming to be distributed worldwide. This has opened up access to people and markets that were unreachable in the past. Online sites provide a variety of opportunities for sharing video. Videos can be distributed to anyone, specific groups or become part of a channel. It is possible to view SD, HD, and even 3D video on many of these websites. It is possible to upload videos from a computer and cell phones.

Video-sharing websites allow users to upload, share, and view productions. Some of the sites allow commercials; others don't. Some video-sharing sites are provided by private companies and are aimed at a specific audience. Others, such as YouTube, Vimeo, and Facebook, are mass communication websites, used by millions around the world (Figure 22.1). Facebook has over 500 million active users who can upload videos, and YouTube says that over 14 billion videos were viewed on their site in 2010.

Websites differ greatly on how they manage their content. Some enable producers to upload their programs for free and make them available free to the audience. Other sites may charge a fee to view the video, splitting the fees between the site and the producer (Figure 22.2).

FIGURE 22.2 *Some websites provide videos for a fee, splitting the fees between the site and the program's producers.*

CASE STUDY: GOOGLE NEWS NEAR YOU

As news staff from television stations and newspapers continue to face difficult financial times and reduce their staff and coverage, alternative news resources began appearing. Google, using its free video distribution service, YouTube, launched what they called the "biggest news platform in the world."

Visitors to "Google News Near You" sign into the site, and Google, using the visitor's Internet address, is able to determine their geographical location. The viewers then have the ability to look through all of the news stories within a 100-mile radius that have been uploaded onto YouTube. As more news outlets participate, Google hopes to shrink the radius to make the stories much more local.

Google has invited more than 25,000 news sources to become content providers. The term "news source" is defined pretty loosely. While any newspaper, radio station, and television can contribute stories, so can about anyone else. Providers include schools, churches, and almost anyone else who wants to take the time to upload "news."

Google is working hard to make this service profitable for themselves, as well as content providers, by splitting the revenue from all ads that appear with each specific video. Although the revenue is small, Google believes that, as they build the audience, the payout will significantly increase. With an audience of over 100 million viewers a month, content providers have a chance to have their material seen by an audience that may not have found them otherwise.

The questions that remain include: (1) Can this be monetized? (2) Will the audience actually trust the content providers?

Festivals and Competitions

Video competitions and "film" festivals can also be a very productive way of getting your project seen. Productions are often exhibited in front of a live audience, providing a way for general viewers and distributors to see the production.

LIVE DISTRIBUTION (OR MOSTLY LIVE . . .)

Live Online Distribution

The Internet and mobile phones have incredibly changed the live distribution of content in other ways. In the past, there were lots of layers of bureaucracy (gatekeepers) between the producer and the audience. As mentioned earlier, the cost of live transmission gear was also prohibitive. Today, you can reach a live audience directly, going around the traditional gatekeepers, by using a computer or some cell phones (Figures 22.3 and 22.4).

FIGURE 22.3 *Some companies, such as Ustream, provide cell-phone apps that enable live video transmission to their website, which then makes the live feed available to an audience. This is a cell-phone screenshot of a live video transmission.*

FIGURE 22.4 *Apps allow producers to transmit live or upload directly to Facebook, Twitter, or YouTube from a mobile phone.*

Source: Photo courtesy of Skype qik.

FIGURE 22.5 *Skype, working with a number of hardware manufacturers, has created a "broadcast" version of their popular consumer video software. It is being used by different broadcast programs, such as Jimmy Kimmel Live, to provide live video feeds from outside their studio. The hardware equipment converts the Skype feed into an HD-SDI video output/input.*

Source: Photo courtesy of Skype.

As with video-sharing websites, live video streaming of events can be provided to a specific audience or available to anyone who would like to see it. Video streaming requires substantial compression of the data, which often results in a lower-quality image. However, the quality is rapidly improving. Some live streaming sites allow unlimited viewers, while others are currently limited to one-on-one direct video transmissions (Figure 22.5).

IPTV

Internet Protocol Television (IPTV) utilizes the Internet to provide programming to its audience instead of broadcast or cable television. IPTV is somewhat of a mix of the services by video-sharing sites and live-streaming sites. Programming can be streamed live, video-on-demand, or interactive television (iTV). Many times, these systems are subscriber-based, which requires the payment of fees in order to access the content.

FIGURE 22.6 *IPTV channels, such as Major League Baseball, provide a wide range of services, often interactive, such as live streaming of games, baseball statistics, merchandise, and baseball news.*

For example, Major League Baseball began their own IPTV called MLB.com. They provide baseball information, news, sports columns, and statistics. MLB.com also provides games, video and audio streaming of baseball games, official baseball fan products, and ticket sales (Figure 22.6). Some of their

information is free; other material is only available to subscribers. These sites provide an interactivity that give much more access to the information the audience wants—allowing the viewer to individualize his or her viewing experience (Figure 22.7).

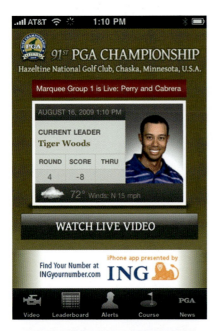
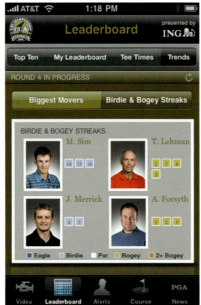

FIGURE 22.7 *The Professional Golf Association (PGA) provides an app for cell phones that allows viewers to customize their viewing. This includes receiving alerts when a specific player is at the tee. The apps enable you to follow a specific person without watching everything, as well as provide updated statistics, show the holes, or even watch a specific hole.*

3G/4G Transmission

A growing method of video transmission is the use of 3G/Wi-Fi and 4G cellular connections. In order to obtain a high quality, these units have to leverage multiple 3G/4G connections, called bonding. The advantage to these systems is that they are becoming incredibly small (Figure 22.8).

Summary

Television production personnel are having to learn how to use the new distribution media as they continue to emerge and converge. Producers need to know how to attract an online audience, including how to use social media to get the word out.

FIGURE 22.8 *HD video can be transmitted live utilizing multiple "bonded" connections of 3G/4G. The transmission gear has become incredibly small.*

Source: Photo/illustration courtesy of LiveU.

INTERVIEW WITH A PROFESSIONAL: TRIPP CROSBY

What kind of things do you put on the Internet? Since 2006, I've had two YouTube channels, multiple websites, blogs, and a podcast. A portion of my professional success relies on the response I get to my original content, so I pay close attention to the numbers. I can tell you what it's like to get anything from a few hits to a few million hits. On a bad day, I get zero blog comments and a steep dip on the line graph that tells me how many people have visited. On my worst day ever, I had to take down a post because it was too offensive or just way off brand. On a good day, I'll post a video, get tens of thousands of views on YouTube, hundreds of comments on my blog, and lots of tweets with a forwarded link. My best day ever was when YouTube featured our sketch "things you can't do when you're not in a pool" on their homepage, and we instantly got millions of views, phone calls from Hollywood, and eventually a reward from iTunes. I've learned that I can't predict an online audience as much as I'd like to be able to. There are certainly ways to increase the odds that content will be somewhat viral (such as featuring a celebrity), but the real way to increase views is to consistently add new content and build a slowly but steadily growing audience.

FIGURE 22.9
Tripp Crosby, producer, writer, and director.

How do you build an audience for your video? The best way I've found to announce new content is through Twitter and Facebook. Ideally, people will subscribe to my podcast, my blog, and my YouTube channel, but to get new viewers I have to keep annoying people who have decided to be my "friend."

continued

Interview with a Professional: Tripp Crosby—*continued*

Why do you create for the Web? The Web is the future for content creators. Sort of. It's really exciting to be able to create something and instantly make it public, but it's really daunting to think about how many other people are doing the same thing. Even though advertisers pay for some of my online material, I still make my living on content that I create for clients. I know there are some folks out there who are rolling in money that they make online, but they are still the exception. So why do I put my content online? Because for me, hits equal credibility. Let's say I make 100 dollars from AdSense for a video on YouTube that got 300,000 views. This hardly pays for lunch on a shoot. But the video got 300,000 views! Now I have a great selling point for my next client who needs to know his 1,000-person event will enjoy what I create. One day, I'd like to be fully sustained by advertising. Anyone would agree that it's more fun to create content for an audience than a client. For now, the Internet is my free marketing campaign.

What changes do you see in the future? I do not believe that the Internet will ever replace television. Instead, I think they will be one and the same. In 10 years, everyone will still have a rectangular something hanging from their wall, and they'll have some sort of portable handheld device with a keyboard. It's just that both will tune into content through the same channels. In other words, you won't need an ISP and a cable provider. Did I say 10 years? I meant one year.

What kind of challenges do you face with video and the Web? Just deciding where to put content is challenging. What blog template works best for my content? YouTube or Vimeo? Should I have my own "site?" How am I going to track my views? My subscribers? Then comes the most challenging aspect of putting content online—making good content that people want to consume. Or even more challenging—making content that people want to consume and then pass along. It doesn't matter if your initial audience is 75 or 75,000. If one view equals more than one view, you have succeeded and your audience is growing. Lastly, one of the keys to building an online audience is being consistent with updating your content online. An Internet audience is like having a girlfriend that doesn't like you as much as you like her. As long as you keep your content updated, they will stick around for more, but as soon as you stop contributing they will move on. Also, it takes a lot of work to get them back.

Tripp is the founder of Green Tricycle Studios. Besides having videos on MTV and on YouTube with over 5 million hits, his clients have included Chick fil A, AOL, Wal-Mart, and InComm.

REVIEW QUESTIONS

1. List three ways of getting your video to an audience.
2. What is the difference between live online video and IPTV?
3. What are the challenges with getting your program on broadcast television?

GLOSSARY

3D: Images that appear to have three dimensions, height, width, and depth.

3D TV: Televisions that are designed specifically for showing 3D programming. Viewers must watch the images through high-precision 3D glasses, which open and close the left and right shutters in synchronization with the alternating images.

4K: Also known as "Ultra High Definition," "Ultra HD," or SHV-1, 4K is defined as at least 3,840 horizontal and at least 2,160 vertical pixels, with an aspect ratio of at least 16 x 9.

8K: Also known as "Super Hi-Vision" and SHV-2, 8K is 16 times more detailed than HD, with 33 million pixels compared to HD's 2 million. The 8K system also has three-dimensional 22.2 surround sound, with three vertical layers of front speakers, two layers of surround speakers, and two subwoofers for bass frequencies.

720p: This HDTV format has 720 scan lines and uses progressive scanning. The 720p format is best for fast-moving motion scenes.

1080i: This HDTV format has 1,080 scan lines and uses an interlaced scanning system. The 1080i format has a sharper image than the 720p format.

1080p: This HDTV format has 1,080 scan lines and uses a progressive scanning system. It is often referred to as *full HD*.

Acoustics: Higher-frequency sound waves travel in straight paths and are easily deflected and reflected by hard surfaces. They are also easily absorbed by porous fibrous materials. Lower-frequency sound waves (below 100 Hz) spread widely, so they are not impeded by obstacles and are less readily absorbed. As sound waves meet nearby materials, they are selectively absorbed and reflected; the reflected sound's quality is modified according to the surfaces' nature, structures, and shapes.

Action line (axis of action line/eyeline): The imaginary line along the direction of the action in the scene. Cameras should shoot from only one side of this line.

AGC (Automatic Gain Control): Circuitry in some cameras that automatically adjusts the audio level (higher or lower) to a preset average level.

Ambient sound: The background sounds that are present when shooting a production.

Analog (analogue) recording: Analog systems directly record the variations of the video and audio signals. They have a tendency to deteriorate when dubbing copies and can be recorded only on tape.

Animation (image composition): Refers to when video images give the audience the same emotional response that you had while shooting.

Aperture: The opening in the lens that lets light into the camera.

Arc: A camera move that moves around the subject in a circle, arc, or horseshoe path.

Aspect ratio: Aspect ratio can be defined as the proportion of width to height of the screen. Televisions have an aspect ratio of 4:3 or 16:9. It is important to keep the aspect ratio in mind when designing sets so that they fit the screen.

Audio filters: Audio filters may be used to reduce background noises (traffic, air conditioners, wind), or compensate for boomy surroundings.

Audio mixer: A unit used to select, control, and intermix audio sources. It may include filter circuits and reverberation control. It is usually operated by the audio mixer (it's a job title as well as the name of the board), also known as A-1.

Audio sweetening: The process of working on the program sound after the video portion is completed; also called a dubbing session or track laying.

Autofocus: A process by which some lenses automatically focus on the subject.

Axis of action line: Often also called the action line or eyeline, this is an imaginary line along the direction of the action in the scene. Cameras should shoot from only one side of this line.

Backlight control: When there is more light in the background than on the subject, some cameras use a backlight control button that opens up the iris an arbitrary stop or so above the auto-iris setting in order to improve the subject's exposure.

Barn doors: These metal flaps are usually attached to the top, bottom, and sides of the light in order to shape the beam.

Base light: The minimal amount of light that allows the camera to see the subject.

Batten: The bar to which studio lights are connected.

Bidirectional microphone: This microphone can hear equally well both in front and back but is deaf on either side of it.

Black level: The intensity of black in the video image.

Black stretch control: Camera circuitry that makes shadow detail clearer and improves tonal gradation in darker tones.

Bonded cellular: The merging of multiple cellular channels in order to increase the quality of the transmitted video.

Boom pole: A pole that is used to hold a microphone close to a subject. Often referred to as a mic boom.

Breakdown sheet: An analysis of a script, listing all of the production elements in order of the schedule.

Bug: A small graphic usually located in one of the corners of the video screen. The bug usually identifies the network or may include scoring information for sports.

Camcorder: A camera with a built-in video recorder.

Camera blocking (stumble-through): The initial camera rehearsal, coordinating all technical operations, and discovering and correcting problems.

Camera control unit (CCU): Equipment that controls the camera from a remote position. The CCU is part of the setup and adjustments of the camera: luminance, color correction, aperture.

Camera script: The camera script adds full details of the production treatment to the left side of the "rehearsal script" and usually also includes the shot numbers, cameras used, positions of camera, basic shot details, camera moves, and switcher instructions (if used).

Camera setup: Adjusting the controls within the camera's circuitry.

Cathode ray tube (CRT): CRT televisions send an electron beam through a vacuum tube to a phosphor-coated screen. These "tube" televisions are large and bulky.

Character generator (CG): A generic name for any type of television graphic creation equipment.

Character makeup: The *makeup* emphasis is on the specific character or type that the actor is playing. By facial reshaping, remodeling, and changes in hair, the subject may even be totally transformed.

Charge-coupled device (CCD): An image sensor used in most video cameras.

Chromakey: Using a production switcher and this technique, the director can replace a specific color (usually blue or green) with another image source (still image, live video, prerecorded material).

Clapboard (clapper or slate): Shot at the beginning of each take to provide information such as film title, names of director and director of photography, scene, take, date, and time. Primarily used in dramatic productions.

Clip: A video segment.

Close-up shot (CU): The CU shot encourages the viewer to concentrate on a specific feature. When shooting people, it is used to emphasize emotion.

CMOS: See *Complementary metal-oxide semiconductor*.

Color bar generator: Color bars provide a consistent reference pattern that is used for matching the video output of multiple cameras. They are also used to obtain the best-quality image on a video monitor. This test signal is composed of a series of vertical bars of standard colors (white, yellow, cyan [blue-green], green, magenta [red-purple], red, blue, black). A color bar generator actually creates the pattern electronically. However, it is possible to use a printed color chart, as long as it has been cared for and is not faded.

Complementary metal-oxide semiconductor (CMOS): An image sensor that consumes less power, saving energy for longer shooting times.

Composition: The goal of composition is to create an image that is attractive or that captures and keeps the audience's attention and effectively communicates the production's message.

Compressor/expander: Deliberately used to reduce or emphasize the audio dynamic range (i.e., the difference between the quietest and loudest sounds).

Condenser microphone: A high-quality microphone that can be very small and is generally powered by an in-board battery, phantom power, or a power supply.

Context (image composition): The content of the image should allow the viewer to understand the subject better. Compose the shot in such a way that it includes a background, or foreground, that adds additional information or context to the image.

Continuity: Making sure that there is consistency from one shot to the next in a scene and from scene to scene. This continuity includes the talent, objects, and sets. An example of a continuity error in a production would be when one shot shows the talent's hair combed one way and the next shot shows it another way.

Contrast: The difference between the relative brightness of the lightest and darkest areas in the shot.

Control room: Sometimes known as a gallery, an area in a studio where the director controls the television production. Although the control room equipment may vary, they all include video and audio monitors, intercoms, and a switcher.

Convergence: In 3D, when a human looks at two overlaid images, one image seen by the left eye and one seen with the right eye, and perceives one image with depth, convergence is when the two images converge on a single point in space.

Convertible camera: This type of camera generally starts out as a camera "head." A variety of attachments can then be added onto the camera head, including different kinds of lenses, viewfinders, and recorders, to suit a specific production requirement.

Coordination meetings: A forum for all parties involved in the production to share ideas, communicate issues that may affect other areas, and ensure that all details are ready for the production.

Corrective makeup: Reduces less-pleasing facial characteristics while enhancing more attractive points.

Cover shot: A video clip that is used to cover an edit so that the viewers do not know that the edit occurred.

Coverage: The term used by filmmakers that refers to repeating a scene from enough different angles to ensure that a seamless performance can be maintained throughout multiple takes.

Crab: See *Truck*.

Crash start: Takes us straight into the program, which appears to have begun already.

Crawl: The movement of text horizontally across the television screen.

Credit roll: Continuous information moving vertically into the frame, and passing out at the top.

Credits: The text that lists and acknowledges those appearing in and contributing to the program.

CU: See *Close-up shot*.

Cue card: The talent may read questions or specific points from this card, which is positioned near the camera. Generally, it is held next to the camera lens.

Cut: The "cut" or "take" is the most common transition when editing. It is an instantaneous switch from one shot to the next.

Cutaway shot: These shots are used to cover edits when any sequence is shortened or lengthened. Generally, it is a shot of something outside of the current frame.

Cyclorama (cyc): The cyclorama (or cyc, pronounced "sike") serves as a general-purpose detail-free background. It can be neutral, colored with lights, or have no light (black).

Dead surroundings: When area surfaces are very sound-absorbent, the direct sound waves strike walls, floor, ceiling, and furnishings, and are largely lost. Only a few weak reflections may be picked up by the microphone.

Deep focus: Deep focus, or large depth of field, is when everything in the shot is clearly in focus.

Depth of field (DoF): The distance between the nearest and farthest objects in focus.

Dichroic filters: These filters produce three color-filtered images corresponding to the red, green, and blue proportions in the scene.

Diffusion material: Can be attached to the front of a light in order to reduce the intensity of the light beam.

Digital recording: A digital system regularly samples the waveforms and converts them into numerical (binary) data. This allows many generations of copies to be made without affecting the quality of the image. Digital systems also allow the data to be recorded on forms of media other than tape, such as hard disks and flash memory.

Digital video effect (DVE): Digital video effect equipment, working with the switcher, is used to create special effects between video images. A DVE could also refer to the actual effect.

Digital zoom: Zooming is achieved by progressively reading out a smaller and smaller area of the same digitally constructed image. The image progressively deteriorates as the digital zoom is zoomed in.

Digitizing: Converting the audio and video signals into data files. This term is used when transferring video footage from a camera (or other video source) to a computer.

Directional microphone: This type of microphone can hear sounds directly in front of it.

Dissolve: A gradual transition between two images. A dissolve usually signifies a change in time or location.

Dolly (truck/track): (1) The action of moving the whole camera and mount slowly toward or away from the subject. (2) A platform with wheels that is used to smoothly move a camera during a shot.

Drag: The variable friction controls located on a tripod head that steady the camera's movements.

Dress rehearsal (dress run): The goal of the dress rehearsal is to time the wardrobe and makeup changes.

Dry blocking (walk-through): Actors perform, familiarizing themselves with the studio settings while the studio crew watch, learning the format, action, and production treatment.

DSLR: A still camera (digital single lens reflex camera) that shoots video, allowing the photographer to see the image through the lens that will capture the image.

Dutch: Tilting the camera is called a "Dutch" or a "canted" shot. This movement increases the dynamics of the shot.

DVE: See *Digital video effect.*

Dynamic microphone: A rugged, low-maintenance microphone that is not easily distorted.

Dynamic range: The range between the weakest and loudest sounds that can be effectively recorded by a recording device.

ECU or XCU: Extreme close-up shot.

Electronic field production (EFP): EFP cameras are used for non-news productions such as program inserts, documentaries, magazine features, and commercials. EFPs can also be used for multi-camera production.

Electronic newsgathering (ENG): Camcorders generally used for newsgathering. Many times, they are the cameras that are equipped with a microphone and camera light and are used to shoot interviews and breaking news.

Ellipsoidal: A sharply focused/defined spotlight.

ELS: Extreme long shot.

Empirical production method: The empirical method is where instinct and opportunity are the guides.

Equalizer: An audio filter that can boost or reduce any segment of the audio spectrum.

EXT: The abbreviation used to signify an external location on a script.

Eye-level shots: Provide an image that is roughly at the eye level of the talent (in a studio show) or the average viewing audience.

Eyeline (line): Where people appear to be looking, or line of sight.

Fact sheet/rundown sheet: Summarizes information about a product or item for a demonstration program, or details of a guest for an interviewer.

Fade: A gradual change (dissolve) between black and a video image. Usually defines the beginning or end of a segment or program.

Fill light: A flood light that reduces the shadows made by the key light.

Filmic space: Intercuts action that is concurrent at different places.

Filmic time: Editing technique that tightens up the pace of a production by leaving out potentially boring portions of the scene when the audience interest could wane.

Filter wheel: Filter wheels are often fitted inside the video camera, just behind the lens. The typical wheel is fitted with a number of different correction filters and may include a 5,600 K (daylight), 3,200 K, and neutral-density filters.

FireWire: Also known as IEEE 1394 or iLink; a method for connecting different pieces of equipment, such as cameras, drives, and computers, so that they can transfer large amounts of data, such as video, quickly and easily.

Fishpole: See *Boom pole*.

Flash memory: Flash cards can store large amounts of digital data but have no moving parts. This makes them durable and able to work in a variety of temperatures, and allows data to be easily transferred into a nonlinear editor.

Flats: Freestanding background set panels.

Flood lighting: Scatters in all directions, providing a broad, nondirectional light.

Floor plan (staging plan or set plan): A rough plan of the staging layout that usually begins with drawing potential scale outlines of settings, including their main features— windows, doors, stairways. Ensure that there is enough room for cameras, sound booms, and lighting.

Focal length: An optical measurement: the distance between the optical center of the lens and the image sensor, when you are focused at a great distance such as infinity. It is measured in millimeters (mm) or inches.

Focus puller: The person responsible for keeping the camera in focus using a follow-focus device.

Focus zone: See *Depth of field*.

Foley: Creating sound effects that can be used to replace the original sounds such as hoof beats or footsteps.

Follow focus: This technique requires the camera operator to continually change the focus as the camera follows the action.

Format: The show format lists the items or program segments in a show in the order in which they are to be shot. The format generally shows the durations of each segment and possibly the camera assignments.

Fresnel light: An unfocused spotlight. It is lightweight, less expensive than an ellipsoidal, and has an adjustable beam.

F-stop: A setting that regulates how much light is allowed to pass through the camera lens by varying the size of the hole the light comes through.

FU: FU is the abbreviation used on scripts for a fade-up (from black to a video signal).

Full script: Includes detailed information on all aspects of the production. This includes the precise words that the talent/actors are to use in the production.

Gaffer: The head of the electrical department, many times in charge of lighting on a television set.

Gain: (1) (video) Amplification of the camera video signal, usually resulting in some video noise. (2) (audio) Amplification of the audio signal.

Gamma: Adjusts the tone and contrast of a video image.

Gel: Colored flexible plastic filters used to adjust the color of the lights.

Genlock: Adjusting the timing of the local sync pulses to get them into step with remote (nonsync) source.

Goals: Broad concepts of what you want to accomplish with the program.

Graphic equalizer (shaping filter): Device with a series of slider controls; allows selected parts of the audio spectrum to be boosted or reduced.

Grip clamps: Clamps designed to allow a light to be attached to almost anything.

Group shot (GS): A director's command to frame the entire group into the shot.

Handheld camera (HH): A camera that is held by a person and not supported by any type of camera mount.

Handheld microphone: Generally refers to any microphone held in the hand, used to pick up human speech.

Hand properties (props): Any items that are touched and handled by the talent during the production. These could include a pen, dishes, a cell phone, or silverware.

Hard disk drive (HDD): Used for recording digital video images; can be built into the camera or attached to the outside of the camera.

HDSLR: A high-definition (HD) version of a DSLR (see *DSLR*).

Headroom: The space from the top of the head to the upper frame.

High angle: When the camera is positioned higher than the subject.

High-definition television (HDTV or HD): Video formats that currently use a range from 720 to 1,080 lines of resolution are considered to be high definition.

Hue: Refers to the predominant color (e.g., blue, green, or yellow).

Image magnification (i-mag): Video on large television screens next to a stage in order to help the viewers see the stage action.

Incident light (direct lighting): The measurement of the amount of light falling on a subject.

INT: The abbreviation used on a script to signify an internal location.

Intellectual property (IP): Property (music, video, etc.) whose owners have been granted certain legal exclusive rights.

Intercom: A wired or wireless communication link between members of the production crew.

Interlaced scanning: The television's electron scans the odd-numbered lines first and then goes back and "paints" in the remaining even-numbered lines.

IPTV: Internet Protocol Television (IPTV) utilizes the Internet to provide programming to its audience instead of broadcast or cable television.

Iris: The adjustable diaphragm of the lens. This diaphragm is adjusted to be open or closed based on the amount of light needed to capture a quality image.

Isolated (ISO): When all the cameras are connected to the switcher as before, the ISO (or isolated) camera is also continuously recorded on a separate recorder.

iTV: Interactive television, or iTV, refers to online programming that allows the viewer to make choices about how he or she watches an event.

Jib: A counterbalanced arm that fits onto a tripod that allows the camera to move up, down, and around.

Jog: Playing the video on a recorder/player frame by frame.

Jump cut: Created when the editor cuts between two similar shots (two close-ups) of the same subject.

Key light: The main light, usually a spotlight, that reveals the shape and surface features of the subject.

Lavalier microphone: These small microphones clip on the clothing of the talent and provide fairly consistent, hands-free audio pickup of the talent's voice.

LCD: See *Liquid-crystal display*.

LED: The abbreviation for a light-emitting diode light bulb.

LED light panel: A camera or studio light that is made from a series of small LED bulbs.

Lighting plot: A lighting plot shows where each light will be placed on the set.

Limiter: A device for preventing loud audio from exceeding the system's upper limit (causing overload distortion), by progressively reducing circuit amplification for louder sounds.

Line level: The audio signal generated by a non-microphone device such as a CD player.

Linear editing: The copying, or dubbing, of segments from the master tape to another tape in sequential order.

Liquid-crystal display (LCD): These flat-screen displays work by sending variable electrical currents through a liquid crystal solution that crystallizes to form a quality image.

Live surroundings: Signifies an area, usually a room, that contains predominantly hard surfaces, which strongly reflect sound, creating an echo.

Logging: Loggers view the footage and write down the scene/take numbers, the length of each shot, time code, and descriptions of each shot.

Long shot (LS or wide-angle shot): The long shot is used to help establish a scene for the viewer.

Low angle: When the camera is positioned lower than the subject.

Lower third (L/3rd): A graphic that appears in lower third of the television screen. Traditionally, this contains biographical information.

LS: See *Long shot.*

Luminance: The brightness of the image—how dark or light it appears.

Macro: Some lenses include a macro setting, a lens capable of extreme close-ups.

Medium shot (MS): A shot close enough to show the emotion of the scene but far enough away to show some of the relevant context of the event.

Mic level: The audio level of a signal that is generated by a microphone.

Minimum focused distance (MFD): The closest distance a lens can get to the subject. With some telephoto lenses, the MFD may be a few yards. Other lenses may be 1/4 inch.

Modular set: Designed in a number of components, these sets can be easily assembled, disassembled, and stored.

Monaural (mono): This single track of audio is limited; its only clue to distance is loudness, and direction cannot be conveyed at all.

Monitors: Monitors were designed to provide accurate, stable image quality. They do not include tuners and may not include audio speakers.

Monopod: A one-legged camera support.

MS: See *Medium shot.*

Multi-camera production: Signifies that two or more cameras were used to create a television production. Usually the cameras are switched by a production switcher.

Narrow-angle lens: See *Telephoto lens.*

Natural sound (NAT): The recording of ambient or environmental sounds on location.

Nonlinear editing: The process whereby the recorded video is digitized (copied) onto a computer. Then the footage can be arranged, rearranged, and special effects added, and the audio and graphics can be adjusted using editing software.

Normal lens: The type of lens that portrays the scene approximately the same way a human eye might see it.

Notch filter (parametric amplifier): A filter that produces a very steep peak or dip in a selected part of the audio spectrum, such as to suppress unavoidable hum, whistle, or rumble.

NTSC (National Television System Committee): The television system traditionally used by the United States and Japan. It has 525 scan lines.

Objective camera: This role is that of an onlooker, watching the action from the best possible position at each moment.

Objectives: Measurable goals, something that can be tested for to confirm that the audience did understand and remember the key points of the program.

Omnidirectional microphone: This type of microphone can pick up audio equally well in all directions.

Opening titles: Introduce the show to the audience.

Open set: Can be created by carefully grouping a few pieces of furniture in front of the wall. Even as little as a couch, low table, table lamp, potted plants, a screen, chair, and stand lamp can suggest a complete room.

Optical zoom: Uses a lens to maintain a high-quality image throughout its zoom range.

Outline script: Usually provides the prepared dialogue for the opening and closing and then lists the order of topics that should be covered. The talent will use the list as they improvise throughout the production.

Outside broadcast (OB): Also known as a *remote production*, an OB takes place outside of the studio.

Pace: The rate of emotional progression. A slow pace suggests dignity, solemnity, contemplation, and deep emotion; a fast pace conveys energy, excitement, confusion, and brashness.

PAL (phase-alternating line): The color television system widely used in Europe and throughout the world. It was derived from the NTSC system but avoids the hue shift caused by phase errors in the transmission path by reversing the phase of the reference color burst on alternate lines. It has 625 lines of resolution.

Pan head (tripod head): Enables the camera to tilt and pan smoothly. Variable friction controls (drag) steady these movements. The head can also be locked off in a fixed position. Tilt balance adjustments position the camera horizontally to assist in balancing the camera on the mount.

Pan shot: When the camera pivots to the left or right with the camera pivoting on a camera mount.

Patch panel/jackfield: Rows of sockets to which the inputs and outputs of a variety of audio units are permanently wired. Units may be interconnected with a series of plugged cables (patch cords).

Pedestal: (1) An adjustable camera support that has wheels. These are normally used in a studio. (2) Can refer to the black level of a video image shown on a waveform monitor. (3) Also can refer to the action of raising a camera higher.

Per diem: Refers to a set amount of money paid per day, above the normal pay, to a worker to cover living expenses when traveling.

Photographic lighting: See *Three-point lighting*.

Pickup shot: If an error is made during the shooting of a scene, this is created by changing the camera angle (or shot size) and retaking the action from just before the error was made.

Planned production method: Organizes and builds a program in carefully arranged steps.

Plasma: A high-quality, thin, flat-panel screen that can be viewed from a wide angle.

Point-of-view (POV) camera: These small, sometimes robotic cameras can be placed in positions that give the audience a unique viewpoint.

Postproduction: Editing, additional treatment, and duplication of the project.

Preamplifier: An amplifier used to adjust the strength of audio from one or more audio sources to a standard level (intensity). It may include source switching and basic filtering.

Preliminary script/writer's script: Initial submitted full-page script (dialogue and action) before script editing.

Preview monitor: This video monitor, located in the control room's monitor wall, is used by the director to preview video before it goes on air.

Prime lens (primary lens): A fixed coverage, field of view, or focal length.

Program monitor (on-air monitor): Shows the actual program that is being broadcast or recorded.

Progressive scanning: This sequential scanning system uses an electron beam that scans or paints all lines at once, displaying the total picture.

Prop (property): An item used as a part of a film or television scene. This could include a set decoration or an item handled by the talent.

Prosumer equipment: Sometimes known as *industrial equipment*; a little heavier-duty, sometimes employs a few professional features (such as interchangeable lenses on a camera), but may still have many of the automatic features that are included on the consumer equipment.

Public domain: Music and lyrics published in 1922 or earlier are in the public domain in the United States. No one can claim ownership of a song in the public domain; therefore, public domain songs may be used by anyone. Sound recordings, however, are protected separately from musical compositions. There are no sound recordings in the public domain in the United States.

Quick-release mount: Attached to the camera and fits into a corresponding recessed plate attached to the tripod/pan head. Allows the camera operator to quickly remove or attach the camera to the camera mount.

RAW: Uncompressed data from the sensor of a digital camera.

Reflected light metering: The light bouncing off of the subject. A handheld light meter or a camera's built-in meter is used to measure by aiming directly at the subject.

Rehearsal script: Usually includes the cast/character list, production team details, and rehearsal arrangements. There is generally a synopsis of the plot or storyline, location, time of day, stage/location instructions, the action, dialogue, effects cues, and audio instructions.

Remote survey (recce): A preliminary visit to a shooting location.

Remote truck (OB van): A mobile television control room that is used when away from the studio.

Reverberation: Device for increasing or adjusting the amount of echo accompanying a sound.

RGB (red, green, blue): The three primary colors used in video processing.

Roll: The movement of text up or down the video screen.

Running order: The order that the scenes or shots will be shown in the final project, which may differ greatly from the shooting order.

Safe title area: The center 80 percent of the screen where it is safe to place graphics.

Sampling rate: Measures how often the values of the analog video signal are converted into a digital code.

Saturation: The chroma, purity, and intensity. It effects its richness or paleness.

Scene: Covers a complete continuous action sequence.

Scoop: A simple floodlight. It is inexpensive, usually not adjustable, lightweight; does not have a sharp outline.

SECAM (*Séquentiel couleur à mémoire*, Sequential Color with Memory): Video system used by France and many countries of the former USSR.

Shooting order: The order that the scenes or shots may be shot using the video camera, which may differ greatly from the running order.

Shot sheet (shot card): A list, created by the director, of each shot needed from each individual camera operator. The shots are listed in order so that the camera operator can move from shot to shot with little direction from the director.

Shotgun microphone: A highly directional microphone used to pick up sound from a distance.

Show format: The show format lists the items or program segments in a show, in the order they are to be shot. It may show durations, who is participating, or shot numbers.

Single-camera production: One camera is used to shoot the entire segment or show.

Site survey: A meeting of the key production personnel at the proposed shooting location. A survey allows them to make sure that the location will meet their production needs.

Situation comedy (sitcom): A television program that has a storyline plot (situation) and is a humorous drama.

SMPTE (Society of Motion Picture and Television Engineers): This international professional organization has developed over 400 standards, practices, and engineering guidelines for audio, television, and the film industry.

Soft light: Provides a large amount of diffused light.

Spotlight: A highly directional light.

Stage props: The furniture on the set: news desks, chairs, couches, tables.

Standard-definition television (SDTV, SD): Standard definition signifies television formats that have 480–576 lines.

Standby: Alerting the talent to stand by for a cue.

Stereo sound: Uses two audio tracks to create an illusion of space and dimension.

Stereographer: A person who operates a stereo 3D camera.

Stick mic: See *Handheld microphone*.

Storyboard: A series of rough sketches that help visualize and organize your camera treatment.

Straight makeup: A basic compensatory makeup treatment, affecting the talent's appearance to a minimum extent.

Streaming: Programming that is shown live or transmitted from a video sharing site to a computer.

Stretch: Telling the talent to go more slowly (meaning that there is time to spare).

Studio: Indoor locations designed to handle a variety of productions, with a wide-open space equipped with lights, sound control, and protection from the impact of weather.

Studio plan: The basis for much of the organization, showing the studio's permanent staging area with such features and facilities as exits, technical supplies, cycloramas, and service and storage areas.

Subjective camera: See *Point-of-view (POV) camera*.

Subtitles: Identify people and places.

Super-cardioid microphone: A super-cardioid (or highly directional) pickup pattern that is used for extremely selective pickup, to avoid environmental noises, or for distance sources.

Surround sound: Instead of the one channel for mono or the two channels for stereo, 5.1 surround has six discrete (distinct, individual) channels. Can provide a sense of envelopment when mixed correctly.

Switcher (vision mixer): A device used to switch between video inputs (cameras, graphics, video players).

Symbolism (image composition): Using images that have hidden or representational meaning to the viewer.

Synopsis: An outline of the characters, action, and plot; helps everyone involved in the production understand what is going on.

Take: See *Cut*.

Tally light: A light usually found on the front of the camera and in the viewfinder. The front tally light is to let the talent know that the camera is recording. The back tally light lets the camera operator know when his or her camera has been chosen to be recorded by the director.

Teaser: Showing dramatic, provocative, intriguing highlights from the production before the opening titles. The goal is to convince the audience to stay for the entire production.

Telephoto lens: A narrow-angle lens that is used to give a magnified view of the scene, making it appear closer. The lens magnifies the scene.

Teleprompter: A device that usually projects computer-generated text on a piece of reflective glass over the lens of the camera. It is designed to allow talent to read a script while looking directly at the camera.

Television receivers: Monitors that include a tuner so that they can display broadcast programs with their accompanying sound.

Three-point lighting: A lighting technique that utilizes three lights (key, fill, and backlights) to illuminate the subject.

Tightening a shot: Zooming the lens in a little bit.

Tilt balance: Adjustments located on the pan head of a tripod to position the camera horizontally and assist in balancing the camera on the mount.

Tilt shot: The camera moves up or down, pivoting on a camera mount.

Time-base corrector (TBC): Equipment that provides automatic compensation for synchronizing inaccuracies on replay or imperfect sync pulses from mobile cameras.

Time code (TC): A continuous time signal throughout the tape, showing the precise moment of recording.

Timeline: Includes multiple tracks of video, audio, and graphics in a nonlinear editing system.

Treatment: A film treatment, or script treatment; it is more than an outline of the production and less than a script. It is usually a detailed description of the story that includes other information such as how it will be directed.

Triangle lighting: See *Three-point lighting*.

Trimming: Cutting frames off of a shot to make it shorter.

Tripod: A camera mount that is a three-legged stand with independently extendable legs.

Tripod arms (pan bars): These handles attach to the pan head on a tripod or other camera mount to accurately pan, tilt, and control the camera.

Truck (crab, track): The truck, trucking, or tracking shot is when the camera and mount move sideways (left or right) with the subject.

Ultra HD (Super Hi Vision or UHD): Ultra HD includes 4K and 8K digital video formats.

Vectorscope: An oscilloscope that is used to check the color accuracy of each part of the video system (cameras, switcher, recorder). Incorrect adjustments can create serious problems with the color quality. Ideally, the color responses of all equipment should match. *Color bars* are usually recorded at the beginning of each videotape to check color accuracy.

Video gain: The amplification of the camera's video signal, usually resulting in some video noise.

Video-sharing sites: Online sites that enable producers to upload programming so that it can be seen by an audience.

Videotape: Tape has been the traditional means of recording video images. However, it is being replaced by hard drives and flash cards.

Viewfinder: Monitors the camera's picture; allows the camera operator to focus, zoom, and frame the image.

Virtual set: Set that uses a blue or green seamless background, chromakeying the computer-generated set into the scene. Most virtual sets employ sophisticated tracking software that monitors the camera's movements so that as the cameras zoom, tilt, pan, or move in any other way, the background moves in a corresponding way.

Voiceover (VO): Commentary over video.

Waveform monitor: An oscilloscope that is designed to monitor a video signal. This ensures that all colors will be correctly recorded.

White balance: The process of calibrating a camera so that the light source will be reproduced accurately as white; the most common technique of color-balancing a camera.

Wide-angle lens: Shows us a greater area of the scene than is normal. The subject looks unusually distant.

Wide shot (WS): See long shot (LS or wide-angle shot).

Wild track: General background noise.

Wild track interviews: A wild track interview usually means that while images are being shown of a person doing something (sawing wood, for example), the voice of the person shown in the images is heard being interviewed, like a voice-over.

Wipe: A special effect transition between two images. Usually shows a change of time, location, or subject. The wipe adds a bit of novelty to the transition but can easily be overused.

XCU: Extreme close-up.

XLR: A professional audio connector. Usually, but not limited to, three pins.

XLS: Extreme long shot.

Zebra: An indicator included on some video cameras in the viewfinder. Allows camera operators to evaluate the exposure of the image in the viewfinder by showing all overexposed segments of the scene with stripes.

Zone focus: Camera operators are focused on a portion of the scene. When the subject comes into that area, the camera has been prefocused to make sure that the action is sharp.

Zoom lens: A lens that has a variable focal length.

INDEX